AF322734

Mathematical Theory of
Adaptive Control

Interdisciplinary Mathematical Sciences – Vol. 4

Mathematical Theory of Adaptive Control

Vladimir G. Sragovich
Russian Academy of Science, Russia

Translator
I. A. Sinitzin
Russian Academy of Science, Russia

Editor
J. Spaliński
Warsaw University of Technology, Poland

Assistant Editors
Ł. Stettner and **J. Zabczyk**
Polish Academy of Sciences, Poland

World Scientific

NEW JERSEY · LONDON · SINGAPORE · BEIJING · SHANGHAI · HONG KONG · TAIPEI · CHENNAI

Published by

World Scientific Publishing Co. Pte. Ltd.

5 Toh Tuck Link, Singapore 596224

USA office: 27 Warren Street, Suite 401-402, Hackensack, NJ 07601

UK office: 57 Shelton Street, Covent Garden, London WC2H 9HE

British Library Cataloguing-in-Publication Data
A catalogue record for this book is available from the British Library.

Interdisciplinary Mathematical Sciences — Vol. 4
MATHEMATICAL THEORY OF ADAPTIVE CONTROL

Copyright © 2006 by World Scientific Publishing Co. Pte. Ltd.

ISBN 981-256-371-7

Printed in Singapore by World Scientific Printers (S) Pte Ltd

… to my teachers,
Professors of Moscow University,

Aleksandr Khintchine
and
Abram Plesner

PREFACE

The primary notions of control theory are those of a *controlled object*, a *control aim* and a *control algorithm* (*strategy*). Both a Markov chain and an ordinary or stochastic differential equation with controls entering into its description can be considered as control problems.

We choose the controls so that the controlled object has certain desired properties, called the control aims. For example, a functional defined on the states of a Markov chain may be required to be extreme or the solutions of the given equation should be stable in some sense. Solving a control problem means finding a strategy (an algorithm) giving the choice rules of the controls to achieve the control aim given beforehand.

For many decades control theory was based on the assumption that the controlled object was known exactly within the framework of its mathematical description (model). For example, if the mathematical model of the considered object is the linear difference equation of order n

$$x_t + a_1 x_{t-1} + a_2 x_{t-2} + \cdots + a_n x_{t-n} = b_1 u_{t-1} + \cdots + b_m u_{t-m} + \psi(t)$$

where x_t is the state of the object, u is the control, $\psi(t)$ is the external disturbance (or noise), then the values of the coefficients (a_i, b_i) are supposed to be known and the states x_t to be observed at each moment t. Moreover, either an explicit form of the function $\psi(t)$ or the probabilistic characteristics of the noise $\psi(t)$ are supposed to be known in the deterministic or stochastic cases respectively. We call the theory of control based on these assumptions *classical control theory*.

However, in many applied engineering problems *a priori* we do not have this information about the controlled object. This has led to the creation of *adaptive control theory*. There are three possible approaches.

The first consists of employing the missing data as soon as they arrive during the control process. The second approach is based on controlling the object given incompletely and searching missing information simultaneously. This approach gives the identification method connecting the estimation procedures of the unknown characteristics of the object with the control methods of classical theory. This method has a wide use. The third approach consists of constructing algorithms of control not requiring detailed knowledge about the object.

Due to successful development, especially of the last approach, adaptive control theory may be regarded as an independent discipline. According to the general concept of adaptive control, instead of working with the incomplete mathematical model of the controlled object, we need to find a class (a collection) of mathematical models containing the model that we are interested in. Hence, the control aim stated in advance refers to no concrete object but to all objects from the specified class. The strategy (the control algorithm) being designed must apply to all objects from the given class. For this reason the algorithms appearing in adaptive theory are more difficult than those in classical theory.

We would like to emphasize three distinctive features of this book in comparison with other books dedicated to the same topic. First, it is the wide range of objects studied (in order of increasing complexity): discrete processes of automata type (inertia-free), the process generated by recurrent procedures, minimax problems, finite Markov chains (with both observable and unobservable states), Markov and semi-Markov processes, discrete time stationary processes, linear difference stochastic equations, ordinary differential equations (we may call this *deterministic adaptive theory*) and, finally, stochastic Ito equations. The controlled objects listed above are mainly stochastic, and hence belong to *controlled stochastic processes*.

The second feature of this book is the detailed description of the research of the Eastern School of adaptive control which has not been easily accessible to the western reader.

The third feature is the formal definition of *adaptive control strategy* which has been given for the first time. This notion is used throughout the present volume. This can be stated as follows.

Let $\mathcal{K}$ be a class of controlled objects (controlled random processes) and let Z denote a control aim defined for all objects from $\mathcal{K}$. Finally, let Σ be a set of strategies which apply to all objects from $\mathcal{K}$.

Then a strategy from Σ that secures the attainment of the aim Z for every object from $\mathcal{K}$ is called an *adaptive strategy*. The goal of adaptive theory (probably unreachable) is to obtain necessary and sufficient conditions for the existance of an adaptive strategy for every collection $\mathcal{K}, Z$ and Σ above.

The purpose of the present volume is twofold. On the one hand, for the mathematically well-trained students of the appropriate specialities the book may serve for a text-book on adaptive control theory. On the other hand, the author hopes that even the specialists will find an inspiration here for their own research.

Many results deserving attention could not have been included in the main text of the book due to constraints on the book's volume. Therefore, to the author's regret, some significant results have been put into appendix — *Comments and Supplements*.

The readers should have a good knowledge of undergraduate mathematics. Nevertheless, most chapters begin with sections containing all necessary information (without proofs) to be used.

Bibliography is divided into two parts. The first one (General References) contains the list of the auxiliary citations. The second part (Special References) presents the original scientific works which form the basis for our consideration. This part is supplemented by some interesting works but, unfortunately, the author had no possibility to review them in detail. As mentioned above the brief survey of them is given in the *Comments and Supplements*. While composing the bibliography the following rule was used. If the results obtained by some authors are cited in a monograph then the readers will be referred to this monograph only.

To reference the text the following scheme is used. In every chapter the sections are numbered successively by two digits. The first of them denotes the chapter number, for example, (3.2) refers to Sec. 2 in Chap. 3. Each section has a separate numeration of equations, theorems, lemmas and so on consisting of one number only. The references to an item from another chapter (or section) are given completely (for example, Theorem 1 from Sec. 1, Chap. 2).

A substantial part of the book has been written in close contact with the authors of the appropriate results. Whether they are post-graduates, colleagues or friends of author is mentioned in the comments to chapters. Their advice was very useful. Here, the author would like to especially mention and to express many thanks to Professor Vladimir A. Brusin and Professor Aleksandr S. Poznyak as well as to Dr. Eugenij S. Usachev.

The author is grateful to the Committee of Scientific Research in Warsaw (Poland) for providing the financial support to complete this work and to translate the manuscript from Russian into English. I express once more my sincere gratitude to Professor Lukasz Stettner.

Vladimir G. Sragovich

EDITOR'S NOTE

This monograph is the magnum opus of Professor Vladimir G. Sragovich — one of the leaders of the Russian School of Adaptive Control. The book offers a very broad treatment of the subject and gives a comprehensive guide to the literature, with a bibliography of almost 200 entries.

We regret that the author, who passed away in 2000, could not take part in the preparation of the final version of this book.

Jan Spaliński, Łukasz Stettner, Jerzy Zabczyk
Warsaw, Poland, March 2005

(Edited by *Jan Spaliński*, with assistance from *Łukasz Stettner* and *Jerzy Zabczyk*.)

CHAPTER 1

BASIC NOTIONS AND DEFINITIONS

In this chapter the basic definitions and notions of control theory such as a mathematical model, a controlled random process, an observable process, a strategy and a control aim are considered. They serve as the basis for the definition of the adaptive strategies. Some auxiliary notions are considered as well. The general approach to the synthesis of the adaptive strategies and their properties is discussed. Specific features of classic and adaptive control are also discussed.

1.1. Random Processes and Systems of Probability Distributions

We shall begin with the definition of a probability space $(\Omega, \mathfrak{F}, \mathbf{P})$ where Ω is the space of elementary events, $\mathfrak{F}$ is the σ-algebra of measurable subsets from Ω, i.e. a class of the subsets of Ω closed with respect to complements, products and countable sums of these sets and, finally, $\mathbf{P}$ is a probability measure, i.e. it is a non-negative, countably-additive function defined on $\mathfrak{F}$ (i.e. $\mathbf{P}\{A\} \geq 0$ for any $A \in \mathfrak{F}$ and for disjoint sets A_i from $\mathfrak{F}$ we have $\mathbf{P}\{\bigcup_1^\infty A_i\} = \sum_1^\infty \mathbf{P}\{A_i\}$ and $\mathbf{P}\{\Omega\} = 1$.)

Let $(X, \mathfrak{X})$ be a pair consisting of a measurable space X and a σ-algebra of measurable subsets $\mathfrak{X}$. In the most interesting cases the space X is Euclidean, i.e. $X = \mathrm{R}^l$, $l = \dim \mathrm{R}^l$. A measurable mapping $\xi : \Omega \to X$ is called a *random variable* (r.v. for short). It means that the inverse image of any measurable set from $\mathfrak{X}$ belongs to $\mathfrak{F}$ under the mapping ξ or, symbolically, $\xi^{-1}(M) \in \mathfrak{F}$ for all $M \in \mathfrak{X}$. If Ω is a topological space we can consider the smallest σ-algebra $\mathfrak{B}_\Omega$ containing all open sets from Ω. It is called the *Borel σ-algebra* and any r.v. ξ which is measurable with respect to $\mathfrak{B}_\Omega$ is called a *Borel random variable*. The measure $\mathbf{P}$ on the probability space $(\Omega, \mathfrak{F}, \mathbf{P})$ defines the probabilities of the events pertaining to the r.v. ξ. For example, the event $\xi \in M$ $(M \in \mathfrak{X})$ occurs with probability $\mathbf{P}\{\xi \in M\} = \mathbf{P}\{\omega : \xi \in M\}$. It is convenient to define a measure on R^1 for the scalar r.v. ξ by using the distribution function $F(x) = \mathbf{P}\{\xi(\omega) \leq x\}$. For the multi-dimentional r.v. this can be done in a similar way.

Using the measure $\mathbf{P}$ the mathematical expectation of ξ is defined as the Lebesgue integral

$$\mathbf{E}\xi = \int_\Omega \xi(\omega)\mathbf{P}\{d\omega\}.$$

If $\xi \in R^1$ we can rewrite this formula as follows

$$\mathbf{E}\xi = \int_{-\infty}^{\infty} x\, dF(x).$$

1

Let the variable t be interpreted as time. We shall distinguish discrete and continuous time. In the first case the parameter t is running over the set $\{0, 1, 2, \ldots\}$ (sometimes $\{0, \pm 1, \pm 2, \ldots\}$) but in the second one t belongs either to a finite interval $[0, T]$ or to the half-axis $[0, \infty)$.

A family of r.v. $\xi_t(\omega)$ in continuous or discrete time is called a *random process* (r.p. for short). The r.p. will be called *non-terminating* if it is defined on an unbounded interval.

For a fixed ω, $\xi_t(\omega)$ is a function of time t which is called a *trajectory* or a *path* of this process. For a fixed t, $\xi_t(\omega)$ is a r.v. which represents the value of this process at time t. The trajectories $\{\xi_t(\omega), \omega \in \Omega\}$ will form a function space and a space of sequences when t is continuous and discrete respectively.

The probabilistic properties of a r.p. are defined by the measure $\mathbf{P}$ given on the appropriate probability space $(\Omega, \mathfrak{F}, \mathbf{P})$. This can be defined in various ways. At first we shall consider the construction of the scalar r.p. by using the family of finite-dimensional distribution functions associated with $\xi_t(\omega)$

$$\mathbf{P}\{\omega : \xi_{t_1}(\omega) \le x_1, \ldots, \xi_{t_m}(\omega) \le x_m\} \overset{\text{def}}{=} F_{t_1 \cdots t_m}(x_1, \ldots, x_m)$$

for all positive integers m and $t_1 < \cdots < t_m$. The distributions from this family have the following consistency properties:

1. For any $x_1, \ldots, x_m \in R^1$, $m < l$ and $t_1 < t_2 < \cdots < t_m < \cdots < t_l$

$$F_{t_1 t_2 \cdots t_l}(x_1, \ldots, x_m, \infty, \ldots, \infty) = F_{t_1 t_2 \cdots t_m}(x_1, \ldots, x_m);$$

2. If $i_1, \ldots, i_m$ is any transposition of the index-set $1, 2, \ldots, m$ then

$$F_{t_{i_1} t_{i_2} \cdots t_{i_m}}(x_{i_1}, \ldots, x_{i_m}) = F_{t_1 t_2 \cdots t_m}(x_1, \ldots, x_m).$$

Properties 1 and 2 are sufficient for the existence of a unique probability measure coinciding with the distributions $F_{t_1 t_2 \cdots t_m}(x_1, \ldots, x_m)$ given on the cylindrical sets (Kolmogorov Theorem). Moreover, there is a probability space $(\Omega, \mathfrak{F}, \mathbf{P})$ and r.p. $\xi_t(\omega)$ defined on it (where t is running over the same set from which $t_1, \ldots, t_m$ are taken) which corresponds to this family so that

$$\mathbf{P}\{\omega : \xi_{t_1}(\omega) \le x_1, \ldots, \xi_{t_m}(\omega) \le x_m\} = F_{t_1 t_2 \cdots t_m}(x_1, \ldots, x_m). \tag{1}$$

The probability measure $\mathbf{P}$ can be constructed in a standard way, i.e. first it is defined on the cylindrical sets of the form

$$M = \{\omega : \xi_{t_1}(\omega) \le x_1, \ldots, \xi_{t_m}(\omega) \le x_m\}, \quad t_1 < \cdots < t_m \tag{2}$$

by (1) and then it is extended to the σ-algebra generated by these sets. If the sets in (2) are open with respect to a topology of the given function space then the σ-algebra will be a Borel σ- algebra.

On the other hand, an r.p. in continuous time can be defined by using the family of conditional probability measures

$$\mu_t(M|\xi_s, 0 \le s < t) = \mathbf{P}\{\xi_t \in M|\xi_s, 0 \le s < t\}, \quad \forall M \in \mathfrak{X}$$

where the past history of the process up to time t is considered as the condition. Often, these conditional probabilities are assigned by using an increasing flow of the σ-algebras $\mathfrak{F}_t$ (i.e. for any $t_1 < t_2$ the inclusions $\mathfrak{F}_{t_1} \subseteq \mathfrak{F}_{t_2} \subseteq \mathfrak{F}$ take place). The σ-algebra $\mathfrak{F}_t$ is generated usually by the r.p. past history, i.e. by the system of the following sets

$$M_{t,a} = \{\omega : \xi_s(\omega) < a, s \leq t, \ a \in \mathrm{R}^1\}.$$

Then one introduces a family of conditional probabilities $\mu_s(\cdot|\mathfrak{F}_{t_-})$ where $\mathfrak{F}_{t_-}$ means the left-hand limit of the σ-algebras, i.e. $\mathfrak{F}_{t_-} = \lim_{s \uparrow t} \mathfrak{F}_s$.

In the discrete time case the r.p. can be defined by using a family of finite-dimensional distribution functions $F_{t_1 \cdots t_m}(\cdot)$ or by a family of conditional probabilities $\mu_{t+1}(M|\xi_0, \xi_1, \ldots, \xi_t)$. Further the abbreviation $\xi^t \stackrel{\text{def}}{=} (\xi_0, \xi_1, \ldots, \xi_t)$ (the past history of the process ξ_t up to time t) will be used to write down this conditional probability as $\mu_{t+1}(M|\xi^t)$. In what follows we shall assume that the above mentioned conditional probabilities satisfy the following conditions:

1. For any past history ξ^t the function $\mu_{t+1}(M|\xi^t)$ is a probability measure with respect to argument M;
2. For a fixed M this function is measurable with respect to the variables $\xi_0, \ldots, \xi_t$.

Using the family of measures $\{\mu_t\}$ one can write down the representations of the finite-dimensional distributions, i.e. an r.p. is given. In what follows we shall prefer to define an r.p. by means of a family $\{\mu_t\}$ which defines the measure uniquely on the state space of the process ξ_t, i.e. on the space of sequences $(\xi_0, \xi_1, \ldots, \xi_t, \ldots)$. Now the existence problem for an r.p. with given characteristics arises.

Let two sequences of measurable spaces $(\Omega_t, \mathfrak{F}_t)$ and $(X_t, \mathfrak{X}_t)$, $t = 1, 2, \ldots$ be given. A probability measure $\mathbf{P}_1$ and a conditional measure $\mathbf{P}(\cdot|\omega_1, \ldots, \omega_{t-1})$ are defined on $(\Omega_1, \mathfrak{F}_1)$ and $(\Omega_t, \mathfrak{F}_t)$ respectively. For any $A \in \mathfrak{F}_t$ let the function $P(A|\omega_1, \ldots, \omega_{t-1})$ be Borelian with respect to the arguments $\omega_1, \ldots, \omega_{t-1}$. Let us put for $A_j \in \mathfrak{F}_j, \ j \geq 1$

$$P_t(A_1 \times \cdots \times A_t) = \int_{A_1} \mathbf{P}_1(d\omega_1) \int_{A_2} \mathbf{P}_2(d\omega_2|\omega_1) \cdots \int_{A_t} \mathbf{P}_t(d\omega_t|\omega_1, \ldots, \omega_{t-1}).$$

Under these conditions the existence problem of a r.p. given on the space $(\Omega, \mathfrak{F}) = \prod_{t \geq 1}(\Omega_t, \mathfrak{F}_t)$ with the paths from $(X, \mathfrak{X}) = \prod_{t \geq 1}(X_t, \mathfrak{X}_t)$ and with the probabilistic characteristics given beforehand is solved by the following theorem:

Theorem 1. (Jonescu Tulcea) *In the space $(\Omega, \mathfrak{F})$ there exist an unique probability measure $\mathbf{P}$ and a process $\xi = (\xi_1(\omega), \xi_2(\omega), \ldots)$ such that*

$$\mathbf{P}\{\omega : \omega_1 \in A_1, \ldots, \omega_t \in A_t\} = \mathbf{P}_t(A_1 \times \cdots \times A_t),$$

$$\mathbf{P}\{\omega : \xi_1(\omega) \in A_1, \ldots, \xi_t(\omega) \in A_t\} = \mathbf{P}_t(A_1 \times \cdots \times A_t)$$

for all $t, \ A_j \in \mathfrak{F}_j.$

Sometimes we have to consider the scalar functions (the functionals) $\varphi_t = \varphi(\xi^t)$ defined on the paths of the r.p. which take values from some measurable space $(X, \mathfrak{X})$. These functions are measurable mappings $\varphi : X^t \to$ R. The sequence φ_t is also an r.p. for which it can be found, as a matter of principle, the family of the conditional distributions (or the finite-dimensional distribution functions) provided one is known for the original process. But it is rarely that such calculations can be done in an explicit form. For the mathematical expectation $W(t) = \mathbf{E}\varphi_t$ we have

$$W(t) = \int_{X^t} \varphi(x_1, \ldots, x_t)\mu(dx_t|x^{t-1})\mu(dx_{t-1}|x^t) \cdots \mu(dx_1).$$

It is often necessary that the parameters specifying the distribution of the r.v. (for example, in the case of the binomial distribution it is the probability of the success denoted by p usually; in the case of the normal distribution they are the mathematical expectation and the variance denoted by (m, σ) respectively) be included in the notation of this distribution. This is a standard situation in the theory of the random processes. Such parameters are indicated in the notations of the corresponding conditional probabilities or the distribution functions, i.e. $\mu_t^{(\theta)}$ or $F_{t_1,\ldots,t_m}^{(\theta)}$, or, generally speaking, $P^{(\theta)}$. In control theory importance is attached to the situation when there exists a time-varing variable, called a control and usually denoted by u_t, such that the probabilistic characteristics of the process depend on it. We shall now discuss this in more detail.

Let, in addition to $(X, \mathfrak{X})$, a measurable space $(U, \mathfrak{U})$ be given. The points u from U are called the *controls*. Let the conditional probabilities have the form

$$\mu_{t+1}(M|\xi^t, u^t), \quad t \geq 1, \quad M \in \mathfrak{X}.$$

and the following conditions are implemented:

1. The functions μ_{t+1} are the probability measures on X with respect to the first argument for all ξ^t and u^t.
2. For any $M \in \mathfrak{X}$ they are measurable with respect to $(\xi^t, u^t) \in X^t \times U^t$.
3. The functions μ_{t+1} depend on u_t essentially.

We shall call the conditional measures from the family $\{\mu_t, t \geq 1\}$ submitted to these conditions the *controlled conditional measures* (*distributions*). Such families given on the space $(\Omega, \mathfrak{F})$ define a class of r.p. taking the values from $(X, \mathfrak{X})$ with the controls from $(U, \mathfrak{R})$. To select some r.p. it is necessary either to fix in advance the sequence of controls or to give the rules of their choice in the course of the r.p. evolution.

Definition 1. The family of conditional distributions $\{\mu_t, t \geq 1\}$ or, which is the same, the class of r.p. associated with it is called the *model of the controlled object*.

Such models will be denoted by ξ, ζ or η.

1.2. Controlled Random Processes

First we shall assume that time is discrete. Let a measurable space $(\Omega, \mathfrak{F})$ be the space of the elementary events, $(X, \mathfrak{X})$ be the *state space* and $(U, \mathfrak{U})$ be the *space of controls* (or *actions*). We suppose that a model of the controlled object $\{\mu_t, t \geq 1\}$ is given as well. The r.p. state will be denoted by x_t, i.e. x_t is the value of the r.p. generated by this model at time t. The state x_t may be accessible to observation and measurement but this does not always occur. We can often judge about it only by the circumstantial evidence or by partial observations. Indeed, it is true in the case of many technological, physical and chemical processes where the direct measurement of all components is impossible. By reason of this we are forced to introduce a variable z_t to denote data observed at time t.

We shall consider z_t as a process given onto a measurable space $(Z, \mathfrak{Z})$ which is called the *space of observations*. The evolution of the process z_t is defined by the family of conditional measures $\{\nu_t, t \geq 1\}$ on Z where the function $\nu_t = \nu(H|x^t, z^{t-1}, u^{t-1})$, $H \in \mathfrak{Z}$ is measurable with respect to the variables included into the condition. The generation of the observations up to a moment t, i.e. $z^t = \{z_s, s \leq t\}$ is called the *history of the observable process* up to the moment t. Another name is an *information image of the model*.

The space Z can be a subspace of X. It means that the object is partially observed, i.e. a piece of information about the states of the process has been lost. If the spaces X and Z have no common points we shall have to judge about the states of the model by using indirect information only. Finally, it often turns out that $X = Z$ (of course, then $\mathfrak{X} = \mathfrak{Z}$). In this case it is said that the model is *completely observable*. Otherwise, it is called *partially observable* (or *with incomplete information*).

Finally, we shall define one more family of conditional measures on the space of the controls U.

Definition 1. We shall call the conditional measure on $\mathfrak{U}$

$$\sigma^{(t)} \overset{\text{def}}{=} \sigma^{(t)}(N|z^t, u^{t-1}), \quad N \in \mathfrak{U}, \ t \geq 0.$$

a *control choice rule* at time t.

In the singular case the randomized rule $\sigma^{(t)}$ turns into a deterministic one which is represented by the measurable mapping $\sigma^{(t)} : Z^t \times U^{(t-1)} \to U$. Hence the rule $\sigma^{(t)}$ points to the control u_t which must be chosen by using the informational image of the model up to time t, i.e. the sequence $u_0, z_1, u_1, \ldots, z_{t-1}, u_{t-1}, z_t$ in the stochastic or deterministic way. Such laws are called the *non-anticipated* ones.

Definition 2. A family of the choice rules of the controls $\sigma = \{\sigma^{(t)}, t \geq 1\}$ is called a *strategy of the control* or just a *strategy*.

The terms "strategy", "control algorithm" and "control law" are synonyms. In terms of automatic control theory a "strategy" is the feedback under which the disconnected control system is closed.

Let $\Sigma = \{\sigma\}$ denote a non-empty set of the "admissible" strategies. The choice of Σ depends on the structure of the model. The "admissibility" means that there are some specific circumstances differentiating one model from another which should be taken into account. We shall denote a strategy for the model $\{\mu_t, t \geq 1\}$ with the observations $\{\nu_t, t \geq 0\}$ by $\sigma_{\mu\nu}$.

Now we can formally define a controlled process in discrete time.

Let a family of conditional probabilities $\{\mu_t, \nu_t, \sigma^{(t)}\}$, $t \geq 0$ be given on the measurable spaces $(X, \mathfrak{X})$, $(Z, \mathfrak{Z})$, $(U, \mathfrak{U})$ of the states, the observations and the controls respectively. Here the strategy σ belongs to the set of the strategies Σ. We shall restrict ourselves to the case of topological spaces X, Z, U. Moreover, for all applications of the proposed theory it is sufficient to suppose that these spaces are complete separable metric spaces. It is quite reasonable to assume that the conditional probabilities μ_t, ν_t, $\sigma^{(t)}$ are the Borelian functions (with respect to the arguments noted into the conditions). It does not restrict the application field of our theory. According to Jonescu Tulcea Theorem there exists a three-dimensional r.p. (x_t, z_t, u_t) whose paths of which belong to the countable product $(X^\infty \times Z^\infty \times U^\infty, \mathfrak{X}^\infty \times \mathfrak{Z}^\infty \times \mathfrak{U}^\infty)$ of the state space, the observation space and the control space, respectively. The corresponding probability measure on the space $(\Omega, \mathfrak{F})$ is unique.

Definition 3. The sequence $\zeta = \{\mu_t, \nu_t, \sigma^{(t)}, t \geq 1\}$ formed by the family of the controlled conditional distributions μ and ν and the class of the admissible strategies $\Sigma_{\mu\nu}$ is called the *controlled random process* (CRP for short).

The spaces X, Z, U are defined by the measures μ and ν respectively. We shall denote the CRP under a fixed strategy σ by $\zeta(\sigma)$.

Any strategy $\sigma \in \Sigma_{\mu\nu}$ defines a random process $\zeta(\sigma)$ pertaining to the r.p. class given by the family μ.

Let us describe the evolution of a CRP at some fixed strategy from $\Sigma_{\mu\nu}$. At the initial moment $t = 0$ our model is in a state x_0 which generates an observation z_0. Taking into account this observation a control u_0 is calculated. At the moment $t = 1$ the model passes into a new state x_1 which produces a new observation z_1 and then a new control u_1 is calculated by using the data (z_0, u_0, z_1). This new control leads to the new observation and state at time $t = 2$ respectively. Using the past history $(z_0, u_0, z_1, u_1, z_2)$ a control u_2 is defined and so on. So step by step the controlled random process is progressed. At each time t it consists of three components $i_t = (x_t, z_t, u_t)$. For the non-terminating processes a result of the control is the infinite sequence of the triplets $(i_0, i_1, \ldots, i_t, \ldots)$ that represents the trajectory of the process in $X^\infty \times Z^\infty \times U^\infty$. We can consider separately the trajectory of the model x_t in the states space X and the trajectories z_t and u_t in the spaces Z and U respectively.

Definition 4. The finite collection $i^t = (i_0, i_1, \ldots, i_t)$ is called the *history of the controlled process* up to time t.

As mentioned above in the important case when $X = Z$ and $x_t = z_t$ one speaks of an observable process. Then the description of the controlled process is simplified.

Let the initial values x_0, u_0 and strategy σ be given. Then the mathematical expectations $E^{(\sigma)}_{x_0,u_0} \varphi_t$ and the higher moments of the function $\varphi = \varphi(x^t, u^{t-1})$ defined on the trajectories of the controlled process can be calculated (if they exist) by the evident formulae.

The deterministic models under the deterministic observations are often met in control problems. It is clear that they fall under the general conception of controlled random processes.

We shall end the list of the main notions of control theory with the notion of a *control aim* which so far had no abstract definition. Instead examples of some concrete goals are usually stated. The control aim is to provide the model with some desirable property that takes place under some strategies but not under others. The control algorithms (the strategies) serve to attain it.

We shall consider some deterministic models as examples of the control goals. As a general rule such a model is represented by a difference equation or a differential one. It is often required that its solution be dissipative or stable (in the Lyapunov sense) or asymptotically stable. Such aims are called the *stabilizational aims*. The *optimization aims* are more complicated. They are associated with some *objective functions* $W(\sigma)$ defined on the set of admissible strategies $\Sigma_{\mu,\nu}$. It is required to find an optimal strategy σ_0 that maximizes this function, i.e. $W(\sigma_0) = \max_{\sigma \in \Sigma} W(\sigma)$ or an ε-optimal strategy σ_ε under which the inequality $W(\sigma_\varepsilon) > \sup_{\sigma \in \Sigma} W(\sigma) - \varepsilon$ holds. Sometimes this inequality is written in the form $W(\sigma_\varepsilon) > (1 - \varepsilon) \sup_{\sigma \in \Sigma} W(\sigma)$. The optimal strategy and the extreme value depend on the given set of the admissible strategies.

The objective function is often additive. So if T is a finite number then

$$W(\sigma) = \sum_{t=1}^{T} \varphi(x_t, u_{t-1})$$

or

$$W(\sigma) = \lim_{T \to \infty} T^{-1} \sum_{t=1}^{T} \varphi(x_t, u_{t-1})$$

in the infinite time case.

Another widespread goal consists of satisfying the inequalities $\varphi(x^t, u^{t-1}) < 0$, $t \geq 1$. Not only the stabilizational aims but also many optimizational aims are its particular cases. The goals mentioned above can refer to "global" aims connected with the non-terminating paths of the model. The "local" aims may be rather important as well. For example, they are the transition from an initial state x_0 into a state $\tilde{x}$ for time T or the optimal high-speed problem, i.e. starting from x_0 it is needed to reach $\tilde{x}$ in minimal time under some restrictions on the controls.

Let us now consider the stochastic controlled models. The main aims of the control are divided into two groups. The first of them is composed of so-called *strong aims* referred directly to the paths. They have a probabilistic sense. Here some examples of such aims are adduced. It is required to provide the implementation of the following inequalities

$$\varphi(x^t, u^{t-1}) < 0, \quad \forall\, t \geq 1$$

or of equality

$$\lim_{T\to\infty} T^{-1} \sum_{t=1}^{T} \varphi(x_t, u_{t-1}) = \max_{\sigma} W(\sigma)$$

which take place for almost all paths, i.e. they are true with probability one.

The strong aims are reached under heavy conditions on the considered models. As a general rule the model must be ergodic.

Finally, the second group is composed of the *weak aims*. They refer not to a function defined on the paths of the model but to the average characteristics defined by the chosen strategy. Let us consider the main types of strategies before we write down the explicit formulae of the mathematical expectations for the considered functions under these strategies.

Program strategies are the simplest. They do not depend on the evolution of the model. Such strategies are sequences of unconditional distributions $\varkappa_t$ defined onto the space of controls U fixed in advance. If these distributions are singular then the controls will be some functions of time $u_t = f(t)$ which form the sequence $u_0, u_1, \ldots, u_t, \ldots$. The program strategies are convenient particularly in the technical applications. Various applied problems of control theory were solved by using them.

Stationary strategies are formed by using the identical rules. If the time t is running over all integers ($t = 0, \pm 1, \pm 2, \ldots$) then $\varkappa_t = \varkappa$ for all t. But if the initial moment is fixed, i.e. $t = 0, 1, 2, \ldots$ then $\varkappa_t = \varkappa$ beginning at $t = h \geq 0$. In the simplest case when $X = Z$ the considered strategies have the form $\varkappa(\cdot\,|\,x_{t-h}, x_{t-h+1}, \ldots, x_t; u_{t-h}, u_{t-h+1}, \ldots, u_{t-1})$ $\left(\varkappa\big(\cdot\,\big|\,x_{t-h}^t u_{t-h}^{t-1}\big)\right.$ for short). In other words, the rules are different up to time $t = h$ but after this moment they coincide.

Definition 5. The number h is called the *memory depth* of the stationary strategy.

The *stationary program strategy* consists of using the distribution $\varkappa$ only. In the deterministic case the control $u_t = u$ is repeated time after time.

Let us write down the representation of the mathematical expectation of the function $\varphi_t = \varphi(x^t, u^{t-1})$ under a program strategy having the form $(u_0, u_1, u_2, \ldots)$. We have

$$W(u^{t-1}) \stackrel{\text{def}}{=} \mathbf{E}\varphi_t$$

$$= \int_{X^{t+1}} \varphi(x_0, \ldots, x_t; u_0, \ldots, u_{t-1})\mu(dx_t\,|\,x^{t-1}, u^{t-1}) \cdots \mu(dx_0\,|\,u_0). \tag{1}$$

Following this $\mathbf{E}\varphi_t$ is a measurable function of u. If the space U is topological then the conditional probabilities $\mu(\cdot|x^t, u^t)$ will be continuous with respect to u for all t and the integral from (1) converges uniformly. Then the functions $W(u^t)$ are continuous with respect to all of its arguments jointly. If the program strategy is randomized then $\mathbf{E}\varphi_t$ will not be a function but only a number. Indeed (again for the sake of simplicity we assume that $X = Z$)

$$
\begin{aligned}
W(t) &\overset{\text{def}}{=} \mathbf{E}\varphi_t \\
&= \int_{X^{t+1} \times U^t} \varphi(x_0, \ldots, x_t; u_0, \ldots, u_{t-1}) \mu(dx_t | x^{t-1}, u^{t-1}) \cdots \mu(dx_0 | u_0) \\
&\quad \times \sigma^{(t-1)}(du_{t-1}) \cdots \sigma^{(0)}(du_0).
\end{aligned}
$$

Definition 6. A strategy generated by a sequence of rules which have memory depth equal to one, i.e. by the distributions $\sigma^{(t)}(\cdot|x_t)$ or $\sigma^{(t)}(\cdot|z_t)$ given on U is called a *Markov strategy*.

For the Markov strategy the current value of the model or the observation is important only. The deterministic Markov strategy is formed by using the functions of either $f_t(x_t)$ type or $f_t(z_t)$ type.

Definition 7. A strategy generated by a distribution $\sigma(\cdot|x)$ (i.e. at every instant of time the controls are chosen according to the same rule) is called a *stationary Markov strategy*.

Finally, a combination of the notions mentioned above leads to the *simple strategies*, i.e. the *stationary deterministic Markov strategies* consisting in using the same function f of the current state, i.e. $f(x_t) = u_t$.

This notion is used especially often since the solution of many problems can be obtained by means of the simple strategies. Then the mathematical expectation for any function can be calculated by analogy with the previous formula

$$
\begin{aligned}
\mathbf{E}\varphi_t &= W(t) \\
&= \int_{X^{t+1}} \varphi(x_0, \ldots, x_t; f(x_0), \ldots, f(x_t)) \mu(dx_t | f(x_t)) \cdots \mu(dx_0 | f(x_0)).
\end{aligned}
$$

For any strategy (non-stationary and randomized) depending on the whole history of the model evolution the measure generated by it on the set of paths leads to the expression

$$
\begin{aligned}
\mathbf{E}_\sigma \varphi_t &= W(\sigma, t) \\
&= \int_{X^{t+1} \times U^t} \varphi(x^t, u^t) \prod_{i=0}^{t} \mu_i(dx_i | x^{i-1}, u^{i-1}) \sigma^{(i)}(du_i | x^i, u^{i-1}),
\end{aligned}
$$

where x^{-1}, u^{-1} should be treated as a formal notation of the absent variables. The partially observable models are given in a more complicated way. The analogues of the integral considered above have the rather cumbersome form. For this reason the appropriate formulae are omitted.

Now we can finally point out the typical aims of the control. It is assumed that the class Σ of the admissible strategies is chosen so that $\mathbf{E}_\sigma \varphi_t$ should be finite for all t. The following functions $\varphi_t = \varphi(x_t, u_{t-1})$ are the most common ones. Their role is connected with the fact that sometimes we can draw a conclusion about a current state x_t of the model knowing the value of some numerical characteristic of the state (for example, the function φ_t). The next aim is quite reasonable with respect to functions

$$\lim_{t \to \infty} \mathbf{E}_\sigma \varphi_t = \sup_{\sigma \in \Sigma} W(\sigma) = \bar{W}.$$

Here $W(\sigma)$ is the limiting objective function obtained by passing to the limit, i.e. $W(\sigma) = \lim_{t \to \infty} W(\sigma, t)$. This aim is called *asymptotic optimality*. In connection to the above we shall cite the appropriate terminology. So φ_t is the *reward* at time t, $\mathbf{E}_\sigma \varphi_t$ is the *average reward* at time t under the strategy σ, $W(\sigma)$ is the *limiting reward* and, at last, $\bar{W}$ is the *maximum limiting average reward*. We are often forced to restrict to approximate optimality. For a fixed $\varepsilon > 0$ it is required to find a strategy σ_ε such that the inequality

$$\lim_{t \to \infty} \mathbf{E}_{\sigma_\varepsilon} \varphi_t > \bar{W} - \varepsilon$$

holds. Such an aim is called *ε-optimality* (in the weak sense).

The weakest optimal aims are concerned with the Cesaro averages such as

asymptotic optimality

$$\lim_{T \to \infty} T^{-1} \sum_{t=1}^{T} \mathbf{E}_\sigma \varphi_n = \bar{W}$$

and *ε-optimality*

$$\lim_{T \to \infty} T^{-1} \sum_{t=1}^{T} \mathbf{E}_\sigma \varphi_t > \bar{W} - \varepsilon.$$

Stabilization problems can sometimes refer to the weak aims. For instance, the asymptotic stability in the mean square sense can be reasonably interpreted as the minimization problem (with respect to the limit) of the function $\mathbf{E}_\sigma \|x_t\|^2$, i.e. $\lim_{t \to \infty} \mathbf{E}_\sigma \|x_t\|^2 = 0$.

The other aims of control are the fulfilment of the "goal inequalities" such as $a(t) \le W(\sigma, t) \le b(t)$ under some given numerical sequences $a(t), b(t)$. The more general form of such inequalities is connected with a family of functions $(\varphi^{(1)}, \ldots, \varphi^{(l)})$. It is required that the inclusions

$$\left(W^{(1)}(\sigma, t), \ldots, W^{(l)}(\sigma, t) \right) \in G$$

where $W^{(i)}(\sigma, t) = \mathbf{E}_\sigma \varphi_t^{(i)}$, $i = 1, \ldots, l$, $G \in \mathrm{R}^l$ take place for all $t \ge t_0$.

The majority of well-known aims of control may be reduced to aims of this kind. For example, the optimizational aims (in the weak sense) or the stability problems (in the last case the goal is the fulfilment of the inequality $E_\sigma \|x_t\|^2 \leq \gamma$) are considered as such aims.

We shall now discuss methods of solving control problems for some types of models.

A *process with independent values* (PIV for short) defined by a family of controlled conditional distributions of the form

$$\mu_{t+1}(\cdot | x^t, u^t) = \mu_{t+1}(\cdot | u_t),$$

is the simplest kind of CRP. This means that only the last of the controls used appears in the condition. In this case the class of the admissible strategies Σ consists only of the sequences of conditional distributions on U. In the physical point of view these processes are inertia-free and independent on the previous evolution. Sometimes it is more convenient to write down PIV in the form $x_t(u_{t-1})$.

Homogeneous processes with independent values (HPIV for short) have fundamental importance. Their conditional probabilities do not depend on time, i.e. $\mu_t(\cdot | u) \equiv \mu(\cdot | u)$. The measurable function $\varphi(x_t, u_{t-1})$ defined on the HPIV path is also a HPIV. The name of these processes is derived from the fact that the program strategy inverses HPIV into a sequence of the independent random variables identically distributed under $u_t \equiv u$. As we shall soon see the control strategy for HPIV has such a form. In the deterministic case, i.e. when the measures μ are degenerate the HPIV will be represented by a single-valued function of the form

$$x_t = g(u_{t-1}),$$

i.e. it is the optimization standard object. The more complex strategy transforms the HPIV. For example, the simple strategy $\sigma = \{f(x)\}$ turns it into a homogeneous Markov process with the transition function $\mu(M | f(x))$. The strategies of the general form having large or increasing memory depth transform the HPIV into some Markov process or, generally speaking, into a random process of general form.

The control aims for PHIV are usually double, i.e. the optimization problem and the achievement of a given level. They are formulated in the terms of "average rewards" for one step.

For the sake of simplicity we shall not touch the function $\varphi(x_t, u_{t-1})$ but we shall consider the scalar model only, i.e. $X = R^1$. Then the average reward at time t under a control u is equal to

$$W(u) = \int_{-\infty}^{\infty} y \mu(dy | u).$$

The average reward at the same time under a strategy $\sigma = \{\sigma^{(j)}\}$ is equal to

$$\mathbf{E}_\sigma x_t = \int_{-\infty}^{\infty} \cdots \int_{-\infty}^{\infty} W(u_{t-1}) \prod_{j=1}^{t} \mu(dx_j | u_{j-1}) \sigma^{(j)}(du_{j-1} | x^{j-1}, u^{j-2}).$$

The following estimate is evident

$$\mathbf{E}_\sigma x_t \leq \max_u W(u).$$

We shall use the same symbols $W(u)$ and $W(\sigma)$ for the mathematical expectation at a fixed moment and for the limiting reward considered as a function given on Σ respectively. They differ by the arguments u or σ only. Having this in mind we shall obtain the obvious inequality

$$W(\sigma) \leq \max_u W(u).$$

Here the equality will take place if the control $u_0 = \arg\max W(u)$ is applied for all t, i.e. the stationary strategy $\sigma_0 = \{u_0, \ldots, u_0, \ldots\}$ is used. Then

$$\sup_\sigma W(\sigma) = \max_u \bar{W}(u).$$

Applying the strong law of the large numbers to the sequence $x_0, x_1, \ldots, x_k, \ldots$ obtained of the independent identically distributed random variables, we shall obtain

$$\lim_{T \to \infty} T^{-1} \sum_{t=1}^{T} x_t = \bar{W} \quad \text{a.s.}$$

For a task to be solved on the achievement of the given average level it is necessary to find a strategy σ^* such that the equality $W(u) = \alpha$ (where α is a number from the domain of the function W) takes place for all time t. The solution is obvious, namely, it is needed to calculate the root of the equation $W(u) = \alpha$.

It is also simple to give methods of solving other control problems by HPIV. Those methods are based on the assumption that all characteristics of the process are known exactly. Before we had used the average reward only.

There is another interpretation of the HPIV notion (in so-called "wide sense") which, in turn, means that

(a) for the fixed $u_1, \ldots, u_{t-1}$ the r.v. $x_{t+1}(u)$ does not depend on the r.v. $x_i(u_{i-1})$, $i \leq t$;

(b) $\mathbf{E}x_t(u) = W(u)$;

(c) $\sup_t \mathbf{E}x_t^2(u) < \infty$.

Let us consider the next class of processes called *controlled Markov chains*. These are specified by the 5-symbol collection $C = \{X, U, P^{(u)}, \bar{p}, \zeta)\}$ where $X = \{x_1, \ldots, x_m\}$, $U = \{u_1, \ldots, u_k\}$ are the sets of states and controls respectively, the stochastic matrices $P^{(u)} = (p_{ij}^{(u)})$ are formed by the controlled one-step transition probabilities from x_i to x_j under the control u, i.e. $p_{ij}^{(u)} = \mathbf{P}\{x_i \overset{u}{\to} x_j\}$, $\bar{p} = (p_1, \ldots, p_m)$ is an initial distribution and, finally, $\zeta = \zeta(x, u, \omega)$ is a numerical r.v. denoting the reward in the state x under the applied control u. The average rewards $r_x^u = \mathbf{E}\zeta(x, u)$ are required to be finite. The choice of the admissible strategy leads to the "evolution" of the Markov chain, i.e. the state x_t and the reward $\zeta_t = \zeta(x_t, u_t, \omega)$ become some functions of time. Unless otherwise stated these functions (the paths) will be assumed non-terminating.

The HPIV considered above are controlled Markov chains with a single state.

In the theory of controlled Markov chains one usually considers the discounted reward

$$W_\beta(\sigma, \bar{p}) = \sum_{t=0}^{\infty} \beta^t \mathbf{E}_{\sigma,\bar{p}} \zeta_t, \quad 0 \le \beta < 1,$$

or the limitting average (for one step) reward

$$W(\sigma, \bar{p}) = \lim_{T \to \infty} T^{-1} \sum_{t=0}^{T} \mathbf{E}_{\sigma,\bar{p}} \zeta_t, \quad 0 \le \beta < 1$$

as the objective functions. It is required to find an optimal strategy maximizing one of these functions. The structure of the optimal strategy is well known and simple. There are only k^m different functions defined onto X which take the values from U and generate the simple strategies. These functions form the set Σ. Linear programming provides the necessary calculation tools to find the optimal strategies. Let us state this problem for the ergodic chains having the communicating states only, i.e., there is a positive probability that for a finite number of transitions each state x_i can be reached from any state x_j. We shall restrict ourselves here to the case of the second objective function; the indexes i, j and l correspond to the states and to the controls respectively.

$$\sum_{i,l} r_i^l x_i^l \to \max;$$

$$\sum_{l} x_j^l - \sum_{i,l} p_{ij}^l x_i^l = 0, \quad j = 1, \dots, m;$$

$$\sum_{i,l} x_i^l = 1;$$

$$x_i^l \ge 0, \quad i = 1, \dots, m; \ l = 1, \dots, k.$$

Here the quantities x_i^l mean the choice probabilities of the control u_l in the state x_i^l or, in other words, the optimal strategy is being searched in the class of randomized strategies. The analysis shows that for any i there exists the unique $l(i)$ such that $x_i^{l(i)} > 0$ and $x_i^l = 0$ for $l \ne l(i)$. The solution of this problem points to

(a) the optimal control law $u(x_i) = u_{l(i)}$;
(b) the limiting probabilities of the states under the optimal strategy $\pi_i(\sigma_{\mathrm{opt}}) = x_i^{l(i)}$;
(c) maximum of the objective function

$$W(\sigma_{\mathrm{opt}}) = \sum_{i=1}^{m} r_i^{l(i)} x_i^{l(i)}.$$

According to the strong law of large numbers for the ergodic Markov chains not only the objective condition concerned with the average rewards but the stronger equality

$$\lim_{T \to \infty} T^{-1} \sum_{t=1}^{T} \zeta_t(u_{t-1}) = \max_{\sigma} W(\sigma), \quad \text{a.s.}$$

holds. The limit value on the left-hand side of the last equality does not depend, naturally, on the initial state of the chain.

For the non-ergodic controlled Markov chains the problem of linear programming is more complicated, i.e.

$$\sum_{i,l} r_i^l x_i^l \to \max;$$

$$\sum_{i,l} (\delta_{ij} - p_{ij}^l) x_i^l = 0, \quad j = 1, \ldots, m;$$

$$\sum_l x_i^l - \sum_{i,l} (\delta_{ij} - p_{ij}^l) y_i^l = p_j; \quad j = 1, \ldots, m;$$

$$x_i^l, y_i^l \geq 0, \quad i = 1, \ldots, m; \quad l = 1, \ldots, k.$$

Here δ_{ij} is the Kronecker symbol, p_i is an initial distribution, the variables x_i^l and y_i^l correspond to states from the ergodic classes and to all states respectively. Here we do not go into details. This problem with 2^{mk} unknown variables can be solved by one of the numerical methods of linear programming. Frequently some variant of the simplex method is used.

As seen from what has been said, to solve the optimization control problems of Markov chains it is required to know information about all chain parameters, i.e. the collection of $m^2 k$ controlled transition probabilities p_{ij}^l must be known exactly. Similarly, for HPIV information about $\mu(\cdot | u)$ is required. This *a priori* information may be replaced by the explicit form of the average reward $W(u)$.

The greatest difficulties arise for the *partially observable Markov chains* differing from the above-mentioned ones by the structure of the admissible strategies. They are formed by the laws $\sigma^{(t)} = \sigma(\cdot | z^t, u^{t-1})$ where z_t signifies the observable variable. It can be:

(1) a reward ζ_t;
(2) a "pseudostate" $z \in Z = \{z_1, \ldots, z_n\}$, z_j being observed with probability q_{ij} in the state x_j. The numbers q_{ij} form the stochastic matrix Q;
(3) the pair (ζ_t, z_t).

For the partially observable chains the main control problem is still unsolved.

Now we shall give another description of Markov chains that has an abstract form. To that end, we shall use the branch of mathematical logic called the theory of automata. We shall start with *deterministic finite automata*. Let three finite sets $X = \{x_1, \ldots, x_m\}$, $Y = \{y_1, \ldots, y_k\}$, and $S = \{s_1, \ldots, s_n\}$ be given. They are called the alphabets of the input signals, the output signals and the states respectively.

We shall denote a transition function by $\varkappa$, $\varkappa : X \times S \to S$ and an output function by λ, $\lambda : X \times S \to Y$ but s_0 is an initial state.

Definition 8. The collection

$$\mathfrak{A} = \{X, S, Y; s_0, \varkappa, \lambda\}$$

is called a *Mili automaton.*[a]

Clearly it is treated as a one-one mapping of the semi-group of the words $G_X = \{x_{i_1} x_{i_2} \ldots x_{i_p}, \forall p, i_1, \ldots, i_p\}$ from the alphabet X into that $G_Y = \{y_{j_1}, \ldots, y_{j_m} \forall q, j_1, \ldots, j_q\}$ from the alphabet Y. It is natural to consider this mapping as a time-varying one. So starting from some initial state s_0 the automaton receives the input symbols $x_{i_1}, \ldots, x_{i_t}$ which are transformed by the transition function $\varkappa$ into the corresponding states $s_{\nu_1}, s_{\nu_2}, \ldots, s_{\nu_t}$ and by the output function λ into the sequence of the output signals $y_{j_1} y_{j_2}, \ldots, y_{j_t}$ having the same length. The initial segments of the input word correspond one-one to that of the output word. This mapping is called an *automaton mapping.* In mathematical logic automata characterize a class of algorithms.

Moor automata differs from Mili ones by the output function λ which depends on the state only, i.e. in this case $y = \lambda(s)$. It means that for Moor automaton the state space can be decomposed into a sum of disjoint subsets, i.e. $S = S^{(1)} + \cdots + S^{(k)}$ where the same output signal y_j corresponds to all states from $S^{(j)}$. With this difference, however, both types of automata are equivalent, i.e. they realize the same mapping.

The inner logic of automata theory and its applications have led to the creation of more complicated types of automata realizing more general automaton mappings. The probabilistic (stochastic) automata were the first of them.

Definition 9. The collection

$$\mathfrak{A} = \{X, S, Y, \bar{p}, \Pi(x), \mu\}$$

is called a *stochastic Moor automaton.*

Here $\bar{p} = (p_1, \ldots, p_m)$ is the initial distribution of the automaton states, $\Pi(x) = \big(\pi_{ij}(x)\big)$ is the stochastic matrix of transition probabilities from the state s_i to the state s_j under the input signal x, $\mu = \mu(y|s)$ is the conditional distribution onto Y under the automaton state s. The stochastic Mili automaton can be defined in a similar way.

It is easy to see that the stochastic Moor automaton (and Mili one as well) are controlled Markov chains. Indeed, the input signals x are the controls, the states of automaton serve as the states of the chain, the matrices $\Pi(x)$ are the analogue of the controlled conditional transitional probabilities $\mathbf{P}\{s(t+1)|s(t), x(t)\}$, the initial distribution $\bar{p}$ has the same meaning in both cases and, finally, the output

---

[a]The *non-initial automata* are used often. Their initial state is not specified.

signals y are the abstract analogue of the rewards to appear with the probabilities $\mu(y|s)$. In the case of numerical rewards one may not introduce the set Y and the conditional distribution μ. Instead, it is enough to given the family of r.v. $\zeta(s,\omega)$ or $\zeta(s,x,\omega)$ which means the presence of the corresponding conditional distribution. Under such an agreement the automaton can be written in the form that does not differ from the notation of a Markov chain (if, additionally, we write $U = \{u\}$ instead of $X = \{x\}$)

$$\mathfrak{A} = \{S, U, \bar{p}, \Pi(x), \zeta(s,x)\}.$$

Automata with variable structure whose matrices $\Pi_t(x)$ and the r.v. ζ_t depend on time t are regarded as another type of automata. They will also be useful for us in the future. These automata can be reduced to the common ones but with some infinite S.

The terminology of automata theory often has a number of advantages. One of them is that an automaton can be naturally interpreted as a control algorithm. It is true in technique. For example, the controlling calculators and discrete devices of automatic control represent by itself the finite deterministic automata.

The strategy realized by finite automata can be directly applied not only to HPIV but to Markov chains as well. As we shall see later the infinite automata can provide attainment of the aims of control for the most complicated models. The process of the control of the model represented in the form of an automaton A_μ by means of an automaton A_σ can be visually illustrated as an interaction of two automata. The first one A_μ sends the sequence of states of the model x_t or the observable values z_t to the automaton A_σ and the latter replies to it by the sequence of the controlling signals u_t. Such a system denoted by $A_\mu \otimes A_\sigma$ realizes the one-one mapping.

As the last example of control theory problem we shall choose the *linear-quadratic problem* (LQP). In the simplest form it is connected with the linear difference equation with constant coefficients

$$x_t + a_1 x_{t-1} + \cdots + a_n x_{t-n} = b_1 u_{t-1} + \cdots + b_m u_{t-m} + \xi_t, \quad t \geq 0,$$

or, in another form, with the system of the first order equations

$$y_{t+1} = Ay_t + Bu_t + \eta_t, \quad t \geq 0. \tag{2}$$

The numbers (a_i, b_i) and matrices A, B are supposed to be known, ξ_t and η_t are the additive noises, i.e. the r.p. of some nature. Let us consider Eq. (2) supposing that y_t, $\eta_t \in \mathrm{R}^l$, $u_t \in \mathrm{R}^m$. We assume that η_t is a sequence of independent, identically distributed r.v. where $E\eta_t = 0$ and the matrix R of their second moments is finite. The initial value y_0 is supposed to be a r.v. with the distribution P and with the matrix of the second moments R_0 but $\mathbf{E}y_0 = m$. We shall choose the function

$$W(\sigma,\mathbf{P}) = \sum_{t=0}^{T-1} \left[y_t^T Q_1 y_t + u_t^T Q_2 u_t \right] + y_T^T Q_0 y_T,$$

defined on the finite time interval $[0, T]$ as an objective function. Here Q_0 and Q_1 are some non-negative matrices, Q_2 is positive definite. The aim is to minimize the non-negative function $W(\sigma, \mathbf{P})$. The solution of this problem has been carefully studied. The optimal control law is linear (by the state), deterministic and non-stationary. It has the form

$$u_t = K(t)x_t$$

where the amplification matrix $K(t)$ can be expressed in terms of the numerical parameters of the equation and of the objective function, namely,

$$K(t) = [Q_2 + B^T S(t+1)B]^{-1} BS(t+1)A.$$

Here the non-negative definite matrices $S(t)$ are the solution of the Riccati matrix recurrent system

$$S(t+1) = A^T S(t)A + Q_1 - A^T S(t)\left[Q_2 + B^T S(t+1)B\right]^{-1} BS(t)A$$

with the following boundary condition (on the right end) $S(T) = Q_0$. This system will be solved **before** the solution of the optimization problem is found. The minimal value of the objective function is equal to

$$\bar{W}(T) \stackrel{\text{def}}{=} \min_{\sigma} W(\sigma, T) = m^T S(0)m + \mathbf{sp}\, S(0)R_0 + \sum_{n=0}^{T-1} \mathbf{sp}\, S(n+1)R_1.$$

If the observations are incomplete, i.e. $z_t = Ly_t + \zeta_t$ we shall have to use additional considerations to obtain the useful signal y_t from the accessible information z_t. With this aim in view the "Kalman filter" is used.

We remark that solving the control problem for the new model again requires full information about characteristics of the linear difference equations, i.e. it is necessary to know $\mathbf{E}y_0$, the matrix R for the noise η_t, and five matrices A, B, Q_0, Q_1, Q_2. For the values of the amplification matrix $K(t)$ to be calculated this *a priori* information is needed. Otherwise, if there is a lack of *a priori* information or inaccuracies into the description of the model then the linearly-quadratic problem cannot be solved and it remains only to sympathize with the designer of such a control system.

Let us return to the controlled model with the strategy. Let the model and the strategy be represented by automata A_μ and A_σ correspondingly and it is assumed that the spaces of the states X, the observations Z and the controls U are some measurable (and may be topological) spaces. Supposing that $X = Z$ we shall accept the controlled Markov process A_μ with the state space X and with the control space U as a model. The properties of the process are specified by the controlled transition function $\mu(\cdot|x, u)$. We shall choose the stochastic Moor automata $A_\sigma = \{X, S, Y; \Pi(x), q\}$ as a class of admissible strategies Σ_μ. As arranged above an interaction between A_μ and A_σ is denoted by $A_\mu \otimes A_\sigma$.

We shall connect with the object $A_\mu \otimes A_\sigma$ an *associated Markov process* (C, P) where $C = X \times S$ is the state space of this process but its transition function P is given by

$$P(U|c) = \int_{X \times U} \pi_{ij}(w)\mu(dw|x, u)q(du|s_i)$$

where $U = M \times S$ is a subset of C, $M \in \mathfrak{X}$, the pair $c = (x, s_i)$ being the previous state of the associated Markov chain. It is easy to write down the transition probability from c into a set $M \times \hat{S}$ where $\hat{S} \subset S$. This process is homogeneous in time. Its paths are the sequences $(x_0, s_0; x_1, s_1; \ldots; x_t, s_t, \ldots)$. For the discrete spaces X and U the integrals in the above mentioned formula are replaced by the sums and we can speak about the transition from the state $s' = (x', s')$ into $s'' = (x'', s'')$. If the spaces X and S are finite then (C, P) will become the *associated Markov chain*. It is interesting to know the conditions under which it is regular, i.e. it is ergodic without cyclic subclasses. Here we shall point to one sufficient condition, namely,

> *the chain is regular but the automaton is strongly tied up and it has no cyclic states.*[b]

The regular chains have positive limiting probabilities. It enables us to calculate the limiting mathematical expectation of the reward (if it is defined for this chain). The form of the limiting average reward depends on the strategy A_σ used.

Let us consider the special case of the controlled HPIV. In this case the structure of the associated process (C, P) is simple enough. The set $C = S$, where S is the state set of this automaton, serves for the state set of the process whose transitions are regulated by the matrix $P = (p_{ij})$

$$p_{ij} = P(s_i \to s_j) = \int_{X \times U} \pi_{ij}(w)\mu(dw|u)q(du|s_i).$$

This process is again homogeneous. If the PIV is considered as a model (a lack of homogeneity takes place) or automaton A_σ has changeable structure then the process (C, P) will be non-homogeneous. If the set S is finite, we shall again call (S, P) a chain. It is regular under the condition stated above and there exists a positive stationary distribution $\pi_1, \ldots, \pi_{|s|}$ which does not depend on the initial state. The limiting average reward (HPIV is scalar) is equal to

$$W(\xi, A) = \sum_{l=1}^{k} W(u_l)\tilde{\pi}_l$$

where $\tilde{\pi}_l = \sum_j \pi_j q(u_l|s_j)$ are the stationary probabilities of the choice of the control u_l and $W(u) = \int x\mu(dx|u)$ is the average reward under the control u.

---

[b]The strong tie-up means that from each state of the chain it can pass into any other. The state of automaton will be called a cyclic if the greatest common divisor of the lengths of the input sequences which transform this state into itself with a positive probability is greater than one.

If the initial state is s_0 and the automaton A realizes the strategy $u_1, \ldots, u_k$ then the mathematical expectation of the reward at time t will be given by

$$W(\xi, A, s_0, t) = \sum_{l=1}^{k} W(u_l) \sum_{j} p_{s_0 j}^{(t-1)} q(u_l | s_j).$$

As known from the theory of Markov chains

$$\lim_{t \to \infty} W(\xi, A, s_0, t) = W(\xi, A)$$

with exponential convergence rate. (It is defined as the second largest absolute value of the eigenvalues of the matrix P.)

Our consideration has been concerned with both controlled models and control problems in discrete time. Fundamental difficulties prevent any formal and correct definition of controlled processes in continuous time. Therefore such definitions are not stated. Nevertheless, this will not pose us problems since from the whole set of such processes we shall confine ourselves to these topics:

(1) the semi-Markov countable processes;
(2) the ordinary differential equations of the form

$$\dot{x} = Ax + Bu + h(t)$$

where $h(t)$ is the deterministic or stochastic additive noise;
(3) the stochastic differential Ito equations

$$dx_t = (Ax_t + Bu_t)dt + C dw_t$$

where w_t signifies a Wiener process.

For all above-mentioned cases the structure of optimal and admissible strategies is well known.

1.3. Definition of Adaptive Control

The control problems described above are characterized by the presence of complete information about the model required for their solution. We have to know not only the model structure but all functions and constants entering into the description of this model. For short we shall name the control theory worked out for such cases as the *classical control theory*. For a long time this theory has been working and has reached an advanced stage of development and has proved its significance in the many practical applications. Let us turn to another situation initiated mainly by practical interest but (partially) also by theoretical one. We shall suppose now that *a priori* information about the model is incomplete. More precisely, we shall assume about the model and the appropriate controlled random process (CRP) that some common properties of their structure are known only. Let a class $\mathcal{K}$ of the CRP be given only.

Definition 1. By *adaptive control theory* we mean the part of control theory devoted to the study of the whole class of control processes $\mathcal{K}$, instead of a particular process, due to the lack of complete information.

We shall begin our consideration with the class $\mathcal{K} = \{\xi\}$ of the CRP each element of which is characterized by the triplet $(\mu_t, \nu_t, \Sigma_{\mu,\nu}, t \geq 0)$, i.e. there is a collection of conditional distributions (μ_t) and (ν_t) and a set of admissible strategies $\Sigma_{\mu,\nu}$. The control aim, related to any process from $\mathcal{K}$ is given. It is supposed that this class contains an infinite number of elements and the intersection $\cap_{\mu\nu}\Sigma_{\mu,\nu}$ of all admissible strategies for CRP from $\mathcal{K}$ is non-empty. It is rather convenient to specify the class $\mathcal{K}$ by using an auxiliary parameter θ belonging to some parameter set Θ. Having this in mind both numerical and other (unknown) characteristics of the conditional distributions (μ_t) and (ν_t) are combined in a collection denoted by the single symbol θ. To emphasize the dependence of the considered distributions upon parameter θ their notations are supplied with a corresponding symbol, i.e. we shall write $(\mu_t(\theta), \nu_t(\theta))$, $t \geq 0$. Then instead of CRP ξ and $\mathcal{K}$ we can write $\xi(\theta)$ and $\mathcal{K}(\Theta)$. Here the parameter set Θ is determined exactly. The advantages of such a notation consist in the obvious description of an "*a priori* uncertainty" set that differentiates the given class of the CRP from the others. Now we shall give examples of parameterization for some classes of CRP.

1. HPIV ξ with the two-element state space $X = (1; -1)$ and the finite set of the controls $U = (u_1, \ldots, u_k)$. This process is completely determined by the collection of probabilities $q_j = \mathbf{P}\{x = 1|u_j\}$, $j = 1, \ldots, k$ which are considered as k-dimensional parameter.
2. The class of linear difference equations $x_{t+1} = Ax_t + Bu_t + h_t$. Here the parameters are the elements of the matrix pair $\theta = (A, B)$.
3. The class of Markov chains $(S, P^{(u)}, U)$ with matrices of transition probabilities $P^{(u)}$, $u \in U$ which form the parameter θ. If the rewards are given on the chain then their average values must be included in θ.

If an aim contains the mathematical expectations of the functions φ_t then, without reservation, they are assumed to exist and to be finite. Such a statement as "the strategy σ has led CRP ξ to the aim" means that the result of the interaction between ξ and σ generates the process that has the properties declared into the definition of the considered aim. We formulate the main definition below:

Definition 2. An admissible strategy (for all processes from $\mathcal{K}$) which leads any process from $\mathcal{K}$ to the given aim C is called an *adaptive control* (or *adaptive strategy*) with respect to the class $\mathcal{K}$ under the aim C.

The "adaptability" of a strategy signifies that it is intended not only for the individual CRP from $\mathcal{K}$ but for all processes entering into $\mathcal{K}$. In the course of the control process we do not know, generally speaking, what process from $\mathcal{K}$ concretely is under the control. In the adaptive control theory the aims are the same as in the classical one. Indeed, practice puts forward the aims which are indifferent to our knowledge about the controlled object. A customer of a control system does not usually worry about the design knowledge of the particular features of the model. He is, generally, interested in the final result only.

In this connection two questions arise. Firstly, whether the control system can be designed under a lack of *a priori* information? But also what the suppositions must be done for it to be realized? Secondly, how (by what means) can the adaptive strategies be constructed? The most of the remaining part is devoted to the answer the first question. Now we would like to give a short answer the second one.

The classification of the adaptive strategies contains three main points:

(1) *the identification strategies*;
(2) *the direct strategies*;
(3) *the searching strategies.*

An identification strategy consists of a combination of two operations simultaneously. The first of them is an evaluation of the unknown parameters of the model (we have denoted them by symbol θ united as the scalars and the vectors as the matrices and the points of some function spaces and so on). The second operation is the calculation of the controls by using the received estimates and according to the choice rules forming the required strategy. Apparently, there are two rather serious restrictions on the identification approach. Namely,

(1) the "good" (converging) estimates of the parameter θ have to exist;
(2) for known θ we must know how (by what method?) an optimal strategy is found.

These suppositions are not trivial. To come nearer to the given aim we have to choose the controls by using the rule which corresponds to the current value of the estimate. Unfortunately, this causes a deterioration of the estimates quality. In this case the convergence of these estimates to the true values of the parameters may be lost. Therefore, it is necessary to "mar" the controls, for example, to randomize them so that the optimal controls provided the achievement of the goal for CRP should appear more often. Doing so we have to choose the "incorrect" controls with a positive probability though it moves the controlled process from the desirable course. Due to the appearing deflexions we can obtain necessary information which gives the appropriate estimates of the parameters. Along with the explicit identification which consists of using the estimates converging to the true values of the parameters the partial (indirect) identification is used as well. In the last case the parameter θ is estimated approximately, i.e. only to such an extent which is necessary for the given problem to be solved (to attain one's aim). The indirect identification methods produce the estimate of the parameter θ belonging to such a region of the parameter space Θ where the control aim can be attained at least approximately.

When constructing an adaptive strategy the use of the identification approach is grounded on two stages. At the first stage the convergence of the estimates $\hat{\theta}_t$ of the parameter θ to the true value of this parameter as $t \to \infty$ should be proved and then, at the second one, it is necessary to prove that the control aim is reached since it does not result, generally speaking, from the convergence of the estimates.

The direct strategies can be applied when the structure of the choice rules for the controlled models from a given class of the CRP is known. For the sake of definiteness we shall suppose that such a strategy is stationary and it is generated by a deterministic rule $h(x_{t-h}^t, u_{t-h}^t; \varkappa(\theta))$ where $\varkappa(\theta)$ will denote the parameter determining the rule h if the model is specified by the parameter θ.[c]

The direct approach ignores the dependence of the rule h on the model parameter θ. The parameter $\varkappa$ of the rule h serves for the unknown parameter instead of the parameter θ and the problem is to find the "proper" estimates $\hat{\varkappa}_t$ of this parameter by using the observations of the process x_t (or by using the process z_t in the case of partial observations). If the function $h(\cdot, \cdot; \varkappa)$ is continuous with respect to $\varkappa$ then the convergence $\hat{\varkappa}_t \to \varkappa$ as $t \to \infty$ will imply the convergence of $h(\cdot, \cdot; \hat{\varkappa}_t)$ to the true law $h(\cdot, \cdot; \varkappa(\theta))$. It remaines to make sure that the control aim would be achieved. An application of the direct strategies usually requires stronger restrictions to the CRP class in contrast with the identification strategies and, besides, the proofs of the corresponding assertions are more complicated.

We shall adduce now an example of a direct strategy based on a wayfaring (a fluctuation) on the given set of rules. Such a wayfaring arises when the structure of the optimal control is well-known in advance and, moreover, is simple enough. For example, in the case of control problems by the HPIV class, the optimal strategy is to repeat unlimitedly the same optimal action u_{opt}. Let the control set $U = \{u_1, \ldots, u_k\}$ be finite and neither the measure $\mu(\cdot|u)$ of the process nor the average reward $W(u)$ associated with the control aim be known. We shall use randomized rules of control choice $\bar{p}_t = (p_t^{(1)}, \ldots, p_t^{(k)})$ where $p_t^{(i)} = P\{u(t) = u_i\}$. The set of all such rules forms the $k-1$-dimensional simplex $S = \{\bar{p} : \sum_1^k p_i = 1, p_i \geq 0, \, \forall i\}$. Under lack of information about the controlled process from the CPIV class, the vector rules $\bar{p}$ are transformed in the course of control so that they converge to the top of the simplex $e_j = (0, \ldots, 0, 1, 0, \ldots, 0)$ corresponding to the optimal action u_j. To that end, a family of operators $T^{(x)}$ depending on the observed value x of the process received in response to the previous action is constructed on the given simplex. The operators $T^{(x)}$ are represented as some automata or some recurrent procedures. So, $\bar{p}_{t+1} = T^{(x)}\bar{p}_t$. Then the sequence of rules $\bar{p}_1, \ldots, \bar{p}_t, \ldots$ generates a random wayfaring on the simplex S. The "adaptability" of this direct strategy

---

[c] In case, for example, of a linear quadratic problem the control law has the following form $u_t = -Kx_t$. The elements of the amplifier matrix $K(t)$ can be expressed by using the well-known procedure in terms of the initiative parameters entering into the description of the task or, more exactly, in terms of five matrices entering both into the control law and into the minimized function [see Sec. 2, Chap. 1].

takes place on the condition that the top set of this simplex corresponding to the optimal action $u_{\mathrm{opt}} = u_{j_0}$ is the absorbing set for the random wayfaring.

The searching strategies differ from the above-mentioned strategies because the search for the required rule is carried out in the set of **all** deterministic rules, i.e. among the rules having the following form $[\{h(x)\},\ \{h(x^{(1)}, x^{(2)})\},\ \dots,$ $\{h(x^{(1)}, \dots, x^{(n)})\}, \dots]$ where $\{h(x, \dots)\}$ stands for the set of all admissible functions having the necessary number of arguments belonging either to the state space X or to the observation space Z. The control of the CRP consists of constructing the optimal strategy using increasing memory depth (or with infinite depth when time is running over the set $\{\dots, -1, 0, 1, \dots\}$). The construction difficulties of the searching strategies, even in the case of discrete spaces X, U, are concerned with the choice of search direction in the set of all competitive rules having an increasing number of arguments. An enumeration of all finite-valued functions with the increasing number of finite-valued arguments is considered as another problem. In practice such an enumeration must be effective, i.e. the function's number defines its explicit form by using either the tables or the formulae which can be used immediately. For this reason the searching method interpreted as a wayfaring on the set of complex functions is used for the most difficult problems but it has no applied importance up to now. This method demonstrates the principal mathematical resolvability of the problem considered. In the present monograph the use of searching strategies is limited to Chaps. 6–8. The identification and direct strategies can be realized in practice without special difficulties. These two types of strategies afford a basis for the application of adaptive control theory in industry, communication and elsewhere.

From what has been outlined above some conclusions about the common features of adaptive strategies can be drawn. First of all, the most typical property is non-stationarity. Next, these strategies are non-Markovian, i.e. they depend not only on the last state of the model but on the more distant past history. Finally, to control the non-deterministic objects it is necessary to use randomization, i.e. the control choice rules are some probability distributions defined on U. Note, however, that there are control problems with the strategies of elementary form, for example, $u(t) = Cx(t)$. Unfortunately, such problems are exceptions.

The adaptive control has a fundamental difference from the classical one. It consists of an uncertainty, as a general rule, of the moment when the control aim gets "near-by". We shall explain this by two examples given below.

First, we shall consider the weak aims for the models $\mu(\theta)$ where $\theta \in \Theta$ is a parameter which differentiates one model from another. Let $W_\theta(\sigma)$ be an objective function, $\bar{W}_\theta = \sup_\sigma W_\theta(\sigma)$. It is required that the following inequality $\lim_{t \to \infty} E_\sigma \varphi_t > \bar{W}_\theta - \varepsilon$ should hold (φ_t is some function defined on the paths of the model). For a given model θ it is possible, in principle, to calculate either the first moment $t^*(\theta)$ when the inequality $E_\sigma \varphi_{t^*(\theta)} > \bar{W}_\theta - \varepsilon$ holds or the upper estimate of this moment (and also to estimate the convergence rate or some other characteristic of the transition process). As indicated in the corresponding notation this moment depends on the model considered. In adaptive theory we have another situation. In the case of lack of *a priori* information about the model the magnitude $t^*(\theta)$ can take any possible value and, as a matter of fact, is unbounded on the class $\mathcal{K} = \{\mu_\theta, \theta \in \Theta\}$. Therefore, the estimates of the convergence rate must be uniform on the class $\mathcal{K}$.

In the case of strong aims the situation is more complicated. If, for example, it is required to ensure the implementation of the inequality

$$\lim_{T \to \infty} T^{-1} \sum_{t=1}^{T} \varphi_t > \bar{W}_\theta - \varepsilon, \quad \text{a.s.}$$

then we shall take an interest in the random moment $\tau(\theta)$ starting from which the following inequality

$$\frac{1}{T} \sum_{t=1}^{T} \varphi_t > \bar{W}_\theta - \varepsilon$$

always holds. The moment $\tau(\theta)$ is non-Markov, i.e. it is not adapted to the information provided by the past history of the process and, therefore, is unobserved. In other words, at every moment of control we have no confidence that the ε-optimal adaptive strategy "has understood" the situation and one "has adjusted" to the proposed model and it has already used almost the optimal rules.

Thus appearance of the non-Markov moments $\tau(\theta)$ and other unknown characteristics such as $t^*(\theta)$ forms an uncertainty entering into the description of our model. The "uncertainty" is the subject-matter of such branches of mathematics as game theory and statistics. The distinction between a decision process in statistics and one in adaptive control theory is based on the fact that in the first case the observations are performed over a finite interval of time (it is fixed beforehand or it ends at some Markov moment) and, hence, it may lead to errors with a positive probability but in the second case the decisions of the adaptive strategies after the moments $\tau(\theta)$ or $t^*(\theta)$ are faultless. In the control problems by the stochastic models it means that we have to consider such problems on unlimited time intervals.

The subject matter of adaptive control theory consists in finding sufficient conditions for existence of adaptive strategies for various aims and classes of CRP. These conditions, with a few exceptions, have constructive character and they can be realized in practice without difficulties. For theoretical and practical purposes it is important to know the necessary conditions for adaptive strategies to exist. However, up to now they are known only for a few classes of the CRP. Heuristic considerations together with experience enable us to state some hypothetic assumptions about the necessary conditions. Consider the following:

A. The control aim is attainable, in principle, with respect to the considered class of the CRP.

B. A class of the CRP must not contain the "very" non-homogeneous in time (non-stationary) processes.

C. In the course of time the influence of the distant controls on the process vanishes.

Here are some intuitive arguments in their favour.

Condition **A** is self-evident but non-trivial. In optimization problems for Markov chains the accessibility of the limiting average reward maximum is the necessary and sufficient condition for an adaptive strategy to exist.

Condition **B** arises in conjunction with the fact that a non-stationary CRP usually has non-stationary optimal strategies. Then in the course of control it has to search not only the optimal choice rules of controls but also to guess a law of their change (or "readjusting" to be able to pass from one rule to the other.) It is easy to give examples of classes of very simple CRP for which adaptive strategies do not exist.

Condition **C** rules out the possibility that incorrect controls chosen on the initial interval of control make the control aim unattainable. We shall illustrate this by an example. Let a class $\mathcal{K}$ of CRP have the set of controls $U = \{u', u''\}$. For any CRP from $\mathcal{K}$ it is required to maximize the average reward $W(u_0, u_1, \ldots, u_t)$ calculated with respect to the measure generated by a program strategy. Let the class $\mathcal{K}$ also have the property that for each CRP from $\mathcal{K}$ the functions $W(u', u_1, \ldots, u_t)$, $t \geq 1$ have the same unknown sign but the functions $W(u'', u_1, \ldots, u_t)$ have the opposite one. It is clear that if the CRP is unknown then we cannot choose the proper initial control and, hence, the given aim is not always achieved.

1.4. Learning Systems

The essence of the definition of adaptive strategy given in the previous section is very broad. For a CRP in discrete time we would like to consider the adaptive strategies from another point of view, namely, to represent them in the form of automata and, hence, to produce the universal constructive definition. We shall begin from a notion of the learning system represented, in a general case, by an infinite automaton.

Let $\sigma^{(l)}$ be the control choice rule, i.e. a probability measure defined on U such that $(z^l, u^{(l-1)}) \in Z^{l+1} \times U^l$, where l is an integer that denotes the memory depth. In the singular case each deterministic rule is just a function $h(z^l)$ since the controls may be successively eliminated.

Definition 1. An object

$$\mathcal{E}_\sigma = (Z, \sigma^{(l)}, U)$$

which is represented as the stationary strategy generated either by the distribution $\sigma^{(l)}$ or by the function h is called an *elementary controlling system* $\mathcal{E}_\sigma$ (or $\mathcal{E}_h$).

The observations z_t of the controlled process x_t and the controls u_t are considered as the input and output of $\mathcal{E}_\sigma$ respectively. At times $t - 1$ and t the controls are calculated by the pre-histories

$$(z_{t-l}, \ldots, z_{t-1}; u_{t-l}, \ldots, u_{t-2}) \quad \text{and} \quad (z_{t-l+1}, \ldots, z_t; u_{t-l+1}, \ldots, u_{t-1})$$

respectively.

In controlling the CRP the system $\mathcal{E}_\sigma$ generates the next value of the process x_t (it is defined by the conditional distribution of the model μ_t) by using the control u_{t-1}. Then this value is transformed into the observable value z_t. It is the input signal of the system $\mathcal{E}_\sigma$ and, further, according to the rule $\sigma^{(l)}$ the next control u_t is generated and so on.

Let $\mathcal{D}_l$ denote the set of all rules with the memory depth l, i.e. the conditional distributions on U at the pre-histories belonging to $Z^l \times U^{l-1}$. Let us also put $\mathcal{D}_\infty = \cup_l \mathcal{D}_l$. We shall choose a non-empty set of admissible rules $\mathcal{D} \subseteq \mathcal{D}_\infty$. Let us now consider the family $\tilde{\mathcal{E}}$ of all elementary control systems corresponding in a one-to-one fashion to the rules from $\mathcal{D}$, i.e.

$$\left(\mathcal{E}_\sigma \in \tilde{\mathcal{E}} \right) \leftrightarrow (\mathcal{E}_\sigma = (Z, \sigma, U), \sigma \in \mathcal{D})$$

or, symbolically,

$$\tilde{\mathcal{E}} = (Z, \mathcal{D}, U).$$

We introduce one more notion. Let $\mathfrak{R}$ be a measurable space (further it will be a metric one but now this is not important) and ξ be a CRP. Let a sequence of measurable functions Ψ_t defined on the observable trajectory z_t and on the controls u_t with values from $\mathfrak{R}$ be given, i.e. $\Psi_t = \Psi(z_0, \ldots, z_t; u_0, \ldots, u_{t-1}) : Z^{t+1} \times U^t \to \mathfrak{R}$. The distributions of these variables are generated by the distributions of the model (μ_t), the observations (ν_t) and the rules $(\sigma^{(t)})$ used.

Definition 2. The sequence Ψ_t is called a *statistic* of the CRP ξ.

We shall explain the sense of these notions below.

We use the notation T_{Ψ_t}, $t \geq 0$, to denote a mapping of $\mathcal{D}$ into $\mathcal{D}$, i.e. $T_{\Psi_t} : \mathcal{D} \to \mathcal{D}$ for all $t \geq 0$. The set T_{Ψ_t}, $t \geq 0$ may be considered as a family of mappings depending on two parameters, namely, Ψ and t. In addition, we shall suppose that under the mapping T_{Ψ_t} a rule having the memory depth not more than $t - 1$ is associated with a rule having the memory depth not more than t. At the initial moment we use the rules having memory depth 0, i.e. they do not depend on the past history of the process. Due to correspondence between the sets $\tilde{\mathcal{E}}$ and $\mathcal{D}$ we can consider T_{Ψ_t} as a mapping of $\tilde{\mathcal{E}}$ into $\mathcal{D}$. At time $t = 0$ we have "the initial system" $\mathcal{E}_0 = (Z, \sigma^{(0)}, U)$ with the constant rule $\sigma^{(0)}$ not depending on the past.

Definition 3. The object

$$\mathcal{L} = [\tilde{\mathcal{E}}; T_{\Psi_t}, \ t \geq 0]$$

is called a *learning system*.

As seen from this definition a learning system is a generalized automaton with infinite sets of input and output signals. The detailed notation of such a system is

$$\mathcal{L} = [Z \times U, S, U; \mathcal{E}_0; T_{\Psi_t}]$$

where $Z \times U$ stands for the input alphabet, U denotes the output one, $S = (\mathcal{D}, \mathfrak{R}, Z^\infty \times U^\infty)$ is the state space, $\mathcal{E}_0$ is the initial state. The time-varying function T_{Ψ_t} is the function of the transitions and the output function is defined by the current state.

To find the structure and the function character of the learning system we shall consider its interaction with the CRP ξ. For the sake of simplicity, we assume that our model is completely observable, i.e. $X = Z$ and $x_t = z_t$. Otherwise, describing the state x_t we have to add "which is followed by appearance of the observation z_t with the probability defined by the conditional distribution ν_t". Let us apply at time $t-1$ the control choice rule $\sigma_{t-1} \in \mathcal{D}$ which has been formed by using the history of the control process from $X^{t-1} \times U^{t-2}$. Then at the next moment t in accordance with the distribution $\mu(\cdot|x^{t-1}, u^{t-1})$ the value x_t appears and, subsequently, the statistics $\Psi_t = \Psi(x^t, u^{t-1})$ can be calculated. It generates the mapping T_{Ψ_t} and the previous rule σ_{t-1} is replaced by $\sigma_t = T_{\Psi_t}\sigma_{t-1}$ which, in turn, generates the new control u_t. Thus the new collection $(\sigma_t, x^t, u^t, \Psi_{t+1})$ is formed and at the next moment $t+1$ the process is repeated again. The above is a repetition, practically word for word, of the process of controlling ξ under the strategy σ given in Sec. 2. Hence the notions of a learning system and a strategy coincide.

The operation of a learning system in the course of control is included in the presence of the set of admissible rules $\mathcal{D}$, memorizing the past history (x^t, u^t), calculating of statistics Ψ_t and producing of the next mapping T_{Ψ_t}. In the general case it has to remember the whole path, i.e. the point of the space $X^\infty \times U^\infty$ but often it is enough to remember the points from $X^l \times U^l$ (if the memory depth is equal to l). Hence the structure of the set S includes the admissible rules, statistics and the history of the control process. In the typical cases the statistics Ψ_t are calculated recursively but sometimes it is necessary to remember the whole past history. Among the widespread statistics we shall emphasize two.

1. The value of the process, i.e. $\Psi_t = x_t$. Then $T_{\Psi_t} = T_{x_t}$;
2. The arithmetic mean of the "rewards", i.e. $\Psi_t = t^{-1} \sum_{i=1}^t g(x_i)$.

The task of statistics is to give a method basing on which we choose the rules σ to have desirable behavior of the model. Then appearance of a sequence $x_0, x_1, \ldots, x_m$ at the input of the system implies the transformation of the initial rule σ_0 into $\sigma_m = T_{\Psi_m} T_{\Psi_{m-1}} \cdots T_{\Psi_1} \sigma_0$. We emphasize that the stochastic nature of the input sequence implies that of the output one. This means that the sequence of rules σ_t is a random process in the set $\mathcal{D}$ or, in other words, the CRP ξ generates some random wayfaring on $\mathcal{D}$. The achievement of the aim means the "purposefulness" of this wayfaring and its "aspiration" to use the most profitable rules. For the deterministic models the situation becomes much more simple.

It remains to explain why a learning system is a Moor automaton. This is because its output signal u_t is determined by the state of this system, i.e. by the rule σ_t acting at that moment and by the memory contents or, more exactly, by some point (x^t, u^{t-1}) of $X^\infty \times U^\infty$. The simplest example of a learning system is the *stochastic learning model* (SLM for short) which has appeared in mathematical modelling of the behaviorism conception in psychology. Because of this the words "behavior", "learning", "adaptation" are used in SLM. This model interacts with an "environment", receiving "reactions" from it in response to "stimuli" sent to the model.

Let the sets of stimuli $X = \{x_1, \ldots, x_m\}$ and reactions $U = \{u_1, \ldots, u_k\}$ be finite. The appearance of a stimulus from the environment is determined by the conditional distributions $\mu(x|u)$ or, in other words, the environment is a HPIV. We now describe the operation of this SLM. The elements of the sets X and U are, strictly speaking, its input and output signals. The admissible rules are the points of the simplex $S = \{p_1, \ldots, p_k : \sum_1^k p_i = 1, \ p_i \geq 0, i = 1, \ldots, k\}$ under the given Euclidean metric. These points are called the *behavior* of the SLM. Each of them is the stochastic vector whose jth coordinate p_j is interpreted as the reaction probability u_j. The number of the last input stimulus as the statistics which defines the collection of the mappings T_{Ψ_t} is chosen. All rules have the memory depth equal to one. The transformation of the behavior $\bar{p}$ for one step is given by the formula

$$T^{(x)}\bar{p} = \alpha_x \bar{p} + (1 - \alpha_x)\bar{q}_x, \quad 0 \leq \alpha_x < 1$$

where $\bar{q}_x = (q_1(x), \ldots, q_k(x))$ is a stochastic vector. If the same stimulus x arrives at the input of the SLM during n successive steps then the mapping $T_{x_n} = \left(T^{(x)}\right)^n$ will have the form

$$T_{x_n}\bar{p} = \alpha_x^n \bar{p} + (1 - \alpha_x^n)\bar{q}_x.$$

The vector $\bar{q}_x$ is the fixed point of this mapping, the parameter α_x pointing the convergence rate of $T_{x_n}\bar{p}$ to $\bar{q}_x$. Hence if the environment sends the SLM the same stimulus x then this system will "learn" the behavior $\bar{q}_x$. The interaction between the environment and the SLM consists of alternating the different stimulus and reactions. Therefore some k-dimensional random process $\bar{p}_t$ corresponds to the input sequence x_t. Under the stated assumptions about the environment, $\bar{p}_t$ is a Markov process and it is possible to prove that its probability distributions converge weakly to some limiting distribution. In some cases this distribution can be calculated. Notice that in the interaction between the environment and the SLM modeled above the notion of aim is absent. The problem of supplying the environment with some properties is not formulated, the scheme being intended for the modelling of some psychological phenomena only.

We introduce a constructive definition of "adaptive control".

The CRP class $\mathcal{K}$ is considered and an control aim is given as well.

Definition 4. A learning system leading each CRP from $\mathcal{K}$ to the given aim is called an *adaptive control* with respect to the class $\mathcal{K}$ under the given aim.

The difference of this formulation from the definition stated in Sec. 3 consists of pointing the universal automaton structure for CRP in discrete time. This structure has many advantages making the realization of the adaptive control system easy to access. This realization may be done in the form of programs, special devices and so on. The control algorithm often happens to be more clear due to automaton structure. It is important that in Moor automata the output signals (the controls) depend on the current states only but not on the past history. If the learning system realizes adaptive control then the random wayfarings on the set of admissible control rules $\mathcal{D}$ will gain a purposeful character. In the case of optimization aims either the random wayfarings lead to the absorbing subset of the optimal rules (belonging to $\mathcal{D}$) or the optimal rules (or almost such) are chosen with increasing frequency. In all cases when we can interpret the process of control as a random wayfaring on $\mathfrak{D}$ it is convenient and reasonable to reformulate the original aim into the goal given for the wayfaring on the set of rules $\mathcal{D}$.

1.5. Bayesian Approach on a Finite Interval

Under the condition of absence or incompleteness of *a priori* information about the model, a control problem on a finite time interval is sometimes treated as an adaptive one but in another point of view than that stated above. In this connection we shall consider a problem of controlling a Markov process with discrete parameter. On the time interval $[1, T]$ a class $\mathcal{K}_\theta$ of controlled Markov processes is given with the state space X and the control space U which are supposed to be Euclidean. The transition functions of these processes are denoted by $\mu(\cdot|x, u; \theta)$ where the parameter $\theta \in \Theta$ characterizes the concrete process and it is running over, for the sake of definiteness, the real axis, i.e. $\Theta = \mathrm{R}^1$. The states x_t are supposed to be observed. At each moment t a reward $\varphi_t = \varphi(x_t, u_{t-1})$ represented by a continuous bounded function on $X \times U$ is given. The total reward per time T is $V_\theta = \sum_1^T \varphi_t(x_t, u_{t-1})$. It is required to maximize the objective function $W_\theta(\sigma) = E_\sigma V_\theta$. In the classical case, i.e. when the transition function μ is known such a problem can be solved by classical methods, for example, by the dynamic programming method without difficulty. But in the adaptive situation the values of parameter θ are unknown. Having the observations $x_1, x_2, \ldots$ we have to construct an optimal strategy maximizing $W_\theta(\sigma)$, i.e. to find the collection of control choice rules $(\sigma_0, \sigma_1, \ldots, \sigma_{T-1})$ which maximize $W_\theta(\sigma)$.

To construct the optimal control we shall use the Bayesian approach to the problems with unknown parameters. The main hypothesis is the following:

H: There exists *a priori* distribution of the parameter θ denoted by $F(\theta)$ that is supposed to be known.

For simplicity, we shall assume that the distribution $F(\theta)$ has the density $f(\theta)$. Then according to Bayes' Theorem the posterior distribution densities of the parameter θ denoted by $f(\theta)^{(1)}, \ldots, f(\theta)^{(T-1)}$ are calculated by using the observations of the process. We assume that the joint distributions appearing below and the rules σ_j have the densities which differ by the notations of their arguments.

The rules which form the optimal strategy are constructed by means of dynamic programing methods. We shall start with the last rule $q_{T-1}(u_{T-1}|x^{T-1})$ that represents the density of the conditional distribution of σ_{T-1}. To that end, we shall write the last summand (taken before averaging with respect to the past history) of the objective function

$$\mathbf{E}(\varphi_T|x^{T-1}) = \int \varphi_T(x_T, u_{T-1})p(x_T, u_{T-1}|x^{T-1})dx_T du_{T-1}$$

for any past history x^{T-1}. In this integral the conditional density is defined by the following formula

$$\begin{aligned}
p(x_T, u_{T-1}|x^{T-1}) &= q(u_{T-1}|x^{T-1})p(x_T|u_{T-1}, x^{T-1}) \\
&= q(u_{T-1}|x^{T-1}) \int p(x_T|u_{T-1}, x_{T-1}; \theta)f^{(T-1)}(\theta|x^{T-1})d\theta
\end{aligned}$$

where the second factor under the integral symbol stands for the *a posteriori* distribution density of the unknown parameter on the step $T-1$. By using Bayes' formula this density is calculated successively from the moment $t=1$ with the prior density $f^{(1)}(\theta)$ up to the moment $t = T - 1$. And so, the function being maximized is the following

$$\mathbf{E}(\varphi_T|x^{T-1}) = \int \Psi_T(u_{T-1}, x^{T-1})q(u_{T-1}|x^{T-1})du_{T-1}$$

where

$$\Psi_T(u_{T-1}, x^{T-1}) = \int \varphi_T(x_T, u_{T-1})p(x_T|u_{T-1}, x_{T-1}, \theta)f^{(T-1)}(\theta|x^{T-1})dx_T\, d\theta.$$

Let u^*_{T-1} denote the value of the argument u_{T-1} where the function Ψ is maximum, i.e.

$$u^*_{T-1} = \mathbf{arg\ max}\ \Psi_T(u_{T-1}, x^{T-1}).$$

It is clear that the function $q(u_{T-1}|x^{T-1})$ must be the "δ-function" concentrated at the point u^*_{T-1}. In other words, the optimal rule at the moment $T - 1$ is deterministic, namely,

$$u^*_{T-1} = h_{T-1}(x^{T-1})$$

and it is determined uniquely by the past history of the control process.

We shall now consider the second rule from the end. We require that, together with the rule h_{T-1} obtained, it gives the maximum of the function $\mathbf{E}(\varphi_{T-1} + \varphi_T|x^{T-2})$ for any past history x^{T-2}. Likewise to the previous we have

$$\mathbf{E}(\varphi_{T-1}|x^{T-2}) = \int \Psi'_T(u_{T-2}, x^{T-2})q(u_{T-2}|x^{T-2})du_{T-2}$$

where

$$\Psi'_T(u_{T-1}, x^{T-2}) = \int \varphi_{T-1}(x_{T-1}, u_{T-2}) p(x_{T-1}|u_{T-2}, x^{T-2}, \theta)$$
$$\times f^{(T-2)}(\theta|x^{T-1}) dx_{T-1}\, d\theta.$$

We account again that the posterior distribution density $f(\theta|x^{T-2})$ was calculated. From the equalities

$$\mathbf{E}(\varphi_T|x^{T-2}) = \mathbf{E}\left(\mathbf{E}(\varphi_T|x^{T-1})|x^{T-2}\right),$$

$$\mathbf{E}(v_T|x^{T-2}) = \int v_T p(x_{T-1}|x_{T-2}, u_{T-2}; \theta) q(u_{T-2}|x^{T-2}) du_{T-2}\, dx_{T-1}$$

it follows

$$\max \mathbf{E}(\varphi_{T-1} + \varphi_T|x^{T-2})$$
$$= \max \int \left[\Psi'_T + \int v_T p(x_{T-1}|x_{T-2}, u_{T-2}) dx_{T-1}\right] q(u_{T-2}|x^{T-2}) du_{T-2}$$

where $v_T = \max_{t\in[0,T]} \mathbf{E}\varphi_t$. We can find the rule at the moment $T-2$ which provides the maximum to the function written down in the brackets. This rule is again deterministic and $u^*_{T-2} = h_{T-2}(x^{T-2})$ is an optimal control. Hence the density $q(u_{T-2}|x^{T-2})$ is again the "δ-function" concentrated at the point u^*_{T-2}. Analogously, we can pass from $T-2$ to $T-3$ and so on. As a result of this we define the rules $h_{T-1}, h_{T-2}, \ldots, h_1$ forming the optimal strategy. This strategy is deterministic and non-Markovian (at each moment t the control depends on the whole past history). One of the peculiarities of this approach consists of arranging the calculations: while the estimates of the parameter and the posterior densities are calculated successively from the moment $t = 1$ to $t = T$, the control choice rules are determined in the opposite order.

Let us analyze the sense of this approach. The supposition about the existence of *a priori* distribution of the parameter θ which means its stochastic nature is **not equivalent** to the absence of information about its value for the concrete controlled process. The original control aim (that is to maximize the objective function $W_\theta(\sigma)$ for any θ) is substituted by another one, namely, it is necessary to provide the maximum to the next function $\tilde{W}(\sigma) = \mathbf{E}_\theta W_\theta(\sigma)$. The additional integration which changes the aim is done with respect to the measure not having any relation to the situation under our consideration, i.e. to the type of the extreme problem and to the class of the controlled processes. Therefore the so-called "optimal Bayesian strategy" does not provide the original maximum to the function $W_\theta(\sigma)$ but only a smaller value depending on the prior density $f^{(0)}(\theta)$ chosen arbitrary. In our consideration this "imperceptible" substitution of the aim has been made at the end of the second paragraph where the function $W(\sigma)$ has appeared instead of the original function $W_\theta(\sigma)$. Next, the choice of $f^{(0)}(\theta)$ has neither reasonable nor natural grounds. Hence one cannot hope to receive even ε-optimality since it is impossible to choose the *a priori* density concentrated in a neighborhood of the true value of the unknown parameter. We have to "spread" the prior distribution over the whole space Θ. The "maximum" value of the objective function and the "optimal" strategy depend on the "spreading" method.

We are forced to conclude that the described symbiosis of the Bayesian approach (to estimate the characteristics of the process) and dynamic programming method (to calculate the optimal strategies) **is not** the adaptive control because it does not guarantee that the chosen aim of the control will be attained. It is possible to reformulate the aim, namely, it is required to maximize $\mathbf{E}_\theta W_\theta(\sigma)$ by using the Bayesian approach. It is doubtful that such a distinct demonstration both of the limited nature and of the discrepancy of the problem to the essence of the matter could arise out of interest in it. It would be necessary to study the sensitivity of the attainable maximum with respect to the chosen prior distribution and to compare the efficiency of the Bayesian approach with the other statistical methods.

The unattainability of the aim under the Bayesian approach may be associated with the fact that the time interval of control is finite. In all known cases the stochastic adaptive control problems are posed and solved on an infinite time interval. Otherwise, there is a problem to determine the real aim of the control. In the stochastic control problems posed on an infinite time interval when the optimizational aim is connected, for example, with maximization of the limiting average reward

$$W(\sigma) = \lim_{t\to\infty} t^{-1} \sum_{n=1}^{t} \mathbf{E}_\sigma \varphi_n(x_n, u_{n-1})$$

the Bayesian approach under some restrictions gives in the limit the true value of the parameter, i.e. the posterior distribution converges to "δ-function" concentrated at the true point. Then the influence of the incorrect controls chosen at first will disappear ("be smoothed") in the course of time and, consequently, the maximum of the function $W(u)$ will be reached exactly or approximately.

CHAPTER 2

REAL-VALUED HPIV WITH FINITE NUMBER
OF CONTROLS: AUTOMATON APPROACH

We study automata (deterministic and stochastic, with variable, fixed, finite or infinite structure) regarded as the adaptive algorithms of controlling the real-valued HPIV and their optimization. The ε-optimality proves to be a typical realized aim. To achieve asymptotic optimality we need either infinity of automata or changeability of structure.

2.1. Formulation of the Problem

In this chapter we shall consider the HPIV[a] with a finite set of controls $U = \{u_1, \ldots, u_k\}$ as the controlled objects. So, the controlled conditional distributions are defined by the conditional measures $\mu(\cdot|u_i) = \mu_i(\cdot)$, $i = 1, \ldots, k$. The state spaces are subsets of the real numbers. We shall now list the most common examples:

a. a finite set $X = \{x_1, \ldots, x_l\}$ is mainly met in the binary form ($l = 2$), i.e. $X = \{-1, 1\}$ or $X = \{0, 1\}$. The value $x^+ = 1$ is called a gain (or encouragement) and the value $x^- = -1$ (or $x^- = 0$) is called a loss (or punishment). These values occur with probabilities

$$q_i = \mathbf{P}\{x^+|u_i\}, \quad p_i = 1 - q_i = \mathbf{P}\{x^-|u_i\}.$$

If $x^- = -1$ and the control u_i (or the "action" of automaton) is used then the average reward will be equal to

$$W(u_i) = q_i - p_i = 2q_i - 1.$$

In case $x^- = 0$ it will be equal to $W(u_i) = W_i = q_i$. We shall always use the first case (i.e. $x^- = -1$). Then the probabilities of the values x^+, x^- can be expressed in terms of the average reward

$$q_i = \frac{1 + W_i}{2}, \quad p_i = \frac{1 - W_i}{2}.$$

We shall say that the HPIV is *binary* if $X = \{x^+, x^-\}$;

---

[a]Recall that the abbreviation HPIV means a "homogeneous random process with independent values".

33

b. a finite interval $X = [a, b]$ of values of the HPIV is used in control algorithms in one of two ways:

(1) operating with these values directly;
(2) transforming them into the binary form by using a probabilistic transformation. Namely, the values x^+ and x^- correspond to the value x of the process with probabilities $(x - a)/(b - a)$ and $(b - x)/(b - a)$ respectively. Then the full probability of x^+ appearing under the control u_i is equal to $q_u = (1 + W(u))/2$ where $W(u) = \int_a^b z\mu(dz|u)$ is the average reward under the control u_i.

c. an unbounded set (e.g. $X = (-\infty, \infty)$ or $X = [0, \infty)$) of the values of the HPIV requires some additional assumption: the average reward $W(u) = \int_{-\infty}^{\infty} z\mu(dz|u)$ should be finite for all $u \in U$. This case can be considered using two approaches: either operating with the values of the process directly or transforming them preliminarily to a finite interval (for example, by using the mapping $\arctan x$).

Assuming that the observations of the control processes are the same, we take a collection of automata $\mathcal{A}$ to have one or another form as the set of the admissible strategies Σ. These automata, called Moor automata, have input alphabet X and output one U. Let us agree to denote the mathematical expectation with respect to the measure generated by the strategy, i.e. by the automaton $\mathcal{A}$ by $\mathbf{E}_{\mathcal{A}}f$. We shall consider the following types of automata:

(α) ordinary finite automata, either deterministic or stochastic with fixed transition functions;
(β) finite automata with variable structure, i.e. their transition functions (or probabilities) change in time according to a given law;
(γ) infinite automata.

We shall now state the aims of control for a class $\mathcal{K}$ of the HPIV.

Let $W_\xi = \max_u W_\xi(u)$ be the maximum average reward of a process ξ.

Definition 1. A control u_{opt} minimizing the function $W_\xi(u)$ is called an optimal control (or an *optimal action of automaton*).

If there are several optimal controls then $I = I_{\mathrm{opt}} = \{i_1, \ldots, i_\varkappa\}$ will denote the index-set of such controls.

The weak optimizational aims are concerned with the mathematical expectations of the process ξ and have two forms, i.e. the first one corresponds to the usual limit

$$\lim_{t \to \infty} \mathbf{E}_{\mathcal{A}} x_t = \bar{W}_\xi = \max_u W_\xi, \quad \forall \xi \in \mathcal{K} \tag{1}$$

and the second one to the limit in the Cesaro sense

$$\lim_{T \to \infty} T^{-1} \sum_{t=1}^{T} \mathbf{E}_{\mathcal{A}} x_t = \bar{W}_\xi, \quad \forall \xi \in \mathcal{K}. \tag{2}$$

These aims are called the "asymptotic optimality". Their approximate versions are called the "ε-optimality". The fulfilment of one of the following inequalities is required

$$\lim_{t\to\infty} \mathbf{E}_{\mathcal{A}} x_t \geq \bar{W}_\xi - \varepsilon, \qquad \lim_{T\to\infty} T^{-1} \sum_{t=1}^{T} \mathbf{E}_{\mathcal{A}} x_t \geq \bar{W}_\xi - \varepsilon, \quad \forall \xi \in \mathcal{K} \tag{3}$$

for the fixed $\varepsilon > 0$. The strong optimizational aims have the form of the strong law of large numbers. One of these aims, called the asymptotic optimality, consist in fulfilling the equality

$$\lim_{T\to\infty} T^{-1} \sum_{t=1}^{T} x_t = \bar{W}_\xi, \quad \text{a.s. } \forall \xi \in \mathcal{K}, \tag{4}$$

accurately, but another one

$$\varliminf_{T\to\infty} T^{-1} \sum_{t=1}^{T} x_t \geq \bar{W}_\xi - \varepsilon, \quad \varepsilon > 0 \quad \text{a.s.} \tag{5}$$

is called the ε-optimality. This can be written in another form

$$\frac{1}{T} \sum_{t=1}^{T} x_t \geq \bar{W}_\xi - \varepsilon$$

for all $T > \tau_\varepsilon(\xi, \omega)$ where $\tau_\varepsilon(\xi, \omega)$ is, generally speaking, some non-Markovian moment such that $\mathbf{P}\{\tau_\varepsilon(\xi, \omega) < \infty\} = 1$. The strong aims mentioned above refer to all ξ from the class $\mathcal{K}$.

The next aim ("expediency") stands separated from the others. It consists of fulfilling the inequality

$$\varliminf_{t\to\infty} \mathbf{E}_{\mathcal{A}} x_t \geq \frac{1}{k} \sum_{i=1}^{k} W_\xi(u_i), \quad \forall \xi \in \mathcal{K}, \tag{6}$$

which is a consolation in the cases when the aims mentioned above cannot be attained. This aim means that the automaton $\mathcal{A}$ has an ability to "earn" no less than in the case of the "thoughtless" equiprobable item-by-item examination of all possible actions.

We note one more traditional aim of the stochastic control theory, i.e. it is required to maximize the discounted reward

$$W_\xi(\mathcal{A}, \beta) = \sum_{n=0}^{\infty} \beta^n \mathbf{E}_{\mathcal{A}} x_n, \quad 0 < \beta < 1.$$

Naturally, in this series the first summands play the main role. Under lack of information about the controlled process it is inevitable that in the initial stage the control algorithm (here we mean the automaton $\mathcal{A}$) makes incorrect decisions, i.e. these first summands take values which differ essentially from the maximum ones. For this reason the maximization of the discounted reward is not considered as the aim in the adaptive approach.

The subjects of this chapter are:

1. to study the possibilities to achieve the aims (1)–(6) with the help of the finite automata and others depending on complexity of their structure (in particular, depending on increasing the number of states);
2. design principles of infinite optimal automata.

2.2. Optimal Properties of Finite Automata

Let $\mathcal{A} = (X, S, U; \Pi(x))$ be a finite non-initial probabilistic Moor automaton with the transition matrix $\Pi(x)$ and decomposition

$$S = S^{(1)} + \cdots + S^{(k)} \tag{1}$$

of the state space S into non-intersecting non-empty sets. We assume that the same output signal (action) u_j corresponds to all states of $S^{(j)}$. This automaton is the control algorithm for any element ξ from the class of the HPIV with the same pairs of finite sets (X, U).[b] Their interaction is denoted by $\mathcal{A}_\xi \otimes \mathcal{A}$. It is described by the associated Markov chain (S, P) (see the end of Sec. 2, Chap. 1). This chain can be decomposed, generally speaking, into the ergodic classes and the set of inessential states. Let $S_1, \ldots, S_h, S_0$ be the subsets of S corresponding to this decomposition. The main demand is:

> *each set S_j should contain the elements from all sets $S^{(i)}$ of the decomposition (1),*

in other words, all actions $u_1, \ldots, u_k$ can be used. We form h automata $\mathcal{A}_j = (X, S_j, U; \Pi_j(x))$, $j = 1, \ldots, h$ where $\Pi_j(x)$ denotes the transition probabilities matrix for the jth ergodic class. All these matrices are induced by the initial matrices $\Pi(x)$ and the process ξ. Automata $\mathcal{A}_j$, $j = 1, \ldots, h$, are called *sub-automata* of the automaton $\mathcal{A}$. They are ergodic, i.e. their associated Markov chains (S_j, P_j) are ergodic.

For an ergodic automaton the limiting probabilities of the states denoted by π_l exist in the Cesaro sense

$$\lim_{T \to \infty} T^{-1} \sum_{t=1}^{T} p_{il}^{(t)} = \pi_l(\xi, \mathcal{A}).$$

They are positive and do not depend on an initial state. If the ergodic sub-automaton is regular the limiting probabilities will exist in the ordinary sense, i.e.

$$\lim_{T \to \infty} p_{il}^{(t)} = \pi_l(\xi, \mathcal{A}) > 0.$$

While using finite automata as the adaptive strategies for the HPIV, it is enough to consider the case of the ergodic automata. Indeed, in finite time a.s. the automaton enters a state corresponding to the ergodic sub-automata (it will be there either

---

[b] We recall that each HPIV with finite sets X and U can be interpreted as some stochastic automaton with one state $\mathcal{A}_\xi$.

from the very beginning or will move there from the inessential states) and remains there for ever. So, the original automaton realizes the strategy corresponding to this sub-automaton. The rest of the states and sub-automata associated with them do not take part in the control. Hence they may be ignored.

Let us turn to the optimizational aims stated in Sec. 1. We start with the asymptotic optimality represented by (1), (2), (4).

Theorem 1. *The finite automata* $\mathcal{A} = (X, S_j, U; \Pi(x))$ *are not asymptotically optimal with respect to the HPIV with the same sets* X *and* U *and any fixed control* $u \in U$ *as the optimal ones.*

Proof. The average reward of automaton $\mathcal{A}$ at time t when the HPIV ξ with an initial state s_0 is under the control is equal to

$$\mathbf{E}_{\mathcal{A}} x_t = W(\xi, \mathcal{A}, s_0, t) = \sum_{j=1}^{k} W(u_j) \sum_{(i \in S^{(j)})} p_{s_0 i}^{(t)}$$

where $p_{s_0 i}^{(t)}$ is the probability of $s_0 \to s_i$ transition for t steps and the symbol $(i \in S^{(j)})$ denotes the index set of the states the action u_j. Let

$$\tilde{\pi}_l(\xi, \mathcal{A}) = \sum_{j \in S^{(l)}} \pi_j(\xi, \mathcal{A}) > 0$$

denote the limiting probability of the action u_l of automaton $\mathcal{A}$ in controlling the HPIV ξ. Then, for the ergodic automaton we have

$$W(\xi, \mathcal{A}) = \lim_{T \to \infty} T^{-1} \sum_{t=1}^{T} \mathbf{E}_{\mathcal{A}} x_t = \sum_{l=1}^{k} W(u_l) \tilde{\pi}_l(\xi, \mathcal{A}).$$

Since $\tilde{\pi}_l(\xi, \mathcal{A}) > 0$ and $W(\xi, \mathcal{A}) < \max_l W(u_l)$, the automaton $\mathcal{A}$ is not asymptotic optimal. $\square$

In a similar manner it is easy to make sure that the regular automata are not asymptotically optimal using the usual (not Cesaro) limit

$$\lim_{t \to \infty} \mathbf{E}_{\mathcal{A}} x_t = W(\xi, \mathcal{A}) \leq \max_l W(u_l).$$

As seen from the above, equality (4) from Sec. 1 fails, i.e. the finite automata are not asymptotically optimal in the strong sense.

The finite optimal automata can exist for some special classes of HPIV. For example, for a class of processes with unique optimal action u_{opt} the optimal automaton is obvious. It always does this action. Naturally, such a case should not be considered as the subject of adaptive theory.

Let us now discuss the attainability of the other aims of automaton control for the HPIV stated in Sec. 1. From the formula for the limiting average reward $W(\xi, \mathcal{A}) = \sum_{i=1}^{k} W(u_i) \tilde{\pi}_i(\xi, \mathcal{A})$ it follows that the ε-optimality property of an automaton with respect to a class of HPIV will be possible if the limiting probability $\tilde{\pi}_i(\xi, \mathcal{A})$ of the optimal action becomes near to one for any ξ from the given class. Below we shall see that this implies the ε-optimality in the strong sense. The expediency property of the automaton $\mathcal{A}$ will occur when for any ξ from the given class the following inequality imposing the obvious conditions on the limiting probabilities $\tilde{\pi}_i(\xi, \mathcal{A})$,

$$\sum_{i=1}^{k} W(u_i)\tilde{\pi}_i(\xi, \mathcal{A}) \geq \frac{1}{k} \sum_{i=1}^{k} W(u_i)$$

holds.

Even simple automata can have this property. We adduce an example. Let an automaton $\mathcal{L}$ have two states s', s'' where the action u_1 and u_2 correspond to s', s'' respectively. The input signals are $+1$ and -1. The transition graph is depicted in Fig. 1 where "$\pm$" means the input signals "± 1" respectively.

We denote the probability of "$+1$ " and "-1" appearing under the control u_i by q_i and $p_i = 1 - q_i$ respectively. If the control u_i is used the average reward will be equal to $W_i = q_i - p_i$. The limiting average reward of automaton $\mathcal{A}$ is defined by the formula

$$W(\xi, \mathcal{L}) = W_1 \pi' + W_2 \pi'',$$

where π', π'' are the limiting state probabilities of the state. They can be found by using the equation $\pi = \pi P$ where

$$P = \begin{pmatrix} q_1 & p_1 \\ p_2 & q_2 \end{pmatrix}, \quad q_i = \frac{1 + W_i}{2}, \quad p_i = \frac{1 - W_i}{2}.$$

Using the normalizing condition $\pi' + \pi'' = 1$ we have

$$\pi' = \frac{p_2}{p_1 + p_2}, \quad \pi'' = \frac{p_1}{p_1 + p_2}.$$

Thus,

$$W(\xi, \mathcal{L}) = \frac{p_2 W_1 + p_1 W_2}{p_1 + p_2}.$$

It is clear that if $W_1 \neq W_2$ then $W(\xi, \mathcal{L}) > (W_1 + W_2)/2$, i.e. automaton $\mathcal{L}$ is expedient with respect to the class of all HPIV with two controls.

From the conditions imposed on the limiting behavior of the associated Markov chain it follows that the attainability of the aims pointed to cannot be ensured by an arbitrary automaton. To do this we have to supply automaton with some specific properties. The means for this are double: on the one hand to increase the number

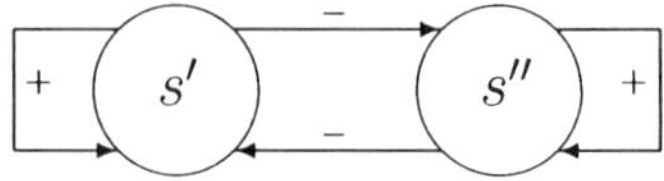

Fig. 1.

of states of the automaton and on the other hand to change the transition function (matrix). We turn to the ε-optimality of the automata leaning upon the principle: "the more profitable actions (ensuring the greater average reward) should be done more often than the less profitable ones". To pass from the intuitive level to strict results we introduce a number of notions and prove some auxiliary assertions.

For the sake of simplicity we assume that the considered HPIV are binary, i.e. their values belong to $X = \{-1, 1\}$. Naturally, the automata are supposed to be binary. So, let the input alphabet X and output one $U = \{u_1, \ldots, u_k\}$ be fixed. According to what has been said in Sec. 1, the conditional probabilities for the HPIV are linear with respect to the average reward, i.e.

$$q_i = \mathbf{P}\{1|u_i\} = \frac{1 + W_i}{2}, \quad p_i = \mathbf{P}\{-1|u_i\} = \frac{1 - W_i}{2}.$$

Let $S^{(j)}$, $j = 1, \ldots, k$ be the elements of the decomposition (1) corresponding to the output signals u_j and $\Pi_j(x)$, $x = x^-$, x^+ be the transition probability matrix into $S^{(j)}$ induced by the original matrices $\Pi(x)$. Let all these subsets have the same amount of states n (this fact will be noted in the notation $S_n^{(j)}$) and "the same structure" or, more precisely, they are isomorphic, i.e. they differ only by notation.[c]

Let $\mathcal{A}_{k,n}$ be a binary automaton with k actions and kn states.

Definition 1. The Moor automaton $\mathcal{A}_{k,n}$ will be called *symmetrical* if the subsets of its state set $S_n^{(j)}$ are isomorphic. Such an automaton will be called a *one-input* one if in any $S_n^{(j)}$ there is a single state the automaton $\mathcal{A}_{k,n}$ passes into *after* changing any action to u_j. It will be called a *one-output* one if in any $S_n^{(j)}$ there is a single state the automaton $\mathcal{A}_{k,n}$ passes into *before* changing any action to u_j.

While interacting the one-input, one-output symmetrical automaton $\mathcal{A}_{k,n}$ with a HPIV ξ, the sojourn time (without interruption) in the set $S_n^{(j)}$ is a random variable $\tau_{j,n}(\omega)$. The set $S_n^{(j)}$ will be called the jth *branch*. If, in addition, the automaton $\mathcal{A}_{k,n}$ is ergodic, then the random variable $\tau_{j,n}(\omega)$ will have finite mathematical expectation (if $\mathcal{A}_{k,n}$ starts from the input state of the branch). The quantity

$$T_{j,n} = \mathbf{E}\tau_{j,n} = T_n(W(u_j)),$$

is called the *average time of action of the control u_j* (or of *sojourning in the branch* $S_n^{(j)}$). The $T_n(W(u_j))$ is generated by the interaction of the HPIV ξ with $\mathcal{A}_{k,n}$ and this fact is represented in its notation. The number n is called the *memory depth* of the branch.

Let us define the rules of changing the output signals. We use two versions of transitions from $S_n^{(i)}$ into $S_n^{(j)}$:

1. *Cyclic* one — deterministic using of all actions in turn

$$S_n^{(i)} \to S_n^{(i+1)}, \quad S_n^{(k)} \to S_n^{(1)}.$$

---

[c]For any i, j the set $S_n^{(i)}$ can be mapped one-to-one on $S_n^{(j)}$ so the matrices $\Pi_i(x)$ and $\Pi_j(x)$ coincide up to, may be, the transposition of their rows.

2. *Equiprobable* one — from any branch $S_n^{(j)}$ with the same probabilities, equal $1/k$, the transition is realized in any branch including the initial one.

Theorem 2. *Let a binary HPIV be controlled by a finite, ergodic, symmetrical, one-input and one-output automaton $\mathcal{A}_{k,n}$. Then for any $i, j \in (1, \ldots, k)$ we have*

$$\tilde{\pi}_j = \frac{T_{j,n}}{\sum_{l=1}^{k} T_{l,n}}, \quad j = 1, \ldots, k,$$

where $\tilde{\pi}_i$ is the limiting probability of the action u_i and, in particular,

$$\frac{\tilde{\pi}_i}{\tilde{\pi}_j} = \frac{T_{i,n}}{T_{j,n}}.$$

Proof. Let τ_n^i, $n \geq 1$, be the successive hitting moments of the automaton into the initial branch. Then this sequence is a renewal process and, obviously, $\mathbf{E}(\tau_n^i - \tau_{n-1}^i) < \infty$. Now, the required result follows from the limiting theorem on the existence of the stationary regime for a renewal process. $\qquad\square$

Corollary 1. *Under the conditions of Theorem 2, the limiting probability and average times of actions are given by the equalities*

$$\tilde{\pi}_j = \frac{T_{j,n}}{\sum_{l=1}^{k} T_{l,n}}, \quad j = 1, \ldots, k.$$

Corollary 2. *Under the condition of Theorem 2, the limiting average reward of the automaton $\mathcal{A}_{k,n}$ is equal to*

$$W(\xi, \mathcal{A}_{k,n}) = \frac{\sum_{i=1}^{k} W_i T_{i,n}(W_i)}{\sum_{i=1}^{k} T_n(W_i)}.$$

in controlling the HPIV ξ.

The following notion is concerned with ε-optimality of automata. It is clear that for the given automaton the value ε is fixed and cannot be made smaller. The value of ε can be decreased only by means of increasing the memory depth n.

Definition 2. A sequence of automata $\mathcal{A}_{k,n}$ which controls a HPIV ξ will be called *asymptotically optimal* if

$$\lim_{n \to \infty} W(\xi, \mathcal{A}_{k,n}) = \max_u W_\xi(u)$$

for any binary ξ.

Sometimes instead of "the asymptotically optimal sequence of automata" one considers "the ε-optimal family of automata". For such families and any $\varepsilon > 0$ there exists n_ε such that

$$W(\xi, \mathcal{A}_{k,n}) > \max_u W_\xi(u) - \varepsilon$$

for any $n > n_\varepsilon$.

The existence condition for families of ε-optimal automata is contained in the following corollary of Theorem 2.

Corollary 3. *A family of automata* $\mathcal{A}_{k,n}$, *which satisfies the conditions of Theorem 2 is optimal in the weak-sense if and only if the relation*

$$\lim_{n\to\infty} \frac{T_n(W')}{T_n(W'')} = 0$$

holds for any W', W'' *such that* $W' < W''$.

Proof. The expression for the limiting average reward from Corollary 2 implies that for the asymptotic optimality of the sequence $\mathcal{A}_{k,n}$ to take place the equalities

$$\lim_{n\to\infty} W(\xi, \mathcal{A}_{k,n}) = \lim_{n\to\infty} \sum_{i=1}^{k} W_i \tilde{\pi}_i(\xi, \mathcal{A}_{k,n}) = \max_u W_\xi(u)$$

should hold, i.e. the limiting probability of the optimal action u_{i_0} tends to one as $n \to \infty$. It means that for a non-optimal action u_i we have

$$\lim_{n\to\infty} \frac{\tilde{\pi}_i(\xi, \mathcal{A}_{k,n})}{\tilde{\pi}_{i_0}(\xi, \mathcal{A}_{k,n})} = 0, \quad i \neq i_0.$$

This correlation, Corollary 2 and the fact that the value of the maximum average reward can be arbitrary imply the result. $\qquad\square$

It remains to ascertain whether a sequence of automata which forms the optimal family exists. We will answer this question by designing such sequences.

The automata of type $\mathcal{D}_{k,n}$ ("deep") are illustrated by the graph in Fig. 2.

Let $S_n^{(j)} = \{s_i^{(j)}, i = 1, \ldots, n\}$ be the jth branch of the automaton (or graph). The input signals $x^+(= +1)$ move the automaton from $s_i^{(j)}$ into $s_n^{(j)}$ but the signals $x^-(= -1)$ move it from $s_i^{(j)}$ into $s_{i-1}^{(j)}$. The passing from $s_1^{(j)}$ is realized either into $s_1^{(j+1)}$ in the case of the cyclic scheme of changing the input signals or into any input state $s_1^{(1)}, s_1^{(2)}, \ldots, s_1^{(k)}$ equiprobably. Hence the input states and output ones of the branch coincide.

Let us define the average sojourn time $T(z)$ in the branch, provided the initial state is s_z, $z = 1, \ldots, n$ and the transitions probabilities from s_z to the right (into s_n) and to the left (into s_{z-1}) are equal to $q = (1 + W)/2$ and $p = (1 - W)/2$ respectively. The value $T_n(W)$ that we are interested in coincides with $T(1)$.

So, the calculation of the average time is reduced to the study of some specific random walk over the integer points of some finite interval. The function $T(z)$ satisfies the following difference equations

$$T(z) = pT(z - 1) + qT(n) + 1, \quad z = 1, 2, \ldots, n$$

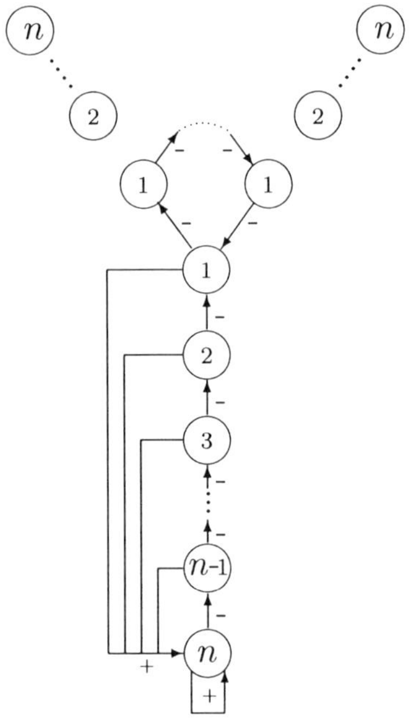

Fig. 2.

with the boundary condition $T(0) = 0$. We have

$$T(1) = qT(n) + 1. \qquad (2)$$

Hence

$$T(2) = pT(1) + qT(n) + 1 = T(1)(1 + p) = T(1)\frac{1 - p^2}{q}.$$

Continuing such calculations for $z = 3, \ldots, n$ we obtain

$$T(z) = T(1)\frac{1 - p^n}{q}, \quad z = 2, \ldots, n.$$

From this and (2) it follows that

$$T(1) = qT(1)\frac{1 - p^n}{q} + 1 = T(1)(1 - p^n) + 1.$$

Therefore

$$T_n(W) = T(1) = p^{-n} = \left(\frac{2}{1 - W}\right)^n. \qquad (3)$$

We shall now consider the extreme values for W. If $W = 1$, then $q = 1$ and only the gains arrive upon the branch and the automaton $\mathcal{D}_{k,n}$ will never leave this branch. If $W = -1$ then $p = 1$ and the automaton stays on the branch only one time due to only losses arriving. For other values $W \in (-1, 1)$ the average time $T_n(W)$ increases exponentially with increasing the memory depth n on the branch. It increases faster than the average reward.

From the explicit representation for the function $T(z)$

$$T(z) = \frac{1 - p^z}{p^n q}, \quad z = 1, \ldots, n$$

it follows that the sojourn time in the branch is maximum for the deepest initial state s_n, i.e. $\max_z T(z) = T(n)$. This fact and the previous formula will be used in Sec. 3.

From (3) it follows that the necessary and sufficient condition of Corollary 3 is fulfilled. Hence the automata $\mathcal{D}_{k,n}$ form the asymptotically optimal sequence with respect to the class of all binary HPIV and, if we add the input convertor (see Sec. 1) then this will be true with respect to the class of the bounded HIPV as well. According to Corollary 3, the limiting average reward of automaton $\mathcal{D}_{k,n}$ is equal to

$$W(n) = W(\xi, \mathcal{D}_{k,n}) = \frac{\sum_{l=1}^{k} W_l (1 - W_l)^{-n}}{\sum_{l=1}^{k} (1 - W_l)^{-n}}.$$

Let us estimate the deviation of $W(n)$ from the maximum reward. Let the latter be equal to W_1 (i.e. u_1 is the optimal action). We have

$$W_1 - W(n) = \frac{\sum_{j=1}^{k} (W_1 - W_j) H_{j,n}}{1 + \sum_{j=1}^{k} H_{j,n}} = \varepsilon_n,$$

where $H_{j,n} = (1 - W_1)^n (1 - W_j)^{-n}$. If not all W_j are equal to W_1 then $\varepsilon > 0$. From this we can find the convergence rate of the maximum average reward to the maximum W_1. To formulate the final result we introduce the notation

$$\mu = \max_{j \geq 2} \frac{1 - W_1}{1 - W_j}, \quad c = \max_j (W_1 - W_j).$$

Then

$$W_1 - W(n) \leq c\mu^n = ce^{-\lambda n}, \quad \lambda = \ln \mu^{-1},$$

i.e. ε_n decreases exponentially fast on the class of the binary HPIV. Moreover, this convergence is uniform on this class.

Note that the automaton strategies $\mathcal{D}_{k,n}$ unlike the other constructions of this section do not use identification of the process under control. The automata "know" only the gains and losses in the form "$+1$ " and "-1" respectively, but their "memory" has the form of a number of states on the branch keeping the number of losses which arrived after the last gain. The characteristics of the process can be restored by using observations of the process of control, i.e. the collection of probabilities $p_1, \ldots, p_k$ must be estimated. To that end, we should fix the sojourn times $\tau_1^{(j)}, \tau_2^{(j)}, \ldots, \tau_N^{(j)}$ on the jth branch and then calculate the empirical average sojourn time

$$\tilde{T}^{(j)} = \frac{1}{N} \sum_{l=1}^{N} \tau_l^{(j)}.$$

Identifying $\tilde{T}^{(j)}$ with its theoretical value $T_{j,n} = p_j^{-n}$ we obtain the required characteristics of the HPIV $p_j \simeq \left(\tilde{T}^{(j)}\right)^{-n}$.

We now define the vast class of *quasi-linear* automata $\mathcal{Q}_{k,n}$.

Their graph is shown in Fig. 3. They again have the star-like form and their branches are linearly ordered. The transition law on a fixed branch is the following. In response to the input signal $x^+(= +1)$ the automaton passes from s_i into either s_{i+1} with probability q_+ or s_{i-1} with probability $p_+ = 1 - q_+$ (the automaton leaves the branch with probability p_+ provided if $s_i = s_1$). Another signal x^- $(= -1)$ implies the same transitions but with different probabilities: from s_i into either s_{i+1} with probability q_- or s_{i-1} with probability $p_- = 1 - q_-$ (it can leave the branch again starting from s_1). The changing of the branches is either cyclical or equiprobable.

The average sojourn time in the branch with depth n when the average reward on it is equal to W takes the value $T(1)$ of the solution of the linear difference equation

$$T(z) = (pp_- + qp_+)T(z-1) + (pq_- + qq_+)T(z+1) + 1, \quad z = 1, \ldots, n \quad (4)$$

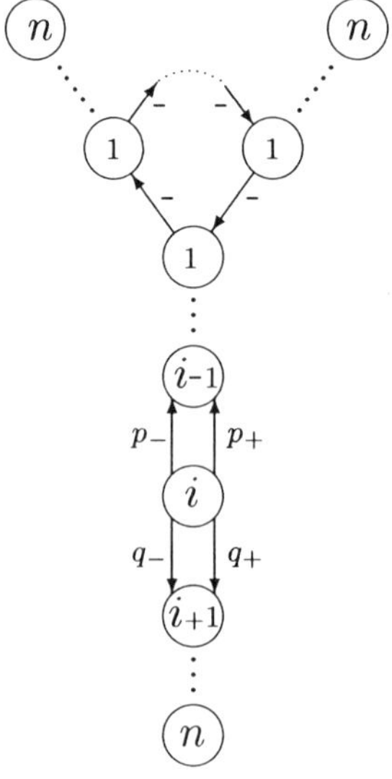

Fig. 3.

with the boundary condition $T(0) = 0$. We put

$$P = pp_- + qp_+, \quad Q = pq_- + qq_+ \quad (P + Q = 1).$$

Then Eq. (4) takes the form

$$T(z) = PT(z - 1) + QT(z + 1) + 1, \quad z = 1, \ldots, n. \tag{5}$$

Let $P \neq Q$. As it is well known, the general solution of (5) is represented by

$$T(z) = c_1 + c_2\lambda^z + \frac{z}{P - Q},$$

where c_1, c_2 are arbitrary constants and $\lambda = P/Q$. From the boundary conditions and (5) it follows that $c_1 = -c_2$ and

$$T(n) = T(n - 1) + P^{-1}$$

at $z = n$. Substituting the values of the solution $T(z)$ at $z = n$ and $z = n - 1$, we can find c_2. As a result of this, we obtain the final form of the solution

$$T(z) = Q\frac{1 - \lambda^z}{(P - Q)^2} + \frac{z}{P - Q}.$$

Hence the average sojourn time in the branch is equal to

$$T(1) = \frac{(Q/P)^n - 1}{Q - P}, \quad P \neq Q.$$

If $P = Q$ then Eq. (5) will take the form

$$T(z) = \frac{1}{2}[T(z - 1) + T(z + 1)] + 1.$$

The function $T(z) = 2nz - z^2$ is its solution. The average sojourn is equal to

$$T_n = T(1) = 2n - 1.$$

We write the limiting average reward for the quasi-linear automata only under the assumptions $P_i \neq Q_i$, $i = 1, \ldots, k$, namely,

$$W(\xi, \mathcal{Q}_{k,n}) = \frac{\sum_{i=1}^{k} W_i((Q_i/P_i)^n - 1)(Q_i - P_i)^{-1}}{\sum_{i=1}^{k}((Q_i/P_i)^n - 1)(Q_i - P_i)^{-1}}. \tag{6}$$

Let us analyze this expression. It is clear that the optimal properties of the automata $\mathcal{Q}_{k,n}$ are defined by the values of the parameters q_+, q_- which characterize the move along the branch accordingly to the input signals.

When increasing the memory depth n unlimitedly the behavior of the value $W(n) = W(\mathcal{Q}_{k,n})$ depends on the value $Q_{\max} = \max_i Q_i$. If $\max_i Q_i = Q_{i_0} \geq 1/2$ then

$$\lim_{n \to \infty} W(n) = W_{i_0}$$

but if $\max_i Q_i = Q_{i_0} < 1/2$, then

$$\lim_{n \to \infty} W(n) = \frac{\sum_{j=1}^{k} W_j (P_j - Q_j)^{-1}}{\sum_{j=1}^{k} W_j (P_j - Q_j)^{-1}}.$$

Hence the attainment of the asymptotic optimality will be possible if the following conditions hold:

(1) $\max_i Q_i = Q_{i_0} \geq 1/2$;
(2) $\max_i W_i = W_{i_0}$.

To verify these we express Q_i in terms of W_i. We have

$$Q_i = p_i q_- + q_i q_+ = W_i \frac{q_+ - q_-}{2} + \frac{q_+ + q_-}{2}.$$

If $q_+ > q_-$, the maximum W_i corresponds to the maximum Q_i, i.e. $Q_{\max} = 2^{-1}(W_{\max}(q_+ - q_-) + q_+ + q_-)$. This means that it will be more preferable to go into the depth of the branch provided we have gain than when we have losses. It remains to satisfy the first condition:

$$Q_{\max} = q_{\max}(q_+ - q_-) + q_- \geq \frac{1}{2}, \quad q_{\max} = \max_i q_i$$

or

$$q_{\max} \geq \frac{1/2 - q_-}{q_+ - q_-}. \tag{7}$$

This inequality links together the parameters of automata $\mathcal{Q}_{k,n}$ and parameters (q_i) of the HPIV. If $q_- \geq 1/2$ the inequality (7) always holds (since its right-hand side is non-positive). Then the automata $\mathcal{Q}_{k,n}$ form the asymptotically optimal sequence. But if $q_- < 1/2$ and $q_+ > 1/2$ then the automata $\mathcal{Q}_{k,n}$ represents the ε-optimal family with respect to the class of binary HPIV subjected to inequality (7).

Let us consider special cases of quasi-linear automata.

The *automata* $\mathcal{K}_{k,n}$ correspond to the values $q_+ = 1$, $q_- = 1/2$. The average time can be found by the general formula with $P = p/2$, $Q = q + p/2$. Then

$$T_n(W) = \frac{2}{1+W} \left[\left(\frac{3+W}{1-W} \right)^n - 1 \right].$$

From Corollary 3, or inequality (7) the ε-optimality of the collection $\mathcal{K}_{k,n}$ follows.

The automata with *linear tactics* $\mathcal{L}_{k,n}$ are defined by $q_- = p_+ = 0$, i.e. $P = P$, $Q = q$ and the average time is equal to

$$T_n(W) = W^{-1} \left[\left(\frac{1+W}{1-W} \right)^n - 1 \right].$$

According to (7) the automata $\mathcal{L}_{k,n}$ form the ε-optimal collection with respect to the class of the HPIV such that $\max_i W_i \geq 1/2$ (or $\max_i q_i \geq 1/2$). The limiting average reward has the form

$$W(\xi, \mathcal{L}_{k,n}) = \frac{\sum_{i=1}^{k}\left[\left(\frac{1+W_i}{1-W_i}\right)^n - 1\right]}{\sum_{i=1}^{k} W_i^{-1}\left[\left(\frac{1+W_i}{1-W_i}\right)^n - 1\right]}.$$

Putting here $n \to \infty$, we find

$$\lim_{n\to\infty} W(\xi, \mathcal{L}_{k,n}) = \begin{cases} \max_i W_i, & \text{if } \max_i W_i \geq 0, \\ \dfrac{k}{\sum_{i=1}^{k} W_i^{-1}}, & \text{if } \max_i W_i < 0, \end{cases}$$

i.e. with respect to the class of all HPIV the expediency is only ensured for the automata $\mathcal{L}_{k,n}$. Indeed, by the inequality relating the arithmetic mean with harmonic one

$$\frac{1}{k}\sum_{j=1}^{k} a_j \geq \frac{k}{\sum_{j=1}^{k} a_j^{-1}}, \quad a_j > 0 \tag{8}$$

(here the equality will take place only if $a_1 = \cdots = a_k$) it follows that for negative W_i

$$\lim_{n\to\infty} W(\xi, \mathcal{L}_{k,n}) = \frac{k}{\sum_{j=1}^{k} W_j^{-1}} \geq \frac{1}{k}\sum_{j=1}^{k} W_j.$$

It may seem that the optimal properties of automata with linear tactics will be better if we separate the output states of a branch from the input ones and put them in the "end" of the branch in the deepest state with the number n (i.e. in state s_n). It turns out that this is not the case. The appropriate automata are called the "automata with hysteresis tactic" $\mathcal{G}_{k,n}$ (at $k = 2$ the graph $\mathcal{G}_{2,n}$ resembles a histeresis loop). The average sojourn time into the branch is the value of the appropriate solution of Eq. (5) at the point $z = n$, i.e.

$$T_n(W) = \begin{cases} \dfrac{1+W}{2W^2}\left[\left(\dfrac{1+W}{1-W}\right)^n - 1\right] - \dfrac{n}{W}, & W \neq 0, \\ n^2, & W = 0. \end{cases}$$

The study of $W(\xi, \mathcal{G}_{k,n})$ with increasing n leads to the same conclusions as for the automata $\mathcal{L}_{k,n}$.

Thus, these examples prove the existence of ε-optimal families of finite automata.

Before now we were paying attention to ε-optimality in the weak sense. It is rather interesting and important to consider ε-optimality in the strong sense. The theorem below illustrates this.

Theorem 3. *Let the ergodic automata $\mathcal{A}_n, n \geq 1$, form the ε-optimal family in the weak sense. Then these automata are ε-optimal in the strong sense.*

Proof. By the assumption there exist limiting probabilities $\tilde{\pi}_j(\xi, \mathcal{A}_n) > 0$ of the actions u_j, $j = 1, \ldots, n$ for any ξ and n. This implies that the function $N_j(t)$, which represents the number of uses of the action u_j up to time t, increases as $t \to \infty$ and, moreover, with probability one the equalities

$$\lim_{T \to \infty} t^{-1} N_j(t) = \tilde{\pi}_j(\xi, \mathcal{A}_n) > 0, \quad j = 1, \ldots, n$$

hold. Let the empirical average reward in time t

$$V_j(t) = t^{-1} N_j(t) \sum_{m=1}^{N_j(t)} x_m^{(j)}$$

correspond to u_j, where $x_m^{(j)}$ is the value of the HPIV to appear in the result of the action u_j. The total average reward in time t can be written in the form

$$V(t) = t^{-1} \sum_{j=1}^{k} N_j(t) V_j(t).$$

Since

$$\mathbf{P}\left\{ \lim_{T \to \infty} \frac{N_j(t)}{t} = \tilde{\pi}_j(\xi, \mathcal{A}_n), \lim_{t \to \infty} V_j(t) = W(u_j), \ j = 1, \ldots, k \right\} = 1,$$

we have

$$\lim_{t \to \infty} V(t) = \sum_{j=1}^{k} \tilde{\pi}_j W(u_j) = W(\xi, \mathcal{A}_n) > \max_j W(u_j) - \varepsilon, \quad \text{a.s.}$$

for any $\varepsilon > 0$. So, there exists a non-Markov moment τ_ε finite a.s. such that $V(t) > \max_j W_j - \varepsilon$ for $t > \tau_\varepsilon$. $\qquad \square$

Hence the constructions of finite automata enumerated above possess the asymptotic optimality property in the strong sense. Let us try to find the best among them. This question is not simple since our automata are already "optimal" (within accuracy up to ε). We leave aside the approach connected with the reliability or simplicity criteria of construction and use as the "rate of converging the average reward $W(\xi, \mathcal{A}_n)$ to the maximum value" which is often considered. For automata with two actions we put

$$\Delta_n = \Delta(\mathcal{A}_n) = \frac{W(\mathcal{A}_n) - W_2}{W_1 - W_2}, \quad W_1 > W_2.$$

Then $\Delta_n \to 1$ as $n \to \infty$ for the ε-optimal family $\mathcal{A}_n$ but for $\mathcal{L}$ this will be the case if $q_1 \geq 1/2$.

Let us compare $\Delta(\mathcal{D}_{2,n})$ with $\Delta(\mathcal{L}_{2,n})$ for the ε-optimal families $\mathcal{D}_{2,n}$ and $\mathcal{L}_{2,n}$ respectively. Since the family $\mathcal{L}_{2,n}$ does not provide the attainment of the aim for all binary HPIV one can expect that $\mathcal{D}_{2,n}$ is always "better" than $\mathcal{L}_{2,n}$, i.e. $\Delta(\mathcal{D}_{2,n}) > \Delta(\mathcal{L}_{2,n})$. Let us verify this conjecture assuming that $q_1 > 1/2$. We have

$$\Delta(\mathcal{D}_{2,n}) = \frac{1}{1 + (p_1/p_2)^n}, \quad \Delta(\mathcal{L}_{2,n}) = \frac{1}{1 + \gamma_n (p_1/p_2)^n},$$

where

$$\gamma_n = \frac{W_1}{W_2} \left(\frac{q_2}{q_1}\right)^n \frac{1 - (p_2/q_2)^n}{1 - (p_1/q_1)^n}.$$

We now consider all possible correlations between $\Delta(\mathcal{D}_{2,n})$ and $\Delta(\mathcal{L}_{2,n})$ in accordance with the value of the "total" reward probability $q_1 + q_2$.

(a) $q_1 + q_2 = 1$.
It is easy to verify that then $\gamma_n = 1$, i.e. $\Delta(\mathcal{D}_{2,n}) = \Delta(\mathcal{L}_{2,n})$ and, hence, both automata are equally "good".

(b) $q_1 + q_2 < 1$.
Then $q_2 < 1/2$ and $p_2 > q_1$, i.e. $W_2 < 0$. Hence

$$\gamma_n \sim \left(\frac{p_2}{p_1}\right)^n \frac{W_1}{-W_2} > 1$$

which implies the advantage of automaton $\mathcal{D}_{2,n}$ over automaton $\mathcal{L}_{2,n}$, i.e. $\Delta(\mathcal{D}_{2,n}) > \Delta(\mathcal{L}_{2,n})$.

(c) $q_1 + q_2 > 1$. Here two cases are possible. Namely,

(1) $q_2 < 1/2$, and since $p_2 < q_1$ we conclude that

$$\gamma_n \sim \left(\frac{p_2}{p_1}\right)^n \frac{W_1}{-W_2} < 1.$$

(2) $q_2 > 1/2$, and since $q_1 > q_2$ we have

$$\gamma_n \sim \left(\frac{q_2}{q_1}\right)^n \frac{W_1}{W_2} < 1.$$

In both cases the automaton $\mathcal{L}_{2,n}$ "wins", i.e. $\Delta(\mathcal{L}_{2,n}) > \Delta(\mathcal{D}_{2,n})$.

The comparison of the automata $\mathcal{D}_{2,n}$ and $\mathcal{L}_{2,n}$ demonstrates the absence of superiority of one of them over the other. For one type of the HPIV the automaton $\mathcal{D}_{2,n}$ has superiority but for another — $\mathcal{L}_{2,n}$, and for yet another they are equivalent. This situation is typical not only in adaptive control theory but in general control theory as well.

2.3. Automata with Increasing Memory

In the previous section the optimizational abilities of the finite automata have been considered and it has been found that these abilities increase when increasing the memory depth. So, the question of efficiency of infinite automata appears. Here we would like to determine whether the potential infinity allows obtaining asymptotic optimality. We shall turn later to actual infinite constructions.

As the fundamental principle of synthesizing the potential infinite automata we choose again, like Sec. 2, the primary choice of profitable actions. We realize this idea for the analogue of the finite automata from the ε-optimal families. Let $\mathcal{A}_{k,n}$ be an automaton from such a collection with cyclic change of actions. We denote a sequence of positive integers by $n_0, n_1, \ldots, n_m \ldots (n_m < n_{m+1})$ and assume that the depth of all branches will change simultaneously from n_{m-1} to n_m after the cyclic round of all branches on the mth cycle is finished. Such a sequence will define a potentially infinite[d] automaton $\mathcal{A}_k[n_m]$ or $\mathcal{A}_k$ provided the sequence n_m is fixed.

Definition 1. The automata $\mathcal{A}_k[n_m]$ are called the *automata with increasing memory*.

The main question of automata theory with increasing memory is to find conditions on integer-valued sequences (n_m) to guarantee the asymptotic optimality of the automata. We shall consider this question when a sequence of ε-optimal automata $\mathcal{D}_{k,n}$ affords for a basis. To state the main results we introduce some notations. Let $g(l) = [l^{2+\delta}]$, $\delta > 0$, and $n(m)$ denote the depth of the branch of the automaton $\mathcal{D}_k$ on the mth cycle of the control.

Theorem 1. *The automaton $\mathcal{D}_k[n_m]$ with increasing memory will be asymptotically optimal if $n(m_1) \leq n(m_2)$ for $n_1 < m_2$ and*

$$\lim_{l \to \infty} [n(g(l+2) - 1) - \alpha n(g(l))] = -\infty$$

for some $\delta > 0$ and any $\alpha > 1$.[e]

Proof. Let us introduce the integer-valued function $\Theta(\nu, T)$ denoting the total number of the uses of the action u_ν up to moment T. Below we will show that if the action u_1 is optimal (i.e. $q_1 > q_i$, $i \geq 2$) then under the conditions of the theorem the equality

$$\lim_{T \to \infty} \frac{\theta(\nu, T)}{\theta(1, T)} = 0, \quad \nu > 1, \quad \text{a.s.} \tag{1}$$

holds. We have

$$\sum_{t=1}^{t} x_i = \sum_{\nu=1}^{k} \sum_{t \leq T} x_{u_0}(u_{t-1} = u_\nu) = \overset{(1)}{\underset{T}{\sum}} + \cdots + \overset{(\nu)}{\underset{T}{\sum}} + \cdots + \overset{(k)}{\underset{T}{\sum}}.$$

In the sums $\sum_T^{(i)}$ the summands represent independent, identically distributed random variables with mathematical expectations $W_i - W(u_i)$ and $W_1 > \max_{i \geq 2} W_i$. Hence

$$T^{-1} \sum_{t=1}^{t} x_i = \frac{\sum_T^{(1)} + \cdots + \sum_T^{(\nu)} + \cdots + \sum_T^{(k)}}{\sum_{\nu \geq 1} \theta(\nu, T)}$$

$$= \frac{\frac{1}{\theta(1,T)} \sum_T^{(1)} + \sum_{\nu \geq 2} \frac{\theta(\nu,T)}{\theta(1,T)} \frac{1}{\theta(\nu,T)} \sum_T^{(\nu)}}{1 + \sum_{\nu \geq 2} \frac{\theta(\nu,T)}{\theta(1,T)}}.$$

---

[d]As seen from the description of functioning the automata $\mathcal{A}$ at any finite moment have finite state space. Only in the limit the number of states is infinite. This is called the *potential infinity*.

[e]For example, the functions $n(m) = [m^c]$, $c > 0$, $n(m) = [\ln m]$, $n(m) = [m^{\ln m}]$ satisfy the conditions of this theorem. It is easy to note that the function $n(m) = 2^m$ does not satisfy the second condition of the theorem. The automaton $\mathcal{D}_k[2^m]$ is not asymptotically optimal.

Making use of (1) and the strong law of the large numbers for a sequence of independent, identically distributed random variables with a finite mathematical expectation we obtain

$$\lim_{t\to\infty} t^{-1} \sum_{n=1}^{t} x_n = \max_i W_i$$

for any binary HPIV.

We turn now to the proof of the key equality (1). We note that in the problem under consideration we cannot restrict ourselves to the investigation of only the first moments of the sojourn time τ_i in the branches of automaton $\mathcal{D}_{k,n}$ with the memory depth n. We will manage to restrict our attention to the second moments. The following estimates

$$\frac{\mathbf{E}\tau_i^2}{(\mathbf{E}\tau_i)^2} \leq \frac{2}{q_i}, \quad i = 1,\ldots,n \tag{2}$$

hold. Their proofs are simple but cumbersome and we will restrict ourselves to short explanations. The second moments $\mathbf{E}\tau_i^2$ are related by the equations similar to ones with respect to the average times, i.e.

$$\mathbf{E}(\tau_i - 1)^2 = p\mathbf{E}\tau_{i-1}^2 + q\mathbf{E}\tau_n^2, \quad \mathbf{E}\tau_0^2 = 0, \qquad i = 1,\ldots,n.$$

They can be written in another form

$$p\mathbf{E}\tau_{i-1}^2 - \mathbf{E}\tau_i^2 + q\mathbf{E}\tau_n^2 = 2T_i - 1, \quad i = 1,\ldots,n.$$

Beginning with the first equation we express successively $\mathbf{E}\tau_i^2$ in terms of $\mathbf{E}_n^2$. Finally, by using the last equation we shall find $\mathbf{E}\tau_2^2$. This process leads to explicit formulae for $\mathbf{E}\tau_i$ from which the estimations (2) follow. For the sake of completeness we shall say that the estimations (2) are the best possible in the following sense

$$\sup_{1\leq i\leq n<\infty} \frac{\mathbf{E}\tau_i^2}{(\mathbf{E}\tau_i)^2} = \frac{2}{q}.$$

This relation can easily be verified by proving that

$$\lim_{n\to\infty} \frac{\mathbf{E}\tau_i^2}{(\mathbf{E}\tau_i)^2} = \frac{2}{q}.$$

Hence, for variance of the r.v. τ_i we have

$$\frac{\mathbf{E}(\tau_i - \mathbf{E}\tau_i)^2}{\mathbf{E}\tau_i^2} = \frac{\mathbf{E}\tau_i^2 - (\mathbf{E}\tau_i)^2}{\mathbf{E}\tau_i^2} \leq \frac{2-q}{q}.$$

Let $\tau(\nu, m)$ be the number of uses of the action u_ν on the mth cycle of the control. As known from Sec. 2 the mean of this r.v. is equal to

$$T(\nu, m) = \mathbf{E}\tau(\nu, m) = p_\nu^{-n_m}.$$

From this it immediately follows that

$$\left.\begin{array}{c}\displaystyle\lim_{h\to\infty}\frac{\sum_{m=g(h)}^{g(h+2)-1}\mathbf{E}\tau(2,m)}{[g(h+1)-g(h)]\mathbf{E}\tau(1,g(h))}=0,\\[3ex]\displaystyle\lim_{h\to\infty}\frac{\mathbf{E}\tau(2,g(h+2)-1)}{\mathbf{E}\tau(1,g(h))}=0\end{array}\right\} \tag{3}$$

if the action u_1 is better than u_2 ($p_2 > p_1$) and the conclusion of Theorem 1 holds. $\qquad\qquad\qquad\qquad\qquad\qquad\qquad\qquad\qquad\qquad\qquad\qquad\qquad\quad\square$

Lemma 1. *Under the conditions of Theorem 1 for $p_2 > p_1$ the following relationship holds:*

$$\lim_{h\to\infty}\frac{\sum_{m=g(h)}^{g(h+2)-1}\tau(2,m)}{\sum_{m=g(h)}^{g(h+2)-1}\tau(1,m)}=0\quad\text{a.s.}$$

Proof. First, we shall verify the limiting correlations

$$\lim_{h\to\infty}\frac{\sum_{m=g(h)}^{g(h+2)-1}\tau(1,m)}{[g(h+1)-g(h)]\mathbf{E}\tau(1,g(h))}\geq\frac{1}{2}\quad\text{a.s.,}\tag{4}$$

$$\lim_{h\to\infty}\frac{\sum_{m=g(h)}^{g(h+2)-1}[\tau(2,m)-\mathbf{E}\tau(2,m)]}{[g(h+1)-g(h)]\mathbf{E}\tau(1,g(h))}=0\quad\text{a.s.}\tag{5}$$

From these relations the assertion of the lemma follows. Indeed, taking into account

$$\frac{\sum_{m=g(h)}^{g(h+2)-1}\tau(2,m)}{\sum_{m=g(h)}^{g(h+2)-1}\tau(1,m)}=\left[\frac{\sum_{m=g(h)}^{g(h+2)-1}[\tau(2,m)-\mathbf{E}\tau(2,m)]}{[g(h+1)-g(h)]\mathbf{E}\tau(1,g(h))}\right.$$

$$\left.+\frac{\sum_{m=g(h)}^{g(h+2)-1}\mathbf{E}\tau(2,m)}{[g(h+1)-g(h)]\mathbf{E}\tau(1,g(h))}\right]\frac{[g(h+1)-g(h)]\mathbf{E}\tau(1,g(h))}{\sum_{m=g(h)}^{g(h+2)-1}\tau(1,m)},$$

and (3) we obtain the required result. So, let us prove the inequality (4). To that end, we introduce a family of independent, identically distributed r.v. $\tau'(1,m)$ which will mean the operating time of the control u_1 on the mth cycle provided the memory depth of the branch is equal to $n(g(h))$. It is clear that $\tau'(1,m)\leq\tau(1,m)$ for $m\in[g(h),g(h+1)-1]$. By Chebyshev inequality and (2) we obtain

$$\mathbf{P}\left\{\frac{\sum_{m=g(h)}^{g(h+2)-1}\tau(1,m)}{[g(h+1)-g(h)]\mathbf{E}\tau(1,g(h))}\leq\frac{1}{2}\right\}$$

$$\leq\mathbf{P}\left\{\frac{\sum_{m=g(h)}^{g(h+2)-1}[\tau'(1,m)-\mathbf{E}\tau'(1,m)]}{[g(h+1)-g(h)]\mathbf{E}\tau(1,g(h))}\leq-\frac{1}{2}\right\}$$

$$\leq\frac{4\mathbf{E}\tau^2(1,g(h))}{\left(\mathbf{E}\tau(1,g(h))\right)^2[g(h+1)-g(h)]}\leq\frac{8}{q_1(g(h+1)-g(h))}.$$

Because

$$\sum_{h=1}^{\infty} \frac{1}{g(h+1) - g(h)} < \infty,$$

Borel–Cantelli Lemma shows that the event standing under the first symbol $\mathbf{P}$ in the inequalities written down above occurs only a finite number of times. This implies the inequality (4).

For any $\varepsilon > 0$, we have

$$\mathbf{P}\left\{ \left| \frac{\sum_{m=g(h)}^{g(h+2)-1} [\tau(2,m) - \mathbf{E}\tau(2,m)]}{[g(h+1) - g(h)]\,\mathbf{E}\tau(1,g(h))} \right| \geq \varepsilon \right\}$$

$$\leq \frac{\sum_{m=g(h)}^{g(h+2)-1} \mathbf{E}\left[\tau(2,m) - \mathbf{E}\tau(2,m)\right]^2}{[g(h+1) - g(h)]^2 \left(\mathbf{E}\tau(1,g(h))\right)^2 \varepsilon^2}$$

$$\leq \frac{2 - q_1}{\varepsilon^2 q_1} \left(\frac{\mathbf{E}\tau(2,g(h+2) - 1))}{\mathbf{E}\tau(1,g(h))} \right)^2 \frac{g(h+2) - g(h)}{[g(h+1) - g(h)]^2 \mathbf{E}\tau(1,g(h))}.$$

According to the second equality in (3)

$$\sup_{h \geq 1} \left(\frac{\mathbf{E}\tau(2,g(h+2) - 1))}{\mathbf{E}\tau(1,g(h))} \right)^2 < \infty$$

if $p_2 > p_1$. Hence, the convergence of the series

$$\sum_{h=1}^{\infty} \frac{g(h+2) - g(h)}{[g(h+1) - g(h)]^2}$$

implies equality (5). $\qquad\qquad\qquad\square$

We can now prove the key equality without difficulties. The quantities $\theta(1,T)$, $\theta(\nu,T)$ denote the total number of uses of the actions u_1 and u_ν up to moment T respectively with $p_1 < p_\nu$. Let N cycles be completed up to time T and the $N+1$th one start. We choose M such that $g(M+1) < N < g(M+2)$ then

$$\frac{\theta(\nu,T)}{\theta(1,T)} \leq \frac{\sum_{n=1}^{N+1} \tau(\nu,n)}{\sum_{n=1}^{N} \tau(1,n)} \xrightarrow{N \to \infty} 0.$$

The last follows from the conditions of Theorem 1.

Now we elucidate the second condition of Theorem 1. It is easy to note that if $p_1 < p_\nu$ this condition is equivalent to

$$\lim_{m \to \infty} \frac{\mathbf{E}\tau(\nu,g(m+2) - 1)}{\mathbf{E}\tau(1,g(m))} = 0. \tag{6}$$

Indeed, as we known already

$$\frac{\mathbf{E}\tau(\nu,g(m+2) - 1)}{\mathbf{E}\tau(1,g(m))} = \frac{p_1^{n_{g(m)}}}{p_\nu^{n_{g(m+2)-1}}} = p_\nu^{-n_{g(m+2)-1} + \frac{\ln p_1}{\ln p_\nu} n_{g(m)}}.$$

This implies the desired equivalence of (6) to the conditions of Theorem 1, since the number $\alpha = \ln p_1 (\ln p_\nu)^{-1} > 1$ can be arbitrary from the interval $(1, \infty)$.

We shall consider the question of the convergence rate of the average reward over the interval $[0, \, T]$, i.e. of the quantity

$$W_T(\mathcal{D}) = T^{-1} \sum_{t=1}^{T} \mathbf{E} x_t,$$

to its limiting value (maximum). Let us introduce the notation: two sequences of variables α_n, β_n, $n \geq 0$ will be linked by the relation $\alpha_n \asymp \beta_n$ if they are positive and

$$0 < \varliminf_{n \to \infty} \frac{\alpha_n}{\beta_n} \leq \varlimsup_{n \to \infty} \frac{\alpha_n}{\beta_n} < \infty.$$

Theorem 2. *Let an automaton $\mathcal{D}_2([n_m])$ be defined by the function $n_m = [m^c]$, $c > 0$, and $p_1 < p_2$ (i.e. $W_1 > W_2$). If $c = 1$ then*

$$W_1 - T^{-1} \sum_{t=1}^{T} \mathbf{E} x_t \asymp \frac{W_1 - W_2}{T^\mu}, \qquad \mu = 1 - \ln_{p_1} p_2.$$

If $c \neq 1$ then

$$\lim_{T \to \infty} \left[W_1 - T^{-1} \sum_{t=1}^{T} \mathbf{E} x_t \right] T^\mu = \infty$$

with the same μ.

Proof. Using the equalities

$$\sum_{t=1}^{T} \mathbf{E} x_t = \mathbf{E} x(u_1) \mathbf{E}\theta(1, T) + \mathbf{E} x(u_2) \mathbf{E}\theta(2, T),$$

$$T = \mathbf{E}\theta(1, T) + \mathbf{E}(2, T),$$

we have

$$W_1 - T^{-1} \sum_{t=1}^{T} \mathbf{E} x_t = (W_1 - W_2) \frac{\mathbf{E}\theta(2, T)}{T} \asymp (W_1 - W_2) \frac{\mathbf{E}\theta(2, T)}{\mathbf{E}\theta(1, T)}. \tag{7}$$

It remains to estimate the factor $\mathbf{E}\theta(2, T)/\mathbf{E}\theta(1, T)$. We consider the quantity

$$\frac{\sum_{m=1}^{M+1} \mathbf{E}\tau(2, m)}{\sum_{m=1}^{M} \mathbf{E}\tau(1, m)}, \tag{8}$$

where the amount of cycles M is related to T by the equality

$$\sum_{m=1}^{M+1} \mathbf{E}\tau(2, m) = \sum_{m=1}^{M} \mathbf{E}\tau(1, m) = T$$

which, in particular, means that $\lim_{t \to \infty} M = \infty$.

First we shall consider the case of linearly increasing memory, i.e. when $c = 1$, $n_m = m$. We have

$$\sum_{m=1}^{M+1} \mathbf{E}\tau(2,m) = \sum_{m=1}^{M+1} p_2^{-m} \asymp p_2^{-M}$$

and

$$\sum_{m=1}^{M} \mathbf{E}\tau(1,m) \asymp p_1^{-M}.$$

Due to the equality $p_2^M = (p_1^M)^{\ln_{p_1} p_2}$ we find

$$\frac{\mathbf{E}\theta(2,T)}{\mathbf{E}\theta(1,T)} \asymp \frac{T^{\ln_{p_1} p_2}}{T} = T^{-\mu}.$$

It remains to consider the case $c \neq 1$.

1. $c > 1$. Since

$$\sum_{m=1}^{M+1} p_2^{-[m^c]} \asymp p_2^{-M^c} p_2^{-cM^{c-1}}, \qquad \sum_{m=1}^{M} p_1^{-[m^c]} \asymp p_1^{-M^c},$$

we obtained

$$\frac{\mathbf{E}\theta(2,T)}{\mathbf{E}\theta(1,T)} > a p_2^{-cM^{c-1}} T^{-M} \tag{9}$$

for some $a > 0$. But $p_2^{-cM^{c-1}} \to \infty$ as $t \to \infty$, and the assertion of the theorem follows from (7) and (9).

2. $c < 1$. Now, in view of the relations

$$\sum_{m=1}^{M+1} p_2^{-[m^c]} \asymp M^{1-c} p_2^{-M^c}, \qquad \sum_{m=1}^{M} p_1^{-[m^c]} \asymp M^c p_1^{-M^c} \asymp T,$$

we have

$$\frac{\mathbf{E}\theta(2,T)}{\mathbf{E}\theta(1,T)} \asymp M^{(1-c)\mu} T^{-\mu}.$$

This and (7) prove the theorem for $c < 1$ since $M^{(1-c)\mu} \xrightarrow[[t\to\infty]]{} \infty$. $\square$

Note that according to the definition of automaton with variable structure the automata with increasing memory above are not regarded as such.

2.4. δω-Automata and Their Modifications

We consider classes of real-valued HPIV with a finite number of states. The mathematical expectations $W(u) = \int_{-\infty}^{\infty} z\mu(dz|u)$ are supposed to exist. The required aims consist in obtaining the optimal limiting reward by means of using finite automata. The original principle of synthesizing the ε-optimal automata consists of recalculating the choice probabilities of the actions by using the estimates of the average rewards $W(u_1), \ldots, W(u_k)$.

The strategies for a class of the HPIV are represented in the form of a learning system $\mathcal{L}$ with some sets X and U as the input and output signals respectively and with the open $k-1$-dimensional simplex

$$\Sigma = \left\{ (p_1, \ldots, p_k); \sum_{j=1}^{k} p_j = 1, p_j > 0, j = 1, \ldots, k \right\}$$

as the state space. The points $\bar{p} = (p_1, \ldots, p_k) \in \Sigma$ are some probability distributions on U, i.e. p_i is the choice probability of u_i. Let a law $T^{(*)}$ of some walk on Σ be given. Then the state $\bar{p}$ turns out to be a function of time, i.e. $\bar{p} = \bar{p}(t)$. We shall say that $\mathcal{L}$ is in a *δ-optimal regime* at the moment t if the distribution $\bar{p}(t)$ is such that the probability of the optimal action is equal to $1 - \delta$ ($\delta \in (0, 1/2)$) and the sum of the probabilities of the rest of the actions is equal to δ.

Definition 1. A learning system $\mathcal{L}_\delta$ will be called *δ-optimal* with respect to a class $\mathcal{K}$ of HPIV if after a time τ (finite a.s. non-Markov) it enters a δ-optimal regime and stays there for ever for any $\xi \in \mathcal{K}$.

We are going to study the optimizational possibilities of the δ-optimal systems. But first, we define a class of the HPIV to be considered.

Let Π_c, $c > 0$, denote a class of real-valued HPIV such that $|W(u)| \leq c$ for all $u \in U$.

Theorem 1. *For any class Π_c, the family of δ-optimal learning systems $\mathcal{L}_\delta$ forms an ε-optimal family in both strong and weak sense.*

Proof. The distribution $\bar{p}(t)$ in a δ-optimal regime has the form

$$\bar{p}_\delta = (q_1\delta, \ldots, q_{j_0-1}\delta, 1 - \delta, q_{j_0+1}\delta, \ldots, q_k\delta) = \left(p_1^{(\delta)}, \ldots, p_k^{(\delta)} \right),$$
$$W_{j_0} = \bar{W} = \max_i W(u_i)$$

where $q_j > 0$, $\sum_{j \neq j_0} q_j = 1$ and j_0 means the index of an optimal action, if there are several it will mean the index of any of them. The appropriate average reward per unit time is equal to

$$W_\delta = (1 - \delta)\bar{W} + \delta \sum_{j \neq j_0} q_j W(u_j)$$

or

$$W_\delta = \bar{W} - \delta \left(\bar{W} - \sum_{j \neq j_0} q_j W(u_j) \right).$$

Choosing δ properly we immediately obtain the assertion on ε-optimality in the weak sense. The ε-optimality in the strong sense follows from the following properties of the system $\mathcal{L}_\delta$:

(a) $W(\mathcal{L}_\delta) > \bar{W} - \varepsilon(\delta)$;
(b) the distribution $\bar{p}_\delta$ is used for any $t > \tau$.

Let $\varepsilon > 0$ be fixed. We choose δ so small that $\varepsilon(\delta) < \varepsilon/2$, then

$$\mathbf{P}\left\{ \lim_{t\to\infty} t^{-1} N_i(t) = p_i^{(\delta)}, \lim_{t\to\infty} V_i(t) = W(u_i), i = 1, \ldots, k \right\} = 1,$$

where $N_i(t)$, $V_i(t)$ denote the same quantities as in Theorem 3, Sec. 2. This means that for sufficiently large t $(> \tau)$

$$V(t) = \sum_{i=1}^{k} \frac{N_i(t)}{t} V_i > W(\mathcal{L}_\delta) - \frac{\varepsilon}{2} > \bar{W} - \varepsilon, \quad \text{a.s.} \qquad \square$$

Note that there exists no finite automaton which is δ-optimal for all $\xi \in \Pi_c$. Therefore, the associated Markov chain corresponding to the pair ξ, $\mathcal{L}_\delta$ is infinite and, moreover, it is not obliged to be regular.

Let us define a class of $\delta\omega$-automata for which we shall prove δ-optimality. The input signals are the pairs (x_t, u_t), the triplets $s = (\bar{N}, \bar{M}, \bar{p})$ serve as the states where $\bar{N} = (N_1, \ldots, N_k)$ and $\bar{V} = (V_1, \ldots, V_k)$ are k-dimensional vectors whose components are defined in advance and $\bar{p}$ is some distribution on U. As the initial state we take $s_0 = \bar{N} = \bar{V} = 0$, $p_0 = (1/k, \ldots, 1/k)$. The output function is the current distribution $\bar{p}(t)$ on U. It remains to define the transition functions. The components of $\bar{N}$, $\bar{V}$ are altered in an obvious way: at the moment t one recalculates the component which corresponds to the number of the control $U(t - 1)$ at the moment $t - 1$. We shall suppose that $V_i(t) = 0$ until $N_i(t) = 0$.

The transformation technique of the distribution $\bar{p}$ is defined by a pair $(\mathcal{Q}, \mathcal{I})$ where $\mathcal{Q}$ is an operator acting on $\bar{p}$ at the moment to be specified by the law $\mathcal{I}$. We impose the following condition on this rule: $\mathbf{P}\left\{ \nu_t \xrightarrow{t\to\infty} \infty \right\} = 1$ where ν_t denotes the number of transformations of $\bar{p}$ over time t. We shall give examples of such laws.

1. At each moment the vector $\bar{p}$ is transformed with probability $\alpha > 0$ and remains the same with probability $1 - \alpha$.
2. The vector $\bar{p}$ will be transformed after $\min_i N_i(t)$ increases by one.
3. The vector $\bar{p}$ will be transformed after all actions have been chosen at least once.

The operator $\mathcal{Q}$ transforms a stochastic vector $\bar{p}$ into a stochastic one again. It is subject to the following conditions:

1. All components of the vector $\mathcal{Q}\bar{p}$ are bounded from below by a number $\beta \in (0, \delta)$.
2. Let j_0 be the number of the best action at the moment t (i.e. $V_{j_0}(t) = \max_j V_j(t)$). Then $\|\mathcal{Q}\bar{p}_t - e_{j_0}\| \leq \|\bar{p}_t - e_{j_0}\|$, where $e_{j_0} = (0, \ldots, 0, 1, 0, \ldots 0)$ (unit corresponds to the j_0th coordinate) is the j_0th vertex of the simplex Σ.
3. If u_{j_0} has been used as the best action during T_δ actions of the operator $\mathcal{Q}$ then as a result of these actions a δ-regime will be set (the choice probability of the action u_{j_0} is equal to $1 - \delta$).

4. When replacing the best action u_{j_0} by the next one u_{i_0} at the moment of the current transformation the vector $\bar{p}$ is transformed in the following way: the j_0th and the i_0th components change over.

For processes with a single optimal action the last condition can be relaxed. For example, we can return to the original uniform distribution $\bar{p}_0$.

This completes the description of the $\delta\omega$-automata. Note that all states of the associated Markov process corresponding to an interaction between a HPIV and a $\delta\omega$-automaton are transient, i.e. no state is visited twice.

Theorem 2. *The $\delta\omega$-automata are ε-optimal with respect to each class of HPIV Π_c.*

Proof. Note that by properties 1–4 of the operator $\mathcal{Q}$ the $\delta\omega$-automaton enters the δ-regime in finite time a.s., i.e. the probability $1 - \delta$ will be assigned to some action (not necessarily to the optimal one). It remains to verify that beginning from some moment the optimal action will remain the best one. The conditions imposed on $\mathcal{Q}, \mathcal{I}$ imply that $N_j(t) \to \infty$ a.s. for all j. Hence $V_j(t) \xrightarrow{\text{a.s.}} W(u_j)$ as $t \to \infty$ for all j. This implies that there exists a j_0 such that beginning from some moment, the value V_{j_0} will be the leading one among the quantities $V_1(j), \ldots, V_k(t)$. Therefore the control u_{j_0} will be optimal. Then the probability p_{j_0} begins to increase. If there is more than one optimal action and I is the set of their indexes then there will exist a non-Markov moment τ finite a.s. such that for all $t > \tau$ the following inequality

$$\min_{i \in I} V_i(t) > \max_{i \notin I} V_i(t)$$

holds. By condition 4 on the operator $\mathcal{Q}$, the automaton remains in the δ-regime for ever after it has entered it. $\qquad\square$

Using this result it is easy to imagine the functioning of $\delta\omega$- automaton as an adaptive strategy: at first it uses all actions equiprobably and accumulates the empirical estimates of the rewards, but thereafter it defines the most profitable action according to the maximum. At the moment τ_δ the automaton enters a δ-regime which can change its carrier a finite number of times. At some moment the optimal action becomes the carrier, and beginning from the moment τ, this optimal action keeps the maximum probability $1 - \delta$. In any class Π_c (c is positive and fixed) the limiting average reward W_δ approaches to the maximum reward $\bar{W}$ as $\delta \to 0$. We supplement these qualitative considerations with some quantitative results.

We shall consider the properties of the process $\bar{N}(t)$. Its values belong to an integer-valued k-dimensional lattice. In view of the identity $N_1(t) + \cdots + N_k(t) \equiv t$ its components are linearly dependent and increase by one in random order. The probability p_j of increasing $N_j(t)$ satisfies the inequalities $\beta \leq p_j \leq 1 - \delta$. It is clear that these components form semi-martingales. Let us estimate the deviation probabilities of the components $N_j(t)$ from their mathematical expectations.

Lemma 1.

$$\mathbf{P}\left\{\beta \leq \varliminf_{t\to\infty} t^{-1}N_j(t) \leq \varlimsup_{t\to\infty} t^{-1}N_j(t) \leq 1-\delta, \; j=1,\dots,k\right\} = 1.$$

Proof. We shall show that for any $\varepsilon > 0$ the following series

$$\sum_{t=1}^{\infty}\mathbf{P}\{N_j(t) \leq (\beta-\varepsilon)t\}, \quad \sum_{t=1}^{\infty}\mathbf{P}\{N_j(t) \geq (1-\delta+\varepsilon)t\}$$

converge. Then, according to Borel–Cantelli Lemma the events $\{N_j(t) \leq (\beta-\varepsilon)t\}$, $\{N_j(t) \geq (1-\delta+\varepsilon)t\}$ occur a.s. only a finite number times and this proves the lemma. To verify the required convergence it is sufficient to make sure that

$$\mathbf{P}\{N_j(t) \leq (\beta-\varepsilon)t\} \leq e^{-2\varepsilon^2 t}, \quad \mathbf{P}\{N_j(t) \geq (1-\delta+\varepsilon)t\} \leq e^{-2\varepsilon^2 t}. \tag{1}$$

We shall only prove the first inequality. Let $N^*(t)$ denote a sequence of r.v of the form $N^*(t) = \zeta_1 + \cdots + \zeta_t$ where

$$\zeta_n = \begin{cases} 1, & \text{with probability } \beta, \\ 0, & \text{with probability } 1-\beta. \end{cases}$$

Applying Hoeffding theorem[f] to the variables $\zeta_n \in [0,1]$ with $\mathbf{E}\zeta_n = \beta$ we have

$$\begin{aligned}
\mathbf{P}\{N^*(t) \leq (\beta-\varepsilon)t\} &= \mathbf{P}\{N^*(t) - \beta t \leq -\varepsilon t\} \\
&= \mathbf{P}\{N^*(t) - \mathbf{E}N^*(t) \leq -\varepsilon t\} \leq e^{-2\varepsilon^2 t}.
\end{aligned}$$

It remains to note that

$$\mathbf{P}\{N^*(t) \leq -Lt\} \leq \mathbf{P}\{N_j(t) \leq -Lt\}. \qquad \square$$

We now turn to the properties of the components of the vector $\bar{V}(t)$. First, we shall specify the class of HPIV.

Let $\mathcal{K}_c$ denote a subclass of Π_c whose elements have the following property: for each r.v. $\xi(u)$ defined by the measure $\nu(\cdot|u)$ there exist some positive $g(u)$ and H such that

$$\mathbf{E}e^{y\xi(u)} \leq \exp\{g(u)y^2/2\}, \quad 0 \leq y \leq H.$$

Each r.v. ξ may have its own constants g, H.

Lemma 2. *Let u_1 be a single optimal action for a process $\zeta \in \mathcal{K}_c$. Then for any t*

$$\mathbf{P}\left\{V_1(t) \leq \max_{i\geq 2} V_i(t)\right\} \leq (k-1)(t^2+4)e^{-\mu t}, \quad \mu > 0.$$

---

[f]**Theorem.** (Hoeffding) *Let $\eta_1,\dots,\eta_n$ be independent r.v. such that $\eta_i \in [a_i, b_i]$, $i = 1,\dots,n$. Then for any $x > 0$*

$$\mathbf{P}\left\{n^{-1}\sum_{i=1}^{n}(\eta_i - \mathbf{E}\eta_i) \geq x\right\} \leq \exp\left[-\frac{2n^2}{\sum_{i=1}^{n}(b_i - a_i)^2}x^2\right].$$

Proof. We have

$$\mathbf{P}\left\{V_1(t) \le \max_{i \ge 2} V_i(t)\right\}$$

$$= \sum_{j=2}^{k} \sum_{l_1, l_j=0}^{t} \mathbf{P}\left\{V_1(t) \le \max_{i \ge 2} V_i(t) | V_j(t) = \max_{i \ge 2} V_i(t), N_1(t) = l_1, N_j(t) = l_j\right\}$$

$$\times \mathbf{P}\left\{V_j(t) = \max_{i \ge 2} V_i(t) | N_j(t) = l_j\right\} \mathbf{P}\{N_1(t) = l_1, N_j(t) = l_j\}$$

$$\le \sum_{j=2}^{k} \sum_{l_1, l_j=0}^{t} \mathbf{P}\left\{V_1(t) \le V_j(t) | V_j(t) = \max_{i \ge 2} V_i(t), N_1(t) = l_1, N_j(t) = l_j\right\}$$

$$\times \mathbf{P}\{N_1(t) = l_1, N_j(t) = l_j\}. \tag{2}$$

On the right-hand side of (2) we choose a summand to correspond, for example, to $j = 2$, i.e. we consider the r.v. $\xi(u_1)$ and $\xi(u_2)$ defined by the distributions $\mu(\cdot|u_\varkappa)$ with the mathematical expectations $W_\varkappa$, $\varkappa = 1, 2$ and write this in the form

$$\sum_{l_1, l_2=0}^{t} \mathbf{P}\{V_1(t) - V_2(t) \le 0 | N_1(t) = l_1, N_2(t) = l_2\} \mathbf{P}\{N_1(t) = l_1, N_2(t) = l_2\}$$

$$= \Sigma_1 + \Sigma_2.$$

In Σ_1 the sum is taken over all indices l_i, $i = 1, 2$ such that $(\beta - \varepsilon)t \le N_{l_i}(t) \le (1 - \delta + \varepsilon)t$ but in Σ_2 over the rest of them.

Let us estimate the sum Σ_1 from above by using the inequality

$$\Sigma_1 < \sum_{l_1, l_2=(\alpha-\varepsilon)t}^{(1-\delta+\varepsilon)t} \mathbf{P}\left\{l_1^{-1} \sum_{i=1}^{l_1} x_i^{(1)} - l_2^{-1} \sum_{i=1}^{l_2} x_i^{(2)} \le 0\right\}$$

where $x_i^{(\varkappa)}$ is the sample values of the r.v. $\xi(u_\varkappa)$. We shall consider one of the summands on the right-hand side. We put $l_\varkappa = \lambda_\varkappa t$ where $\lambda_\varkappa$ is fixed and such that $\beta - \varepsilon \le \lambda_\varkappa \le 1 - \delta + \varepsilon$ (it is obvious that $\varepsilon < \min(\beta, \delta)$), and introduce the centred r.v. $z^{(\varkappa)} = x^{(\varkappa)} - W_\varkappa$. Then

$$\mathbf{P}\left\{l_1^{-1} \sum_{i=1}^{l_1} x_i^{(1)} - l_2^{-1} \sum_{i=1}^{l_2} x_i^{(2)} \le 0\right\}$$

$$\le \mathbf{P}\left\{\lambda_1^{-1} \sum_{i=1}^{l_1 t} z_i^{(1)} - \lambda_2^{-1} \sum_{i=1}^{\lambda_2 t} z_i^{(2)} \le (W_1 - W_2)t\right\}.$$

After the substitution

$$\zeta_j = \frac{z_j^{(1)}}{\lambda_1}, \quad j = 1, \ldots, \lambda_1 t, \qquad \zeta_{\lambda, t+j} = -\frac{z_j^{(2)}}{\lambda_2}, \quad j = 1, \ldots, \lambda_2 t,$$

we obtain a family of r.v. which satisfy the conditions of the Petrov Theorem.[g]
Putting $M = W_1 - W_2$, $G = (g_1/\lambda_1 + g_2/\lambda_2)t$ and changing $\lambda_\varkappa$ by the least value
we apply this theorem to the sum $S_t = \sum_{j=1}^{(\lambda_1+\lambda_2)t} \zeta_j$. Then

$$\mathbf{P}\left\{ l_1^{-1} \sum_{i=1}^{l_1} x_i^{(1)} - l_2^{-1} \sum_{i=1}^{l_2} x_i^{(2)} \leq 0 \right\} = \mathbf{P}\{S_t \leq Mt\} \leq e^{-\nu t}$$

where

$$\nu = \begin{cases} \dfrac{(\beta - \varepsilon)M^2}{2(g_1 + g_2)}, & \text{if } M \leq (g_1 + g_2)H, \\[2mm] \dfrac{HM}{2}, & \text{if } M > (g_1 + g_2)H. \end{cases}$$

So, each summand in the sum Σ_1 was estimated. To estimate the whole sum
we put

$$M = \min_{j \geq 2}(W_1 - W_2), \quad \tilde{\nu} = \min\left(\frac{(\beta - \varepsilon)M^2}{2(g_1 + g_2)}, \frac{1}{2}HM \right).$$

Then

$$\Sigma_1 < \sum_{l_1, l_2 = (\beta - \varepsilon)t}^{(1-\delta+\varepsilon)t} \mathbf{P}\left\{ l_1^{-1} \sum_{i=1}^{l_1} x_i^{(1)} - l_2^{-1} \sum_{i=1}^{l_2} x_i^{(2)} \leq 0 \right\} \leq t^2 e^{-\tilde{\nu} t}. \tag{3}$$

It remains to estimate Σ_2. Making use of inequality (1) and Lemma 1 we obtain

$$\Sigma_2 \leq \mathbf{P}\{N_2(t) \leq (\beta - \varepsilon)t\} + \mathbf{P}\{N_1(t) \geq (1 - \delta + \varepsilon)t\}$$
$$+ \mathbf{P}\{N_2(t) \leq (\beta - \varepsilon)t\} + \mathbf{P}\{N_1(t) \geq (1 - \delta + \varepsilon)t, N_2 \leq (\beta - \varepsilon)t)\}$$
$$\leq 4e^{-2\varepsilon^2 t}. \tag{4}$$

Combining (3) and (4) we see that

$$\Sigma_1 + \Sigma_2 < t^2 e^{-\tilde{\nu} t} + 4e^{-2\varepsilon^2 t} \leq (t^2 + 4)e^{-\nu t}$$

where $\nu = \min(\tilde{\nu}, 2\varepsilon^2)$. By analogy with this inequality we obtain the estimate

$$\mathbf{P}\{V_1(t) \leq V_j(t)\} \leq (t^2 + 4)e^{-\nu t}.$$

From this and (2) the desired result follows. $\square$

---

[g]**Theorem.** (Petrov) *Let $\eta_1, \ldots, \eta_n$ be independent r.v. and suppose there exist positive constants
H and $g_1, \ldots, g_n$ such that*

$$\mathbf{E}e^{y\eta_i} \leq \exp(g_i y^2/2)$$

for $y \in [0, H]$. Put $G = g_1 + \cdots + g_n$. Then

$$\mathbf{P}\left\{ \sum_{i=1}^{n} \eta_i \geq x \right\} \leq \begin{cases} \exp\left(-\dfrac{x^2}{2G} \right), & \text{if } 0 \leq x \leq GH, \\[2mm] \exp\left(-\dfrac{Hx}{2} \right), & \text{if } GH \leq x. \end{cases}$$

Corollary 1.

$$\mathbf{P}\left\{\min_{i\in(i_1,\ldots,i_n)} V_i(t) \leq \max_{j\notin(i_1,\ldots,i_n)} V_j(t)\right\} < ae^{-\lambda t}, \quad a,\lambda > 0$$

for any HPIV from $\mathcal{K}_c$ with optimal actions $u_{i_1},\ldots,u_{i_n}$.

We obtain more complete information on the behavior of the $\delta\omega$-automata by considering the properties of the following r.v : τ'_δ is the first hitting time of the automaton in the $\delta\omega$-optimal regime (Markov moment), τ''_δ is the last hitting time in this regime (non-Markov moment). They are related by the inequalities

$$\bar{\tau} \leq \tau'_\delta \leq \tau''_\delta \leq \tau_\Pi + \tau_\delta \tag{5}$$

where, in the simplest case of a single optimal action u_1,

$$\bar{\tau} = \min\left\{t : V_1(t) > \max_{j\geq 2} V_j(t)\right\},$$

$$\tau_\Pi = 1 + \max\left\{t : V_1(t) \leq \max_{j\geq 2} V_j(t)\right\}.$$

In what follows we shall assume that the rule $\mathcal{I}$ which defines the moments when the operator $\mathcal{Q}$ is used is such that the random lengths Δ of the time intervals between the successive uses of this operator *have finite moments* $\mathbf{E}\Delta^n$, $n \geq 1$.

Theorem 3. *If a $\delta\omega$-automaton and a control of HPIV from $\mathcal{K}_c$ is given then all moments of the r.v. τ'_δ and τ''_δ are finite.*

Proof. By Corollary 2 we have

$$\mathbf{P}\{\bar{\tau} = t\} = \mathbf{P}\left\{V_1(s) \leq \max_{i\geq 2} V_i(s), s < t, V_1(t) > \max_{i\geq 2} V_i(t)\right\}$$

$$\leq \mathbf{P}\left\{V_1(t-1) \leq \max_{i\geq 2} V_i(t-1)\right\} \leq ae^{-\nu(t-1)}.$$

Hence

$$\mathbf{E}\bar{\tau}^n = \sum_{t=1}^{\infty} t^n \mathbf{P}(\bar{\tau} = t) < a\sum_{t=1}^{\infty} t^n e^{-\nu(t-1)} < \infty, \quad n = 1, 2, \ldots,$$

i.e. all moments of $\bar{\tau}$ are finite. The assertion on τ_Π can be proved by similar arguments. Now, from (5) and the condition on the rule $\mathcal{I}$ the assertion follows. $\quad\square$

Let us define a subclass of $\delta\omega$-automata to be used later on. They are defined by the specific form of the operator Q which depends on two numerical parameters θ and $\hat{p}$ such that

$$0 < \theta < 1, \quad \frac{]\theta k[}{k} < \hat{p} \leq 1 - \delta < 1, \quad \delta < 1/2,$$

where k is the number of controls, $]r[$ means the least integer no less than r.

At the moments pointed by the rule $\mathcal{I}$ the variational series

$$V_{i_1} > V_{i_2} > \cdots > V_{i_k}$$

is formed by using the empirical rewards $V_1(t), \ldots, V_k(t)$. We put here the symbol ">" but if there exist groups of the same V_i then they are arranged in order of increasing their indices. The first application of the operator Q picks the group of the "best" actions $u_{i_1}, \ldots, u_{i_{]\theta k[}}$ and assigns them the total probability $\hat{p}$ instead of the previous one $\frac{]\theta k[}{k} < \hat{p}$. This new probability is distributed equally between the actions picked uniformly. The supplementary probability $1 - \hat{p}$ is divided in equal parts between the rest of the actions as well. Let the selected group of the actions keep its position at the head of the variational series

$$V_{j_1} > \cdots > V_{j_{]\theta k[}},$$

where $(j_1, \ldots, j_{]\theta k[})$ is some transposition of the indices $(i_1, \ldots, i_{]\theta k[})$ by the moment of the next transformation. From this group of actions the proportion θ of better ones is picked over again, i.e. $u_{j_1}, \ldots, u_{j_{]\theta k[}}$ and the total probability $\hat{p}$ is distributed between them. The repetition of this procedure leads, in the long run, to the best action u_ν that the probability $\hat{p}$ is assigned to. So, the distribution of the probabilities of the actions is the following

$$\bar{p} = \left(\underbrace{\frac{1-\hat{p}}{k-1}, \ldots, \frac{1-\hat{p}}{k-1}}_{\nu-1}, \hat{p}, \frac{1-\hat{p}}{k-1}, \ldots, \frac{1-\hat{p}}{k-1} \right).$$

If at the moment of the next transformation the action u_ν is the leader again, the automaton will pass into the δ-regime, i.e. it will have the distribution

$$\bar{p}_\delta = \left(\underbrace{\frac{\delta}{k-1}, \ldots, \frac{\delta}{k-1}}_{\nu}, 1-\delta, \frac{\delta}{k-1}, \ldots, \frac{\delta}{k-1} \right).$$

Having additional information about the class of the HPIV the properties of the operator Q may be worked out in detail. For example, if u_{opt} is unique, the following variants will be possible:

(1) to return from the current distribution $\bar{p}$ to the original uniform distribution $\bar{p}_0$;
(2) to increase θ^{-1} times the available group of $\varkappa$ best actions;
(3) to assign the probability $\hat{p}$ (or $1 - \delta$, provided the moment occurs) to a new group to consist of the $\varkappa$ actions (without changing the number of the member of the group).

The $\delta\omega$-automata defined here are called automata of $\mathcal{G}$-type ("group"). We denote them by $\mathcal{G}(k, \theta, \hat{p}, \delta, \mathcal{I})$.

The quantitative estimates of the optimizational properties of the $\delta\omega$-automata can be based on the different tests, i.e. the total reward till the final hitting of the δ-optimal regime, the average time of the first (or the last) hitting in the δ-optimal regime and so on. It is rather difficult to obtain such estimates in an analytical way and this can be done only in special cases. Almost always such estimates can be obtained by computer simulation. However we note that in the adaptive theory the "uniform estimates" are of interest, i.e. such estimates which hold on the whole class of controlled models. The individual estimates which refer to a concrete process from the class are not useful because we do not know which process concretely is under the control.

We shall use the design principles of $\delta\omega$-automata which are ε-optimal with respect to the classes of the HPIV to synthesize the asymptotically optimal automata. The aim of control defined as

$$\mathbf{P}\left\{ \lim_{t\to\infty} t^{-1} \sum_{n=1}^{t} x_n = \max_i W_i \right\} = 1$$

is obviously equivalent to the one stated in terms of the choice probabilities of the action u_t at the moment t

$$\mathbf{P}\left\{ \lim_{t\to\infty} \sum_{i\in I} p_i(t) = 1 \right\} = 1 \tag{6}$$

where I denotes the index set of the optimal actions. This equality should take place for each process $\xi \in \mathcal{K}$. When a unique optimal action u_{j_0} exists the appropriate vertex $e_{j_0} = (0, \ldots, 0, 1, 0, \ldots, 0)$ of the simplex Σ will be the unique absorbing point of the random walk corresponding to the interaction between ξ and the automaton. We shall consider the automata whose state sets contain the vectors $\bar{N}$, $\bar{V}$, $\bar{p}$. Condition (6) refers to the components of the vector $\bar{p}$. We rewrite this condition in terms of $\bar{N}$:

$$\mathbf{P}\left\{ \lim_{t\to\infty} t^{-1} \sum_{i\in I} N_i(t) = 1 \right\} = 1. \tag{7}$$

The pairs $(\bar{N}, \bar{V})$ serve as the states of the simplest automaton construction. Its output function is defined by an integer $r \geq 1$ and a sequence of positive integers n_m, $(n_m < n_{m+1})$. The controls u_i (referred to as actions as before) are chosen in two alternative regimes:

(1) each element from $\{u_1, \ldots, u_k\}$ is used r times in succession;
(2) during n_m times the best action (corresponding to the maximum empirical rewards $V_i(t)$) is applied.

This automaton is denoted by $\mathcal{M}$.

Theorem 4. *The automaton $\mathcal{M}$ is asymptotically optimal with respect to the class of all scalar HPIV.*

Proof. By construction we have $N_j(t) \xrightarrow[t\to\infty]{} \infty$ and, hence,

$$\mathbf{P}\left\{ \lim_{t\to\infty} V_i(t) = W_i, \ i = 1, \ldots, k \right\} = 1.$$

So, beginning from some moment τ the quantity $V_{j_0}(t)$ will become greater than all other V_j if u_{j_0} belongs to the group of the optimal actions. Beginning from this moment, the choice frequency of the optimal actions approaches one (i.e. $N_{j_0}/t \xrightarrow[t\to\infty]{} 1$) but the frequencies of the other actions approach to 0. This implies that the total average reward in time t

$$\mathbf{V}(t) = \sum_{l=1}^{k} \frac{N_l(t)}{t} V_l(t)$$

converges to $\max W_i = \bar{W}$ a.s. $\qquad\square$

The *automata* $\mathcal{G}_\delta$ are similar to $\mathcal{G}$. The difference consists of the changeability of δ, i.e. instead of the constant number δ we have a sequence δ_n such that

$$0 < \delta_n < 1/2, \quad \delta_n \downarrow 0, \quad \sum_{n=1}^{\infty} \delta_n = \infty. \tag{8}$$

On reaching the δ-optimal regime at the moments defined by the rule $\mathcal{I}$ we replace the value δ_n obtained by the smaller one δ_{n+1}, i.e. the probabilities δ_{n+1} and $\delta_{n+1}/k - 1$ are assigned to the best action or the optimal one and the rest of them respectively. If the rule $\mathcal{I}$ consists of changing the leader of the variation series at the moment when an action which has been used most rarely so far is applied then instead of δ_n we shall write $\delta_{n(t)}$ where $n(t) = \min_i N_i(t)$. Then, in conditions (8) the divergence of the series is not required. We denote this version of automaton $\mathcal{G}_\delta$ by $\mathcal{G}_{\delta_n}$.

Theorem 5. *The automata $\mathcal{G}_\delta$ (in particular, $\mathcal{G}_{\delta_n}$) are asymptotically optimal with respect to the class of all scalar HPIV.*

Proof. After getting the automaton $\mathcal{G}_\delta$ in the δ-regime the choice probabilities of other than the best actions became equal to $\delta_t/(k - 1)$. According to (8) the series composed of these numbers diverges. This implies that the number of uses of these actions tends to infinity, i.e. $\lim_{t\to\infty} N_j(t) = \infty$ for all j. In view of $V_j(t) \xrightarrow[t\to\infty]{\text{a.s.}} W_j$ convergence in finite time, one can find an optimal action which will be used with frequency approaching one as $t \to \infty$. For the automata $\mathcal{G}_{\delta_n}$ it is obvious that $n(t) \xrightarrow[t\to\infty]{\text{a.s.}} \infty$. Otherwise the probabilities of all actions would be bounded from below by some positive constants but then $N_j(t) \xrightarrow[t\to\infty]{\text{a.s.}} \infty$. This leads to a contradiction. $\qquad\square$

Definition 2. *An automaton $\mathcal{SA}$ is defined by the recurrent procedure*

$$\bar{p}_{t+1} = \bar{p}_t + \frac{1}{t+1} \left[\bar{\psi}(t) - \bar{p}_t \right], \tag{9}$$

where the components of the vector $\bar{\psi}(t)$ are defined as follows

$$\psi_j(t) = \begin{cases} 1, & \text{if } V_j(t) = \max_i V_i(t), \\ 0, & \text{if } V_j(t) < \max_i V_i(t). \end{cases}$$

If there are several maximum estimates then $\psi_j(t) = 1$ only for one index j (for example, for the least one).

Theorem 6. *The automaton $\mathcal{SA}$ is asymptotically optimal with respect to the class of scalar HPIV with unique optimal action.*

Proof. For almost all $\omega \in \Lambda = \{\omega : \lim_{t \to \infty} N_i(t) = \infty, \ i = 1, \ldots, k\}$ we have $V_i(t) \xrightarrow[t \to \infty]{} W_i$. Hence $V_{i_0}(t) > V_i(t)$ for all sufficiently large t where $i_0 \ (\neq i)$ is the index of the optimal action. Then the transformation (9) increases the probability $p_{i_0}(t)$ and decreases the rest of them. For $i \neq i_0$, $t > t_0$ we have $(a(t) = (1+t)^{-1})$

$$p_i(t+1, \omega) = (1 - a(t))p_i(t, \omega) = \sum_{l=t_0}^{t}(1 - a(l)) \cdots (1 - a(t_0))p_i(t_0, \omega).$$

The divergence of $\sum a(t)$ implies that $\prod_{l=t_0}^{t}(1 - a(l)) \xrightarrow[t \to \infty]{} 0$, i.e. $p_i(t, \omega) \xrightarrow[t \to \infty]{} 0$ for $i \neq i_0$. Hence $p_{i_0}(t, \omega) \xrightarrow[t \to \infty]{} 1$.

It remains to show that $\mathbf{P}\{\Lambda\} = 1$. To that end we shall prove the divergence of $\sum_{t=1}^{\infty} p_i(t, \omega)$, $i = 1, \ldots, k$ with probability one. Indeed, by (9),

$$\bar{p}(t+1) = \frac{t}{t+1}\bar{p}(t) + \frac{1}{t+1}\bar{\psi}(t),$$

i.e. $p_i(t+1) \geq t/(t+1)p_i(t)$ and equality takes place if and only if $\psi_i(t) = 0$. This implies the divergence of the series in question.

Let the automaton $\mathcal{SA}$ controlling the HPIV be given. We put

$$\eta_t^{(i)}(\omega) = \begin{cases} 1, & \text{if } u(t) = u_i, \\ 0, & \text{if } u(t) \neq u_i. \end{cases}$$

Then $N_i(t) = \sum_{l=1}^{t} \eta_l^{(i)}$. We have

$$\mathbf{E}(\eta_t^{(i)}|\mathcal{F}_{t-1}) = \mathbf{P}\{u(t) = u_i|x^t, u^{t-1}\} = \mathbf{P}\{u(t) = u_i|s(t)\} = p_i(t, \omega).$$

In this sequence of equalities the second one holds since the past history (x^t, u^{t-1}) defines the state $s(t)$ uniquely. Thus, the conditional mathematical expectations of $\eta_t^{(i)}$ are equal to the choice probabilities of u_i at time t. By the lemma on series of r.v.[h] the divergence of the series $\sum_{t=1}^{\infty} p_t(\omega)$ implies that of $\sum_{t=1}^{\infty} \eta_t^{(i)}$ and this means that $\mathbf{P}\{\lim_{t \to \infty} N_i(t) = \infty\} = 1$, i.e. $\mathbf{P}\{\Lambda\} = 1$. $\qquad \square$

The following construction is similar to the previous ones.

Let $\mathcal{FP}$ denote an automaton whose states are represented by the triplets $(\bar{N}, \bar{V}, \bar{p})$.

---

[h]**Lemma.** *Let $\xi_t(\omega)$ $(0 \leq \xi_t \leq B < \infty)$ be a sequence of r.v. measurable with respect to a sequence of σ-algebras $\mathcal{F}_t$. Let $p_t(\omega) = \mathbf{E}(\xi_t|\mathcal{F}_{t-1})$, $t \geq 1$. Then the series $\sum_{t=1}^{\infty} \xi_t(\omega)$, $\sum_{t=1}^{\infty} p_t(\omega)$ converge a.s. simultaneously.*

Let

$$V(t) = t^{-1} \sum_{i=1}^{k} N_i(t) V_i(t)$$

be the empirical average reward over time t and $\bar{\varphi}(t) = (\varphi_1(t), \ldots, \varphi_k(t))$, where

$$\varphi_i(t) = \begin{cases} V_i(t) - V(t), & \text{if } V_i(t) > V(t), \\ 0, & \text{otherwise.} \end{cases}$$

We define a "norm" by the relation $\varphi = \sum_{i=1}^{k} \varphi_i$ and a transformation of the vector $\bar{p}$ by the equality

$$\bar{p}(t+1) = Q\bar{p}(t) = \frac{\bar{p}_t + \bar{\varphi}(t)}{1 + \varphi(t)}.$$

In studying the properties of the automata $\mathcal{FP}$ we shall restrict ourselves to the case of two actions, the action u_1 being optimal ($W_1 > W_2$). Beginning from the moment t_0 let the inequality $V_1(t) - V_2(t) \geq v > 0$ ($t > t_0$) hold. Then

$$p_2(t+1) = \frac{p_2}{1 + \varphi(t)} = \frac{p_2(t)}{1 + t^{-1}N_2(t)[V_1(t) - V_2(t)]} \leq \frac{p_2(t)}{1 + t^{-1}vN_2(t)},$$

and iterating this correlation $T = t - t_0$ times we obtain

$$p_2(t_0 + T) \leq \frac{p_2(t_0)}{\prod_{s=0}^{T}(1 + v(t_0 + s)^{-1}N_2(t_0 + s))}.$$

From the divergence of the product in the denominator as $t \to \infty$ it follows that $p_2(t) \to 0$, i.e. the choice probability of the optimal action u_1 tends to one.

We shall define the following measurable sets in the probability space:

$$\Lambda_\varkappa = \left\{ \omega : \lim_{t \to \infty} N_\varkappa(t, \omega) = \infty \right\}, \quad \varkappa = 1, 2.$$

Then $\lim_{t \to \infty} V_\varkappa(t) = W_\varkappa$ a.s. on $\Lambda_\varkappa$. We shall show that $N_1(t)$ and $N_2(t)$ cannot tend to infinity simultaneously. To prove this we suppose that $\Lambda_1 \cap \Lambda_2 \neq \emptyset$. Then $V_\varkappa(t) \to W_\varkappa$ for almost all points of this intersection and, hence, beginning from some moment the inequality $V_1(t) > V_2(t)$ holds and the probability $p_1(t)$ approaches one. To study the convergence of $\sum_{t=1}^{\infty} p_2(t)$ we use the Raabe test: this series will converge if for all $t > t'$

$$t\left(\frac{p_2(t)}{p_2(t+1)} - 1\right) = N_2(t)[V_1(t) - V_2(t)] > 1.$$

Indeed, this inequality holds since we have $N_2(t) \to \infty$ and $V_1(t) - V_2(t) \to W_1 - W_2 > 0$ a.s. on $\Lambda_1 \cap \Lambda_2$. So, the series $\sum_{t=1}^{\infty} p_2(t)$ converges. It contradicts the assumption that $N_2(t) \not\to \infty$ on the set $\Lambda_1 \cap \Lambda_2$.

Thus, the automaton $\mathcal{FP}$ cannot choose every action infinitely often. This fact is the reason that the automaton $\mathcal{FP}$ is not asymptotically optimal with respect to the class of the scalar HPIV with $U = \{u_1, u_2\}$. Indeed, for any however large T the probability of the event $\{V_1(t) < V_2(t), \, t \leq T\}$ that is "interpreted" by the automaton as the indication of optimality of u_2 (but not u_1) is positive.

2.5. Automata with Formed Structure

The strategies realized by the infinite optimal automata which we have studied above are non-identificational. The full identification consists of estimating the conditional distributions $\mu(\cdot|u)$. But in the previous constructions (except automata with increasing memory) there exists some identification element in estimating the average rewards $W(u)$. It is of interest to construct automaton algorithms of adaptive control without this element of identification but which should be asymptotically optimal. The present section is devoted to this problem.

Let $X = \{-1, +1\}$ be the binary input alphabet and $U = \{u_1, \ldots, u_k\}$ be the output alphabet, $\Sigma = \left\{ \bar{p} = (p_1, \ldots, p_k) : \sum_{j=1}^{k} p_j = 1, p_j \geq 0, j = 1, \ldots, k \right\}$ be the $k-1$-dimensional simplex and $T^{(x)}$ be some transformation of the simplex into itself that depends on the input signal x. From these elements we form the non-initial Moor automaton $\mathcal{G} = (X, \Sigma_k, U; T^{(\cdot)})$ with the continual state space Σ_k and transition function $T^{(x)}$, i.e. its state at the moment $t+1$ is defined by the equality $s(t+1) = T^{(x_t)}s(t)$ and its output signals at the moment t are defined by the family of probabilities $p_1(t), \ldots, p_k(t)$ where $p_i(t) = \mathbf{P}\{u(t) = u_i\}$ is the component of the vector $\bar{p}(t)$ (the state of automaton at the moment t). The similarity of such automata with the stochastic learning models is obvious. However, the constructions being considered here, which are called the *automata with formed structure*, have the following feature concerned with the sequence of states $\bar{p}(0), \bar{p}(1), \ldots, \bar{p}(t), \ldots$, i.e. it is the random process (on simplex Σ), and for the automaton $\mathcal{G}$ to be asymptotically optimal this process should be absorbed in the vertex e_{i_0} of the simplex Σ_k corresponding to the optimal action u_{i_0} for any binary HPIV from some class $\mathcal{K}$.

Hence the optimizational abilities of the automaton $\mathcal{G}$ depend on the form of the transformation $T^{(x)}$ (transition function).

Theorem 1. *There exists an automaton with formed structure $\mathcal{G}$ which is asymptotically optimal with respect to the class $\mathcal{K}_{1,k}$ of binary HPIV with k controls among which only one is optimal.*

Proof. This theorem will be obvious if we find at least one transformation $T^{(x)}$ which generates a random process on Σ absorbed in the vertex of the simplex corresponding to a single optimal action for any HPIV from $\mathcal{K}_{1,k}$. We define such a transformation as follows

$$\bar{p}(t+1) = \bar{p}(t) + \frac{a(t)(1+x_t)}{2}[e(u(t)) - \bar{p}(t)], \quad \bar{p}(1) > 0 \tag{1}$$

where $e_i = (\overbrace{0, \ldots, 0}^{i-1}, 1, 0, \ldots, 0)$, $e(u) = \sum_{i=1}^{k} e_i \chi(u(t) = u_i)$ is the vertex of the simplex corresponding to the action $u(t)$ at the moment t, $x(t)$ is a value of the input signal at the moment t (1 or -1) but $a(t)$ is the sequence of the "steps" of the recurrent procedure (1). Hence the transformation (1) changes a state of the automaton only in the case of arriving at the input "encouragement" ($+1$) but in case of "penalty" (-1) this state remains the same. It is easy to note that the sequence of states $\bar{p}(t)$ is a Markov process on Σ. To judge the convergence of this process it remains to impose the following conditions on the numerical sequence $a(t)$:

$$\left.\begin{array}{c} 0 < a(t) < 1, \quad \displaystyle\sum_{t=1}^{\infty} a(t) = \infty, \\[4mm] \varlimsup_{t\to\infty} \dfrac{a(t)}{\prod_{l=1}^{t}(1 - a(l))} = c < \dfrac{(q_{i_0} - q_{i_1})p_{i_0}}{q_{i_0}} \end{array}\right\} \tag{2}$$

where q_{i_0} and q_{i_1} are the gains probabilities for the optimal (u_{i_0}) action and the subsequent $\left(u_{i_1} = \arg\max_{i\neq i_0} W(u_i)\right)$ one respectively. For the class $\mathcal{K}_{1,k}$ we have $q_{i_0} > q_{i_1}$. Hence the adaptability of the automaton $\mathcal{G}$ is proved. $\qquad\square$

Theorem 2. *Let the procedure* (1) *realize the control of a HPIV from* $\mathcal{K}_{1,k}$ *with steps* $a(t)$ *obeying conditions* (2). *Then the sequence of vectors* $\bar{p}(t)$ *converges both with probability one and in the mean square sense to the vertex* e_{i_0} *of the simplex* Σ *corresponding to the optimal action* u_{i_0}.

Proof. From the initial condition $p_i(1) > 0$, $\forall i$, and the form of the procedure it follows that $p_{i_0} > 0$ a.s. for all t.

We define the r.v.

$$L_t = \frac{1 - p_{i_0}(t)}{p_{i_0}(t)}$$

and the sequence of σ-algebras $\mathcal{F}_t = \sigma(u_1, x_1; \cdots; u_{t-1}, x_{t-1})$. With probability one we have

$$\mathbf{E}(L_{t+1}|\mathcal{F}_t) = \sum_{i=1}^{k} \mathbf{E}(L_{t+1}|\mathcal{F}_t; u(t) = u_i)p_i(t)$$

$$= \sum_{i=1}^{k} \left[\frac{1 - p_{i_0}(t)}{p_{i_0}}p_i + \frac{1 - p_{i_0}(t) - a(t)(\delta_{i_0,i} - p_{i_0}(t))}{p_{i_0}(t) + a(t)(\delta_{i_0,i} - p_{i_0}(t))}q_i\right]p_i(t)$$

$$= L_t\left[p_{i_0}p_{i_0}(t) + \frac{(1 - a(t))q_{i_0}}{a(t) + p_{i_0}(t)(1 - a(t))}p_{i_0}^2(t) + \sum_{i\neq i_0} p_i p_i(t)\right.$$

$$\left. + \frac{1 - p_{i_0}(t)(1 - a(t))}{(1 - p_{i_0}(t))(1 - a(t))}\left(1 - p_{i_0}(t) - \sum_{i\neq i_0} p_i p_i(t)\right)\right]$$

$$= L_t\left[1 + \frac{a(t)}{1 - a(t)}p_{i_0} - \frac{a(t)\sum_{i\neq i_0} p_i p_i(t)}{(1 - p_{i_0}(t))(1 - a(t))}\right.$$

$$\left. + \frac{a^2(t)q_{i_0}}{(1 - a(t))[a(t) + p_{i_0}(t)(1 - a(t))]}\right]$$

$$\leq L_t\left[1 - a(t)\frac{q_{i_0} - q_{i_1}}{1 - a(t)} + \frac{a^2(t)q_{i_0}}{(1 - a(t))(a(t))} + p_{i_0}(t)(1 - a(t))\right]$$

where the inequality appears because

$$\sum_{i \neq i_0} p_i p_i(t) \geq q_{i_1}(1 - p_{i_0}(t)).$$

We also have

$$p_{i_0}(t) \geq p_{i_0}(1) \prod_{h=1}^{t-1}(1 - a(h)).$$

And so,

$$\frac{a^2(t)q_{i_0}}{(1 - a(t))\left[a(t) + p_{i_0}(t)(1 - a(t))\right]} \leq \frac{a^2(t)q_{i_0}}{(1 - a(t))\left(a(t) + p_{i_0}(t)\prod_{l=1}^{t-1}(1 - a(l))\right)}$$

$$\leq \frac{a(t)q_{i_0}}{1 - a(t)}\left[1 + \frac{p_{i_0}(t)}{a(t)}\prod_{l=1}^{t}(1 - a(l))\right]^{-1}$$

$$\leq \frac{a(t)q_{i_0}}{1 - a(t)}\left[1 + p_{i_0}(c^{-1} + o(1))\right]^{-1}.$$

Here the last inequality holds by (1). Hence we finally have the inequality

$$\mathbf{E}(L_{t+1}|\mathcal{F}_t) \leq L_t\left[1 - \frac{a(t)}{1 - a(t)}\left(q_{i_0} - q_{i_1} - \frac{q_{i_0}}{1 + p_{i_0}(1)c^{-1}} + o(1)\right)\right]. \tag{3}$$

The last condition in (2) and the martingale lemma[i] imply that $L_t \xrightarrow[t \to \infty]{\text{a.s.}} 0$. Therefore $p_{i_0}(t) \xrightarrow[t \to \infty]{\text{a.s.}} 1$. From the dominated convergence theorem it follows that the
convergence in the mean square sense takes place as well. $\qquad\square$

Note that the conditions (2) require knowing the probabilities q_{i_0} and q_{i_1}. This
prevents from reaching the aim on the whole class $\mathcal{K}_{1,k}$. Hence the conditions (2)
take place for the sequences $a(t)$ of the form

$$a(t) = \frac{a}{t + b},$$

where $a \in (0, 1)$, $b > a - 1$.

---

[i]**Lemma.** (Martingale lemma) *Let a sequences of r.v. η_n, θ_n (≥ 0) be measurable with respect to
the σ-algebras $\mathcal{F}_n$, $\mathbf{E}\eta_1 < \infty$, $\sum_{n=1}^{\infty} \mathbf{E}\theta_n < \infty$ and*

$$\mathbf{E}(\eta_{n+1}|\mathcal{F}_n) \leq (1 - a(t) + b(t))\eta_t + \theta_n, \quad n \geq 1, \quad a.s.$$

where $a(t)$, $b(t)$ are non-negative and such that $\sum_{t=1}^{\infty} a(t) < \infty$, $\sum_{t=1}^{\infty} b(t) < \infty$. Then

$$\mathbf{P}\left\{\lim_{n \to \infty} \eta_n = 0\right\} = 1.$$

This immediately follows from the easily verified inequality

$$\left(\frac{1+b}{n+b-a+1}\right)^a \geq \prod_{l=1}^{n}(1-a(l)) \geq \left(\frac{b-a}{n+b}\right)^a. \tag{4}$$

We can now prove Theorem 1 quickly. The choice of steps in the form $a(t) = a/(t+b)$ ensures that the procedure (1) will be asymptotically optimal with respect to the class $\mathcal{K}_{1,k}$. Therefore the automaton $\mathcal{G}$ with formed structure is adaptive.

Information on the convergence rate of the sequence $\bar{p}(t)$ to the vertex e_{i_0} of the simplex Σ_k is given below.

Theorem 3. *Under the conditions of Theorem 2 and for $a(t) = a/(t+b)$, $b > a > 0$, $a < 1$ the following estimate*

$$\mathbf{E}p_{i_0}(t) \geq 1 - A\frac{1-p_{i_0}(1)}{p_{i_0}}\left(\frac{t+b-a+1}{b-a+1}\right)^{-a(q_{i_0}-q_{i_1})}, \quad t \geq 0$$

holds. Here

$$A = \exp\left\{\frac{a^2 p_{i_1}}{p_{i_0}(1)(1-a)(b-a)}\right\}.$$

Proof. We take the mathematical expectation on both sides of (3) and use the last inequality in (4). Iterating the inequality (4) we obtain

$$\mathbf{E}L_t \leq \mathbf{E}L_1 \prod_{l=1}^{t}\left[1 - \frac{a}{l+b-a}\left(q_{i_0} - q_{i_1} - \frac{1-q_{i_0}}{1+p_{i_0}(b-a)^a(l+b)^{1-a}}\right)\right].$$

Using inequality $\ln(1-x) \leq -x$ $(0 < x < 1)$ we see that

$$\mathbf{E}L_t \leq \mathbf{E}L_1 \exp\left\{\sum_{l=1}^{t}\frac{a}{l+b-a}\left(q_{i_0} - q_{i_1} - \frac{c_1}{1+p_{i_0}c_2(l+b)^{1-a}}\right)\right\}$$

$$\leq \mathbf{E}L_1 \exp\left\{\frac{a^2 p_{i_1}}{(1-a)(b-a)p_{i_0}}\right\}\left(\frac{b-a+1}{t+b-a+1}\right)^{a(q_{i_1}-q_{i_0})}.$$

The required estimate for $\mathbf{E}p_{i_0}(t)$ follows from the obvious inequality $L_t \geq 1 - p_{i_0}(t)$. $\square$

Thus, the probabilities $p(t)$ converge at a snail pace. It is possible to find the upper estimate for the convergence rate (in the mean square sense) of the current reward to the limiting value. For the sake of completeness we write down this estimate

$$\overline{\lim_{t\to\infty}} \, t^\gamma \mathbf{E}\left(t^{-1}\sum_{i=1}^{t} x_i - W_{i_0}\right)^2 \leq C(1-\gamma/2)^{-2}$$

where

$$\gamma = a(q_{i_0} - q_{i_1}), \quad C = A(q_{\max} - q_{\min})^2 \frac{q_{i_0}(1)(b-a+1)^\gamma}{p_{i_0}(1)}.$$

2.6. Asymptotic Optimality of Automata with Variable Structure

All optimal automata considered above have either an infinite set of states or a set of states increasing unboundedly. What are the possibilities the finite automata with variable structure (FAVS)? The heuristic consideration in favour of a positive answer is the following. The associated Markov chain corresponding to the interaction between a FAVS $\mathcal{A}$ and a HPIV ξ denoted by $\mathcal{A} \otimes \xi$ is non-homogeneous. Therefore the current probabilities for some state $p_i(t)$ may tend to zero as $t \to \infty$. As seems *a priori* it is possible to find some transition functions and to alternate them in time so that for any HPIV the limiting probabilities of the optimal actions equal to one. We are going to demonstrate the existence of such constructions of FAVS restricting ourselves, for the sake of simplicity, to the case of an automaton with two actions.

Let $\mathcal{K}_4$ denote the Moor automaton with binary input and output alphabet and $S = \{s'_1, s'_2; s''_1, s''_2\}$ be the states of the given automaton.[j] The transitions between the states are shown in Fig. 4.

Let the automaton $\mathcal{K}_4$ interact with the HPIV ξ. If the action u_i is used then the win $x^+ = 1$ and the loss $x^- = -1$ will appear with the probabilities q_i, p_i respectively. For two possible transition laws I and II we define the following transition probability matrices

$$
P_I = \begin{pmatrix} p_1 & q_1 & 0 & 0 \\ 0 & 1 & 0 & 0 \\ 0 & 0 & p_2 & q_2 \\ 0 & 0 & 0 & 1 \end{pmatrix}, \quad
P_{II} = \begin{array}{c} \\ s'_1 \\ s'_2 \\ s''_1 \\ s''_2 \end{array}
\begin{array}{c} \begin{array}{cccc} s'_1 & s'_2 & s''_1 & s''_2 \end{array} \\
\begin{pmatrix} 0 & 0 & 1 & 0 \\ 1 & 0 & 0 & 0 \\ 1 & 0 & 0 & 0 \\ 0 & 0 & 1 & 0 \end{pmatrix} \end{array}.
$$

Let us define an alternation law of these matrices. We agree to denote the l-multiple iterations of a matrix P by the symbol $[P]^l$. Then the alternation technique of the

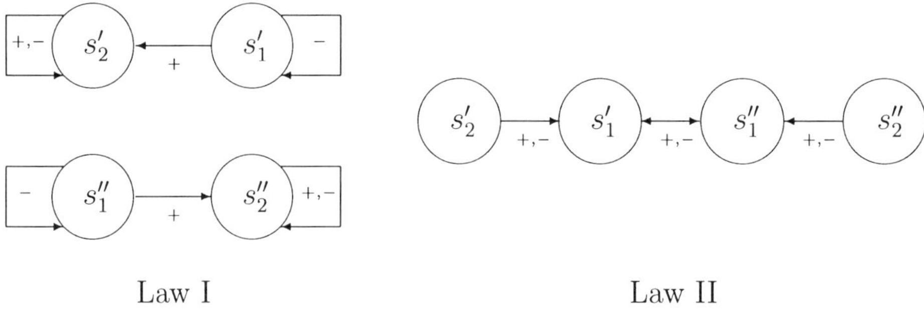

Fig. 4.

[j]In the states s'_1, s'_2 the automaton performs the action u_1 but in the states s''_1, s''_2 the action u_2.

transition laws of the automaton $\mathcal{K}_4$ can be written symbolically as follows

$$\prod_{m=1}^{\infty} [[P_I]^{n(m)} P_{II}]$$

where m denotes the number of stage of control and $n(m)$ is the duration of this stage. The structure of automaton $\mathcal{K}_4$ is described completely.

It remains to formulate the optimizational aim of the control. It is convenient to represent it in the form

$$\lim_{t \to \infty} \mathbf{E} x_t = \max W(u_i). \tag{1}$$

Theorem 1. *For the automaton $\mathcal{K}_4$ to be optimal it is necessary and sufficient that the following conditions*:

1. $\lim_{m \to \infty} n(m) = \infty$,

2. $\sum_{m=1}^{\infty} p^{n(m)} = \infty, \forall p \in (0, 1)$

be satisfied.

Proof. We denote the probabilities to get in the state s_i' (s_i'') at the moment t and the probability of the action u_l at the moment t by $p_i'(t)$ $(p_i''(t))$ and $b_l(t)$ respectively. Then

$$b_1(t) = s_1'(t) + s_2'(t), \quad b_2(t) = s_1''(t) + s_2''(t).$$

We define the numerical sequence t_m in a recurrent way

$$t_0 = 0, \quad t_m = t_{m-1} + n(m) + 1, \quad m \geq 1.$$

Change of branches (or actions) occurs only at the moments t_m. Therefore the function $b_l(t)$ is piece-wise constant, i.e. $b_l(t) = b_l(t_m)$ for $t_m \leq t < t_{m+1}$. The optimizational aim of the control (1) is equivalent (for $q_1 > q_2$) to

$$\lim_{m \to \infty} b_2(t_m) = 0. \tag{2}$$

Change of branch on the mth stage occurs only after $n(m)$ penalties in succession. So,

$$b_2(t_m) = \left(1 - p_2^{n(m)}\right) b_2\left(t_{m-1}\right) + p_1^{n(m)} b_1\left(t_{m-1}\right) \tag{3}$$

and, in view of $b_1(t_m) + b_2(t_m) = 1$, we obtain

$$b_2(t_m) = p_1^{n(m)} + \left(1 - p_1^{n(m)} - p_2^{n(m)}\right) b_2(t_{m-1}). \tag{4}$$

Now, let us prove the necessity. The necessity of the first condition is obvious. Indeed, if there were a subsequence $m_j \leq m_0 < \infty$, $\forall j$, then

$$b_2(t_{m_j}) = \left(1 - p_2^{n(m_j)}\right) b_2(t_{m_j - 1}) + p_1^{n(m_j)} b_1(t_{m_j - 1})$$

$$\geq q_2 b_2(t_{m_j - 1}) + p_1^{n(m_0)} b_1(t_{m_j - 1}) \geq \min\left\{q_2, p_1^{n(m_0)}\right\} > 0$$

and equality (2) would be false. We consider now condition 2. Suppose this fails for some $p > 0$. Then for any p_1, $p_2 < p$ we have $\sum_{m=1}^{\infty} \left(p_1^{n(m)} + p_2^{n(m)}\right) < \infty$. According to the necessity of the first condition proved above there exists an $N < \infty$ such that $1 - p_1^{n(m)} - p_2^{n(m)} > 0$ for all $m > N$. Hence

$$\prod_{j=N+1}^{\infty} \left(1 - p_1^{n(j)} - p_2^{n(j)}\right) > 0.$$

From (4) it follows that

$$b_2(t_m) \geq \left(1 - p_1^{n(m)} - p_2^{n(m)}\right) b_2(t_{m-1}).$$

According to (3) all $b_2(t_m) > 0$ except, may be, $b_2(t_0)$. Hence, for $m > N$ we have

$$b_2(t_m) \geq b_2(t_N) \prod_{j=N}^{m} \left(1 - p_1^{n(j)} - p_2^{n(j)}\right).$$

Putting $m \to \infty$ we find that condition (2) holds.

The necessity is proved. We shall now prove sufficiency. By condition 1 we conclude that for any fixed $\varepsilon > 0$ there exists $L = L_\varepsilon < \infty$ such that

$$p_1^{n(m)} + p_2^{n(m)} < 1, \quad \frac{p_1^{n(m)}}{p_1^{n(m)} + p_2^{n(m)}} < \varepsilon, \quad p_1^{n(m)} < \varepsilon \tag{5}$$

for all $m \geq L_\varepsilon$ (here the last inequality follows from the preceding ones). We want to prove that for any fixed $\varepsilon > 0$

$$\overline{\lim_{m \to \infty}} \, b_2(t_m) \leq 2\varepsilon. \tag{6}$$

If this is the case then this inequality and the arbitrariness of ε will imply the required equality (2). Hence, the theorem will be proved.

The probability $b_2(t_{m-1})$ where $m \geq L_\varepsilon$ satisfies one of two inequalities

$$(\alpha) \; b_2(t_{m-1}) > \frac{p_1^{n(m)}}{p_1^{n(m)} + p_2^{n(m)}}, \quad (\beta) \; b_2(t_{m-1}) \leq \frac{p_1^{n(m)}}{p_1^{n(m)} + p_2^{n(m)}}.$$

From the first of them it follows that

$$b_2(t_m) = p_1^{n(m)} + \left(1 - p_1^{n(m)} - p_2^{n(m)}\right) b_2(t_{m-1})$$

$$= b_2(t_{m-1}) + p_1^{n(m)} - \left(p_1^{n(m)} + p_2^{n(m)}\right) b_2(t_{m-1})$$

$$< b_2(t_{m-1}) + p_1^{n(m)} - \frac{p_1^{n(m)} \left(p_1^{n(m)} + p_2^{n(m)}\right)}{p_1^{n(m)} + p_2^{n(m)}} = b_2(t_{m-1}),$$

i.e. $b_2(t_m)$ is a decreasing function of m. The inequality (β) means that $b_2(t_{m-1}) < \varepsilon$. Then, for the next value we have

$$b_2(t_m) = (1 - p_2^{n(m)})b_2(t_{m-1}) + p_1^{n(m)}b_1(t_{m-1}) \leq 2\varepsilon.$$

The function $b_2(t_m)$ can have two types of behavior: in the first case $b_2(t_m)$ decreases as m increases until entering an ε-neighborhood of zero that, according to the last inequality, from which it does not depart further than 2ε; in the second one this function is bounded from below $\inf_m b_2(t_m) = a > 0$, i.e. $b_2(t_m) \geq a + \varepsilon$ for all finite m.

Now we prove that the second type of behavior is impossible. From this inequality (6) will follow. Let

$$b_2(t_m) \geq a + \varepsilon \geq a + \frac{p_1^{n(m)}}{p_1^{n(m)} + p_2^{n(m)}} \tag{7}$$

for all $m \geq L$. By (4) we have

$$b_2(t_m) \leq p_1^{n(m)} + b_2(t_{m-1}) - \left(a + \frac{p_1^{n(m)}}{p_1^{n(m)} + p_2^{n(m)}}\right)\left(p_1^{n(m)} + p_2^{n(m)}\right)$$

$$\leq b_2(t_{m-1}) - a\left(p_1^{n(m)} + p_2^{n(m)}\right).$$

Summing these inequalities over $m \in [L, N]$ we see that

$$b_2(t_N) \leq b_2(t_L) - a \sum_{m=L}^{N} \left(p_1^{n(m)} + p_2^{n(m)}\right).$$

Due to condition 2 inequality (7) fails for all $m \geq L$. This proves that inequality (6) holds. Hence the theorem is proved. $\qquad\square$

The function $n(m) = [\ln \ln m]$, $m \geq 2$, is an example of a function which ensures the divergence of the series $\sum_{m=1}^{\infty} p^{(m)}$, $p < 1$.

CHAPTER 3

STOCHASTIC APPROXIMATION

We discuss applications of recurrent procedures of stochastic approximation to adaptive control problems of HPIV. The aims considered consist in maximizing the average reward and keeping it at some given level. The necessary and sufficient conditions for convergence are stated. From the conditions obtained follows the extension of such procedures to more general (than HPIV) classes of random processes. Some asymptotic properties are discussed as well.

The extension of the ideas of stochastic approximation to searching a conditional extremum for vector HPIV is also considered.

3.1. Formulation of the Problem

The recurrent procedures serve as the most convenient means of numerical calculations. For the main problems such as solving a system of equations or searching an extremum of a function they are preferable due to simplicity of their realization and the absence of necessity to store a large data base. The advantages of recurrent procedures have appeared at the same time as calculation problems having a statistical nature.

Let $W(u)$ be a scalar function of the argument $u \in \mathrm{R}^1$ which can be calculated for any u with some random error ξ. We need to find the solution of the equation $W(u) = w$ provided it exists. If the function W is non-decreasing we shall use the following recurrent (or iterative) correlation

$$u_{t+1} = u_t - \gamma(t)[W(u_t) - \zeta_t - w], \quad u_0, \ t \geq 0 \qquad (1)$$

where u_0 is an initial value, $\gamma(t)$ is a "step", index t ("time" means the number of the current calculation. This correlation represents the *Robinson–Monro Procedure* (RMP for short). Its convergence can be substantiated on a heuristic level if the errors (called noises or hindrances) without constant drifts and with finite variance (i.e. $\mathbf{E}\zeta_t = 0$, $\mathbf{E}\zeta_t^2 < \infty$) are independent and if the steps $\gamma(t)$ satisfy

$$\gamma(t) > 0, \quad \sum_{t=1}^{\infty} \gamma(t) = \infty, \quad \sum_{t=1}^{\infty} \gamma^2(t) < \infty.$$

We should formulate the notion of convergence. Having in mind that it has a probabilistic character we shall further write about convergence $\lim_{t \to \infty} u_t = u_*$ (u_* is a root of the equation to be solved) both with probability one and in the mean square sense.

77

If the function $W(u)$ is non-increasing then we should change the sign in (1) before $\gamma(t)$.

In multi-dimensional case, when $W(u) = (W^{(1)}(u), \ldots, W^{(n)}(u))$, $u = (u^1, \ldots, u^n)$ and a system of either n algebraic equations or transcendental ones with n unknown variables is under consideration, RMP keeps its form. The fact that we can find a basis for an RMP can be applied to find the extremum of some unimodal differentiable function $W(u)$. With this aim in view we should find the root of its derivative $W'(u)$. However it is impossible to use it directly. We shall use its approximation $\Delta W(u) = (2\delta)^{-1}[W(u + \delta) - W(u - \delta)]$. To that end, let us introduce a sequence $\delta(t)$ vanishing to zero as $t \to \infty$. By analogy with (1) (at $w = 0$) we obtain the correlation

$$u_{t+1} = u_t - \gamma(t) \left[\frac{W(u_t + \delta(t)) - W(u_t - \delta(t))}{2\delta(t)} \right], \tag{2}$$

which is called the *Kiefer–Wolfowitz Procedure* (KWP for short).

In the multi-dimensional case the equality (2) contains the estimate of the gradient of the function $W(u)$, i.e. the vector with the components

$$\frac{W(u_t^i + \delta(t)^i) - W(u_t^i - \delta(t)^i)}{2\delta(t)^i}, \qquad i = 1, 2, \ldots, n,$$

where $u_t^i + \delta^i(t) = (u_t^1, u_t^2, \ldots, u_t^i + \delta(t)^i, u_t^{i+1}, \ldots, u_t^n), i = 1, 2, \ldots, n$. If $W'(t)$ or grad $W(u)$ is continuous, we can rewrite (2) in the form

$$u_{t+1} = u_t - \gamma(t)[\operatorname{grad} W(u_t) + \varkappa_t + \zeta_t] \tag{3}$$

where $\varkappa_t = \varkappa(u_t, \delta(t))$ means a decreasing systematic error.

The procedures (1) and (2) can be varied depending on the concrete conditions of the problem under study. Quite often, a "truncation" is used, i.e. after defining (in the scalar case) the admissible interval $[u', u'']$ of varying the argument u one can put $u_{t+1} = u'$ or $u_{t+1} = u''$ if the right-hand side of the equality (1) or (2) is less than u' or grater than u'' respectively. In the multi-dimensional case the projection on the closed convex set of admissible values of the arguments is used.

It is common practice to unite the procedures (1) and (2) and their modifications under the name of *stochastic approximation*. This theory consists of studying the convergence conditions, the convergence rate and some questions of asymptotic behavior.

It is reasonable to use the ideas and methods of stochastic approximation for adaptive control problems. One result of this kind has been described in the previous chapter. We now make a more systematic investigation. The processes with continuous spaces of states and controls (first of all the HPIV) are the object of our attention.

So, let an HPIV with the space of states and controls $\mathbf{R}^n$ ($n \geq 1$) and finite average rewards

$$W(u) = \mathbf{E}x(u) = \int_{-\infty}^{\infty} z\mu(dz \mid u),$$

be given. These rewards can be vector, i.e. they have $n \geq 2$ components

$$\bar{W}(u) = (W^{(1)}(u), \ldots, W^{(n)}(u)).$$

The usual conditions imposed on the measure $\mu(\cdot|u)$ imply the measurability of the average reward with respect to u. Its continuity is guaranteed by an additional assumption, namely, $\mu(\cdot|u)$ is continuous in u and the integral defining $W(u)$ converges uniformly.

The following aims of control are of paramount importance.

1. To solve the equation (or the system of equations)

$$W(u) = w$$

where w belongs to the domain of $W(u)$. The root of this equation will be denoted by u_*.
2. To maximize the average reward, i.e. to find u_{opt} such that

$$W(u_{\mathrm{opt}}) = \max_u W(u).$$

In the classical statement of the control aim (i.e. when the function $W(u)$ is known) it is necessary to solve the following problem: to find u_* or u_{opt} in advance, i.e. before beginning the control process. Then the required strategy turns out to be stationary. Such strategies transform the HPIV into a sequence of independent random variables. According to the strong law of large numbers we have

$$\lim_{t \to \infty} t^{-1} \sum_{l=1}^{t} W(u_l) = W(u_*) = w, \quad \text{a.s.}$$

$$\lim_{t \to \infty} t^{-1} \sum_{l=1}^{t} W(u_l) = W(u_{\mathrm{opt}}) = \max_u W(u), \quad \text{a.s.}$$

The solution is unlikely to remain simple in the adaptive version. Proceeding from the stochastic approximation stated above we use the recurrent procedures inspired by it which deal with the controls u_t and HPIV values $x_t(u_{t-1})$. To solve the equation $W(u) = w$ we shall use RMP

$$u_{t+1} = u_t - \gamma(t)[x_{t+1}(u_t) - w], \quad t \geq 1 \tag{4}$$

if the function $W(u)$ is non-decreasing, and

$$u_{t+1} = u_t + \gamma(t)[x_{t+1}(u_t) - w], \quad t \geq 1 \tag{5}$$

if this function is non-increasing, some u_0 being taken as initial. The sequence $\gamma(t)$ is positive and converges to zero as $t \to \infty$. Moreover, it satisfies the equality

$$\sum_{t=1}^{\infty} \gamma(t) = \infty$$

which serves as the condition of achieving the true root u_* at any initial value u_0.

The sequence of controls $u_1, \ldots, u_t, \ldots$, generated by the interaction between algorithms (4) or (5) and the HPIV is represented as a random walk on the set of controls. It must ensure the convergence to the required limit u_* in one of the probabilistic senses. We note that the sequence $\{u_t\}$ is Markovian since $x_t(u(t-1))$ represents the Markov process but u_t is its linear transformation.

The $u_t \xrightarrow[t\to\infty]{\text{a.s.}} u_*$ convergence means that for any $\varepsilon > 0$ there exists a non-Markovian moment τ_ε such that $\mathbf{P}\{\tau_\varepsilon < \infty\} = 1$ and

$$|u_t - u_*| < \varepsilon$$

for all $t > \tau_\varepsilon$. If $W(u)$ is continuous, this convergence implies $\lim_{t\to\infty} W(u_t) = w$ a.s. Moreover, the equality

$$\lim_{t\to\infty} t^{-1} \sum_{n=1}^{t} x_n = w$$

holds with probability one. We now turn to the optimization problem of the average reward for the HPIV. We consider first the scalar HPIV with unimodal (maximum is reached at the point u_{opt}) continuously differentiable function $W(u)$. To realize KWP we have to construct some estimate of $W'(u)$ which, obviously, requires the values of the process when the controls $u_t + \delta(t)$ and $u_t - \delta(t)$ are used. Then KWP takes the form

$$u_{2t+2} = u_{2t} + \frac{\gamma(t)}{\delta(t)} \left[x_{2t+1}(u_t + \delta(t)) - x_{t+2}(u_t - \delta(t)) \right] \tag{6}$$

with appropriate initial value u_0. From the various conditions imposed on $\gamma(t)$ and $\delta(t)$ we choose the following:

(a) $\gamma(t) > 0, \delta(t) > 0, t \geq 1; \lim_{t\to\infty} \delta(t) = 0$;
(b) $\sum_{t=1}^{\infty} \gamma(t) = \infty, \sum_{t=1}^{\infty} \left[\gamma(t)\delta(t) + \gamma^2(t)\delta^{-2}(t) \right] < \infty$.

In particular, these imply $\lim_{t\to\infty} \gamma(t) = 0$. The following sequences satisfy the conditions stated above:

$$\gamma(t) = \frac{g}{t}, \quad \delta(t) = \frac{d}{t^\delta}, \quad g, \ d > 0, \quad 0 < \delta < \frac{1}{2}.$$

The convergence with probability one of the KWP (6) implies

$$\lim_{t\to\infty} W(u_t) = W(u_{\text{opt}}) = \lim_{t\to\infty} t^{-1} \sum_{n=1}^{t} x_n, \quad \text{a.s.}$$

which means that the aim of control is attained with respect to the appropriate class of HPIV.

The study of asymptotic properties of the optimizational procedure of stochastic approximation should be based on the following three factors.

(1) the "steps" $\gamma(t)$;
(2) the function $W(u)$, i.e. the average reward (or the regression function);
(3) the noise ζ_t or, which is the same, the controlled random process x_t.

First of all we are interested in the conditions about the noise (or random process x_t) which allow to control a wider class of random processes by using the stochastic approximation methods. At the same time it is necessary to define more exactly the structure of the corresponding classes of the HPIV. The study of the influence of the steps $\gamma(t)$ and function $W(u)$ will be less important in our investigation.

3.2. Convergence Conditions of Stochastic Approximation Procedures

Having in mind the convergence either with probability one or in L_p sense (mainly in the mean square sense) we shall try to find the most general necessary and sufficient conditions of convergence of RMP and KFP. Using no terminology of control theory but only the statistical one (of solving a problem on the regression function $W(u)$ in the presence of noise ζ_t) we shall study RMP ($u_t \in \mathrm{R}^n$, $W : \mathrm{R}^n \to \mathrm{R}^n$)

$$u_{t+1} = u_t - \gamma(t)[W(u_t) + \xi_t]. \tag{1}$$

A simple heuristic consideration suggests that $u_t \xrightarrow[t\to\infty]{} u_*$ (u_* is the root of the equation $W(u) = 0$) with the strong law of large numbers for the noise $\xi_t \in \mathrm{R}^n$

$$\lim_{t\to\infty} \gamma(t) \sum_{n=1}^{t} \xi_n = 0. \tag{2}$$

The explicit statement of this relation must include some restrictions on $W(u)$ and $\gamma(t)$. The function $W(u)$ is always supposed to be non-decreasing and its rate of increasing is at most linear at infinity. The sequence of steps $\gamma(t)$ satisfies the restrictions

$$\gamma(t) \geq 0, \quad \sum_{t=1}^{\infty} \gamma(t) = \infty, \quad \lim_{t\to\infty} \gamma(t) = 0.$$

Under these preconditions which are inadequate to give formal proofs we state the convergence hypothesis:

For the convergence of RMP (1) *either with probability one or in the L_p ($p \geq 2$) sense to take place it is necessary and sufficient that the strong law of the large numbers* (2) *hold in the same sense*, i.e.

$$\{u_t \to u_*\} \Leftrightarrow \left\{ t^{-1} \sum_{n=1}^{t} \xi_n \to 0 \right\}.$$

Using the examples we can show that the correlation (2) understood as the convergence in the probability sense, is neither necessary nor sufficient for convergence of RMP. We will not discuss the corresponding examples here.

Passing to the exact formulation of the hypothesis we note that the necessary convergence conditions differ from the sufficient ones for RMP. We begin with the necessary condition always supposing that the root of the equation $W(u) = 0$, $u_* = 0$ is being searched.

Theorem 1. *If the function $W(u)$ is non-decreasing and continuous in a neighborhood of the point $u = 0$ and $\gamma(t) = t^{-1}$ then from the convergence $\lim_{t\to\infty} u_t = 0$ a.s. (or in the L_p sense, $p \geq 2$) it follows that*

$$\lim_{t\to\infty} t^{-1} \sum_{n=1}^{t} \xi_n = 0$$

in the same sense.

Proof. Multiplying both parts of the equation

$$u_{t+1} = u_t - t^{-1}[W(u_t) + \xi_t]$$

by t and summing over t from 1 to T, we have

$$Tu_{T+1} = \sum_{t=1}^{T} u_t - \sum_{t=1}^{T} W(u_t) - \sum_{t=1}^{T} \xi_t$$

or

$$u_{T+1} = T^{-1} \sum_{t=1}^{T} u_t - T^{-1} \sum_{t=1}^{T} W(u_t) - T^{-1} \sum_{t=1}^{T} \xi_t. \tag{3}$$

By $u_T \to 0$, the properties of the Cesaro averages and the continuity of $W(u)$ at zero we obtain

$$\lim_{T\to\infty} T^{-1} \sum_{t=1}^{T} u_t = 0, \quad \lim_{T\to\infty} T^{-1} \sum_{t=1}^{T} W(u_t) = 0.$$

This and (3) imply the assertion of the theorem. $\qquad\square$

It is surprising that this theorem does not require the fulfilment of the following conditions: $\mathbf{E}\xi_t = 0$ and $\mathbf{E}\xi_t^2 < \infty$ which may seem inevitable.

The weak assumptions on $W(u)$ are related with the assumption that the procedure converges. The converse must include stronger restrictions on $W(u)$. Several types of sufficient conditions are known. We start with a special result on the convergence of RMP for the simplest regression function: the linear function $W(u) = \alpha u$, $\alpha > 0$. Such functions satisfy the assumptions of all theorems on sufficiency. The problem about the root of the equation $W(u) = 0$ is solved by the recurrent procedure

$$u_{t+1} = u_t(1 - \alpha t^{-1}) + t^{-1}\xi_t$$

under the condition $t^{-1} \sum_{t=1}^{t} \xi_n \to 0$. Then $\lim_{t\to\infty} u_t = 0$. This follows from the following lemma.

Lemma 1. *If $\alpha > 0$ and $\lim_{t\to\infty} t^{-1} \sum_{n=1}^{t} \xi_n = \xi_0$ then $\lim_{t\to\infty} u_t = \xi_0/\alpha$.*

From the general sufficient conditions of Theorem 2 another proof follows.

We turn now to the converse with respect to Theorem 1. The sufficient conditions for convergence have non-coinciding forms which differ one from another by the requirements imposed upon the regression function and steps $\gamma(t)$. For the sake of simplifying the statements we only consider the case $\gamma(t) = t^{-1}$ for the RMP (1) assuming that $W : \mathrm{R}^n \to \mathrm{R}^n$, $W(u_*) = 0$ and $W(u)$ is continuous.

Theorem 2. *The equality*

$$\lim_{t \to \infty} t^{-1} \sum_{n=1}^{t} \xi_n = 0, \quad \text{a.s. (or in the } L'_p \text{ sense, } p \geq 2)$$

will imply $\lim_{t \to \infty} u_t = u_*$ *in the same sense if at least one of the following conditions holds:*

(α) *the function W satisfies the conditions*

$$\|W(u') - W(u'')\| \leq L\|u' - u''\|, \quad W(0) = 0, \quad L > 0,$$
$$(u, W(u)) \geq q\|u\|^2, \qquad\qquad\qquad q > 0;$$

(β) *there exists a twice continuously differentiable function $V(u) : \mathrm{R}^n \to \mathrm{R}^n$ such that for any $u \neq u_*$*

$$V(u) > 0, \qquad\qquad V(0) = 0;$$
$$\lim_{\|u\| \to \infty} V(u) = \infty, \quad (\operatorname{grad} V, W) \geq 0;$$

(γ) *the following inequalities*

$$\|W(u)\| \leq a\|u - u_*\| + b, \qquad\qquad a, b > 0,$$
$$(u - u_*)^T W(u) \geq H(\|u - u_*\|), \quad \lim_{x \to \infty} x^{-1} H(x) > 0$$

hold, where $H(x)$, $x \geq 0$, is a continuous, strictly increasing function such that $H(0) = 0$.

We have no possibility to produce a detailed proof and restrict ourselves to some remarks about the conditions (α), (β) and (γ). Each of them means (explicitly or not) the presence of a Lyapunov function. In the case (γ) only its existence is required without specifying its form but in the case (α) it has the quadratic form $v(u) = \|u\|^2$. The corresponding inequalities in these conditions are equivalent to the requirement of assymptotic stability of the trivial solution of the following differential equation

$$\dot{u}_t = -W(u_t),$$

associated with the RMP as its continuous analog.

In the optimization problems of stochastic approximation the following two cases are possible.

(1) The gradient of the regression function can be calculated directly with the error ξ_t.

(2) It is required to find the increment of the function $W(u)$ (with some errors as well).

In the first case Theorems 1, 2 remain in force with $W'(u)$ instead of $W(u)$. In the other we operate with the formula (3) from Sec. 1 which is based on trial steps. We write that formula in the form

$$u_{t+1} = u_t - t^{-1}[\operatorname{grad} W(u_t) + \varkappa_t(u_t) + \xi_t].$$

Under the natural assumption about $\varkappa_t = \|\varkappa_t(u_t)\|$, namely,

$$\sum_{t=1}^{\infty} t^{-1}\varkappa_t < \infty$$

Theorems 1, 2 are valid. Hence, in all cases the convergence hypothesis is true.

In conclusion we discuss the following question: for what types of random sequences ξ_t does the strong law of large numbers hold? We restrict ourselves again to the convergence either with probability one or in the mean square sense. Let us give some examples.

A. The sequence ξ_t formed by the independent, identically distributed random variables with $\mathbf{E}\xi_t = 0$. Then the Kolmogorov Theorem ensures the fulfilment of the equality

$$\mathbf{P}\left\{\lim_{T\to\infty} T^{-1}\sum_{t=1}^{T}\xi_t = 0\right\} = 1. \tag{4}$$

Hence the structure of the class of HPIV the RMP allows us to solve the problem at the level $w = W(u_*)$, under the appropriate conditions on W.

B. The sequence ξ_t is stationary in the wide sense with $\mathbf{E}\xi_t = 0$ and the covariance function $R(t, s) = \mathbf{E}\xi_t\xi_s$ is such that

$$|R(t, s)| \le k\frac{t^{\alpha} + s^{\beta}}{1 + |t - s|^{\beta}}, \quad k > 0,\ 0 \le 2\alpha < \beta < 1.$$

According to Cramer–Leadbetter Theorem it satisfies (4). If the process has bounded spectral density[a] $f(\lambda)$ then the convergence in the mean sense (with respect to L_2) $t^{-1}\sum_{n=1}^{t}\xi_n \to 0$ will take place. Taking into account the estimation $|f(\lambda)| \le f_0 < \infty$ this fact can be verified by a simple calculation.

C. The sequence ξ_t is obtained by summation with the finite memory M

$$\xi_t = \sum_{n=1}^{M} A_n\eta_{t-n}$$

---

[a]It means that the covariance function R can be represented in the form

$$R(n) = \frac{1}{2\pi}\int_{-\pi}^{\pi} f(\lambda)e^{-i\lambda n}d\lambda,$$

and the spectral density $f(\lambda)$ is supposed to be bounded.

of independent, identically distributed random variables (or random vectors) η_t under the constant coefficients (or matrices) $A_1, \ldots, A_M$. In the more general case $\{\eta_t\}$ is a martingale-difference (i.e. $\mathbf{E}(\eta_t | \mathcal{F}_{t-1}) = 0$, $t \geq 0$, a.s.) with respect to a flow of σ-algebras $\mathcal{F}_t$, with bounded second moments. The correlation (4) is obvious. Indeed,

$$t^{-1} \sum_{m=1}^{t} \xi_m = t^{-1} \sum_{m=1}^{t} \sum_{n=1}^{M} A_n \eta_{m-n} = \sum_{n=1}^{M} A_n t^{-1} \sum_{m=1}^{t} \eta_{m-n} \xrightarrow[t \to \infty]{} 0,$$

because

$$t^{-1} \sum_{m=1}^{t} \eta_{m-n} \xrightarrow[t \to \infty]{} 0$$

for any n by the strong law of large numbers.

D. The sequence ξ_t represented by the outputs of a linear *stable* filter

$$\sum_{i=0}^{L} A_i \xi_{t-i} = \sum_{j=0}^{N} B_j \eta_{t-j}.$$

On the right-hand side we have a martingale-difference with bounded second moments. The elements of ξ_t can be written explicitly as follows

$$\xi_t = \sum_{k=1}^{t} C_{t-k} \eta_k + \theta_t$$

where θ_t is some linear function of the previous values of η_t (i.e. η_k, $k \leq t$). This function increases as $t^\alpha \lambda^t$, $\lambda \in (0,1)$. The coefficients (matrices C_t) have the growth order $t^\alpha \lambda^t$ as well.

In a similar way as in example **C** we see that the limit $\lim_{t \to \infty} t^{-1} \sum_{n=1}^{t} \xi_n$ exists (a.s. and in the mean sense) and equals zero.

E. We start with some definitions concerning a random process ξ_t defined onto a probability space $(\Omega, \mathfrak{F}, \mathbf{P})$. Let $\mathfrak{F}_t = \sigma(\xi_t)$ denote the σ-algebra generated by the r.v. ξ_t. Let $\mathfrak{F}_{\leq t}$ and $\mathfrak{F}_{\geq t}$ denote also the σ-algebras generated by the collections $\{\ldots, \xi_{t-1}, \xi_t\}$ and $\{\xi_t, \xi_{t+1}, \ldots\}$ respectively. Let $A \in \mathfrak{F}_{\leq t'}$, $B \in \mathfrak{F}_{\geq t}$, $t' < t$.

Definition 1. The scalar function

$$\alpha(t) = \sup_{A \in \mathfrak{F}_{\leq 0}, B \in \mathfrak{F}_{\geq t}} |\mathbf{P}\{AB\} - \mathbf{P}\{A\}\mathbf{P}\{B\}|$$

is called the *strong mixing coefficient* and the function

$$\beta(t) = \sup_{A \in \mathfrak{F}_{\leq t'}, B \in \mathfrak{F}_{\geq t'+t}} |\mathbf{P}\{A\} - \mathbf{P}\{A|B\}|$$

is called the *uniform-strong mixing coefficient.*

A stationary sequence ξ_t has the property of the *uniform-strong mixing (strong mixing)* if $\lim_{t \to \infty} \beta(t) = 0$ ($\lim_{t \to \infty} \alpha(t) = 0$). The first of these notions represents the particular case of the second one. For the sake of brevity we shall call such sequences just the "sequences with mixing". In addition, we suppose that $\mathbf{E}\xi_t = 0$ and $\mathbf{E}\xi_t^2 < \infty$ for all t. The properties of such sequences are well known.

We are now going to find the conditions when the equality

$$\lim_{t \to \infty} t^{-1} \sum_{n=1}^{t} \xi_n = 0$$

takes place with probability one and (or) in the mean sense. By the Toeplitz Theorem it is enough to make sure that the series of r.v. $\sum_{n=1}^{\infty} \xi_n/n$ converges in the same sense.

We give one example of solving this problem in the mean sense. We have

$$\mathbf{E}\left(\sum_{n=1}^{t} n^{-1}\xi_n\right)^2 = \sum_{n=1}^{t} n^{-2}\mathbf{E}\xi_n^2 + \sum_{(n,m):n \neq m}^{t} (nm)^{-1}\mathbf{E}\xi_n\xi_m = S_1(t) + S_2(t).$$

The following estimate (where $\sigma_t^2 = \mathbf{E}\xi_t^2$) is well-known in the theory of random sequences with mixing

$$\|\mathbf{E}\xi_t\xi_s\| \leq c_1 \alpha^a(t-s)\sigma_t^2, \quad \|\mathbf{E}\xi_t\xi_s\| \leq c_2 \beta^b(t-s)\sigma_t^2, \quad 0 < a, b \leq 1/2.$$

So, supposing

$$\sigma_t \leq L, \quad \alpha(t) = O(t^{-\delta}), \quad \beta(t) = O(t^{-\delta}), \quad L, \delta > 0$$

as $t \to \infty$, we derive the convergence $S_1(t) \xrightarrow[t \to \infty]{} 0$, and $S_2(t) \xrightarrow[t \to \infty]{} 0$.

We leave the investigation of other versions of this problem to the reader. Once more we stress that the purpose of our study is whether the strong law of large numbers takes place for the considered class of random sequences.

The main conclusion is that the application sphere of stochastic approximation considered as a means of control (i.e. a strategy) goes far beyond the scope of the HPIV.

3.3. Survey of Asymptotic Properties of Stochastic Approximation Methods for HPIV

We consider the adaptive problem of attaining the given average reward for the processes of the HPIV-type.

Let ξ_t be a HPIV with $\mathbf{R}^n$ ($n \geq 1$) as the state and control spaces for which there exists a continuous average reward $W(u)$ for all $u \in \mathbf{R}^n$. We shall use the RMP as a control algorithm always supposing $W(u)$ satisfies at least one of the conditions (α), (β), (γ) in Theorem 2 from the previous section.

Let us list some limiting properties of RMP. We start with the estimates of the convergence rate of this procedure with probability one and in the mean square sense. For simplicity we assume that the function $W(u)$ does not increase in a neighborhood of the root and the RMP has the form

$$u_{t+1} = u_t - at^{-1}[x_{t+1}(u_t) - w], \quad t \geq 1. \tag{1}$$

Theorem 1. *Let a HPIV ξ_t satisfy the following conditions:*

(1) *There exist $n \times n$-matrix $\Gamma = \Gamma^T > 0$ and number $\lambda > 0$ such that*

$$(\Gamma W(u), u - u_*) \leq -\lambda(\Gamma(u - u_*), u - u_*), \quad 2a\lambda > 1;$$

  for all u.
(2) $\mathbf{E}x_t^2(u) \leq g(1 + \|u\|^2)$;
(3) *the initial value u_0 has a finite second moment.*

Then

$$\lim_{t \to \infty} t^r \|u_t - u_*\|^2 = 0 \quad \text{a.s.} \quad \forall r \in (0, 1),$$

$$\mathbf{E}\|u_t - u_*\|^2 = O(t^{-1}).$$

Moreover, we can prove convergence in the distribution sense and show that the random sequence u_t is asymptotically normal. This means as usual that the distributions of the random variable $\sqrt{t}(u_t - u_*)$ approach (converge weakly) to the normal distribution $N(0, d)$ with zero mathematical expectation and variance d as $t \to \infty$.

We consider again the RMP (1) and suppose that the convergence $u_t \xrightarrow[t \to \infty]{\text{a.s.}} u_*$ takes place.

Theorem 2. *Let a HPIV satisfy the conditions:*

(1) *The function $W(u)$ has the form*

$$W(u) = B(u - u_*) + \delta(u)$$

  where B is a constant matrix and $\delta(u) = o(\|u - u_*\|)$ as $u \to u_*$, the matrix $A = aB + 2^{-1}I$ being stable;
(2) *the covariance matrix*

$$C(u) = \mathbf{E}(x_t(u) - W(u))(x_t(u) - W(u))^T$$

  is finite and there exists

$$\lim_{u \to \infty} C(u) = S_0;$$

(3) *for some $h > 0$*

$$\lim_{R \to \infty} \sup_{\|u - u_*\| < 0} \int_{\|x_t(u) - W(u)\| > R} \|x_t(u) - W(u)\|^2 \mathbf{P}\{d\omega\} = 0.$$

Then

$$\sqrt{t}(u - u_*) \sim N(0, S),$$

where

$$S = a^2 \int_0^\infty e^{Az} S_0 e^{Az} dz.$$

Both theorems point to the convergence rates of the RMP in the mean sense, in the distribution sense and in the a.s sense. These rates are characteristic for statistics, i.e. they have the order $t^{-1/2}$.

Until now it has been supposed the equation $W(u) = 0$ has a unique solution. It would be interesting to study in detail the behavior of the control sequences u_t when the considered equation has several solutions. For example, whether we can hope that u_t enters set $B = \{u : W(u) = w\}$.

We assume that the set B consists of a finite number of connected components and put $U_\varepsilon(B) = O_\varepsilon(B) \cup \{u : \|u\| < \varepsilon^{-1}\}$ where $O_\varepsilon(B)$ denotes the ε-neighborhood of the set B.

Theorem 3. *Suppose there exists a function $V(u) \geq 0$ with bounded and continuous second derivatives such that for $\varepsilon > 0$*

$$\lim_{\|u\| \to \infty} V(u) = \infty, \qquad \sup_{u \in U_\varepsilon(B)} (\mathrm{grad}\ V(u), W(u)) < 0,$$

$$\mathbf{E}x_t^2(u) \leq c(1 + V(u)).$$

Then the sequence u_t converges a.s. either to one of the points of the set B or to the boundary of one of its connected components.

Under some restrictions on the HPIV it is possible to narrow the limiting set of sequence u_t in comparison with Theorem 3.

Let us define a class of HPIV for which the KWP ensures maximization of the average reward $W(u)$. This class includes the HPIV such that

(a) the function $W(u)$ has a unique maximum were $W'(u) > 0$ and $W'(u) < 0$ on the left and on the right respectively;

(b) there exists a number $d > 0$ such that

$$(W'(u))^2 + \mathbf{E}x_t^2(u) \leq d(1 + u^2).$$

for all u.

The KWP possesses many interesting and important properties of which we consider only the convergence rate in the mean square sense calculated by using the value $s(t) = (\mathbf{E}[u_t - u_{\mathrm{opt}}]^2)^{1/2}$. It turns out that for large t its asymptotic properties are defined by the smoothness of $W(u)$ in a neighborhood of the point u_{opt}.

If the derivative $W'(u)$ is continuous and some power functions are chosen as $\gamma(t)$, $\delta(t)$ then

$$s(t) = \begin{cases} O(t^{-\frac{1}{2}+\gamma}), & \gamma > 1/4, \\ O(t^{-\gamma}), & \gamma < 1/4. \end{cases}$$

If there exists a continuous second derivative $W''(u)$ in a neighborhood of u_{opt} then

$$s(t) = \begin{cases} O(t^{-\frac{1}{2}+\gamma}), & \gamma \geq 1/6, \\ O(t^{-2\gamma}), & \gamma < 1/6, \end{cases}$$

i.e. the growth order of $s(t)$ is equal or close to $t^{-1/3}$. For analytic and symmetric (in a neighborhood of u_{opt}) function $W(u)$ we have

$$s(t) = O(t^{-1/2+\gamma})$$

for all $\gamma \in (0, 1/2)$.

We now give information on normality of the sequence u_t. The result depends again on the smoothness of $W(u)$. It turns out that either $t^{-1/2+\gamma}(u_t - u_{\mathrm{opt}})$ or $t^{1/2}\delta(t)(u_t - u_{\mathrm{opt}})$ is asymptotically normal with zero mean and variances which can be expressed in terms of the numerical characteristics of the HPIV.

These (and a number of other) facts can be extended to the multi-dimensional case when $u \in \mathbb{R}^n$.

3.4. Calculation of the Conditional Extremum

The simplicity of the recurrent procedures above motivates us to extend the application sphere of stochastic approximation methods. We have in mind other (then mentioned) types of problems and aims of control and also more general classes of controlled processes. One of these extensions was made in connection with automata having the formed structure (Sec. 5, Chap. 2) where the randomized control choice rules $\bar{p} = (p_1, \ldots, p_k)$ have been used. In Sec. 5, Chap. 5 these procedure will be considered for Markov chains. So, we now formulate the general principles of constructing the recurrent procedures of stochastic approximation. Having in mind further applications we turn to problems with a finite set of controls $U = \{u_1, \ldots, u_k\}$ and shall use randomized rules $\bar{p}$ which are probability distribution on U. This means that the rules $\bar{p}$ are identified with the points of the $k - 1$-dimensional simplex $\Sigma_k = \{\bar{p} : \bar{p} = (p_1, \ldots, p_k), \sum_{j=1}^{k} p_j = 1, \ p_i \geq 0, \ i = 1, \ldots, k\}$. We seek transformations of these rules in the form

$$\bar{p}_{t+1} = \bar{p}_t - a(t)\bar{g}(x_{t+1}, u_t, \bar{p}_t), \quad t \geq 0 \tag{1}$$

where $g(\cdot)$ is a vector specifying a "move" of the distributions $\bar{p}_t$.

The following conditions on the steps $a(t)$ have appeared in all procedures considered above: $a(t) \geq 0, \ \forall t; \sum_{t=1}^{\infty} a(t) = \infty$. In the optimization problems the vector $\bar{g}$ is constructed by using the function $W(\bar{p})$ whose extremum must be found. But this function is unknown and we have only the controls being sent and the responses of the HPIV obtained in reply to them. The procedure is based only on the current information about the process. Two operations are natural:

(1) Define the vector g as

$$g = \operatorname{grad} W(p) = (\partial W/\partial p_1, \ldots, \partial W/\partial p_n);$$

(2) represent the derivative $\partial W/\partial p_i$ in the form having probabilistic sense or, more precisely, write it down as

$$\frac{\partial W}{\partial p_i} = \mathbf{E}h_i(x_{t+1}, u_t, \bar{p}_t)$$

where $h_i(\cdot, \cdot, \cdot)$ is a known function.

If both steps are realized, then the original procedure will have the form (1) where the function $g = h = (h_1, \ldots, h_k)$ is called the *stochastic gradient*. The proof that this procedure leads to the required aim can be done by using the Lyapunov function $L(\bar{p})$ specified as follows

(a) $L(p) > 0$, $\forall p \in \Sigma_k$, $\bar{p} \neq \bar{p}_*$, $L(\bar{p}_*) = 0$;
(b) $L(\bar{p})$ is continuously differentiable.

The point $\bar{p}_*$ is defined as $\bar{p}_* = \lim_{t \to \infty} p_t$ which has the desired property. The proof of Theorem 2 in Sec. 5, Chap. 2 represents an example of using the Lyapunov function to prove the existence of this limit. In this case the right-hand side of the recurrent correlation is constructed so that it always belongs to the simplex Σ_k, i.e. $\bar{p} \in \Sigma_k$ for all t. There exist problems where this inclusion may fail. Then the procedure (1) has no sense and we have to modify it.

Let $\Sigma_k(\varepsilon)$, $\varepsilon > 0$ denote the subsimplex of the simplex Σ_k:

$$\Sigma_k(\varepsilon) = \left\{ \bar{p} : \bar{p} = (p_1, \ldots, p_k), \ \sum_{j=1}^{k} p_j = 1, \ p_i \geq \varepsilon, \ i = 1, \ldots, k \right\},$$

and $\pi_\varepsilon(\cdot)$ be a projector on $\Sigma_k(\varepsilon)$, i.e. a mapping which associates with any vector $\bar{p} \in \mathbf{R}^k$ the vector $\tilde{p} \in \Sigma_k(\varepsilon)$ such that the distance between $\tilde{p}$ and $\bar{p}$ is minimal. The procedure

$$\bar{p}_{t+1} = \pi_{\varepsilon_{t+1}}(\bar{p}_t - a(t)h(x_{t+1}, u_t, p_t)) \tag{2}$$

is called a *projective procedure*. Here the size of the subsimplex $\sum_k(\varepsilon)$ increases as $\varepsilon_t \to 0$. The vector $\bar{p}_t$ belongs to the interior of Σ_k. The necessity of using the projective procedures usually arises in the case of a non-binary process.

We shall apply these general considerations to one more problem of adaptive control.

We consider a class to consist of the HPIV being understood in the extended sense (see Sec. 2, Chap. 1). At that, any process is $l + 1$-dimensional, i.e. $x_t = (x_t^{(0)}, x_t^{(1)}, \ldots, x_t^{(l)})$. Let $\bar{W}(\bar{p}) = (W_0(\bar{p}), W_1(\bar{p}), \ldots, W_l(\bar{p}))$ be the vector of average rewards. We need to find the "conditional optimal" strategy σ^0 such that

$$W_0(\bar{p}) \to \min, \quad W_i(p) \leq 0, \quad i = 1, \ldots, l.$$

It turns out that in this case one can reach the strong aim, i.e. (here the notation $V_j(t) \overset{\text{def}}{=} t^{-1} \sum_{n=1}^{t} x_n^{(j)}$ is used)

$$\varlimsup_{t\to\infty} V_0(t) = \min, \qquad \varlimsup_{t\to\infty} V_i(t) \le 0, \quad i = 1,\ldots,l, \quad \text{a.s.} \tag{3}$$

The attainment of aim (3) is called the *conditional extremum problem*. In the classical statement it can be reduced to the following linear programming problem

$$W_0(\bar{p}) = \sum_{j=1}^{k} w_j^{(0)} p_j \to \min, \qquad W_i(p) = \sum_{j=1}^{k} w_j^{(i)} p_j \le 0, \quad i = 1,\ldots,l \tag{4}$$

with the constraint $\bar{p} \in \Sigma_k$. Here $w_j^{(i)}$ will denote the mathematical expectation of the ith component of the HPIV if the control u_j is used. It is easy to see that the problems (3) and (4) are equivalent for the class of the HPIV (in the extended sense).

Let us turn to the adaptive version of the conditional extremum problem when the characteristics of the controlled HPIV are unknown. One approach to solve the problem is the method of Lagrange multipliers.

We form the Lagrange function

$$\mathcal{L}(\bar{p}, \bar{\lambda}) = W_0(p) + \sum_{i=1}^{l} \lambda(i) W_i(\bar{p})$$

where the vector $\bar{\lambda} = (\lambda(1), \lambda(2), \ldots, \lambda(l))$ has non-negative components ($\bar{\lambda} \in R_+^l$ for short). One of the conditions for the existence of a solution $\bar{p}_*$ of problem (3) can be formulated with the help of the function $\mathcal{L}$ as follows.

It is necessary and sufficient that for some vector $\bar{\lambda}_* \ge 0$ the point $(\bar{p}_*, \bar{\lambda}_*)$ be a saddle point of the function $\mathcal{L}$ for all $\bar{p} \in \Sigma_k$ and $\bar{\lambda} \ge 0$, i.e.

$$\mathcal{L}(\bar{p}_*, \bar{\lambda}) \le \mathcal{L}(\bar{p}_*, \bar{\lambda}_*) \le \mathcal{L}(\bar{p}, \bar{\lambda}_*). \tag{5}$$

It remains to find a method of searching for the coordinates of the saddle point $(\bar{p}_*, \bar{\lambda}_*)$. In view of the bilinearity of the function $\mathcal{L}$ we cannot use stochastic approximation directly because of the non-stability of the procedures (this can be shown by a simple example[b]). To avoid difficulties it is necessary to replace the original problem by a "close" one (to regularize the original one). In our case the replacement consists of adding summands to the function so it is strongly convex with respect

---

[b]For instance, let us consider the following gradient procedure:

$$\begin{cases} x_n = x_{n-1} - \gamma_n y_{n-1}, \\ y_n = y_{n-1} + \gamma_n x_{n-1}, \end{cases}$$

of searching the saddle point of the function $f = xy$. If $\sum_{n=1}^{\infty} \gamma_n^2 = \infty$, this procedure will not be stable since $x_n^2 + y_n^2 = (1 + \gamma_n^2)(x_{n-1}^2 + y_{n-1}^2) \to \infty$ as $n \to \infty$ but $(0,0)$ is the saddle point.

to $\bar{p}$ (at any admissible $\bar{\lambda}$) and concave with respect to $\bar{\lambda}$ (at any admissible $\bar{p}$). The regularized Lagrange function $\mathcal{L}$ is defined by the equality

$$\mathcal{L}_\delta(\bar{p}, \bar{\lambda}) = \mathcal{L}(\bar{p}, \bar{\lambda}) + \frac{\delta}{2}(\|\bar{p}\|^2 - \|\bar{\lambda}\|^2), \qquad \delta > 0$$

where $\bar{p} \in \Sigma_k(\varepsilon)$, $\varepsilon \in [0, 1/k)$. Obviously, it possesses the required properties and, hence, it has a unique saddle point $(\bar{p}_*(\varepsilon, \delta), \bar{\lambda}_*(\varepsilon, \delta))$on the set $\Sigma_k(\varepsilon) \times \mathrm{R}_+^l$. Let us consider the question of closedness of saddle points of the regularized and original Lagrange functions. First we state the following condition on the parameters of regularization.

Condition A. Let

$$\varepsilon_n \in [0, k^{-1}), \quad \delta_n > 0, \quad \lim_{n \to \infty} \delta_n = 0, \quad \lim_{n \to \infty} \varepsilon_n \delta_n^{-1} = \varkappa < \infty.$$

This condition will be used in two auxiliary lemmas. We put $\bar{p}_*(n) = \bar{p}_*(\varepsilon_n, \delta_n)$, $\bar{\lambda}_*(n) = \bar{\lambda}(\varepsilon_n, \delta_n)$.

Lemma 1. *If condition* **A** *holds then*

$$\lim_{n \to \infty} (\bar{p}_*(n), \bar{\lambda}_*(n)) = (\bar{p}_*, \bar{\lambda}_*),$$

where $\bar{p}_*$, $\bar{\lambda}_*$ are defined by (5) and may depend on $\varkappa$.

Proof. For all $\bar{p} \in \Sigma_k(\varepsilon), \bar{\lambda} \in \mathrm{R}_+^l$ we have $\mathcal{L}_{\delta_n}(\bar{p}_*(n), \bar{\lambda}) \leq \mathcal{L}_{\delta_n}(\bar{p}, \bar{\lambda}_*(n))$. Taking into account the notation

$$e = (1, \ldots, 1), \quad \bar{\lambda} = \bar{\lambda}_*, \quad \bar{p} = \bar{p}_* - \varepsilon_n(k\bar{p}_* - e), \tag{6}$$

we can write

$$\|\bar{p}_*(n)\|^2 + \|\bar{\lambda}_*(n)\|^2 \leq \|\bar{p}_*\|^2 + \|\bar{\lambda}_*\|^2 - 2k\varepsilon_n\delta_n^{-1}\left[\mathcal{L}(\bar{p}_*, \lambda_*(n)) - \mathcal{L}(k^{-1}e, \bar{\lambda}_*(n))\right],$$

i.e. the sequence $(\bar{p}_*(n), \bar{\lambda}_*(n))$ is bounded. Let $(\bar{p}_*(n_i), \bar{\lambda}_*(n_i))$ be any subsequence converging to $(\tilde{p}, \tilde{\lambda})$ as $i \to \infty$. It is clear that $(\tilde{p}, \tilde{\lambda})$ is the saddle point of $\mathcal{L}(\bar{p}, \bar{\lambda})$ on the set $\Sigma_k \times \mathrm{R}_+^l$. Hence the lemma will be proved if the saddle point is unique. Otherwise, we can choose any point $(\bar{p}_*, \bar{\lambda}_*)$ satisfying conditions (5). Since the function $\mathcal{L}(\cdot, \cdot)$ is convex with respect to $\bar{p}$ and concave with respect to λ, we have

$$(\nabla_p \mathcal{L}_{\delta_n}(\bar{p}_*(n), \bar{\lambda}_*(n)), \bar{p}_* - \bar{p}) - (\nabla_\lambda \mathcal{L}_{\delta_n}(\bar{p}_*(n), \bar{\lambda}_*(n)), \bar{\lambda}_* - \bar{\lambda}) \leq 0.$$

From this and (6) it follows that

$$k\frac{\varepsilon_n}{\delta_n}\left[\mathcal{L}(\bar{p}_*, \bar{\lambda}_*(n)) - \mathcal{L}(k^{-1}e, \bar{\lambda}_*(n)) - \mathcal{L}(\tilde{p}, \bar{\lambda}_*(n)) + \mathcal{L}(k^{-1}e, \tilde{\lambda}_*)\right]$$

$$+ (\bar{p}_*(n), \bar{p}_*(n) - \tilde{p}) + (\bar{\lambda}_*(n), \bar{\lambda}_*(n) - \tilde{\lambda}) \leq 0$$

where $\tilde{p} = (p_*(n) - \varepsilon_n e)(1 - k\varepsilon_n)^{-1} \in \Sigma_k$. It remains to find the limit along the subsequence (n_i). We have

$$k\varkappa\left[\mathcal{L}(1/ke, \lambda_*) - cL_*) - \mathcal{L}(k^{-1}e, \tilde{\lambda})\right] + (\tilde{p}, \tilde{p} - \bar{p}_*) + (\tilde{\lambda}, \tilde{\lambda} - \tilde{\lambda}_*) \leq 0,$$

i.e. the saddle point $(\tilde{p}, \tilde{\lambda})$ points to the unique minimum $(\bar{p}_*, \bar{\lambda}_*)$ of the quadratic form

$$\|\bar{p}\|^2 + \|\bar{\lambda}\|^2 - 2\varkappa k \sum_{i=1}^{l} \lambda_i W_i(k^{-1}e)$$

on the closed convex set of saddle points. Hence

$$\lim_{n \to \infty} \bar{p}_*(n) = \tilde{p}, \qquad \lim_{n \to \infty} \bar{\lambda}_*(n) = \tilde{\lambda}.$$

The following lemma has a "technical" character. $\qquad\qquad\qquad\qquad\square$

Lemma 2. *Let the condition* **A** *hold. Then there exists a constant $C > 0$ and an integer $N \geq 1$ such that*

$$\|\bar{p}_*(\varepsilon_n, \delta_n) - \bar{p}_*(\varepsilon_m, \bar{\lambda}_m)\|^2 + \|\bar{\lambda}_*(\varepsilon_n, \delta_n) - \bar{\lambda}_*(\varepsilon_m, \delta_m)\|^2$$
$$\leq C\left[|\varepsilon_n - \varepsilon_m| + |\delta_n - \delta_m| + |\varepsilon_n \delta_n^{-1} - \varepsilon_m \delta_m^{-1}|\right] \overset{\text{def}}{=} C I_{n,m}$$

for all n and $m \geq N$.

Proof. It is enough to verify that $\bar{p}_*$ and $\bar{\lambda}_*$ as functions of ε, δ and ε/δ satisfy the Lipschitz condition. We define the following sets

$$M_0 = \left\{(\bar{p}, \bar{\lambda}) : \sum_{i=1}^{k} p_i = 1\right\},$$
$$M_1(i_1, \ldots, i_s) = \{(\bar{p}, \bar{\lambda}) : p_{(i_g)} = \varepsilon, g = 1, \ldots, s\} \cap M_0,$$
$$M_2(i_1, \ldots, i_r) = \{(\bar{p}, \bar{\lambda}) : \lambda_{(j_h)} = 0, g = 1, \ldots, r\} \cap M_0,$$
$$M_3(i_1, \ldots, i_s; j_1, \ldots, j_r) = M_1(i_1, \ldots, i_s) \cap M_2(j_1, \ldots, j_r)$$

where $i_1, \ldots, i_s$, $j_1, \ldots, j_r$ are all possible transpositions of the indices $(1, \ldots, k)$ and $(1, \ldots, l)$ of the controls u_i and factors λ_j respectively. Reindexing these sets we consider the correspondence between the nth of them and the problem Z_n of calculating the coordinates of the saddle point for the function $\mathcal{L}_\delta(\bar{p}, \bar{\lambda})$ $(\delta > 0)$ on this nth set. Let $(\bar{p}(Z_n), \bar{\lambda}(Z_n))$ denote the solution of Z_n. For any $\varepsilon \in [0, 1/k)$, $\delta > 0$ the point $(\bar{p}_*(\varepsilon, \delta), \bar{\lambda}_*(\varepsilon, \delta))$ coincides with the solution of one of the problems Z_n. So, the needed properties of the functions $(\bar{p}_*(\varepsilon, \delta), \bar{\lambda}_*(\varepsilon, \delta))$ will be found if we understand the dependence of the points $(p(Z_n), \lambda(Z_n))$ on the arguments ε, δ for each n. We consider the problem Z_n for the set

$$M_3(1, 1) = \left\{(\bar{p}, \bar{\lambda}) : \sum_{i=1}^{k} p(i) = 1, p(1) = \varepsilon, \lambda(1) = 0\right\}.$$

Consider the system of linear equations

$$\left.\begin{aligned}
\nabla_p \mathcal{L}(\bar{p}, \bar{\lambda}) + \delta \bar{p} - \varkappa_0 e - \varkappa_1 e_1 &= 0, \\
\nabla_\lambda \mathcal{L}(\bar{p}, \bar{\lambda}) + \delta \lambda - \theta_1 e_1 &= 0, \\
p^T e = 1, \quad p(1) = \varepsilon, \quad \lambda(1) &= 0
\end{aligned}\right\} \tag{7}$$

where $e_1 = (1, 0, \ldots, 0)$, $\varkappa_0$, $\varkappa_1$, θ are the Lagrange multipliers in the problem Z_n. The solution of this system, i.e. the representation of the coordinates of the saddle point $(\bar{p}(Z_n), \bar{\lambda}(Z_n))$ and the Lagrange multipliers in terms of ε and δ, has the form

$$\frac{\sum_{i=1}^{k+l}(a_i + b_i \varepsilon)\delta^i}{\sum_{i=1}^{k+l} c_i \delta^i}. \tag{8}$$

The polynomial in the denominator is the determinant of (7). It has no roots at $\delta > 0$, otherwise the saddle point of the function $\mathcal{L}_\delta(p, \lambda)$ on the set $M_3(1,1)$ would not be unique. Making all possible cancellations in the fractions (8) (if needed) we consider their least common denominator. Let it contain a factor δ^r, $r \geq 0$. Then $r = 0$ or $r = 1$. Indeed, substituting the solution (8) in (7) and multiplying the first equation by $\bar{p}(Z_n)$, the second by $\bar{\lambda}(Z_n)$ and summing, we obtain

$$\delta(\|\bar{p}(Z_n)\|^2 + \|\bar{\lambda}(Z_n)\|^2) = \varkappa_0 + \varkappa_1 \varepsilon.$$

Comparing the powers of the polynomials we find that $2r - 1 = r$ or $2r - 1 = r - 1$, i.e. $r = 1$ or $r = 0$. If $r = 0$ then $(\bar{p}(Z_n), \bar{\lambda}(Z_n))$ is the Lipschitz function of ε, δ in the region of their values. If $r = 1$ then we write down the fraction (8) in the form

$$\frac{a_0 + \sum_{i=1}^{k+l} a_i \delta^i + \varepsilon \sum_{i=1}^{k+l} b_i \delta^i}{\sum_{i=0}^{k+l-1} c_i \delta^i}.$$

The denominator has no roots at $\delta \geq 0$. If $a_0 \neq 0$ at least for one of the components of the $k+l$-vector $(\bar{p}(Z_n), \bar{\lambda}(Z_n))$ the point $(\bar{p}_*(\varepsilon_n, \delta_n), \bar{\lambda}_*(\varepsilon_n, \delta_n))$ is not a solution of the problem Z_n. In the case $a_0 = 0$ for all components of the vector $(\bar{p}(Z_n), \bar{\lambda}(Z_n))$ the last function is Lipschitz in the arguments $\varepsilon, \delta, \varepsilon/\delta$. This holds for all problems Z_n. From this the stated assertion follows. $\qquad\square$

Corollary 1. *Under the conditions of Lemma 2 there exists a constant C such that*

$$\|\bar{p}_*(\varepsilon_n, \delta_n) - \bar{p}_*(\varepsilon_{n+1}, \delta_{n+1})\| + |\bar{\lambda}_*(\varepsilon_n, \delta_n) - \bar{\lambda}_*(\varepsilon_{n+1}, \delta_{n+1})\|$$

$$\leq C\left[|\varepsilon_n - \varepsilon_{n+1}| + |\delta_n - \delta_{n+1}| + |\varepsilon_n \delta_n^{-1} - \varepsilon_{n+1} \delta_{n+1}^{-1}|\right] \overset{\text{def}}{=} C\mathcal{I}_n.$$

We now consider the calculation of the conditional optimal control. This procedure of gradient type is a combination of two methods, namely, of stochastic approximation and Lagrange multipliers. For the gradients with respect to the vectors $\bar{p}$ and $\bar{\lambda}$ we have

$$\nabla_p \mathcal{L}_\delta(\bar{p}, \bar{\lambda}) = \mathbf{E} \left\{ \frac{e(u_n)}{\bar{p}_n^T e(u_n)} \left[x_n^{(0)} + \sum_{j=1}^{l} \lambda_n(j) x_n^{(j)} + \delta p_n^T e(u_n) \right] \Bigg| \bar{p}_n = \bar{p}, \bar{\lambda}_n = \bar{\lambda} \right\},$$

$$\frac{\partial \mathcal{L}_\delta(\bar{p}, \bar{\lambda})}{\partial \lambda(j)} = \mathbf{E}[x_n^{(j)} - \delta \lambda_n(j) | \bar{p}_n = p, \bar{\lambda}_n = \lambda], \quad j = 1, \ldots, l.$$

Here, as in Sec. 5, Chap. 2, we put $e(u) = \sum_{i=1}^{k} e_i \chi(u(t) = u_i)$. The vectors $\bar{p}(t) = (p_1(t), \ldots, p_k(t)) \in \Sigma_k(\varepsilon)$, $\bar{\lambda}_t = (\lambda_t(1), \ldots, \lambda_t(l))$ are the current estimates of the saddle point of the function $\mathcal{L}(p, \lambda)$. To complete the construction of the required procedure it remains to consider the projection operation

$$\bar{p}_{t+1} = \varphi_{\varepsilon_{t+1}} \left[\bar{p}_t - a(t) \frac{e(u_n)}{\bar{p}_t^T e(u_t)} \left\{ x_t^{(0)} + \sum_{j=1}^{l} \lambda_t(j) x_t^{(j)} + \delta p_t^T e(u_t) \right\} \right], \quad (9)$$

$$\bar{\lambda}_{t+1}(j) = \max \left\{ 0, \bar{\lambda}_t(j) + a(t)(x_t^{(j)} - \delta_t \bar{\lambda}_t(j)) \right\}, \quad j = 1, \ldots, l. \quad (10)$$

In these recurrent relations the parameters $a(t)$, δ_t, ε_t are used. For this procedure to be valid they must satisfy some restrictions.

Condition B. *The parameters* $a(t), \delta_t, \varepsilon_t$ *satisfy the following restrictions*

$$a(t) > 0, \ \delta_t > 0, \ \forall t \geq 0, \ \lim_{t \to \infty} \delta_t = 0, \ \varepsilon_t \in (0, 1/k), \ \sum_{t=1}^{\infty} a(t)\delta_t < \infty;$$

there exists $\lim_{t \to \infty} \varepsilon_t/\delta_t = \varkappa < \infty$ *and*

$$\sum_{t=1}^{\infty} \left[a^2(t)/\varepsilon_t + |\varepsilon_t - \varepsilon_{t+1}| + |\delta_t - \delta_{t+1}| + |\varepsilon_t \delta_t^{-1} - \varepsilon_{t+1} \delta_{t+1}^{-1}| \right] < \infty.$$

We can now state the main result.

Theorem 1. *Let condition* **B** *hold. Then the procedure* (9) *converges to the saddle point* $(\bar{p}_*, \bar{\lambda}_*)$ *of the Lagrange function with probability one and, moreover,* $\bar{p}_t$ *converges in the mean square sense as well with respect to the class* $N_{k,l}$ *of the generalized* $l + 1$-*dimensional HPIV with finite number* k *of controls.*

Proof. As before (Sec. 5, Chap. 2) we define the Lyapunov function in the form

$$L_t = \|\bar{p}(t) - \bar{p}_*(t)\|^2 + \|\bar{\lambda}(t) - \bar{\lambda}(t)\|^2$$

where $(\bar{p}_*(t), \bar{\lambda}_*(t))$ is the saddle point of the function $L_{\delta_t}(\bar{p}, \bar{\lambda})$. If we show that $\lim_{t \to \infty} L_t = 0$ then from Lemma 1 the assertion of the theorem will follow. The proof of this relationship is similar to that of Theorem 2 (Sec. 5, Chap. 2). Therefore, we are brief. The first step consists of obtaining the following inequality

$$\mathbf{E}(L_{t+1}|\mathcal{F}_t) \leq L_t + C \left[\mathcal{I}_t \sqrt{L_t} + \mathcal{I}_t^2 + (1 + L_t)\varepsilon_t^{-1} a^2(t) \right]$$

$$- 2a(t) \left[L_{\delta_t}(\bar{p}(t), \bar{\lambda}_*(t)) - L_{\delta_t}(\bar{p}_*(t), \bar{\lambda}_t) + 2^{-1} \delta_t L_t \right]$$

where $\mathcal{I}_t$ has been defined in the corollary of Lemma 2. This inequality is easily derived from the representation of $\nabla_p \mathcal{L}_\delta(p, \lambda)$, the procedure (9), the properties of a projector and Corollary 1. The second step leads to the inequality

$$\mathbf{E}(L_{t+1}|\mathcal{F}_t) \le (1 - 2^{-1}a(t)\delta_t + \varepsilon_t^{-1}Ca^2(t))L_t + C\left[\mathcal{I}_t^2 + \frac{a^2(t)}{\varepsilon_t} + \frac{c\mathcal{I}_t^2}{2a(t)\delta_t}\right].$$

It is derived from the definition of the saddle point and the inequality $2ab \le a^2 + b^2$ (where $a = \sqrt{a(t)\delta_t L_t}$, $b = C(\mathcal{I}_t((a(t)\delta_t)^{-1/2})$. Applying to this inequality the martingale lemma (see foot-note on page 70) we obtain $\lim_{t\to\infty} L_t = 0$ a.s. By the dominated convergence theorem, the convergence of $(p(t), \lambda(t))$ a.s. implies the convergence of $\bar{p}(t)$ in the mean square sense. $\qquad\square$

Hence procedure (9) is the conditional optimal strategy with respect to the class $N_{k,l}$. While using procedure (9) it is convenient to choose the sequences which appear there in the form

$$a(t) = \frac{a}{(t + f)^\alpha}, \quad \varepsilon_t = \frac{e}{(t + g)^\varepsilon}, \quad \delta_t = \frac{\alpha}{(t + h)^\delta}.$$

For the condition **B** to be satisfied we must impose some restriction on the power:

$$\alpha + \delta \le 1, \quad \varepsilon \ge \delta > 0, \quad 2\alpha > 1 - \varepsilon.$$

The estimates of the convergence rate are given below.

Theorem 2. *Under the conditions of Theorem 1 the conditional extremum is attained with probability one and*

$$\overline{\lim_{t\to\infty}} \, t^d \mathbf{E}(W_j(\bar{p}_0) - V_j(t))^2 < \infty, \quad j = 0, 1, \ldots, l,$$

where $\bar{p}_0$ is the solution of the linear programming problem (4) and

$$d = \begin{cases} \min(2\delta, \alpha - 2\delta), & \text{for } \varepsilon = \delta, \\ \min(2\delta, 2(\varepsilon - \delta)), & \text{for } \varepsilon > \delta, \end{cases}$$

with $d \le 2/5$ but the equality $d = 2/5$ takes place only at $\alpha = 4/5$, $\varepsilon = \delta = 1/5$.

CHAPTER 4

MINIMAX ADAPTIVE CONTROL

In this chapter we consider control problems whose aims are obtaining minimax (or maximin) of the appropriate function. For such problems the game interpretation is natural enough, i.e. several players are trying to gain profits at the others expense. The first game below is deterministic but in the following two the players act on a vector-valued HPIV. We consider two statements of the problem: synthesis of a strategy to obtain the game aim (Sec. 1, 3) and analysis leading to find the solution by a family of the finite automata (Sec. 4).

4.1. Games with Consistent Interests

A number of economical and sociological situations leads to the following two-participant game. The first and second player uses controls from the sets $U = \{u\}$ and $V = \{v\}$ respectively. The game consists of the following. The first player (I) chooses a move u_1 and informs the second player (II) about it. Player II makes a move v_1 and informs the first player about it. Thereafter the process repeats. The order of the moves is essential in such a game.

Let us turn to the motivations of the players to choose a given move. For player I it is maximization of his gain $W(u, v)$ and for player II the purpose (or interest) is obtaining desired values for some function $G(u, v)$. For example, it could be either to obtain the maximum value of the function $G(u, v)$ or to secure the inequality $G(u, v) \geq 0$. A set of "expedient moves" $N(u) \subset V$ corresponds to each of such purposes. For the interests of player II stated above these sets have the forms

$$N(u) = \left\{ v : G(u, v) = \max_{v'} G(u, v') \right\}, \quad N(u) = \{ v : G(u, v) \geq 0 \}.$$

It is essential that for a fixed interest of player II these sets are uniquely defined for any u. If player I knows the family of expedient responses $\{N(u), u \in U\}$ exactly, then the choice of his move u will be determined by the size of the reward $W(u) = \inf_{v \in N(u)} W(u, v)$. In this case "solving" the game means to find both the "guaranteed result" $w = \sup_u W(u)$ and a strategy realizing it (maybe accurate within ε). Player I may be informed insufficiently, i.e. instead of expedient responses $\{N(u)\}$ a family $\{N_\theta(u), u \in U, \theta \in \Theta)\}$ where Θ is a set of parameters is given. Then the best guaranteed result is equal to $w(\Theta) = \sup_u \inf_\theta \inf_{v \in N_\theta(u)} W(u, v)$ where the sets $N_\theta(u) = \{ v : G_\theta(u, v) = \max_{v' \in V} G_\theta(u, v') \}$ are assumed to be non-empty. We assume that for any $v \in V$ we can check if it belongs to the given set $N_\theta(u)$. This game is denoted by the symbol Γ_1.

97

So, in each turn player I makes a move and informs player II about it. Next, player II chooses his move in keeping with his desire to maximize the function $G(u, v)$. The first player knows:

(1) the move of his adversary;
(2) the class of the admissible functions $G_\theta(u, v)$;
(3) the behavior of the second player which consists of choosing the response v so his gain function is maximized. However the value of this gain is unknown to the player I.

In the n-th round player I has the knowledge of the history of the game, i.e. the collection of pairs $(u_1, v_1), (u_2, v_2), \ldots, (u_{n-1}, v_{n-1})$. Using this information he tries to achieve the maximum guaranteed result w (maybe accurate within ε). This will be the case if beginning from some moment the participants choose the moves (u_n, v_n) so that $W(u_n, v_n) > \sup_{(u,v)} W(u, v) - \varepsilon$. Thus, the adaptive version of the minimax problem has arisen.

We introduce the necessary notation. We suppose that θ_0 is the true value of the parameter. Next,

$$w(u) = \inf_{v \in N_{\theta_0}(u)} W(u, v), \quad w_0 = \sup_u w(u).$$

Definition 1. The value w_0 is called the *maximum guaranteed result* under complete information of player I.

If player I knows the interests of player II accurately within the class Θ then the best guaranteed result will be equal to

$$w = \sup_u \inf_\theta \inf_{v \in N_\theta(u)} W(u, v) = \sup_u \inf_\theta W_\theta(u).$$

The inequality $w_0 \geq w$ holds but the value w is unknown to player I. We introduce one more notation

$$w(\theta) = \sup_u \inf_{v \in N_\theta(u)} W(u, v).$$

Now we construct optimal game strategies. These depend on the aim. First, we consider the aim:

α_ε. *for all $n \geq N$ the inequalities*

$$W(u_n, v_n) \geq w_0 - \varepsilon$$

hold (here $\varepsilon > 0$ is fixed).

This means that by using information obtained for the first N games about the adversary, player I wants to receive a guaranteed result which is not less than w_0, accurate within ε. Player I does not know the value w_0.

We shall find a way of attaining this aim in the case of a finite parameter set $\Theta = \{\theta_1, \ldots, \theta_n\}$. Let all sets $N_\theta(u)$ be non-empty.

Definition 2. The set Θ is called *distinguishable* if for any pair θ_i, θ_j $(i \neq j)$ there exists a move u such that $N_{\theta_1}(u) \cap N_{\theta_2}(u) = \emptyset$. Otherwise, this set is said to be *undistinguishable*.

Theorem 1. *If the set Θ is distinguishable then for any $\varepsilon > 0$ and $N \geq |\Theta| - 1$ $(= n - 1)$ the aim α_ε will be attainable. If the set Θ is undistinguishable and $a = \sup_u \min\{w_{\theta'}(u) - w(\theta'), w_{\theta''}(u) - w(\theta'')\} < 0$ then the aim α_ε will be unattainable for $0 < \varepsilon < -a$ and any N.*

Proof. 1. Always supposing player I knows the collection of sets $\{N_\theta(u)\}$, he can choose u_i move by move in such a manner that every time one of the values of θ_i be rejected. In $n - 1$ moves the true value of the parameter will be found. This will give the gain function $G_{\theta_0}(u, v)$ of player II as well.

It is clear that if there exists a u_0 such that $N_{\theta'}(u_0) \cup N_{\theta''}(u_0) = \emptyset$ for all θ', θ'' then the aim α_ε will be attainable for any ε and $N \geq 1$.

2. Let the functions $G_{\theta'}$, $G_{\theta''}$ be undistinguished and in the first n turns player II has chosen

$$v_i \in N_{\theta'}(u_i) \cup N_{\theta''}(u_i), \quad i = 1, \ldots, n.$$

This means that before the n plus first game player I knows that the function G is equal to either $G_{\theta'}$ or $G_{\theta''}$. By the conditions of the theorem we have

$$\min\{w_{\theta'}(u) - w(\theta')w_{\theta''}(u) - w(\theta'') + \varepsilon\} < 0,$$

for all u. Hence, in the n plus first game player I cannot guarantee the fulfilment of the inequality $W(u_{n+1}, v_{n+1}) \geq w_0 - \varepsilon$. $\qquad\square$

There are simple examples to demonstrate the non-attainability of the aim α_ε for $N < |\Theta| - 1$ and sufficiently small ε.

We shall now consider another aim of control:

β_ε. *in all turns, except for at most N of them, the inequality*

$$W(u_n, v_n) > w_0 - \varepsilon.$$

should taken place.

Theorem 2. *For the finite parameter set Θ and any $\varepsilon > 0$ the aim β_ε is attainable at $N > |\Theta| - 1$ and is unattainable, generally speaking, at $N < |\Theta| - 1$.*

Proof. We can assume that $|\Theta| = 2$. If the functions G_1 and G_2 are distinguished then player I will choose his first move u_1 so that $N_1(u_1) \cap N_2(u_1) = \emptyset$. Hence he will know the true function G_0. Therefore, in the other games he can guarantee himself the gain being no less than $w_0 - \varepsilon$, i.e. the aim of control will be attained. If G_1 and G_2 are undistinguished then $M(u) = N_1(u) \cap N_2(u) \neq \emptyset$ for any u. Then u_1 should be chosen according to the condition

$$\inf_{v \in M(u)} W(u, v) \geq \sup_u \inf_{v \in M(u)} W(u, v) - \varepsilon.$$

We shall consider two cases: either $v_1 \in M(u_1)$ and then, obviously, $W(u_1, v_1) \geq w_0 - \varepsilon$ or $v_1 \notin M(u_1)$. In the last case player I will determine the true function G_θ. In the next games, if the player does not know the true function yet, he chooses his moves in a similar way. It is easy to see that if $|\Theta| \geq 3$ the arguments will remain in force. It is possible to construct some simple games where the aim β_ε is unattainable for $N < |\Theta| - 1$ and any $\varepsilon > 0$ in. $\qquad\qquad\square$

It is of interest to note two peculiarities of the control problem with the aim β_ε. One is the fact that there is no need to demand that the functions G_θ be distinguished. The other one is that when player I cannot obtain new information about the function G_θ he can assure the very same gain as in the case of full information about the gain function of player II. If player I knows in advance that there is no information about the function G then in each game he can guarantee that his own gain will be no less than $w_0 - \varepsilon$.

Now let the set Θ be infinite. Then the question of attainability of the aim β_ε has a more complicated solution.

Theorem 3. *Let the following conditions hold:*

(1) the set Θ is compact in a complete metric space with metric ρ;
(2) the function $w_\theta(u)$ is lower semi-continuous[a] in $\theta \in \Theta$ for any fixed u;
(3) the function $w(\theta)$ is continuous for $\theta \in \Theta$.

Then for any $\varepsilon > 0$ there exists $N(\varepsilon)$ such that the aim β_ε is attainable at $n \geq N(\varepsilon)$.

Proof. It is sufficient to show that for any $\varepsilon > 0$ there exists a finite decomposition, i.e. sets $\Theta_1, \ldots, \Theta_L$, $\Theta_j \cap \Theta_j = \emptyset$, $i \neq j$ such that $\Theta = \bigcup_{j=1}^{L} \Theta_j$, such that for any $i = 1, \ldots, L$ we have $w(\theta) - w(\Theta_i) \leq \varepsilon$, $\theta \in \Theta_i$. (Here $w(\Theta_i) \stackrel{\text{def}}{=} \sup_u \inf_{\theta \in \Theta_i} w_\theta(u)$.) Making use of conditions 2 and 3 we define a system of open sets in Θ as follows

$$S(u, \varepsilon) = \{\theta : w_\theta(u) > w(\theta) - \varepsilon/2\}.$$

This system forms an open cover of the compact space Θ since for any θ there exists at least one u such that $w_\theta(u) > w(\theta) - \varepsilon/2$. From this system we select a finite cover $S(u_1, \varepsilon), \ldots, S(u_L, \varepsilon)$. Using it we shall construct the decomposition of the set Θ formed by the following disjoint sets:

$$S_1 = S(u_1, \varepsilon), \quad S_2 = S(u_2, \varepsilon) \backslash S_1, \ldots, \quad S_L = S(u_L, \varepsilon) \backslash \bigcup_{j=1}^{L-1} S_j.$$

---

[a] A function $f(x)$ is said to be lower semi-continuous if for any sequence $x_1, x_2, \ldots, x_n, \ldots \to x$ the inequality

$$\varliminf_{n \to \infty} f(x_n) \geq f(x)$$

holds. It is equivalent to the following: the sets $\{x : f(x) < c\}$ are open for any c.

According to condition (3) we can choose $\delta > 0$ so that the inequality $\rho(\theta, \theta') < \delta$ implies $|w(\theta) - w(\theta')| < \varepsilon/2$. We shall write down the set S_j in the form $S_j = \bigcup_{i=1}^{n_j} M_{i,j}$ where the sets $M_{i,j}$ are disjoint and their diameters are less than δ. Then $\Theta = \bigcup_{j=1}^{L} \bigcup_{i=1}^{n_j} M_{i,j}$, $M_{i_1,j_1} \cap M_{i_2,j_2} = \emptyset$ if $(i_1, j_1) \neq (i_2, j_2)$. From the following sequence of inequalities

$$w(M_{i,j}) = \sup_u \inf_{\theta \in M_{ij}} w_\theta(u) \geq \inf_{\theta \in M_{ij}} w_\theta(u_j) \geq \inf_{\theta \in M_{ij}} w(\theta) - \varepsilon/2 \geq \sup_{\theta \in M_{ij}} w(\theta) - \varepsilon$$

for all i, j it follows that, except for at most $n_1 + \cdots + n_L$ games, player I obtains the gain no less than $w(M_{ij_0}) - \varepsilon$ (where j_0 is the index of the set belonging to the decomposition and containing the true objective set of player II), i.e. the aim β_ε is attainable. $\qquad\square$

The conditions of the theorem will be fulfilled if all functions $W(u, v)$, $G_\theta(u, v)$ are continuous on the product of metric compacts $U \times V \times \Theta$.

We shall now state another sufficient condition for the aim to be attained. First, we introduce the necessary notation:

$$\varphi_\theta(u, v) = \sup_{z \in V} G_\theta(u, z) - G_\theta(u, v), \quad N_\theta^\gamma(u) = \{v : \varphi_\theta(u, v) \leq \gamma\}, \quad \gamma \geq 0,$$

$$w_\theta^\gamma(u) = \inf_{z \in N_\theta^\gamma} W(u, z), \qquad\qquad w^\gamma(\theta) = \sup_u w_\theta^\gamma(u).$$

Theorem 4. *In the game above let player II maximize his gain (defined by a function from the family $(G_\theta, \theta \in \Theta)$) and assume the following conditions hold:*

(1) *The sets U, V, θ are compacts in some full metric spaces;*
(2) *the functions $W(u, v)$ and $G_\theta(u, v)$ are continuous on $U \times V \times \Theta$;*
(3) *$\lim_{\gamma \to 0+} w^\gamma(\theta) = w^0(\theta)$ uniformly with respect to θ.*

Then for any $\varepsilon > 0$ there exists $N(\varepsilon)$ such that the aim β_ε is attainable for $n \geq N(\varepsilon)$.

We omit the proof, though it is attractive due to using information about the course of the game more fully than in Theorem 3.

Now we put player I in a more complex situation: in every set (u_n, v_n) let him know only his own gain $W(u_n, v_n)$ but not the move v_n of player II. In this new situation we consider the analogues of the aims $\alpha)$, $\beta)$ (there is no need to formulate them again).

Under these conditions the aim (α) proves to be unattainable even if all functions G_θ are distinguishable in pairs. This may be proved by the following example. Let

$$U = V = [0, 1], \quad W(u, v) = 4v(v - 1)$$

and

$$N_1(u) = \begin{cases} [0, 1/2], & u \in [0, 1/2], \\ 0, & u \in (1/2, 1], \end{cases} \qquad N_2(u) = \begin{cases} 1, & u \in [0, 1/2), \\ [1/2, 1], & u \in [1/2, 1], \end{cases}$$

be the objective sets of player II. (We suppose here that $\theta_1 = 1, \theta_2 = 2$.) Obviously, $N_1(u) \cap N_2(u) = \emptyset$ if $u \neq 1/2$. If $W(u_i, v_i) = 0$ in the first N games, then according to

$$\{1,2\} = \bigcap_{l=1}^{N} \{\theta : \exists v_l \in N_\theta(u_l) : W(u_l, v_l) = 0\}$$

in the $N{+}1$-st game player I will obtain the gain no greater than

$$\sup_u \ \inf_{v \in N_1(u) \cap N_2(u)} 4v(v-1) = -1.$$

However, player I could obtain the gain $w(\theta) = 1$ for both values of the parameter θ. Hence, for $\varepsilon < 1$ the aim α_ε is unattainable. As concerns the aim β_ε, Theorems 2, 3, 4 remain in force.

The game approach described above can be extended from two-participant games to the more complex two-level hierarchical structure. Then we shall say "center" instead of "player I" and "producers" instead of "player II" respectively. The center wants to maximize its own gain $W(u, v_1, \ldots, v_h)$ and the producers try maximizing their private gains $G_i(u, v_i)$, $i = 1, \ldots, h$. Choosing the move u the center informs the producers about it and they choose their own moves independently of each other. Their moves form a collection $(v_1, \ldots, v_h)$. If the center knows exactly the local gains (G_i) then its best guaranteed result in an individual game is equal to

$$w_0 = \sup_u \ \min_{v_1 \in N^{(1)}(u)} \cdots \min_{v_h \in N^{(h)}(u)} W(u, \bar{v})$$

where

$$N^{(i)} = \left\{ v_i \in V_i | G_i(u, v_i) = \max_{z \in V_i} G_i(u, z) \right\}.$$

This game is equivalent to that of the center with one producer having the gain $G(u, \bar{v}) = \sum_{i=1}^{h} G_i(u, v_i)$. It is clear that the preceding analysis can be directly extended to this new situation.

4.2. Some Remarks on Minimax Control of Vector HPIV

In the remainder of this chapter ξ_t denotes a HPIV with values $\bar{x}_t = \left(x_t^{(1)}, \ldots, x_t^{(\nu)}\right) \in \mathbf{R}^\nu$ and controls $\bar{u}_t = \left(u_t^{(1)}, \ldots, u_t^{(r)}\right) \in \mathbf{R}^r$. Each component $u_t^{(j)}$ takes the values from a finite set $U^{(j)} = \{u_{i,j}, i = 1, \ldots, k_j\}$. Such a process is usually defined by the conditional distribution

$$\mu(M|\bar{u}_{t-1}) = \mu\left(x_t^{(i)} \in M_l, l = 1, \ldots, \nu | \bar{u}_{t-1}\right), \quad M = M_1 \times \cdots \times M_\nu.$$

The components are assumed to be independent, i.e.

$$\mathbf{E}(M|\bar{u}) = \prod_{j=1}^{\nu} \mu_j(M_j|\bar{u}).$$

In addition, we assume that the vectors $\bar{x}$ and $\bar{u}$ have the same dimensions, i.e. $\nu = r$.

The classes of the HPIV will be controlled by *decentralized* systems, i.e. by means of the direct product $L = L_1 \times \cdots \times L_\nu$ of learning systems $L_j, j = 1, \ldots, \nu$. We assume that the jth component of the controlled process ξ_t serves as input of the system L_j but its output represents the jth component of the vector of controls u. There is neither interaction nor exchange of information between these learning systems. Each of them is acting as if it were single. However the distributions of components of the HPIV depend on all components of the vector $\bar{u}$. Below, the systems L_j are represented either by recurrent procedures or by finite automata.

Definition 1. A strategy in the form of a decentralized system represented by the direct product of elementary learning systems is called a *game control*.

The operating principle of such systems consists of adopting a decision (or in choosing the next move) by using the input values of the corresponding component of HPIV without knowledge of the successes of the other participants. It means that the moves of the players are independent and the game itself is coalition-free.

Let us formulate some necessary notions of game theory.

Definition 2. A vector $\bar{u}_t = \left(u_t^{(1)}, \ldots, u_t^{(\nu)}\right)$ is called the *party at the time moment t*.

A game is an unbounded sequence of parties. It is accompanied by the sequence of ν-dimensional vectors $\bar{x}_{t+1} = \left(x_{t+1}^{(1)}, \ldots, x_{t+1}^{(\nu)}\right)$ denoting the corresponding gains of the players.

Definition 3. The mathematical expectation of the jth component of HPIV $W_j(\bar{u}) = \int_{-\infty}^{\infty} x \mu_j(dx|\bar{u})$ which is supposed to be finite for all j, u is called the *pay function* of the jth player.

The vector $\bar{W}(\bar{u}) = (W_1(\bar{u}), \ldots, W_\nu(\bar{u}))$ is called the *game pay function*. We enumerate the most typical parties.

Nash point or *equilibrium party* is a party $\tilde{u} = \left(\tilde{u}^{(1)}, \ldots, \tilde{u}^{(\nu)}\right)$ such that

$$W_j(\tilde{u}) \geq W_j\left(\tilde{u}^{(1)}, \ldots, \tilde{u}^{(j-1)}, u^{(j)}, \tilde{u}^{(j+1)}, \ldots, \tilde{u}^{(\nu)}\right), \quad j = 1, \ldots, \nu,$$

i.e. if the jth player changes his move from $\tilde{u}^{(j)}$ to $u^{(j)}$ then he reduces his average gain. Therefore, the equilibrium party is stable in the sense that a reasonable player will not change the appropriate move.

Definition 4. A party in which the minimal component of the vector pay function $\bar{W}(\bar{u}) = (W_1(\bar{u}), \ldots, W_\nu(\bar{u}))$ takes the maximum value is called a *minimax party*.

The problem to find an equilibrium in minimax games can be considered as the aim of control for the vector HPIV.

The investigation of game problems of control for the vector HPIV is useful for understanding and simulating many biological, economical and sociological phenomena. Of course, we cannot discuss this theme here.

4.3. Recurrent Procedure of Searching Equilibrium Strategies in a Multi-person Game

The symbol ξ denotes a HPIV in the extended sense taking the values $x = \left(x^{(1)}, \ldots, x^{(\nu)}\right) \in \mathrm{R}^{\nu}$ with a vector control $u = \left(u^{(1)}, \ldots, u^{(\nu)}\right)$ whose jth component runs over the finite set $U^{(j)} = \{u_1^{(j)}, \ldots, u_{k_j}^{(j)}\}$. The components $x^{(i)}$ are supposed to be independent.

The process is controlled by ν participants (players) independent from each other. Each of them obtains the corresponding component of the vector x as a gain (here we shall talk about "losses") and produces one of the components of the vector u. In other words, the jth player receives the jth component of x and forms the jth component of u. Unlike the usual concept of game theory meaning that all participants know a pay function we now use the adaptive point of view, i.e. the players know only their own moves and losses. So, the adaptive strategies of the players are decentralized and depend only on the information about the course of the game or, in other words, on the course of the coalition-free ν-persons game. The strategy of the jth player will be denoted by $\sigma^{(j)}$. Let us connect with the jth player his average current losses in time T

$$V_T^{(j)} = V_T^{(j)}\left(\sigma^{(1)}, \ldots, \sigma^{(\nu)}\right) = T^{-1} \sum_{t=1}^{T} x_t^{(j)}, \quad j = 1, \ldots, \nu,$$

if the strategy $\left(\sigma^{(1)}, \ldots, \sigma^{(\nu)}\right)$ has been used. The problems of control (or interests of the participants of the game) consist in producing a behavior of every player independent from the other participants so that the stable, "balanced" regime which is similar to the Nash points in the usual games (see Sec. 2) be achieved. More precisely, this means the following. We need to construct independent strategies $\sigma_0^{(1)}, \ldots, \sigma_0^{(\nu)}$ so that

$$\varlimsup_{t \to \infty} \left[V_t^{(j)}\left(\sigma_0^{(1)}, \ldots, \sigma_0^{(j)}, \ldots, \sigma_0^{(\nu)}\right) - V_t^{(j)}\left(\sigma_0^{(1)}, \ldots, \tilde{\sigma}_0^{(j)}, \ldots, \sigma_0^{(\nu)}\right) \right] \leq 0, \text{ a.s.} \quad (1)$$

for all processes from the class of the HPIV described above for all j and any *stationary* program strategy $\tilde{\sigma}^{(j)}$. This requirement is similar to the notion of Nash point but with the following differences:

(1) the move (or action) of the player by the strategy which is not necessarily stationary;
(2) the average losses (over the set of parties) are replaced by real losses on realization.

We associate the control problem of the process ξ considered with some coalition-free ν-person matrix game. It is specified by ν sets of moves $U^{(j)}$ and the numerical matrices $\left(W_j(u^{(1)}, \ldots, u^{(\nu)})\right)$ whose elements give the average losses of the jth player in the party $\left(u^{(1)}, \ldots, u^{(\nu)}\right)$. Let the players use the constant randomized choice rules, i.e. the jth player chooses his ith move with the probability $p_i^{(j)}$ ($i = 1, \ldots, k_j$). The partners act independently from each other. The average pay of this player is equal to

$$R^{(j)}(\bar{p}) = \sum_{i_1 \ldots i_\nu} W(u_{i_1}, u_{i_2}, \ldots, u_{i_\nu}) \prod_{l=1}^{\nu} p_{i_l}^{(l)},$$

$$\bar{p} = \left(\bar{p}^{(1)}, \ldots, \bar{p}^{(\nu)}\right), \quad \bar{p}^{(j)} = \left(p_1^{(j)}, \ldots, p_{k_j}^{(i)}\right).$$

We denote this game by Γ.

The main result is the fact that Γ will have at least one Nash point (equilibrium party) provided all participants use the invariable stationary randomized choice rules. The inequalities to define the Nash points in terms of the game Γ take the form

$$R^{(j)}(\bar{p}_*) \le R^{(j)}(p_*^{(1)}, \ldots, p_*^{(j-1)}, \bar{p}^{(j)}, p_*^{(j+1)}, \ldots, p_*^{(\nu)}), \ \forall\, j.$$

So, we consider the game control problem for the vector HPIV (in extended sense). Inequality (1) serves as the aim of control. The required strategies of the players will be constructed in the form of gradient type recurrent procedures (see Sec. 4, Chap. 3). We shall use the probabilistic notation for the gradient of the average losses[b]

$$\nabla_{\bar{p}^{(j)}} R^{(j)}(\bar{p}) = \mathbf{E}\left\{ \frac{x_t^{(j)}}{e^T(u_t^{(j)})\bar{p}_t^{(j)}} e(u_t^{(j)}) \big| \bar{p}_t = \bar{p} \right\}$$

to express explicitly the strategies in terms of observable variables

$$\bar{p}_{t+1}^{(j)} = \pi(\Sigma_{k_j}(\varepsilon_{t+1}))\left[\bar{p}_t^{(j)} - a(t)\frac{x_t^{(j)}}{e^T(u_t^{(j)})\bar{p}_t^{(j)}} e(u_t^{(j)}) \right], \quad \bar{p}_1^{(j)} \in \Sigma_{k_i}(\varepsilon), \quad j = 1, \ldots, \nu$$

$$(2)$$

where $\pi(\Sigma)$ denotes projecting on the simplex

$$\Sigma_{k_j}(\varepsilon) = \left\{ \bar{p} : \bar{p}_i \ge \varepsilon, \sum_{i=1}^{k_j} p_i = 1 \right\}.$$

For each j the sequence $\bar{p}_1^{(j)}, \bar{p}_2^{(j)}, \ldots, \bar{p}_t^{(j)}, \ldots$ consists of randomized rules formed in the course of the control process. They form the strategy of the jth player $\sigma_0^{(j)} = \{\bar{p}_t^{(j)}, t \ge 1\}$. We should show that these strategies ensure the attainability of the aim (1), i.e. if the j_0th player uses instead of the strategy $\sigma_0^{(j_0)}$ another stationary program strategy then his losses will not decrease.

In procedure (2) the sequences $a(t)$ and ε_t, just as in the previous recurrent procedures, cannot be arbitrary. We shall use the following condition:

Condition C. $0 < a(t+1) \le a(t)$, $\varepsilon \in (0, \max_i k_i^{-1})$, $\lim_{t \to \infty} ta(t) = \infty$

$$\lim_{t \to \infty} (\varepsilon_t + a^{-1}(t)|\varepsilon_t - \varepsilon_{t+1}|) = 0, \quad \sum_{t=1}^{\infty} (t\varepsilon_t)^{-1}a(t) < \infty.$$

---

[b]Recall that $e(u) = \sum_{i=1}^{k} e_i \chi(u = u_i)$ where $u = (u_1, \ldots, u_k)$ and $e_i = (0, \ldots 0, \underbrace{1}, 0, \ldots, 0)$.
$\underbrace{}_{i-1}$

The main result of this section is the following theorem.

Theorem 1. *With respect to the class of vector (ν-dimensional) HPIV (understood in the extended sense) procedure (2) with positive initial vector and parameters satisfying condition* **C** *secures the fulfilment of the following inequalities*

$$\lim_{t\to\infty}\left[V_t^{(j)}\big(\sigma_0^{(1)},\ldots,\sigma_0^{(j)},\ldots,\sigma_0^{(\nu)}\big) - V_t^{(j)}\big(\sigma_0^{(1)},\ldots,\tilde{\sigma}_0^{(j)},\ldots,\sigma_0^{(\nu)}\big)\right] \leq 0$$

for any stationary program strategy $\tilde{\sigma}^{(j)}$ and $j = 1,\ldots,\nu$.

Proof. The proof will be divided into two steps. First we prove that

$$\left.\begin{aligned}
\lim_{t\to\infty}\left[V_t^{(j)}\big(\sigma_0^{(1)},\ldots,\sigma_0^{(\nu)}\big) - t^{-1}\sum_{n=1}^{t}R^{(j)}(\bar{p}_n)\right] &= 0, \\[2ex]
\lim_{t\to\infty}\left[V_t^{(j)}\big(\sigma_0^{(1)},\ldots,\tilde{\sigma}^{(j)},\ldots,\sigma_0^{(\nu)}\big) - t^{-1}\sum_{n=1}^{t}w^{(j)}\big(\bar{p}_n,\tilde{p}^{(j)}\big)\right] &= 0
\end{aligned}\right\} \tag{3}$$

a.s. for all $j = 1,\ldots,\nu$ where

$$w^{(j)}(\bar{p},\tilde{p}) = R^{(j)}\big(p^{(1)},\ldots,p^{(j-1)},\tilde{p},p^{(j+1)},\ldots,p^{(\nu)}\big).$$

The values $w^{(j)}$ denote the losses of the jth player which have deviated from the "Nash party". In the second step we shall check (for $j = 1,\ldots,\nu$ and any stationary rules $\tilde{p}^{(j)}$ forming the strategy $\tilde{\sigma}^{(j)}$) the fulfillment of the following inequalities

$$\varlimsup_{t\to\infty} t^{-1}\sum_{n=1}^{t}\left[R^{(j)}(\bar{p}_n) - w^{(j)}\big(\bar{p}_n,\tilde{p}^{(j)}\big)\right] \leq 0, \quad \text{a.s.} \tag{4}$$

which demonstrate that a deviation from the "Nash party" leads to increasing the losses of the jth player.

Evidently, from (3) and (4) the required assertion follows:

$$\varlimsup_{t\to\infty}\left[V_t^{(j)}\big(\sigma_0^{(1)},\ldots,\sigma_0^{(\nu)}\big) - V_t^{(j)}\big(\sigma_0^{(1)},\ldots,\tilde{\sigma}^{(j)},\ldots,\sigma_0^{(\nu)}\big)\right]$$

$$= \lim_{t\to\infty} t^{-1}\sum_{n=1}^{t}\left[R^{(j)}(\bar{p}_n) - w^{(j)}\big(\bar{p}_n,\tilde{p}^{(j)}\big)\right] \leq 0$$

for all $\tilde{p}^{(j)} \in \sum_{k_j}$ and $j = 1,\ldots,\nu$.

So, it remains to prove (3) and (4). The arguments will be presented briefly since they are similar to those of Sec. 5, Chap. 2 and consist in reducing the original problem to the martingale lemma. Leaving all details to the reader we only give the general description of this idea. Let us consider the first equality in (3). We put

$$L_t = \left(V_t^{(j)}\big(\sigma_0^{(1)},\ldots,\sigma_0^{(\nu)}\big) - t^{-1}\sum_{n=1}^{t}R^{(j)}(\bar{p}_n)\right)^2$$

and introduce the flow of the σ-algebras $\mathcal{F}_t = \sigma\{\tilde{u}_n^{(j)}, x_n^{(i)}(u_{t-1}), i \neq j, x_n^{(j)}(u_n^{(j)}), n = 1, \ldots, t\}$. The elementary arguments lead to the inequalities (for all t)

$$\mathbf{E}(L_t|\mathcal{F}_{t-1}) \leq (1 - t^{-1})L_{t-1} + t^{-1}c_1 + t^{-2}c_2, \quad c_1, c_2 > 0.$$

By the martingale lemma, it implies $L_t \to 0$ a.s. Hence, the first inequality in (3) is proved.

The proof of (4) is more difficult. We introduce the vectors $q_t^{(j)} = (1 - \varepsilon_t k_j)\bar{p}^{(j)} + \varepsilon_t e$ (where $e = (1, 1, \ldots, 1)$). According to the procedure (2), for any $t\ (= 1, 2, \ldots)$ we have (the parentheses $(\cdot, \cdot)$ means the scalar product)

$$\left\|p_{t+1}^{(j)} - q_{t+1}^{(j)}\right\|^2 \leq \left\|\bar{p}_t^{(j)} - q_t^{(j)}\right\|^2 + c_1|\varepsilon_t - \varepsilon_{t+1}| - 2a(t)\left(p_t^{(j)} - q_{t+1}^{(j)}, \frac{x_t^{(j)} e\left(u_t^{(j)}\right)}{e^T\left(u_t^{(j)}\right)p_t^{(j)}}\right)$$

$$+ a^2(t)\left(\frac{x_t^{(j)}}{e^T(u_t^j)p_t^j}\right)^2$$

$$= \left\|p_t^{(j)} - q_t^{(j)}\right\|^2 + c_1|\varepsilon_t - \varepsilon_{t+1}| + c_2 a(t)\varepsilon_{t+1}$$
$$- 2a(t)\left[R^{(j)}(\bar{p}_t) - w^{(j)}(p_t, q_{t+1}^j)\right] + 2a(t)\gamma_t' + a^2(t)\gamma_t'',$$

where we have used the following notation

$$\gamma_t' = R^{(j)}(\bar{p}_t) - w^{(j)}(\bar{p}_t, q_{t+1}^{(j)}) - \left(\bar{p}_t^{(j)} - q_{t+1}^{(j)}, \frac{x_t^{(j)} e\left(u_t^{(j)}\right)}{e^T\left(u_t^{(j)}\right)p_t^{(j)}}\right),$$

$$\gamma_t'' = \left(\frac{x_t^{(j)}}{e^T(u_t^j)p_t^j}\right)^2 \geq 0.$$

Hence

$$n^{-1}\sum_{t=1}^{n}\left[R^{(j)}(\bar{p}_t) - w^{(j)}\left(\bar{p}_t, q_t^{(j)}\right)\right] \leq n^{-1}\sum_{t=1}^{n}\frac{\gamma_t}{a(t)} + n^{-1}\sum_{t=1}^{n}a(t)\gamma_t''$$

$$+ c_3 n^{-1}\sum_{t=1}^{n}\left(\varepsilon_{t+1} + \frac{|\varepsilon_t - \varepsilon_{t+1}|}{a(t)}\right), \quad (5)$$

where

$$\gamma_t = \left\|p_t^{(j)} - q_t^{(j)}\right\|^2 - \left\|p_{t+1}^{(j)} - q_{t+1}^{(j)}\right\|^2 + 2a(t)\gamma_t'.$$

According to the conditions of the theorem the last term on the right-hand side of (5) vanishes as $n \to \infty$ and

$$\sum_{t=1}^{\infty}\mathbf{E}(a(t)\gamma_t')^2 < \infty, \quad \sum_{t=1}^{\infty}t^{-1}a(t)\mathbf{E}\gamma_t'' < \infty.$$

From this it follows that the following series of r.v. $\sum_{t=1}^{\infty}a(t)\gamma_t'$, $\sum_{t=1}^{\infty}t^{-1}a(t)\gamma_t''$ converge a.s. Hence, $\sup_n|\sum_{t=1}^{n}\gamma_t| < \infty$ a.s. This implies that the right-hand side of (5) tends to zero. Indeed, using the lemma on the limit (see Sec. 6, Chap. 5), the

Toeplitz and Kronecker Theorems[c] from the convergence of these series it follows that the right-hand side of (5) approaches zero. This implies inequality (4). $\qquad\square$

Thus, procedure (2) secures the attainability of the aim for the HPIV but the convergence of the rules $\bar{p}_t^{(j)}$ (as $t \to \infty$) is not guaranteed.

Let us clear up the convergence rate in (1) if the parameters $a(t)$, ε_t are taken as follows

$$a(t) = \frac{a}{(t + f)^\alpha}, \quad \varepsilon_t = \frac{e}{(t + g)^\varepsilon}.$$

First of all, for the condition **C** to be satisfied it must be $0 < \varepsilon < \alpha < 1$. On the more comprehensive analysis we arrive at the conclusion that in the inequalities

$$V_t^{(j)}\big(\sigma_0^{(1)}, \ldots, \sigma_0^{(\nu)}\big) - V_t^{(j)} \asymp \frac{1}{n^{\theta - \delta}}$$

(where $\delta > 0$ and symbol $\asymp$ has been defined before Theorem 2 in Sec. 3 Chap. 2) the quantity $\theta = \min(\varepsilon, \alpha - \varepsilon, 1 - \alpha)$. It is not difficult to verify that $\theta < 1/3$, the equality taking place only at $\alpha = 2/3$, $\varepsilon = 1/3$.

4.4. Games of Automata

The previous sections of this chapter have been devoted to constructing and investigating adaptive strategies for minimax control problems. Such an approach may be called direct. Sometimes the opposite situation arises: there is a system of control (strategy) and a class of controlled processes. It is necessary to find what aims of control can be attained by this system on a given class. Here we would like to consider the inverse problem, namely, the analysis of the decentralized system of control in the form of a direct product of finite automata, i.e. games of automata provided the vector HPIV having binary components are under control. Let us turn to the accurate statement of the problem.

We consider a class $Q(\nu, k)$ of HPIV ξ_t with independent components. Their values belong to a finite interval $[a, b]$ and $U = U^{(1)} \times \cdots \times U^{(\nu)}$ is the space of controls, all factors $U^{(j)} = \{u_1^{(j)}, \ldots, u_{k_j}^{(j)}\}$ being finite sets. Using the quantum technique described above, i.e. passing from the process with continuous value space to the binary one we shall assume from now on that the components of HPIV ξ_t are binary. If a control u is used these will take the value to be equal to one (gain) with

---

[c]**Theorem.** (Toeplitz) *Let the sequences $a_n \geq 0$, x_n be such that $0 < b_n = \sum_{i=1}^n a_i \to \infty$ and $x_n \to x_0$ as $n \to \infty$. Then*

$$\frac{1}{b_n} \sum_{i=1}^n a_i x_i \xrightarrow[n \to \infty]{} x_0.$$

Theorem. (Kronecker) *If the sequences $b_n \geq 0$, x_n are such that $b_n \leq b_{n+1} \to \infty$ as $n \to \infty$ and $\sum_{n=1}^\infty x_n < \infty$ then $b_n^{-1} \sum_{i=1}^n b_i x_i \xrightarrow[n \to \infty]{} 0$.*

the probability $q_j(u) = \mathbf{P}\{x^{(j)} = 1|u\} = W_j(u)$ and that to equal to zero with the probability $p_j(u) = \mathbf{P}\{x^{(j)} = 0|u\} = 1 - q_j(u)$ for the jth component respectively.

As the strategies for the processes ξ_t we choose the direct product of ν automata from some ε-optimal family (see Sec. 3, Chap. 2) $\mathcal{A}^{(n)} = \mathcal{A}_1^{(n)} \times \cdots \times \mathcal{A}_\nu^{(n)}$, all *automaton-factors having the same memory depth n.* We agree to use only automata $\mathcal{D}_{k,n}$ and $\mathcal{Q}_{k,n}$ with cyclic and equiprobable change rules of actions (or transition from branch to branch) and their modifications. The latter means that when changing a branch the automaton passes not into the first state s_1 but into the last one s_n having the depth n. So, the considered class $\mathcal{K}$ of automata is formed by using four subclasses $\mathcal{K} = \mathcal{K}_1 \cup \mathcal{K}_2 \cup \mathcal{K}_3 \cup \mathcal{K}_4$ where $\mathcal{K}_1$ consists of the modifications of automata $\mathcal{D}_{k,n}$ and $\mathcal{Q}_{k,n}$ with cyclic change of branches, $\mathcal{K}_2$ of the modifications of the same automata with equiprobable change of branches, $\mathcal{K}_3$ is represented by the automata $\mathcal{D}_{k,n}$ and $\mathcal{Q}_{k,n}$ with cyclic change of branches and $\mathcal{K}_4$ consists of the same automata with equiprobable change of branches.

The family of automata $\mathcal{D}_{k,n}$ is ε-optimal with respect to the class of all binary HPIV but for the collection $\mathcal{Q}_{k,n}$ this is not the case. Generally speaking, there exists a constant $c(\mathcal{A}^{(n)}) \geq 0$ such that the automata from $\mathcal{A}^{(n)}$ form the ε-optimal family with respect to the class of binary HPIV for which $\max_i q_i > c(\mathcal{A}^{(n)})$. According to the notation from Sec. 3, Chap. 2 we have

$$
c(\mathcal{Q}_{k,n}) = \begin{cases} \dfrac{1/2 - p_+}{p_- - p_+}, & \text{if } p_+ < 1/2,\ p_- > 1/2, \\[2mm] 0, & \text{if } p_- > 1/2,\ p_+ + q_- < 1. \end{cases}
$$

For these definite types of the automata we know that $c(\mathcal{L}_{k,n}) = 1/2$ and $c(\mathcal{K}_{k,n}) = 0$. We shall also assume that $c(\mathcal{D}_{k,n}) = 0$. We impose the following constraint on the automata forming the considered learning systems $\mathcal{A}^{(n)} = \mathcal{A}_1^{(n)} \times \cdots \times \mathcal{A}_\nu^{(n)}$ and on the corresponding HPIV

$$
c(\mathcal{A}_i^{(n)}) < W_i(u) < 1, \quad \forall\, i = 1, \ldots, \nu,\ u \in U. \tag{1}
$$

As the investigation tool for the automaton games, i.e. for the pairs $(\xi_t, \mathcal{A}^{(n)})$, we shall take the associated Markov chains $M_n = (S^{(n)}, \mathcal{P}^{(n)})$ having the direct product of the state sets of automata as their state space $S^{(n)} = S_1^{(n)} \times \cdots \times S_\nu^{(n)}$. The elements of the transition matrix $\mathcal{P}^{(n)} = (p_{\bar\lambda\bar\mu})$ where $\bar\lambda$ and $\bar\mu$ are indices of the states $s_{\bar\lambda} = (s_{\lambda_1}, \ldots, s_{\lambda_{\bar\nu}})$ and $s_{\bar\mu} = (s_{\mu_1}, \ldots, s_{\mu_{\bar\nu}})$ from $S^{(n)}$ are calculated according to the formula

$$
p_{\bar\lambda\bar\mu} = \prod_{i=1}^{\nu} \left[q_i(u)p_{\lambda_i\mu_i}(1) + p_i(u)p_{\lambda_i\mu_i}(0)\right].
$$

Here the following notation has been used: $\bar u = \left(u^{(1)}, \ldots, u^{(\nu)}\right)$ is an output signal of the automaton $\mathcal{A}^{(n)}$ in the state $s_{\bar\lambda}$, ($u^{(j)}$ is an output signal of the automaton $\mathcal{A}_j^{(n)}$ in the state s_{λ_j}) and $p_{\lambda_i\mu_i}(0)$, $p_{\lambda_i\mu_i}(1)$ are the elements of the transition probability matrix of the automaton $\mathcal{A}_i^{(n)}$ under the input signals 0 and 1 respectively. The chain M_n has $k_1 k_2 \ldots k_\nu n^\nu$ states.

The assumed constraint (1) guarantees the regularity of the chains M_n. Therefore there exist limiting states probabilities $\pi(s, n) > 0$. In view of the Moor property of automata, the set $S^{(n)}$ may be decomposed into a sum of disjoint subsets $S^{(n)}(u)$, the control u corresponds to every one of them, i.e. $S^{(n)} = \bigcup_u S^{(n)}(u)$. This implies that there exist limiting probabilities (their total number is equal to $\varkappa = k_1 \cdots k_\nu$) of the actions of the automata

$$\pi(\bar{u}, n) = \sum_{s \in S^{(n)}(u)} \pi(s, u).$$

Definition 1. The quantity

$$W_j^{(n)}(\bar{u}) = \sum_{u \in U} W_j(\bar{u}) \pi(\bar{u}, n)$$

is called the *limiting gain* of the jth participant of the game $\mathcal{A}_j^{(n)}$.

We have presented characteristics of classes of controlled objects ξ_t and strategies ΣK as well as the objective vector $\left(W_1^{(n)}(\bar{u}), \ldots, W_\nu^{(n)}(\bar{u})\right)$. The problem consists of finding the aims which can be attained by the strategies from ΣK provided the process ξ_t is under control. Hence, it is necessary to ascertain the asymptotic properties of the objective vector as $n \to \infty$. The form of the limiting gains shows that the problem consists of calculating $\lim_{n \to \infty} \pi(\bar{u}, n)$ or in estimating these limits. Hence the analysis of automata games consists of investigating the sequences of Markov chains M_n. The difficulties connected with unlimited increase of the number of states of these chains will be overcome by means of auxiliary Markov chains which are asymptotically equivalent to the original ones but have a more simple structure. We shall now describe the construction of these chains.

Let $M = (S, \mathcal{P})$ be a Markov chain and $S' \subseteq S$. Let us also consider the states of the chain at the moments of entering the set S'. They form a new chain $M' = (S', \mathcal{P}')$ so-called *restriction* of the chain M to the subset S'. One step in this new chain corresponds to several steps of the original chain M between two hittings into S'. It can be proved that if the chain M is regular then the chain M' will be regular too and the final probabilities of both chains will be related as follows (the upper index points to the chain (M or M') that the appropriate notation refers to)

$$\pi^{M'}(s) = \frac{\pi^M(s)}{\sum_{s_i \in S'} \pi^M(s_i)}, \quad s \in S'. \tag{2}$$

In particular, for any pair of the states $s_i, s_j \in S'$ we have

$$\frac{\pi^M(s_i)}{\pi^M(s_j)} = \frac{\pi^{M'}(s_i)}{\pi^{M'}(s_j)}.$$

Let us define two restrictions of the chain M_n. The first, M_n', is the restriction of M_n to the set $(s_1, \ldots, s_\varkappa)$, $\varkappa = k_1 \cdot k_2 \cdots \cdot k_\nu$ of the "deep states". Each subset

$S^{(n)}(u)$ is represented by the states corresponding to the maximum depth n of each automaton on the branch.

Another restriction, F_n, is to the set $s_l \cup s \cup \Gamma_l$ where s_l is the "deep" state of the subset $S^{(n)}(u_l)$, $s \in S^{(n)}(u)$ is arbitrary and Γ_l is the set of boundary states of $S^{(n)}(u_l)$, i.e. the states from which it is possible to leave this set.

We turn to estimating the transitions and limiting probabilities for these constructions. We shall begin with the subchain F_n. First, we shall make one general remark on the proofs of the subsequent lemmas and theorems. These proofs will become more simple, without losing generality, if we assume that only automata $\mathcal{D}_{k,n}$ and $\mathcal{L}_{k,n}$ take part in the game. Indeed, the chain M_n remains the same if the ith automaton $\mathcal{Q}_{k,n}$ having the parameters (p_+, p_-) and the average gain is replaced with $\mathcal{L}_{k,n}$ and $\tilde{W}_j(u) = p_+ W_j(u) + p_-(1 - W_j(u))$ simultaneously. The letters $c_1, c_2, \ldots$ mean positive constants not depending on n and s.

Lemma 1. *There exists a constant $c_1 > 0$ such that*

$$P^{F_n}_{ss_l} \geq c_1.$$

Proof. Obviously, if all participants of the game are represented by automata $\mathcal{D}_{k,n}$ then one time they will get into the state s_l provided all automata receive the gain at the same time. In this case $c_1 = \prod_{i=1}^{\nu} W_i(u)$. So, we can assume that only the automata $\mathcal{L}_{k,n}$ take part in the game.[d] We now introduce some notation.

Let $g_i(t)$ be a depth of the state of the automaton $\mathcal{A}_i$ at the moment t. Then

$$Q^i_t(t_1, t_2; , l_1, l_2, l_3) \stackrel{\text{def}}{=} \mathbf{P}\{g_i(t_2) = l_3; g_i(t) > l_2, t_1 < t < t_2 | g_i(t_1) = l_1\},$$

$$Q^i_2(t_1, t_2; l) \stackrel{\text{def}}{=} \mathbf{P}\{g_i(t) > l, t_1 < t \leq t_2 | g_i(t_1) = n\}.$$

The following relations are obvious.

$$Q^i_2(t_1, t_2; h) = 1 - \sum_{t=t_1+1}^{t_2} Q^i_1(t_1, t; n, h, h),$$

$$Q^i_2(t_1, t_2; h) = \sum_{l=h+1}^{n} Q^i_1(t_1, t_2; n, h, l).$$

Next, let $\tau_i(l) > 0$ be the first hitting moment of the automaton into either the state with depth $\leq l$ (not taking into account the moment $t = 0$) or into the state with depth n provided the initial state has depth l. The event $B_i(l)$ means that the transition of $\mathcal{A}_i$ from the initial state with depth l into some state with depth n occurs before hitting it into the set of states with depth $\leq l$ (except for the moment $t = 0$ again).

---

[d]The general case requires some insignificant additions. We will not consider them here.

Let μ be such that $l_\mu = \min_{1 \le i \le \nu} l_i$. For any $T > 0$ we have

$$
\begin{aligned}
P_{ss_l}^{F_n} &= \sum_{m=1}^{\infty} \mathbf{P}\{s(m) = s_l, s \notin \Gamma_l, s(t) \neq s, 0 < t < m | s(0) = s\} \\
&\ge \mathbf{P}\{s(N) = s_l, s(t) \notin \Gamma_l, s(t) \neq s \text{ for } 0 < t < T | s(0) = s\} \\
&\ge \prod_{i=1}^{\nu} Q_1^l(0, T; l_i, h_i, n),
\end{aligned}
\tag{3}
$$

where $h_i = 1$, if $i \neq \mu$ and $h_\mu = l_\mu$.

We shall consider the probabilities Q_1^l in detail.

The change of states by each automaton in the course of the game is a random walk over the points $1, 2, \ldots, n$ with a reflecting screen at the point n. Hence, we can use the results obtained for such a walk (or, in other words, the solution of the ruin problem in an inoffensive game) to estimate the quantities $\mathbf{P}\{B_j(l_j)\}$ in the following inequalities $(j = 1, \ldots, \nu)$

$$
Q_1^j(0, T; l_j, h_j, n) \ge \mathbf{P}\{B_j(l_j)\} \sum_{t=1}^{T} \mathbf{P}\{\tau_j(l_j) = t | B_j(l_j)\} Q_1^j(t, T; n, h_j, n).
\tag{4}
$$

For any l_j, according to the solution of the ruin problem, we have

$$
\mathbf{P}\{B_j(l_j)\} = q_j \frac{1 - p_j q_j^{-1}}{1 - (p_j/q_j)^{n-l_j}} \ge q_j - p_j.
\tag{5}
$$

For $h_j \le m \le n$ the combinatorial arguments imply the following inequalities

$$
Q_1^j(t_1, t_2; n, h_j, m) \le \left(\frac{p_i}{q_j}\right)^{n-m} Q_i^j(t_1, t_2; n, h_j, n).
$$

Summing them over m from $h_j + 1$ to n we obtain

$$
\begin{aligned}
Q_1^j(t_1, t_2; n, h_j, n) &\ge \left(1 - \frac{p_j}{q_j}\right) \sum_{m=h_j+1}^{n} Q_1^j(t_1, t_2; n, h_j, m) \\
&= \left(1 - \frac{p_j}{q_j}\right) Q_2^j(t_1, t_2; h_j).
\end{aligned}
\tag{6}
$$

According to the properties of the function Q_1^j obtained we arrive at the inequality

$$
\begin{aligned}
Q_2^j(t_1, t_2; h_j) &\ge 1 - \left(\frac{p_j}{q_j}\right)^{n-h_j} \sum_{t=t_1+1}^{t_2} Q_1^j(t_1, t_2; n, h_j, n) \\
&\ge 1 - (t_2 - t_1) \left(\frac{p_j}{q_j}\right)^{n-t_j}.
\end{aligned}
\tag{7}
$$

We put

$$
a = \max_i \frac{2}{(q_i - p_i)^2}, \qquad b = \max_i \frac{p_i}{q_i}.
$$

Let l_0 be the minimal integer such that $axb^x < 1/2$ for $x > l_0$. Such l_0 exists since $b < 1$. We set $L_0 \subset S^{(n)}(u_L)$ as the subset of states such that the depths of all states of automaton are greater than $n - l_0$ for $s \in L_0$. For $s \in L_0$ we also obtain the estimate $P_{ss_l}^{F_n} \geq \prod_{i=1}^{\nu} q_i^{l_0}$. Otherwise, in (3) we put $T =]a(n - l_\mu)[$ (where $]a[= [a] + 1$). Then for $t \in [1, T)$ we have

$$(T - t) \left(\frac{p_j}{q_j} \right)^{n - h_j} < 1/2, \quad j = 1, \ldots, \nu.$$

Together with (6) and (7) this implies

$$Q_1^j(t, T; n, h_j, n) \geq 1 - \frac{p_j}{q_j}.$$

Substituting this estimation and (5) in (4) we obtain the inequality

$$Q_1^j(0, N; l_j, h_j, n) \geq \frac{(q_j - p_j)^2}{2q_j} \sum_{t=1}^{N} \mathbf{P}\{\tau_j(l_j) = t | B_j(l_j)\}$$

$$= \frac{(q_j - p_j)^2}{2q_j} \mathbf{P}\{\tau_j(l_j) \leq T | B_j(l_j)\}.$$

We shall use the inequality

$$\mathbf{P}\{\tau_j(l_j) \leq T | B_j(l_j)\} \geq 1 - T^{-1} \mathbf{E}(\tau_j(l_j) | B_j(l_j)) \tag{8}$$

to estimate the probabilities on the right-hand side of the previous inequality. In view of the identity

$$\mathbf{E}\tau_j(l_j) = \mathbf{P}\{B_j(l_j)\} \mathbf{E}(\tau_j(l_j) | B_j(l_j)) + \mathbf{P}\{\bar{B}_j(l_j)\} \mathbf{E}(\tau_j(l_j) | \bar{B}_j(l_j)),$$

we have

$$\mathbf{E}(\tau_j(l_j) | B_j(l_j)) \leq \frac{\mathbf{E}(\tau_j(l_j))}{\mathbf{P}\{B_j(l_j)\}}.$$

Using again the solution of the ruin problem we obtain

$$\mathbf{E}(\tau_j(l_j)) = \frac{n - l_j}{q_j - p_j} \frac{1 - p_j/q_j}{1 - (p_j/q_j)^{n - l_j}} - \frac{1}{p_j - q_j} \leq \frac{n - l_j}{q_j - p_j}.$$

Substituting this estimation in (8) and (5), we finally obtain the estimation for the right-hand side of (4)

$$\mathbf{P}\{\tau_j(l_j) \leq N | B_j(l_j)\} \geq 1 - \frac{1}{a(q_j - p_j)^2} \geq \frac{1}{2}.$$

This leads to the following

$$Q_1^j(0, T; l_j, h_j, n) \geq \frac{(q_j - p_j)^2}{4q_j}.$$

Thus, for all $s \in L_0$, according to (3), we have

$$P_{ss_l}^{F_n} \geq \prod_{j=1}^{\nu} \frac{(q_j - p_j)^2}{4q_j}.$$

It remains to write down the value of the constant c_1:

$$c_1 = \min \left(\prod_{i=1}^{\nu} q_i^{l_0}, \prod_{j=1}^{\nu} \frac{(q_j - p_j)^2}{4 q_j} \right).$$

This completes the proof of Lemma 1. $\qquad\qquad\qquad\qquad\qquad\qquad\qquad$ $\square$

Let a binary HPIV be controlled by the automaton $\mathcal{A}$. We suppose that we obtain the unit gain with probability q in reply to the action u. Also, let $T(\mathcal{A}, q)$ be the average operating time of the action u.

Definition 2. The quantity

$$\lambda_j(\bar{u}) = \lim_{n \to \infty} \left[T(\mathcal{A}_j^{(n)}, W_j(\bar{u})) \right]^{1/n}$$

is called the *reduced gain* of the sequence of automata $\{\mathcal{A}_i^{(n)}\}$ under the control u.

The function $\lambda_j(\bar{u})$ characterizes both the process and the controlling automaton. However, it also increases monotonically with $W_j(\bar{u})$. The restrictions on automata stated above can be written as follows

$$\lambda_j(u) < 1 \quad \text{for all } u \quad \text{and} \quad j = 1, \ldots \nu.$$

The reduced gains exist for all automata from the class $\mathcal{K}$. They are equal to

$$\lambda_i(\bar{u}) = (1 - W_i(\bar{u}))^{-1} \qquad\qquad \text{for } \{\mathcal{D}_{k,n}\},$$
$$\lambda_i(\bar{u}) = W_i(\bar{u})(1 - W_i(\bar{u}))^{-1} \qquad \text{for } \{\mathcal{L}_{k,n}\},$$
$$\lambda_i(\bar{u}) = (1 + W_i(\bar{u}))(1 - W_i(\bar{u}))^{-1} \quad \text{for } \{\mathcal{K}_{k,n}\}.$$

Here is the first application.

Lemma 2. *There exist constants c', $c'' > 0$, not depending on n and s, such that*

$$c' \prod_{i=1}^{\nu} [\lambda_i(\bar{u}_l)]^{l_j - n} \le p_{s_l s}^{F_n} \le c'' \prod_{i=1}^{\nu} [\lambda_i(\bar{u}_l)]^{l_j - n}$$

where $l_1, \ldots, l_\nu$ are the depths of the automata $\mathcal{A}_1, \ldots, \mathcal{A}_\nu$ corresponding to the state s.

Proof. Without restricting attention to identical automata we first obtain the upper estimate. To achieve this, we change the transition functions of all automata: if the output signal is equal to zero, the states having the "unit" depth pass into themselves. We denote the Markov chain corresponding to such a new game and its restrictions to the sets $S^{(n)}(\bar{u})$ and $s_l \cup s \cup \Gamma_l$ by M_n^* and $M_{n,\bar{u}}^*$, F_n^* respectively. Obviously, $p_{s_l s}^{\mathcal{F}_n} = p_{s_l s}^{F_n^*}$. Since F_n^* is the restriction of the chain $M_{n,\bar{u}}^*$ we have

$$p_{s_l s}^{F_n} = p_{s_l s}^{F_n^*} \le \frac{\pi^{F_n^*}(s)}{\pi^{F_n^*}(s_l)} = \frac{\pi^{M_{n,\bar{u}}^*}(s)}{\pi^{M_{n,\bar{u}}^*}(s_l)}. \qquad (9)$$

We estimate the limiting distribution of the chain $M_{n,\bar{u}}^*$. Its components $s(t)$ change independently since this chain is formed by ν non-interacting chains each of which characterizes one automaton. Let $\pi_i^{(n)}(l)$ be the limiting probability of the state with depth $l(1 < l < n)$ for the automaton $\mathcal{A}_i^{(n)}$ in such chains. Therefore

$$\pi^{M_{n,\bar{u}}^*}(s_l) = \prod_{j=1}^{\nu} \pi_j^{(n)}(n), \quad \pi^{M_{n,\bar{u}}^*}(s) = \prod_{j=1}^{\nu} \pi_j^{(n)}(l_j).$$

As applied to the automata $\mathcal{L}_{k,n}$ such chains represent a random walk over the points $1, 2, \ldots, n$ with the reflecting screens at the point 1 and n. The limiting probabilities can be calculated as follows

$$\pi_i^{(n)} = \frac{1 - p_i/q_i}{1 - (p_i/q_i)^n} \left(\frac{p_i}{q_i}\right)^{n-1}.$$

For the automata $\mathcal{D}_{k,n}$ we have

$$\pi_i^{(n)}(l) = \begin{cases} q_i p_i^{n-l_i}, & 2 \le l \le n, \\ p_i^n, & l = 1. \end{cases}$$

The upper estimate can be obtained by substituting these expressions in (9).

We turn now to the lower estimate containing the constant c' which depends on the construction of the automata. If the collection of automata consists of $\mathcal{D}_{k,n}$ the required inequality will follow from the fact that in the chain M the probabilities of the $s_l \to s$ transitions for $n - \max l_i$ periods are equal to $\prod_{j=1}^{\nu} q_j p_j^{n-l_j}$. If the automata $\mathcal{L}_{k,n}$ are playing, we obtain a different constant (the general case is not considered). By Lemma 1, $p_{ss_l}^{F_n} \ge c_1$, since this probability is the sum of probabilities of trajectories which begin at s, end in s_l, and do not pass through the other states of the chain F_n. We reverse all such trajectories in time. These and only these reversed trajectories form the transitions from s_l to s in the chain F_n. This defines a one-to-one mapping $s_l \longleftrightarrow s$ between the direct and inverse transition. For the reversed trajectory the "gain" (i.e. $x = 1$) corresponds to the "loss" (i.e. $x = 0$) for the direct one. For the direct trajectory the total number of gains of each automaton is greater than that of losses by $n - l_i$ but for the reversed chain the situation is opposite. Hence the probability of the transition along any trajectory from s to s_l is $\prod_{j=1}^{\nu} (q_j/p_j)^{n-l_j}$ times that along the reversed trajectory from s_l to s. Summing over all trajectories we obtain the final estimate

$$p_{s_l s}^{F_n} = p_{ss_l}^{F_n} \prod_{j=1}^{\nu} \left(\frac{q_j}{p_j}\right)^{n-l_j} \ge c'' \prod_{j=1}^{\nu} [\lambda_j(\bar{u}_l)]^{l_j - n}. \qquad \square$$

We now prove one of the central results of automata game theory.

Theorem 1. *There exist constants* $c_2, c_3 > 0$ *such that*

$$c_2 \pi^{M_n'}(s_l) \le \pi(\tilde{u}_l, n) \le c_3 \pi^{M_n'}(s_l)$$

for all $l = 1, \ldots, \varkappa$ *and all* $n \ge 1$.

Proof. We fix l and show that for all n

$$\pi(\bar{u}_l, n) \geq c_l \pi^{M_n}(s_l), \quad c_l > 0. \tag{10}$$

Another form of this inequality is

$$\sum_{s \in S^{(n)}(\bar{u}_l)} \frac{\pi^{M_n}(s)}{\pi^{M_n}(s_l)} \leq c_l.$$

This reduces the proof to obtaining the upper estimate of $\pi^{M_n}(s)/\pi^{M_n}(s_l)$ for all $s \in S^{(n)}(\bar{u}_l)$. According to the inequality $\pi(s_g) \geq \pi(s_g)p_{gg} + \pi(s_h)p_{hg}$ we have

$$\frac{\pi(s_h)}{\pi(s_g)} \leq \frac{1 - p_{gg}}{p_{hg}}.$$

Hence

$$\frac{\pi^{M_n}(s)}{\pi^{M_n}(s_l)} = \frac{\pi^{F_n}(s)}{\pi^{F_n}(s_l)} \leq \frac{p_{s_l s}^{F_n} + \sum_{s_i \in \Gamma_l} p_{s_l s_i}^{F_n}}{p_{s s_l}^{F_n}}. \tag{11}$$

The upper estimate from Lemma 2 for $s_i \in \Gamma_l$ leads to the inequality $p_{s_l s_i}^{F_n} \leq cb^{n-1}$. Substituting it into (11), together with the upper estimates from Lemmas 1 and 2, we obtain (10). It is easy to see that $\pi^{M_n}(s_l) \leq \pi(\bar{u}_l, n)$. Summing the inequalities (10) we obtain $\sum_{l=1}^{\varkappa} \pi^{M_n}(s_l) \geq C_2$, where $C_2 = (c'')^{-1}$. Putting $C_3 = \max c_l$ and using (2) we finally obtain the required result. $\square$

Thus, it is possible to obtain good estimates of limiting control probabilities $\pi(\bar{u}, n)$ (as $n \to \infty$) by studying the relatively simple Markov chains M_n' with $\varkappa$ states. However, this requires the transitional probabilities of these chains to be known. We need a new notion.

Definition 3. A control (party) $\bar{u}_l = \left(u_l^{(1)}, \ldots, u_l^{(\nu)}\right)$ is called an $(h_1, \ldots, h_r)$-*neighbour of a control* $\bar{u}_m = \left(u_m^{(1)}, \ldots, u_m^{(\nu)}\right)$ *if:*

(α) for $i = h_1, \ldots, h_r$ either $\bar{u}_l^{(i)} \neq \bar{u}_m^{(i)}$ (for the automata from the classes $\mathcal{K}_2$, $\mathcal{K}_3$, $\mathcal{K}_4$) or $\bar{u}_l^{(i)} - \bar{u}_m^{(i)} = 1 (\mathrm{mod}\ k_i)$ (for the automata from the class $\mathcal{K}_1$);

(β) $\bar{u}_l^{(i)} = \bar{u}_m^{(i)}$ for the other values of i.

When applied to the classes $\mathcal{K}_2$, $\mathcal{K}_3$, $\mathcal{K}_4$ this definition is simplified: the control $\bar{u}_l$ is an $(h_1, \ldots, h_r)$-neighbour of $\bar{u}_m$ if the collections $\bar{u}_l$ and $\bar{u}_m$ differ only by the components with the numbers $h_1, \ldots, h_m$. In this case this relation becomes symmetric.

Lemma 3. *There exist constants* $c_4, c_5 > 0$ *such that*

1. for all n, l *and* m $(\neq l)$

$$p_{lm}^{M_n'} \leq c_4 \min[\lambda_i(\bar{u}_l)]^{-n}; \tag{12}$$

2. *if the control $\bar{u}_l$ is an $(h_1, \ldots, h_r)$-neighbour of $\bar{u}_m$ then for all n*

$$p_{ml}^{M'_n} \geq c_5 \prod_{i=1}^{r} [\lambda_{h_i}(\bar{u}_m)]^{-n}. \tag{13}$$

Proof. We first prove (12). To pass from the deep state s_l (which corresponds to the control $\bar{u}_l$) into the deep state s_m, the chain M_n definitely must visit the set of boundary states Γ_l. For this reason we have

$$p_{lm}^{M'_n} \leq \sum_{s \in \Gamma_l} p_{s_l s}^{F_n}.$$

Using the estimate of probabilities on the right-hand side from Lemma 2 we obtain the inequality

$$p_{lm}^{M'_n} \leq c_6 \sum_{j=1}^{\nu} \frac{1}{\lambda_j(\bar{u}_l) - 1} [\lambda_j(\bar{u}_l)]^{-n}$$

which implies (12).

Let us turn to the inequality (13) and assume for the sake of simplicity that $\bar{u}_i$ is neighbour with $\bar{u}_m$ in the hth component. Let $s_1 \in S^{(n)}(\bar{u}_m)$ be the one-depth state of $\mathcal{A}_h^{(n)}$ and $s_2 \in S^{(n)}(\bar{u}_l)$ have depth n in the other automata. Then

$$p_{ml}^{M'_n} \geq p_{s_m s_1}^{F_n} p_{s_1 s_2}^{F_n} p_{s_2 s_l}^{F_n}.$$

From the estimates of Lemmas 1 and 2 it follows that

$$c_1 \leq p_{s_2 s_l}^{F_n}, \quad c\,[\lambda_h(\bar{u}_m)]^{-n} \leq p_{s_1 s_1}^{F_n}.$$

It remains to estimate the probability of the $s_1 \to s_2$ transition. If $\mathcal{A}_h^{(n)}$ belongs to one of the classes $\mathcal{K}_1$, $\mathcal{K}_2$, $\mathcal{K}_4$ then the one-step probability of this transition in the chain M_n will not be less than $p_h/g_h \prod_{j \neq h} q_j$. But if $\mathcal{A}_h^{(n)} \in \mathcal{K}_3$ and $u_l^{(h)} - u_m^{(h)} = d \pmod{k_h}$ then under d losses of the automaton $\mathcal{A}_h^{(n)}$ and the gains of the others the transition $s_1 \to s_2$ occurs for d steps without visiting both Γ_m and the set of the deep states. The probability of this event is equal to $p_h^d \prod_{j \neq h} q_j^d$. This leads to the desired estimate. $\qquad\square$

The results obtained allow answering the principal questions of automata games. In addition, they provide a possibility to calculate the limiting probabilities of the typical parties and to clear up the asymptotic properties (as $n \to \infty$) of automata. We note the non-identical "quality" of the estimates of the limiting probabilities: in Lemma 2 it has been obtained with an accuracy up to some constant but in Lemma 3 the upper and lower estimates differ from each other with accuracy up to the order. However we shall see later that Theorem 1 allows finding the "limiting parties" for a wide class of games.

Definition 4. A sequence of automata $\mathcal{A}_h^{(n)}$ is called *critical* under the control $\bar{u}$ if $\lambda_h(\bar{u}) = \min_i \lambda_i(\bar{u})$. The quantity $\lambda(\bar{u}) = \lambda_h(\bar{u})$ is called the *critical gain* of the control $\bar{u}$.

For any control it may happen that there exist several critical sequences of automata. All automata from such sequences will be called critical as well.

Definition 5. A control (party) $\bar{u}_m$ is called *maximin* (in the automaton sense) if its critical gain is maximum, i.e. $\lambda(\bar{u}_m) = \max_u \lambda(u)$.

A game of automata can have several maximin controls (parties). The introduced notion of the maximin party differs from the one stated above (Sec. 2). They are formulated in different terms: the previous one — by means of pay functions, the new one — in terms of critical gains. The situations when these notions coincide will be discussed later on.

An oriented graph V_Γ with $\varkappa$ vertices can be associated with an automata game Γ: an edge will lead from the vertex i to the vertex j if the control (party) $\bar{u}_j$ is an h-neighbour (at some $i = 1, \ldots, \nu$) with $\bar{u}_i$ and the sequence of automata $\mathcal{A}_h^{(n)}$ is critical under the control $\bar{u}_i$.

Lemma 4. *If a vertex j of the graph V_Γ is attainable from a vertex i and $\lambda(\bar{u}_j) > \lambda(\bar{u}_i)$ then*

$$\frac{\pi(\bar{u}_i, n)}{\pi(\bar{u}_j, n)} \leq c \left[\frac{\lambda(\bar{u}_i)}{\lambda(\bar{u}_j)}\right]^n, \quad c > 0.$$

Proof. Let us introduce the sequence of Markov chains M_n'' with the same state spaces as M_n' and transition probabilities

$$p_{j_1 j_2}^{M''} = \begin{cases} \dfrac{p_{j_1 j_2}^{M'}}{1 - p_{j_1 j_2}^{M'}}, & \text{if } j_1 \neq j_2, \\[12pt] 0, & \text{if } j_1 = j_2. \end{cases}$$

The limiting state probabilities of the chains M_n'' and M_n' are related by

$$\frac{\pi^{M_n'}(s_i)}{\pi^{M_n'}(s_j)} = \frac{\pi^{M_n''}(s_i)\left(1 - p_{jj}^{M_n'}\right)}{\pi^{M_n''}(s_j)\left(1 - p_{ii}^{M_n'}\right)}. \tag{14}$$

By Lemma 3

$$c_6 \lambda^{-n}(\bar{u}_i) \leq 1 - p_{ii}^{M_n'} \leq c_7 \lambda^{-n}(\bar{u}_i), \qquad c_6, c_7 > 0 \tag{15}$$

for all t and n. If the graph V_Γ has an edge from i_1 to i_2 then by (14) and (15) $p_{i_1 i_2}^{M_n''} \geq c_4/c_7 = c$. Hence $\pi^{M_n''}(s_{i_2}) \geq c\pi^{M_n''}(s_{i_1})$. By assumption there exists a sequence of edges of lengths greater than $\varkappa$ which lead from i to j. Hence $\pi^{M_n''}(s_j) \geq c^\varkappa \pi^{M_n''}(s_i)$. The desired assertion follows from this by substituting inequality (15) in (14). $\square$

Corollary 1. *If the vertex j of the graph V_Γ is attainable from the vertex i and $\lambda(\bar{u}_j) > \lambda(\bar{u}_i)$ then $\lim_{n \to \infty} \pi(\bar{u}_i, n) = 0$.*

Let $\mathcal{U}_M$ be the set of all maximin controls (parties) and $\mathcal{B}_M$ be the set of the graph vertices. The main result on automata games from the classes considered is the next theorem.

Theorem 2. *If the set $\mathcal{U}_M$ is attainable from any vertex of the graph V_Γ, then*

$$\lim_{n\to\infty} \sum_{u\in\mathcal{U}_M} \pi(\bar{u}, n) = 1.$$

Proof. We choose an arbitrary non-maximin control $\bar{u}_i$. By assumption there exists a maximin control $\bar{u}_j$ such that the jth vertex of the graph is attainable from the ith one. According to the corollary from Lemma 4 $\lim_{n\to\infty} \pi(u_i, n) = 0$, i.e. the limiting probabilities of all non-maximin parties tend to zero. $\square$

We shall now formulate another criterion for the party to be maximin.

Theorem 3. *If for each party $\bar{u}_i \notin \mathcal{U}_M$ there exists an $(h_1, \ldots, h_r)$-neighbour party $\bar{u}_j$ (which is not necessarily maximin) such that $\lambda(\bar{u}_j) > \prod_{g=1}^{r} \lambda_{h_g}(u_i)$, then*

$$\lim_{n\to\infty} \sum_{u\in\mathcal{U}_M} \pi(\bar{u}, n) = 1.$$

Proof. In the inequality

$$\frac{\pi^{M'_n}(s_i)}{\pi^{M'_n}(s_j)} \leq \frac{1 - p_{jj}^{M'_n}}{p_{ij}^{M'_n}}$$

we substitute the estimate (15) and

$$p_{ij}^{N'_n} \geq c_4 \prod_{g=1}^{r} \lambda_{h_g}^{-n}(\bar{u}_i)$$

from Lemma 2. After simple transformations we arrive at a conclusion that $\lim_{n\to\infty} \pi^{M'_n}(s_i) = 0$. In view of Theorem 1 we obtain that $\lim_{n\to\infty} \pi(\bar{u}_i, n) = 0$. From this the required assertion follows. $\square$

A natural question arises whether in the automata games the choice of the maximin controls is preferable. It turns out that the answer is negative. There exist games (for example, the games of two automata $\mathcal{D}_{2,n}$) in which probability of non-maximin parties increases together with n.

We now discuss attainable aims for games of identical automata. Instead of the gains $\lambda_l(\bar{u})$ we now use the average gains $W_l(\bar{u})$. Let us denote by $\tilde{\mathcal{U}}_M$ the set of all controls (parties) for which the value

$$\max_{\bar{u}} \min_{l} W_l(\bar{u})$$

is achieved. Let V be an oriented graph with $\varkappa$ vertices such that an edge will lead from vertex h to vertex g provided the control $\bar{u}_g$ is β-neighbour with $\bar{u}_h$ and $W_\beta(\bar{u}_h) = \min_l W_l(\bar{u}_h)$. The set of vertices of the graph which correspond to the controls from $\tilde{\mathcal{U}}_M$ is denoted by $\tilde{\mathcal{B}}_M$.

Theorem 4. *If the set $\tilde{\mathcal{B}}_M$ is attainable from any vertex of the graph V then*

$$\lim_{n \to \infty} \sum_{u \in \tilde{U}_M} \pi(\bar{u}, n) = 1.$$

The proof follows immediately from Theorem 2.

According to the stated result the collection of identical automata from the classes considered above provides the greatest possible average gain for the least "lucky" one among them. We emphasize once more that Theorems 2–4 assert only that the total probability of the parties from the set $\tilde{\mathcal{U}}_M$ converges to one (with exponential rate). Until now we have not discussed how the probabilities of the parties $\pi(\bar{u}, n)$, $\bar{u} \in \tilde{\mathcal{U}}_M$ are distributed on $\tilde{\mathcal{U}}_M$ and whether they converge. Having the results on the asymptotic behavior (as $n \to \infty$) of the limiting probabilities of the parties in the game of automata with the memory depth n it becomes possible to describe the average gains.

Definition 6. The quantity

$$W_i^{(n)} = \sum_{\bar{u}} W_i(\bar{u}) \pi(\bar{u}, n)$$

is called the *average gain* of the automaton $\mathcal{A}_i^{(n)}$ with memory depth n.

The above results do not imply, in general, the convergence of the average gains as $n \to \infty$. If the set $\tilde{\mathcal{B}}_M$ contains a single element, i.e. the maximin party $\tilde{\mathcal{U}}_M$ is unique then

$$\lim_{n \to \infty} W_i^{(n)} = W_i(u_M).$$

If in the assumptions of Theorem 2 each of the limiting probabilities has the limit at $\bar{u} \in \tilde{\mathcal{U}}_M$, i.e. $\lim_{n \to \infty} \pi(\bar{u}, n) = \pi_\infty(\bar{u})$, then

$$\lim_{n \to \infty} W_i^{(n)} = \sum_{\bar{u} \in U_M} W_i(\bar{u}) \pi_\infty(\bar{u}).$$

We now consider in detail the case when $\bar{W}_i(\bar{u}) = W(\bar{u}), \forall i$. This is the case if all components of the controlled process have identical distributions. In addition, let the automata taking part in the game be identical. We call such games the *common pay-box games*. The next theorem gives the attainable aim.

Theorem 5. *For the games with common pay-box the automata from the considered class correspond to the solution-party in the limit, i.e.*

$$\lim_{n \to \infty} W^{(n)} = \max_u W(u).$$

Proof. It turns out that all automata on the graph are critical in each party since all reduced gains of all automata are the same. Hence, from any vertex j the edges lead to all vertices corresponding to the controls which differ from $\bar{u}_j$ by the output signal of one automaton. Such graphs are connected, and by Theorem 2 the probability of the set of maximin parties tends to one with increasing the memory depth n. The function $W(u)$ achieves its maximum on this set. $\qquad\square$

Consider Gur's game as an example. Its participants have only two moves u_1 and u_2. From ν participants of the game, let m choose the action u_1. Then the gains of all participants are equal to $W(m/\nu)$. In this game the set of solutions includes all parties in which $m_0 = \theta_0\nu$ participants have chosen the first action. Here $\theta_0 = \operatorname{argmin} W(\theta)$ (this means that $W(\theta_0) = \max_\theta W(\theta)$).

Thus, in the games with common pay-box the automata $\mathcal{A}^{(n)}$ from the considered classes obtain an average income which is close to the maximum one for large n.

It remains to compare the equilibrium situation (in the Nash sense) with the maximin parties. As noted above it is rather difficult to do this in the general case due to different characteristics of such parties. For such a comparison to be successful we should restrict attention to games of identical automata. But even then we find, in general, that the notions of an equilibrium situation and a maximin party cannot be enclosed one in another. In some special games these notions can be connected. We now turn to one such game.

The distribution games are played by ν participants having k identical actions (moves) $u_1, \ldots, u_k$, the gains of which are defined by the "weights" $a_1, \ldots, a_k$ satisfying the conditions $1 \geq a_1 \geq \cdots \geq a_k$. If the participants are distributed so that ν_i have chosen the action ν_i (i.e. $\nu_i \geq 0, i = 1, \ldots, \nu, \sum_{i=1}^{k} \nu_i = \nu$) then each participant which has chosen the action u_i will obtain the gain equal to one with probability a_i/ν_i and zero gain with complementary probability. In other words, the result of party is governed by the collection $(\nu_1, \ldots, \nu_k)$. The parties with the collections $\{\nu_i\}$ such that the inequalities

$$\frac{a_i}{\nu_j} \geq \frac{a_i}{\nu_i + 1}$$

hold at any i, j correspond to the stable situation.

Lemma 5. *Each stable situation in the distribution game is a maximin party.*

Proof. Indeed, let $\bar{u}_l$ be a stable situation and one of the participants, may be even the critical one among them, has changed his action. Then he comes off a loser since

$$\frac{a_i}{\nu_i(\bar{u}_l) + 1} \leq \min_l W_l(\bar{u}_l), \quad i = 1, \ldots, k.$$

In the party $\bar{u}$ let $\nu_{j_0}(\bar{u}) \neq \nu_j(\bar{u}_l)$ at least for one index j. Then there exists j_0 such that $\nu_{j_0}(\bar{u}) \geq \nu_{j_0}(u_l) + 1$. Hence

$$\min_l W_l(\bar{u}) \leq \frac{a_{j_0}}{\nu_{j_0}(\bar{u})} \leq \frac{a_{j_0}}{\nu_{j_0}(\bar{u}) + 1} \leq \min_l W_l(\bar{u}_l).$$

This means that

$$\min_l W_l(\bar{u}_l) = \max \min W_l(\bar{u}).$$

i.e. the party $\bar{u}$ is minimax. $\qquad\square$

There exist distribution games for which the opposite assertion does not hold, i.e. they have maximin parties which are not the equilibrium situations. The following fact is interesting enough.

Theorem 6. *If the identical automata from the classes $\mathcal{K}_2$, $\mathcal{K}_3$ and $\mathcal{K}_4$ take part in the distribution game then*

$$\lim_{n \to \infty} \sum_{\bar{u} \in U_M} \pi(\bar{u}, n) = 1.$$

Proof. By Theorem 2 it is sufficient to verify that on the graph associated with the distribution game the set of vertices $\mathcal{B}_M$ corresponding to the set of parties $\mathcal{U}_M$ is attainable from any vertex not belonging to $\mathcal{B}_M$. We choose an arbitrary party $\bar{u}_i \notin \mathcal{U}_M$ and the maximin one $\bar{u}_m$. Let the automata choosing the action $\bar{u}_g$ be the critical one for $\bar{u}_i$. According to the definition of the party $\bar{u}_M$ either it does not contain the action $\bar{u}_g$ or the automata using this move gains more than the one choosing it in the party $\bar{u}_i$, i.e.

$$\frac{a_g}{\nu_g(\bar{u}_i)} = \min_l W_l(\bar{u}_i) < \min_l W(\bar{u}_M) \leq \frac{a_g}{\nu_g(\bar{u}_M)}.$$

In both cases $\nu_g(\bar{u}_M) < \nu_g(\bar{u}_i)$, and since

$$\sum_{j=1}^{k} \nu_j(\bar{u}_i) = \sum_{j=1}^{k} \nu_j(\bar{u}_M)$$

there exists an h such that $\nu_h(\bar{u}_M) > \nu_g(\bar{u}_i)$.

First assume that the automaton $\mathcal{A}_l$ has chosen the action u_g in the party $\bar{u}_i$ and define a new party $\bar{u}_j$ by replacing the action u_g by u_h in the party $\bar{u}_i$. By definition of the game graph the edge leads from the vertex i to the vertex j. If $\bar{u} \in \mathcal{U}_M$ the attainability of $\mathcal{U}_M$ from the vertex j is proved. Otherwise, the arguments for $\bar{u}_i$ are repeated for $\bar{u}_j$ and so on. This produces a sequence of vertices $i, j, \ldots$ connected by edges. Beginning from the first step the equality

$$\sum_{l=1}^{k} |\nu_l(\bar{u}_j) - \nu_l(u_M)| = \sum_{l=1}^{k} |\nu_l(\bar{u}_j) - \nu_l(\bar{u}_M)| - 2$$

holds. By reason of this each vertex in this sequence appears no more than once. Hence, we enter the set $\mathcal{U}_M$ for a finite number of steps. $\qquad\square$

Note that for the automata from the class $\mathcal{K}_1$ these arguments fail.

CHAPTER 5

CONTROLLED FINITE HOMOGENEOUS MARKOV CHAINS

We study a number of optimization problems of adaptive control of Markov chains with rewards. First, we consider finding the unconditional extremum of the average reward, next, the problems with constraints (conditional extremum), and finally, the minimax problems.

5.1. Preliminary Remarks

Definition 1. An object

$$C = \{X, U, \mathcal{P}^{(u)}, \bar{p}, \zeta\}$$

is called a *controlled Markov chain*. Here $X = \{x_1, \ldots, x_m\}$ and $U = \{u_1, \ldots, u_k\}$ are the sets of states and controls respectively, $\mathcal{P}^{(u)} = (p_{ij}^{(u)})$ is a stochastic matrix of transition probabilities $p_{ij}^{(u)} = \mathbf{P}\{x_i \xrightarrow{u} x_j\} = \mathbf{P}\{x(t) = x_i | x(t-1) = x_j, u(t-1) = u\}$ provided the control $u \in U$ was used, $\bar{p} = (p_1, \ldots, p_m)$ is the initial distribution, and $\zeta = \zeta(x, u, \omega) \in \mathrm{R}^h, h \geq 1$ is the random vector of rewards (if $h = 1$ it will be called just a reward).

The reward $\zeta(x, u, \omega)$ appears in the state x provided the control u was applied. The average reward $r_x^u = \mathbf{E}\zeta(x, u; \omega)$ is assumed to be finite. Sometimes one assumes that every state of the chain is related with some subset $U_i(\subseteq U)$ of controls which may be used when the chain enters the state x_i. This condition is prompted by many applications but for the sake of simplicity we shall assume that $U_i \equiv U$.

The chain will be called *homogeneous* if the matrix $\mathcal{P}^{(u)}$ does not depend on time. Otherwise it is *nonhomogeneous*. Until now only homogeneous chains were considered in adaptive theory.

In classical theory of controlling Markov chains one usually restricts attention to stationary deterministic Markovian strategies (the simplest case). They are defined by a collection of functions $f : X \to U$. If $u = f(x)$ then the control u is applied every time the chain enters the state x. Let $\sum_{\mathrm{SDM}} = \{f\}$ denote the collection of such rules. It contains k^m different elements. The stationary randomized Markov strategies from the set Σ_{SRM} generated by the distributions $p(\cdot|x)$ on the set U and depending on the current state of the process are used as well. The stationary non-degenerate Markov strategies such that $p(u|x) > 0$ at all (x, u) play an important role below. They form the set Σ_+. The stationary non-degenerated strategies are

123

defined by a control choice law γ. This law is represented by some stochastic $m \times k$-matrix with positive elements

$$\Gamma(\gamma) = (p(x|u)).$$

We consider control problems only on the infinite time interval $[0, \infty)$. To state these the objective functions should be introduced: the discounted reward (gain)

$$W_\beta(\sigma, p) = \sum_{n=0}^{\infty} \beta^n \mathbf{E}_{\sigma, p} \zeta_n, \quad 0 < \beta < 1,$$

and the limiting average reward (gain) per "one step"

$$W(\sigma, p) = \lim_{T \to \infty} T^{-1} \sum_{t=0}^{T} \mathbf{E}_{\sigma, p} \zeta_t.$$

Weak aims regarded as a type of unconditional extremum can be written as follows:

find the optimal strategies σ_β or σ^0 such that

$$W_\beta(\sigma_\beta, p) = \max_\sigma W_\beta(\sigma, p), \quad W(\sigma^0, p) = \max_\sigma W(\sigma, p),$$

or, in the more weak form, find the ε-optimal strategies $\sigma_{\beta, \varepsilon}$ or σ_ε^0) ($\varepsilon > 0$ and is fixed) such that

$$W_\beta(\sigma_{\beta, \varepsilon}, p) > \max_\sigma W_\beta(\sigma, p) - \varepsilon, \quad W(\sigma_\varepsilon^0, p) > \max_\sigma W(\sigma, p) - \varepsilon.$$

In the minimization problems the quantities ζ should be interpreted, of course, as losses not gains.

The strong aim regarded as a type of unconditional extremum is formulated for the limiting average (one-step) reward as follows:

find an optimal strategy σ^0 which secures the fulfilment a.s. of the following equality

$$\lim_{T \to \infty} T^{-1} \sum_{t=0}^{T} \zeta_n = \max_\sigma W(\sigma, p).$$

The ε-optimal version of this aim is of interest as well. The same aims can be set for the discount reward.

Optimization problems with constraints have the following form:

find a conditional-optimal strategy $\hat{\sigma}$ such that for the h-dimensional reward vector $\zeta = \left(\zeta^{(1)}, \zeta^{(2)}, \ldots, \zeta^{(h)}\right)$ the following conditions

$$\lim_{T \to \infty} T^{-1} \sum_{t=0}^{T} \zeta_t^{(1)} = \max_\sigma W^{(1)}(\sigma, p) = W^{(1)}(\hat{\sigma}, p),$$

$$\varliminf_{T \to \infty} T^{-1} \sum_{t=0}^{T} \zeta_t^{(j)} \geq g^{(j)}, \quad j = 2, \ldots, h, \quad \text{a.s.}$$

hold, where (g_i) are given numbers and $W^{(i)}(\sigma, p) = \varliminf_{T \to \infty} T^{-1} \sum_{t=0}^{T} \mathbf{E}_{\sigma,p} \zeta_t^{(i)}$ is the limiting average reward of the i-th component of the reward vector ζ.

We now describe minimax problems for Markov chains. There are two players A and B. The set of controls is $U = U' \times U'' = \{(u_i', u_j'') \in U' \times U''\}$. The first component of the pair is specified by player A and the second by player B. Players A and B use the sets of strategies Σ' and Σ'' respectively. Choosing the moves, player A obtains some reward ζ_{ij}, which depends on the state x and the move of the rival made. The quantity of reward of player B is equal to $-\zeta_{ij}$, i.e. we deal with a two-person zero-sum game. This game is complicated by the fact that under the control $u = (u_i', u_j'')$, the Markov chain passes from the current state x to a new one $\tilde{x}$ in such a way that the dependence of the reward ζ on the participants' moves differs from the previous. The aim of player A consists of maximizing his gain and of player B it consists of minimizing his loss. This means that the players have to form the strategies $\sigma_1 \in \Sigma'$ and $\sigma_2 \in \Sigma''$ respectively so that the limiting average reward $W(\sigma_1, \sigma_2, p)$ possesses the property:

$$\max_{\sigma_1} \min_{\sigma_2} W(\sigma_1, \sigma_2; p) = \min_{\sigma_2} \max_{\sigma_1} W(\sigma_1, \sigma_2; p).$$

5.2. Structure of Finite Homogeneous Controlled Markov Chains

We are interested in structure and properties of controlled Markov chains. For this reason we study the strategies regulating their behaviour. Let us first write down the transition probability matrices under the stationary deterministic strategies (from Σ_{SDM}) generated by the functions $f(x)$, i.e. $\sigma = \sigma(f)$. To each function corresponds a control choice law which can be written in vector form $\bar{u}(x) = (u_{l_1}, u_{l_2}, \ldots, u_{l_m})$ the ith component corresponds to the chain state x_i. The transition probability matrix $P(\bar{u}(x)) = \big(p_{ij}(u(x))\big)$ associated with this law has the elements

$$p_{ij}^{(\bar{u}(x))} = p_{ij}(\bar{u}(x_i)) = p_{ij}^{(u_{l_j})}.$$

Let us consider a strategy from Σ_{SRM}. The control choice law generating this strategy can also be represented by the vector $\bar{p}(x) = (p_1(x), \ldots, p_k(x))$ where $\gamma = p_l(x) = \mathbf{P}\{u = u_l | x\}$ is the conditional probability of using the control u_l. For short, such a strategy is written as $\sigma = \sigma(p)$. The transition probability matrices $P(p)$ contain the elements

$$p_{ij}(p) = \sum_{l=1}^{k} p_{ij}^{(u_l)} p_l(x_i).$$

The non-stationary strategies with time-varing laws γ_t generate the matrices $P(\gamma_t)$. This means that the associated (usual) Markov chain is non-homogeneous in contrast to the previous cases.

Definition 1. A controlled Markov chain is called *regular* if for every strategy $\sigma(\bar{u}) \in \Sigma_{\text{SDM}}$ the matrix $\mathcal{P}(\bar{u})$ is regular, i.e. for some $l \geq 1$ all elements of the matrix $\mathcal{P}^l(u)$ are positive.

This definition means that the chain has a single ergodic class without any cyclic subclasses. If a controlled Markov chain is regular then all matrices $\mathcal{P}(\gamma)$ are also regular.

Definition 2. A controlled Markov chain is called **ergodic** if for every strategy $\sigma(\bar{u}) \in \Sigma_{\text{SDM}}$ the matrix $\mathcal{P}(u)$ is ergodic, i.e. it is indecomposable.

In this case the chain has one class of communicating states with several cyclic subclasses. All matrices $\mathcal{P}(\gamma)$ are also ergodic.

Definition 3. A controlled Markov chain is called **connected** if for every ordered pair of states (x_g, x_h) there exists at least one deterministic Markov law $\bar{u}_{gh}$ such that $p_{gh}^{(n)}(u_{gh}) > 0$ for some n.

The least value of n from this definition satisfies $n \leq m - 1$. The realization of the inverse transition from x_h into x_g may require using another law $\bar{u}_{hg}$. The regular and ergodic chains are connected.

Theorem 1. *A controlled Markov chain is connected if and only if there exists a randomized Markov law γ such that the matrix $P(\gamma)$ is indecomposable.*

Proof. 1. First, we assume that the controlled Markov chain is connected. By definition, there exist some intermediate states $x_{j_1}, x_{j_2}, \ldots, x_{j_n}$, such that under the appropriate controls we have

$$p_{gj_1}(u_{gj_1}) > 0, \quad p_{j_1 j_2}(u_{j_1 j_2}) > 0, \quad \ldots, \quad p_{j_n h}(u_{j_n h}) > 0$$

for any x_g, x_h. In the states $x_g, x_{j_1}, \ldots, x_{j_n}$ we assign some positive probabilities to the controls $u_{gj_1}, u_{j_1,j_2}, \ldots, u_{j_n h}$. We repeat this procedure for all pairs of states in such a manner that the sum of their probabilities is equal to one in each state. This construction defines some randomized Markov law γ, and hence some strategy $\sigma(\gamma) \in \Sigma_{\text{SRM}}$. We also form the matrix $P(\gamma)$. It is clear that in the chain $(X, P(\gamma))$ all states communicate. Therefore the matrix $P(\gamma)$ is indecomposable.

2. Let there exist a strategy $\sigma(\gamma) \in \Sigma_{\text{SRM}}$ such that $P(\gamma)$ is indecomposable. Hence, for any x_g, x_h there exists a sequence of states $x_{j_1}, \ldots, x_{j_n}$ such that one can pass from every state x_{jl} into the successive state x_{jl+1} under the action of the proper control, i.e. $x_g \to x_{j_1}, \ldots, x_{j_n} \to x_h$. If there are repetitions in these transitions we

cross them out. We assign some of the controls to the states $x_{j_1}, \ldots, x_{j_n}$. This produces a deterministic Markov law to transfer x_g into x_h. This law can be extended to the other states in an arbitrary manner. $\qquad\square$

This theorem will hold for non-stationary strategies if the control choice probabilities converge (as $t \to \infty$) to positive limits. It follows from the proof that instead of Σ_{SRM} one can consider the set of non-degenerate strategies Σ_+. The controlled Markov chains can essentially differ when choosing different strategies. Even the deterministic laws sometimes generate rather different chains. We give one such result which will be important later on.

Theorem 2. *Let some decomposition of a connected controlled chain into the ergodic classes $E_1, \ldots, E_M$ and class of the inessential states N correspond to a strategy $\sigma(\bar{u}')$. Then for any $j = 1, \ldots, M$ there exists a deterministic strategy $\sigma(\bar{u}'') \in \Sigma_{\mathrm{SDM}}$ dividing the set of states of this chain into an ergodic class E_j and a class of the inessential states $X \backslash E_j$.*

Proof. Let us construct the law $\bar{u}''$ defining the required strategy. On the states of the chosen ergodic class E_j we define the same controls as those defined by the law $\bar{u}'$. We consider the states from $X \backslash E_j$. Among them there exists a subset of states $E^{(1)}$ which enter E_j for one period under the proper controls. These controls are associated with the states of $E^{(1)}$. Next, we choose the subset $E^{(2)}$ consisting of points which enter the set $E^{(1)}$ at some point in time. We assign again the appropriate controls to the states of this subset. Next, we determine the subset $E^{(3)}$ and so on. As a result of this the set $X \backslash E_j$ will be exhausted in a finite number of steps. Otherwise, one could find at least one state from which it would be impossible to pass into the set E_j under any law. This would contradict the connectedness of the chain. $\qquad\square$

Let us now turn to properties of controlled Markov chains when the non-degenerate strategies from the class Σ_+ are applied. First, note that for all non-degenerate laws γ, the location of zeros in the matrices $P(\gamma)$ is the same. Hence, use of these laws leaves the structure of the chains $(X, P(\gamma))$ unchanged: the number of ergodic classes and their structure are the same, in each ergodic class the amount $d \geq 1$ and structure of cyclic subclasses does not change together with γ and the sets of the inessential states coincide.

Let $\sigma(\gamma) \in \Sigma_+$. We say that the state x_i is *followed* by the state x_j if the transition probability $x_i \to x_j$ for n steps $p_{ij}^{(n)} > 0$ at some $n \geq 1$. The states x_i, x_j are called *communicating* if x_j is followed by x_i and x_i is followed by x_j. The set of all communicating states of a controlled Markov chain is called the *connective component*. Its structure and composition does not depend on the prescribed law γ. The strategies not belonging to Σ_+ can decompose the connective component into ergodic classes and the set of inessential states in an arbitrary way. We give an example of a connective component with six states $x_1, \ldots, x_6$ and three controls

u_1, u_2, u_3. The transition matrices are given as follows:

$$P^{(u_1)} = \begin{pmatrix} + & + & 0 & 0 & 0 & 0 \\ + & + & 0 & 0 & 0 & 0 \\ 0 & 0 & + & + & 0 & 0 \\ 0 & 0 & + & + & 0 & 0 \\ + & + & 0 & 0 & + & + \\ 0 & 0 & + & + & + & + \end{pmatrix}, \quad P^{(u_2)} = \begin{pmatrix} + & 0 & 0 & 0 & 0 & 0 \\ 0 & + & 0 & + & 0 & 0 \\ 0 & 0 & + & 0 & + & + \\ 0 & + & 0 & + & 0 & 0 \\ 0 & 0 & + & 0 & + & 0 \\ 0 & 0 & + & 0 & 0 & + \end{pmatrix},$$

$$P^{(u_3)} = \begin{pmatrix} + & 0 & 0 & + & 0 & + \\ 0 & + & 0 & 0 & + & 0 \\ + & 0 & + & 0 & 0 & + \\ 0 & + & 0 & + & 0 & + \\ + & 0 & 0 & + & + & 0 \\ 0 & + & 0 & + & 0 & + \end{pmatrix}$$

where the symbol "+" means some positive number (the numerical values are not important to us.) The deterministic law $(u_1, u_1, \ldots, u_1)$ generates two ergodic classes (x_1, x_2) and (x_3, x_4) and the class of the inessential states (x_5, x_6). The law $(u_2, u_2, \ldots, u_2)$ generates again two ergodic classes $(x_1, x_4), (x_3, x_5, x_6)$ and the inessential state (x_1). The law $(u_3, u_3, \ldots, u_3)$ leads to one ergodic class and the inessential state x_3.

Fix a non-degenerate law γ and consider the Markov chain $(X, P(\gamma))$. Assume that the state x_i has the following property: there exists at least one state x_j such that $p_{ij}^{(l)}(\gamma) > 0$ for some $l \geq 1$ but $p_{ji}^{(h)} = 0$ for all h. In the theory of Markov chains such a state is usually called *inessential*. The set N of all inessential states is called the *inessential set*. It is characterized by the fact that for strategies from Σ_+ the chain leaves it in a finite number of steps with probability equal to one. However, for some degenerate strategies, in particular, for deterministic ones, the set N turns out either to be ergodic or to contain some ergodic subset. Here is an example of such a possibility for the chain with the sets of states $X = \{x_1, x_2, x_3, x_4, x_5\}$ and controls $u = \{u_1, u_2\}$:

$$P^{(u_1)} = \begin{pmatrix} 1 & 0 & 0 & 0 & 0 \\ 0 & 1 & 0 & 0 & 0 \\ + & 0 & + & + & + \\ 0 & + & 0 & + & + \\ + & 0 & + & 0 & + \end{pmatrix}, \quad P^{(u_2)} = \begin{pmatrix} 1 & 0 & 0 & 0 & 0 \\ 0 & 1 & 0 & 0 & 0 \\ 0 & 0 & + & 0 & + \\ 0 & 0 & 0 & + & + \\ 0 & 0 & + & + & 0 \end{pmatrix}.$$

The degenerate strategies of the form $(\bar{p}_1, \bar{p}_2, u_2, u_2, u_2)$ where $\bar{p}_1$, $\bar{p}_2$ are two-dimensional probability vectors corresponding to u_1 and u_2 in the states x_1 and x_2 make the set of states $\{x_3, x_4, x_5\}$ ergodic and even regular. Application of the control u_1 in these states makes the probability to pass into the absorbing states x_1 and x_2 positive. It is easy to find the transition matrices when the set N contains a proper subset which represents an ergodic class under some degenerate strategy.

The main result on controlled Markov chains is the following theorem.

Theorem 3. *The set of states of each finite homogeneous controlled Markov chain can be represented in the form*

$$X = S(1) + S(2) + \cdots + S(L) + N$$

where $S(i)$ are the connective components and N denotes the inessential set. The corresponding transition probability matrix for the strategy $\sigma(\gamma) \in \Sigma_+$ has the following canonical form

$$P(\gamma) = \begin{pmatrix} P_1(\gamma) & & & & \\ & P_2(\gamma) & & & \\ & & \ddots & & \\ & & & P_L(\gamma) & \\ R_1(\gamma) & R_2(\gamma) & \ldots & R_L(\gamma) & R_0(\gamma) \end{pmatrix}$$

where the square matrices $P_1(\gamma), \ldots, P_L(\gamma)$ are indecomposable, the matrices $R_1(\gamma), \ldots, R_L(\gamma), R_0(\gamma)$ are non-negative and not all matrices $R_1(\gamma), \ldots, R_L(\gamma)$ are zero. The chains $(S(i), P_i(\gamma)), i = 1, \ldots, L$ are ergodic with d_i (≥ 1) cyclic subclasses.

Once the definitions of connective components and the inessential set are given, the proof of this theorem does not contain any new ideas in comparison with similar results for the usual Markov chains (without control). This does not change the fact that Theorem 3 is of fundamental importance.

We now turn to the optimization problems for the objective functions $W_\beta(\sigma, p)$ and $W(\sigma, p)$ (see Sec. 1) defined on a finite Markov chain with rewards. The appropriate optimal strategies exist, do not depend on the initial state (or distribution) and are simple (stationary deterministic Markov strategies), i.e. they belong to the set Σ_{SDM}. They can be obtained by known calculation methods which have been discussed in Sec. 2, Chap. 1 in application to the objective function $W(\sigma, p)$. The mentioned methods consist of using either linear programming (with the help of the simplex method) or the iterative method which is, in fact, the dual of the linear programming problem. The latter method will be useful to us, hence we outline it.

Let us fix a Markov law $\bar{u}(x) = (u_{l_1}, u_{l_2}, \ldots, u_{l_m})$, a transition probability matrix $P^{(\bar{u})}$, and a rewards vector $\bar{r}(\bar{u}) = (r_1, \ldots, r_m)$ corresponding to this law. We can now write down the dual problem to the linear programming problem for ergodic chains (see Sec. 2, Chap. 1)

$$W + v_i = \max\left(r_i + \sum_{j=1}^{m} p_{ij}(\bar{u}(x))v_j \right), \quad i = 1, \ldots, m. \tag{1}$$

We are looking for the limiting average reward $W(u)$ and "weights" $v_1, \ldots, v_m$. Because the number of unknown variables is greater by one than the order of the system, one of them may be taken arbitrarily but it is common practice to put $v_m = 0$. Then the system (1) becomes solvable. Its solution is the first step of every iterative cycle. The second step consists of improving this solution, i.e. in seeking a control $\tilde{u}(x)$ such that the quantity

$$r_i + \sum_{j=1}^{m} p_{ij}(u)v_j, \quad i = 1, \ldots, m,$$

with the weights found in the previous step, becomes maximum for any state of the chain. The collection $\tilde{u}(x) = (u_{j_1}, \ldots, u_{j_m})$ serves as the next Markov law. This completes the iterative cycle. In the next cycle we solve system (1) again, where now $W = W(\tilde{u})$ means the limiting average reward under the stationary strategy generated by the law $\tilde{u}$. In the second step one improves the obtained Markov law and so on. After a finite number of cycles the iterative procedure stops and leads to the optimal law $\bar{u}_{\mathrm{opt}}$ supplying the maximum to the limiting average reward $W(\sigma(\bar{u}_{\mathrm{opt}})) = \sup_\sigma W(\sigma)$ equal to the value W.

Let us express the function $W(\sigma, p)$ in explicit form at a strategy $\sigma(\gamma) \in \Sigma_{\mathrm{SRM}}$. First, we compute the mathematical expectation of the reward at the moment t,

$$\mathbf{E}_{\gamma,p}\zeta_t = \bar{p}P^{t-1}(\gamma)r(\gamma), \tag{2}$$

where $\bar{p} = (p_1, \ldots, p_m)$ is a row-vector of the initial state probabilities, $P^{t-1}(\gamma)$ is the transition matrix for $t-1$ steps under a randomized law γ and $r(\gamma) = (r_1^{(\gamma)}, r_2^{(\gamma)}, \ldots, r_m^{(\gamma)})$ is the column-vector of the one-step rewards under the law γ in the states x_i, i.e.

$$r_i^{(\gamma)} = \sum_{l=1}^{k} r_i^l p(u_l|x_i), \quad r_i^l = \mathbf{E}\zeta(x_i, u_l).$$

Quite often one uses the more complex form of reward $\zeta(x_i, x_j; u_l)$. Then the one-step reward under the control u_l is defined by $r_i^l = \sum_{j=1}^{m} r_{ij}^l p_{ij}^l$. From (2) it follows that

$$W(\sigma(\gamma), p) = \lim_{T \to \infty} T^{-1} \sum_{t=1}^{T} \bar{p}P^{t-1}(\gamma)r(\gamma).$$

Since for any stochastic matrix P there exists the Cesaro limit

$$\lim_{T \to \infty} T^{-1} \sum_{t=1}^{T} P^{t-1} = Q,$$

we obtain the desired representation of the objective function

$$W(\sigma(\gamma), p) = \bar{p}Q(\gamma)r(\gamma). \tag{3}$$

If the transition matrix has the form

$$P(\gamma) = \begin{pmatrix} P_1(\gamma) & & & & \\ & P_2(\gamma) & & O & \\ & & O & \ddots & \\ & & & & P_L(\gamma) \\ R_1(\gamma) & R_2(\gamma) & \cdots & R_L(\gamma) & R_0(\gamma) \end{pmatrix}$$

where $P_1(\gamma), \ldots, P_L(\gamma)$ are indecomposable matrices associated with ergodic classes but the matrices $R_1(\gamma), \ldots, R_L(\gamma), R_0(\gamma)$ are associated with the inessential states, then the limiting matrix $Q(\gamma)$ has the following form

$$Q(\gamma) = \begin{pmatrix} Q_1(\gamma) & & & & \\ & Q_2(\gamma) & & O & \\ & O & \ddots & & \\ & & & Q_L(\gamma) & \\ N_1(\gamma) & N_2(\gamma) & \cdots & N_L(\gamma) & N_0(\gamma) \end{pmatrix}.$$

As explained above, for the degenerate strategy $\sigma(\gamma)$ the matrix $N_0(\gamma)$ can be non-trivial and may contain an indecomposable submatrix.

For chains of a special form the expression for the objective function (3) is simpler. Let the chain be regular. Then, as known, there exists $\lim_{T\to\infty} P^T(\gamma) = Q(\gamma)$ representing a matrix with identical rows

$$\bar{\pi}(\gamma) = (\pi_1(\gamma), \ldots, \pi_m(\gamma)), \quad \pi_i(\gamma) > 0, \quad \sum_{i=1}^{m} \pi_i(\gamma) = 1$$

where $\pi_i(\gamma)$, $i = 1, \ldots, m$ are the limiting state probabilities. This leads to the objective function

$$W(\sigma(\gamma), \bar{p}) = (\bar{\pi}(\gamma), r(\gamma)) = \sum_{j=1}^{m} \pi_j(\gamma) r_j(\gamma) \tag{4}$$

that does not depend on the initial distribution $\bar{p}$. Clearly, the extremum of the function W does not depend on $\bar{p}$ for connected chains. If $\sup W(\sigma, x') > \sup W(\sigma, x'')$ then we replace the strategy $\sigma(\gamma)$ by a non-stationary strategy which chooses a non-degenerate law γ_0 and uses it thereafter up to a random moment τ (finite a.s.) until the chain passes into the state x'. This state must belong to one of the ergodic classes into which the strategy $\sigma(\gamma)$ decomposes the connected chain. If $\sigma(\gamma)$ is the optimal strategy, then by Theorem 2 one can pass from any state of the chain into the class pointed out by using only the deterministic laws. Therefore, formula (4) holds though among the limiting probabilities $\{\pi_i(\gamma)\}$ there may be zeros under the appropriate laws γ.

For the chains of a general form which can be decomposed into the ergodic classes $E_1, \ldots, E_L$ by the optimal strategy, the objective function has the form: γ

$$W(\sigma(\gamma), p) = \sum_{j=1}^{L} W(\sigma(\gamma), \bar{p}_j) = q_1 W_1(\gamma) + \cdots + q_L W_L(\gamma)$$

where $\bar{p}_j$ corresponds to the states from E_j, $q_j = \sum_{i \in E_i} p_i$ is the hitting probability of the chain into the class E_i, $W_i(\gamma)$ is the limiting average reward with respect to the class E_i. Below we need some types of strong laws of large numbers for controlled Markov chains with rewards. First, we introduce a useful notion.

Definition 4. A strategy $\sigma = (\gamma_1, \gamma_2, \gamma_3, \ldots)$ is called *asymptotically stationary* if the sequence of laws $\gamma_n = \mathbf{P}\{u_n = u | x'', x^{n-1}, u^{n-1}\}$ converges a.s. to the conditional probability $p(u|x)$ for all x and u.

Theorem 4. *Let $(X, U, P^{(u)}, p, \zeta)$ be a connected controlled Markov chain, σ be an asymptotically-stationary strategy with the limiting Markov law γ_∞ such that the chain $(X, P(\gamma_\infty))$ is ergodic. Then*

$$\lim_{T \to \infty} T^{-1} \sum_{t=1}^{T} (\zeta_t - \mathbf{E}_\sigma \zeta_t) = 0, \quad \text{a.s.}$$

Proof. By assumption, the matrix $P(\gamma_\infty)$ is indecomposable. Hence all matrices $P(\gamma_t)$, beginning from some t, are indecomposable as well, and all chains $(X, P(\gamma_t))$ have the same structure (moreover, their transition probabilities are close to each other.) First, we assume that these chains are regular, i.e. without cyclic subclasses. Then with probability one we have

$$\lim_{T \to \infty} T^{-1} \sum_{t=1}^{T} \zeta_t = \sum_{i=1}^{m} \pi_i(\gamma_\infty) r_i(\gamma_\infty).$$

If the chain has $d \geq 2$ cyclic subclasses we write the matrix $P(\gamma_\infty)$ in the following form (maybe after renumbering its states in an appropriate way)

$$P(\gamma_\infty) = \begin{pmatrix} 0 & \mathcal{P}_1 & 0 & \ldots & 0 \\ 0 & 0 & \mathcal{P}_2 & \ldots & 0 \\ \hdotsfor{5} \\ 0 & 0 & 0 & \ldots & \mathcal{P}_{d-1} \\ \mathcal{P}_d & 0 & 0 & \ldots & 0 \end{pmatrix}.$$

By using standard arguments we arrive at a conclusion that the states of the chain considered at the moment j (mod d) form ergodic classes. Therefore for any $j = 0, 1, \ldots, d-1$ the strong law of large numbers can be applied to the sequence $T^{-1} \sum_{t=1}^{T} \zeta_{td+j}$, i.e. there exist $\lim_{T \to \infty} T^{-1} \sum_{t=1}^{T} \zeta_{td+j}$ and these limits can easily be expressed in terms of numerical characteristics of the jth cyclic subclass. This implies immediately the existence of the limit of the sequence (as $T \to \infty$)

$$\frac{1}{dT} \sum_{t=1}^{dT} \zeta_t = \frac{1}{d} \sum_{j=0}^{d-1} T^{-1} \sum_{t=0}^{T-1} \zeta_{td+1}$$

which can be expressed in terms of the matrix $P(\gamma_\infty)$ and mathematical expectations of the rewards. $\qquad\square$

Let us estimate the convergence rate of the average rewards under the strategy $\sigma(\gamma)$. We consider the regular chain $(X, P(\gamma))$. For the elements of the matrix $P^t(\gamma) = \left(p_{ij}^{(t)}(\gamma) \right)$ we have the following decompositions

$$p_{ij}^{(t)}(\gamma) = \pi_i(\gamma) + \sum_{n=2}^{h} a_n(i, j; \gamma; t) \lambda_n^t(\gamma)$$

where λ_κ are the eigenvalues of the matrix $P(\gamma)$ and $\lambda_1 = 1 > |\lambda_2| \geq |\lambda_3| \geq \cdots \geq |\lambda_n|$, the coefficients $a_n(\ldots; t)$ being polynomials in t. The mathematical expectations of the rewards $\mathbf{E}_{\gamma,p}\zeta_t = \sum_{i,\nu=1}^{m} \mathbf{E}\zeta(x_\nu) p_i p_{i\nu}^{(t)}(\gamma)$ have the limits

$$\lim_{t \to \infty} \mathbf{E}_{\gamma,p}\zeta_t = \sum_{\nu=1}^{m} \mathbf{E}\zeta(x_\nu)\pi_\nu(\gamma) = \mathbf{E}_\infty \zeta,$$

the convergence rate being exponential and being defined by the second eigenvalue λ_2 of the matrix $P(\gamma)$. Similar result holds for the convergence rate of the variables $\eta_T = T^{-1} \sum_{t=1}^{T} \zeta_t$. For the latter, we state a result on the convergence rate in the a.s. sense.[a] Namely, the asymptotic equality

$$T^{-1} \sum_{t=1}^{T} \zeta_t = \mathbf{E}\zeta + o(t^{\delta - 1/2})$$

takes place, where $\mathbf{E}\zeta = \sum_{i=1}^{m} \pi_i \mathbf{E}\zeta(x_i)$ is the limiting mathematical expectation of the random variable $\zeta(x)$ defined on the regular Markov chain, $\delta > 0$ being arbitrary.

We again remind of the possibility to treat the controlled Markov chains with rewards as a stochastic automaton. This interpretation is often convenient and useful.

5.3. Unconditional Optimal Adaptive Control for Finite Markov Chains

Throughout the remainder of this chapter the sets X and U are assumed to be known, the states x_t and rewards ζ_t observed at each moment. The matrices $P^{(u)}$ and characteristics r_x^u of the rewards $\zeta(x, u; \omega)$ are unknown.

The problem of finding extrema of objective functions $W_\beta(\sigma, p)$ and $W(\sigma, p)$ may be considered as an optimization aim of control of a Markov chains both in adaptive theory and in the classical one. In this case for the class $\mathcal{K}$ of controlled Markov chains we have to find a strategy $\sigma_\mathcal{K}$ such that

$$\sup_\sigma W_\beta(\sigma, p) = W_\beta(\sigma_\mathcal{K}, p) \quad \text{or} \quad \sup_\sigma W(\sigma, p) = W(\sigma_\mathcal{K}, p)$$

for all chains $C \in \mathcal{K}$. However, it is impossible to obtain the fulfilment of the first equality (approximately with an accuracy up to $\varepsilon > 0$) under lack of information

---

[a]As known, this type of convergence does not generate a metric in the space of random variables.

about the controlled chain by using only the observations of its evolution. Indeed, from the form of the discounted average reward

$$W_\beta(\sigma, p) = \sum_{t=0}^{\infty} \beta^t \mathbf{E}_{\sigma,p} \zeta_t$$

it follows that starting from the initial moment we should use the optimal laws. Here we cannot expect this. For this reason we exclude the function $W_\beta(\sigma, p)$ from the set of objective functions and in what follows restrict attention to the case of the limiting average (one-step) reward $W(\sigma, p)$.

Both aims stated above and other optimization aims contain the value $W(p) = \sup_\sigma W(\sigma, p)$. We are concerned with the cases when it does not depend on the initial distribution.

Definition 1. A controlled Markov chain with rewards is called *equiprofitable* if $\sup_\sigma W(\sigma, p) = W$.

The chains which are regular and connected are equiprofitable.

We denote a family of all finite homogeneous controlled Markov chains with rewards which have X ($|X| = m$) as the state set and U ($|U| = k$) as the set of controls by $M(X, U) = M(m, k)$. For this family we use the Euclidian metric. Hence, we interpret any chain from $M(m, k)$ as some point of the $km^2 + km$-dimensional space generated by k stochastic matrices $P^{(u)}$ (of order m) and mk mathematical expectations of the rewards $r_i^l = \mathbf{E}\zeta(x_i, u_l)$. We now consider the set of equiprofitable chains $RM(X, U) \subset M(X, U)$. *This set is neither closed nor open in the Euclidean space* $R^{km^2 + km}$. First, we construct an example of a sequence of equiprofitable chains whose limit is not equiprofitable. Let $X = (x_1, x_2, x_3)$, $U = (u_1, u_2)$ and the rewards in the states x_1, x_2 are r_1 but in the state x_3 it is r_2. Let $r_1 > r_2$. The transposition probability matrices are defined as follows

$$P^{(u_1)} = \begin{pmatrix} 1/2 & 1/2 & 0 \\ 1/2 & 1/2 & 0 \\ p & p & 1-2p \end{pmatrix}, \quad P^{(u_2)} = \begin{pmatrix} 0 & 0 & 1 \\ 0 & 0 & 1 \\ 0 & 0 & 1 \end{pmatrix}$$

where $0 < p < 1/2$. For any p this chain is equiprofitable. When $p \to 0$ we obtain the equiprofitable chain with two ergodic classes $E_1 = \{x_1, x_2\}$, $E_2 = \{x_3\}$. The rewards are equal to r_1 and r_2 for the classes E_1 and E_2 respectively. We note that the optimal strategy for the original chains with $p > 0$ consists of repeating the control u_1 unlimitedly.

Second, it is easy to verify that in the complement of the set $RM(X, U)$ there exist converging sequences of non-equiprofitable chains whose limits are equiprofitable.

The notion of equiprofitable chain plays an important role in adaptive control problems. We consider the following two aims of adaptive control:

asymptotic optimality *means constructing a strategy σ^0 ensuring the fulfilment of the equality*

$$\lim_{T\to\infty} T^{-1} \sum_{t=1}^{T} \zeta_t = \sup_{\sigma,p} W(\sigma,p) \quad \text{a.s.} \tag{1}$$

for any chain from the class $\mathcal{K} \subseteq M(X,U)$.

The strategy σ^0 will be called ***asymptotic-optimal***;

*ε-**optimality** means constructing a family of strategies σ_ε ensuring the fulfilment of the equality*

$$\lim_{T\to\infty} T^{-1} \sum_{t=1}^{T} \zeta_t > \sup_{\sigma,p} W(\sigma,p) - \varepsilon \quad \text{a.s.} \tag{2}$$

for any chain from the class $\mathcal{K} \subseteq M(X,U)$.

These strategies σ^0 will be called ε-*optimal*.

The following proposition shows the importance of equiprofitable chains in classical control theory.

Proposition. *If an asymptotic-optimal strategy (or a family of ε-optimal strategies) exists then the chain is equiprofitable.*

Obviously, such a result also holds in the adaptive version of control theory. The reverse assertion about the sufficiency of the equiprofitability condition of the Markov chain for an optimal adaptive strategy to exist is more substantial.

Theorem 1. *For a compact class $\mathcal{K} \subset RM(X,U)$ there exist ε-optimal families of strategies. In other words, for any $\varepsilon > 0$ there exist strategies which secure achievement of the aim (2) for all chains from $\mathcal{K}$.*

Theorem 2. *For the class $RM(X,M)$ there exist asymptotic-optimal strategies which secure the achievement of aim (1) for all equiprofitable Markov chains from $\mathcal{K}$.*

Compactness of the set of Markov chains is understood in the sense of the natural topology of the Euclidean space R^{km^2+km}. Hence, the class $\mathcal{K}$ is bounded and closed in this space. By Theorem 1 the family of ε-optimal strategies can be constructed under stronger assumptions than in the case of asymptotic-optimal strategies.

The proofs of Theorems 1 and 2 we give in the subsequent three sections are constructive (i.e. appropriate strategies are designed). These strategies will be based on different principles. It is possible to construct other types of optimal strategies but this would lead us far away. It remains to describe properties of optimal adaptive strategies when non-equiprofitable chains are under control.

Theorem 3. *If x_0 is an arbitrary initial state of a non-equiprofitable chain (it may belong to any connective component or to an inessential set) then the inequality*

$$\lim_{T \to \infty} T^{-1} \sum_{t=1}^{T} \zeta_t \geq \min_p \max_\sigma W(\sigma, p), \quad \text{a.s.}$$

(or its ε-version) is satisfied.

If the maximum gain is attained (under a proper initial condition!) in the ergodic class belonging to an inessential set then no adaptive strategy allows reaching it. We recall that under the non-degenerate strategies the inessential set is left with probability one in finite time. For this reason for a finite time the chain passes into one of the connective components which can be reached from the inessential set. Hence adaptive control can lead to losses which are never present in classical control.

As noted above (in Sec. 1) for a number of applied problems we are forced to restrict the collections of the controls used depending on the current state of the chain x_t, i.e. to assume that in that state the controls from the subset $U_{x_t} \subset U$ are only used. Such restrictions could be introduced into the adaptive concept without difficulty. However for the sake of exposition we have not taken them into account. We leave it to the interested readers.

5.4. The First Control Algorithm for a Class of Markov Chains (*identificational*)

The schemes of constructing the control algorithms here and in the subsequent sections are the same: first, we design some ε-optimal strategy for a fixed $\varepsilon > 0$, and after that, by the proper passing to the limit, we obtain the asymptotic-optimal strategy.

Here we are concerned with the strategies (or algorithms) of controlling the Markov chains which are based on statistical estimates of their unknown characteristics computed by observing the rewards and the changes of states. We need to estimate the elements $p_{ij}^{(u)}$ of the matrix $P^{(u)}$ and the mathematical expectations of the rewards $r_x^u = \mathbf{E}\zeta(x, u)$. We use two types of non-degenerate randomized Markov laws $\gamma = (p_1, \ldots, p_m)$: the uniform one $\gamma_0 = (1/k, \ldots, 1/k)$ and k laws of the following form $\gamma_l = \left(\frac{\delta}{k-1}, \ldots, \frac{\delta}{k-1}, 1 - \delta, \frac{\delta}{k-1}, \ldots, \frac{\delta}{k-1} \right)$, $\delta \in (0, 1/2)$, $l = 1, \ldots, k$ where $1 - \delta$ corresponds to the lth coordinate. While using such laws the connected chain proves to be ergodic, and visits all states infinitely often (using all controls in every state!).

The estimates $p_{ij}^{(u)}(t)$ of the transition probabilities $p_{ij}^{(u)}$ can be obtained in the usual way:

$$p_{ij}^{(l)}(t) = \frac{N_{ijl}(t)}{N_{il}(t)} \quad \text{for } N_{il}(t) \geq 1$$

where $N_{ijl}(t)$ means the number of transitions on the interval $[0, t]$ from x_i into x_j under the control u_l and $N_{il}(t) = \sum_j N_{ijl}(t)$. If $N_{il}(t) = 0$, then we put $p_{ij}^{(l)}(t) = 0$. Under the condition $N_{il}(t) \to \infty$ this estimate is consistent, i.e.

$$\lim_{t \to \infty} p_{ij}^{(l)}(t) = p_{ij}^{(u_l)} \quad \text{a.s.}$$

for all i, j, l. The estimates $\hat{r}_x^u$ of the mathematical expectations of the rewards are defined as the arithmetic means of the observed rewards $\zeta_t = \zeta(x_t, u_t)$. For connected chains under the laws γ_0 and γ_l these estimates are consistent.

We denote by $\mathcal{I}$ the rule which at the moment τ_n selects the "quasi-optimal" law. The selection can be done either at each moment or in every hn, $h > 1$, with the probability q. At the moment τ_n the optimal (deterministic and Markovian) law $\bar{u}$ is calculated by using the available empirical data (the collection of the matrices $(p_{ij}^l(t))$ and $(\hat{r}(t))$). For this purpose we use the linear programming method or the iterative one. Let the law $\bar{u}$ assign the control u_l to the state x_i. Then we associate the distribution $\gamma_l(x_i)$ of type δ with this state.

On the initial interval of control (up to τ_n) the law γ_0 is applied. This completely defines the identificational algorithm of control which is denoted by AI_δ. The limiting average gain corresponding to it is denoted by WI_δ.

Lemma 1. *For any connected chain from $M(X, U)$ the algorithms AI_δ form an ε-optimal family, i.e. for any $\varepsilon > 0$ and each connected chain there exists a δ_ε such that the inequality*

$$\lim_{T \to \infty} T^{-1} \sum_{t=1}^{T} \zeta_t = WI_\delta > \sup_{\sigma} W(\sigma) - \varepsilon \quad \text{a.s.}$$

holds for all $\delta < \delta_\varepsilon$.

Proof. By consistency of the estimates $p_{ij}^l(t)$, $\hat{r}_i^l(t)$ and finiteness of the set of deterministic Markov laws, the law $\bar{u}(x)$ coincides with the true optimal one from some moment τ on and after that moment it will remain unchangeable. Let σ_δ be a strategy based on infinite iteration of the randomized law which is constructed by using the δ-type distributions γ_l in the way described above from the moment τ on. The obtained Markov chain is ergodic and the limiting average reward $W(\sigma_\delta) = WI_\delta$ does not depend on the initial distribution and is continuous in δ. Hence there exists the number δ_ε declared in the lemma. In view of the ergodicity of the chain the strong law of large numbers holds. $\qquad\square$

From this lemma one interesting result follows.

Theorem 1. *The algorithms AI_δ form an ε-optimal family with respect to any compact class $\mathcal{K} \subset RM(X, U)$.*

Proof. The non-degenerate laws used make the chain pass into one of the connective components (if it has not been there from the very beginning), and the maximum limiting average reward is the same in all components. We fix $\varepsilon > 0$. To every chain $C \in \mathcal{K}$ we put in correspondence a number $\delta_\varepsilon(C)$ such that for $\delta < \delta_\varepsilon(C)$ the algorithm AI_δ ensures the attainability of the aim for the chosen chain C. We denote by $U_{c',a}$ the neighborhood of the chain C' consisting of the chains C such that $|\delta_\varepsilon(C) - \delta_\varepsilon(C')| < a\delta_\varepsilon(C')$ where $0 < a < 1/2$. The set of these neighborhoods forms an open cover of the compact set $\mathcal{K}$, and hence one can find a finite cover $U_{c_1,a_1}, \ldots, U_{c_n,a_n}$ of this set. For all $C \in U_{c_i,a_i}$ we have $\delta_\varepsilon(C) > (1 - a_i)\delta_\varepsilon(C_i)$. Therefore for each $C \in \mathcal{K}$ we have

$$\delta_\varepsilon(C) \geq \min_i(1 - a_i)\delta_\varepsilon(C_i) > \frac{1}{2} \min_i \delta_\varepsilon(C_i) = \delta_\varepsilon(\mathcal{K}).$$

This means that if we choose $\delta < \delta_\varepsilon(\mathcal{K})$ the desired inequality will hold for all $C \in \mathcal{K}$ and for any strategy AI_δ. $\qquad\square$

The arguments above show why ε-optimal strategies are constructed only for compact classes of chains from $RM(X,U)$. Indeed, for every ε, it is not difficult to construct a sequence of chains $C_1, \ldots, C_n, \ldots$ from $RM(X,U)$ such that $\delta_\varepsilon(C_n) \to \infty$ as $n \to \infty$. Therefore it is impossible to obtain ε-optimality for the whole class $RM(X,U)$. To Theorem 1 we add one result about the convergence rate of the class of regular chains.

As noted at the end of Sec. 2 the convergence on the regular class $\mathbf{E}_{\gamma,p}\zeta_t \to \mathbf{E}_\infty\zeta$ has an exponential rate defined by the absolute value of the second eigenvalue λ_2 of the matrix $P(\gamma)$. We shall now extend this result to $\mathcal{K}$, i.e. the compact class of regular chains. Namely, the following estimate holds:

$$|\mathbf{E}_{AI_\delta}\zeta_t - WI_\delta| < O(\lambda^t(\mathcal{K})), \quad 0 < \lambda(\mathcal{K}) < 1.$$

Indeed, let $\lambda(C)$ be the second largest modulus of the eigenvalues the transition matrix under the strategy σ_δ (their total number is equal to k^m). If $\lambda(C') = \rho_{C'}e^{i\phi}$ for the chain C' then the set of chains $U_{C',a}$ satisfying the condition

$$|\lambda(C) - \lambda(C')| < a(1 - \rho_{C'}), \quad 0 < a < 1$$

is called a neighborhood of this chain. The set of such neighborhoods forms a cover of $\mathcal{K}$ from which we choose a finite subcover $U_{C_1,a_1}, \ldots, U_{C_n,a_n}$. For $C \in U_{c_i,a_i}$ we have $|\lambda(C)| < a_i + (1 - a_i)\rho_{C_i}$, and hence

$$|\lambda(C)| < \max_i[a_i + (1 - a_i)\rho_{C_i}] = \lambda(\mathcal{K}) < 1$$

for any regular chain $C \in \mathcal{K}$. Therefore, the required estimate of the convergence rate holds for any $C \in \mathcal{K}$ and for any strategy from AI_δ.

It remains to consider the asymptotic-optimal identificational strategy. With this purpose in view we modify the algorithm AI_δ by allowing the parameter δ to change, and denote it by δ_n. At the current moment τ_n given by the law $\mathcal{I}$ the former value δ_{n-1} is replaced by δ_n. It is assumed that $\delta_n < \delta_{n-1}$, $\lim_{n\to\infty} \delta_n = 0$ and $\sum_{n=1}^\infty \delta_n = \infty$. According to what has been said, the laws γ of type δ are changeable as well. The described strategy that is denoted by AI can be naturally interpreted as the limit (as $\delta \to 0$) of the strategies AI_δ.

Theorem 2. *The algorithm AI is asymptotic-optimal with respect to the whole class of the equiprofitable Markov chains $RM(X,U)$.*

Proof. In finite time the chain enters one of the connective components. In this component the estimates $p_{ij}^{(u)}(t)$, $\hat{r}_x^u$ are consistent and from some non-Markov moment on, the quasi-optimal law $\bar{u}(x)$ coincides with the optimal one $\bar{u}_{\mathrm{opt}}(x)$ and remains unchangeable for ever. According to the construction of the randomized laws γ_n they converge to $\bar{u}_{\mathrm{opt}}(x)$. This means that the strategy AI is asymptotic-stationary, i.e. we can apply Theorem 4 from Sec. 2. If the chain $(X, P(\bar{u}_{\mathrm{opt}})\}$ has only one ergodic class the assertion will be proved. Otherwise, the maxima of the limiting average rewards are identical. $\qquad\square$

Thus, the identificational control algorithms are some of the possible adaptive strategies whose existence was asserted in Sec. 3.

5.5. The Second Control Algorithm for a Class of Markov Chains (*automata*)

To control Markov chains with rewards we use modifications of finite automata designed in Sec. 2, Chap. 2 to control HPIV. The efficiency of such automata is based on the following principle: the more the average reward corresponds to the controls (or the actions), the longer is the time period during which these controls are used. To the chains we should substitute the choice of actions by the choice of Markov laws and use those which bring the greater reward. We are going to discuss one concrete class of automata. For convenience and simplicity we choose the automata $\mathcal{D}_{k,n}$.

We assume that the rewards $\zeta(x,u)$ are bounded and belong to the interval $[0,1]$. The automata $\mathcal{MD}_{k^m,n}$ can be depicted by the graphs of the automata $\mathcal{D}_{k,n}$. Each of them has k^m branches and the law $\bar{u}_j$ corresponds to the jth one (we suppose that these branches are numbered). Any branch has n states. The input state has depth n and the output has depth one. The input signals of the automaton are represented by the pairs (x_t, ζ_t) consisting of the current state and reward. The control $\bar{u}_j(x_t)$ is produced in accordance with the current state x_t, and it assigns to this state the Markov law associated with the given branch. The value of the reward obtained, ζ_t, is used to calculate the empirical reward on the branch, i.e. the value $\eta_t = t^{-1} \sum_{h=1}^{t} \zeta_h$ (here t means fictitious time since it is counted from the moment when the chain has entered into this branch). With the sequence of variables $\eta_1, \eta_2, \ldots$ associate the sequence of independent r.v. $\rho_1, \rho_2, \ldots$ defined as follows:

$$\rho_t = \begin{cases} 1, & \text{with probability } \eta_t, \\ 0, & \text{with probability } 1 - \eta_t. \end{cases}$$

If $\rho_t = 1$ ("encouragement") the automaton will pass from the ith state into the nth one and will not leave it till the value $\rho = 0$ ("penalty") appears. Then the automaton passes into the state $n - 1$. Generally speaking, on the appearence of a penalty the automaton passes from the ith state into to the $i - 1$-st one. If the penalty has appeared in the first state the automaton leaves the branch and passes into another branch that is chosen either equiprobably among k^m branches or cyclically, i.e. from the jth branch into the $j + 1$-st one and from the k^mth into the first one. This completes the description of the automaton $MR(X, U)$.

Now we study the properties of these automata.

The method is the same as in the previous section: first, we show that the automata $MD_{k^m, n}$ form an ε-optimal family for each equiprofitable chain, and then this fact is to be extended to the compact class $\mathcal{K} \subset RM(X, M)$. We also consider some modifications of the automata to give them asymptotic optimality.

Lemma 1. *The automata* $\mathcal{M D}_{k^m, n}$ *are an ε-optimal family for each equiprofitable chain.*

Proof. Without restricting generality we assume that the initial state of the chain belongs to a connective component C. Let some Markov law $\bar{u}(x)$ corresponding to the branch of an automaton decompose the chain C into ergodic classes $E_1, \ldots, E_{g(\bar{u})}$ and a set of inessential states E_0. If initially the branch is in control the chain in the class E_0 then after some finite time this chain will pass into one of ergodic classes. So, we can assume the chain is ergodic with the limiting average reward $W(\bar{u})$.

The probabilities of encouragement and penalty have the form

$$q_t = \mathbf{P}\{\rho_t = 1\} = \mathbf{E}\eta_t, \quad p_t = \mathbf{P}\{\rho_t = 0\} = 1 - \mathbf{E}\eta_t$$

respectively. In the ergodic chain $\lim_{t \to \infty} \eta_t = W_j$ a.s., and apart from this $\lim_{t \to \infty} \mathbf{E}\eta_t = W_j$. Hence for any $\Delta > 0$ and all sufficiently large t ($\geq t_\Delta$ depending on the chain) we have

$$W(\bar{u}) - \Delta \leq \mathbf{E}\eta_t \leq W(\bar{u}) + \Delta.$$

This enables us to estimate the average sojourn time $T_{\bar{u}}^{(n)}$ of the automaton in the branch by using the explicit expression for the average time of the automaton $\mathcal{D}_{k, n}$ obtained in Sec. 2, Chap. 2. Note that the automaton $\mathcal{M D}_{k^m, n}$ changes its state under the action of a sequence of independent binary r.v. with nonconstant encouragement probability. These arguments lead to the estimates

$$(1 - W(\bar{u}) + \Delta)^{-n} \leq T_{\bar{u}}^{(n)} \leq (1 - W(\bar{u}) - \varepsilon)^{-n}$$

which hold for $t > t_\Delta$ for an arbitrary small number Δ. Similar arguments as in case of the HPIV enable us to conclude that the automata $\mathcal{M D}_{k^m, n}$ form an ε-optimal family (in the weak sense), i.e.

$$\lim_{t \to \infty} \mathbf{E}\zeta_t > \sup_\sigma W(\sigma) - \varepsilon_n, \quad \varepsilon_n \to 0, \quad n \to \infty,$$

since the greatest limiting average reward among all values $W(\bar{u})$ corresponds to the optimal law $\bar{u}_{\mathrm{opt}}$, i.e. this law is applied on the average no less than $(1 - W(\bar{u}) - \varepsilon)^{-n}$ times in succession. In view of the ergodicity of the chain we obtain ε-optimality in the strong sense. $\qquad\square$

We give one similar construction of automaton for controlling Markov chains.

The *automata* $MD_{k^m,n,\delta}$ do not deal with deterministic laws $\bar{u}(x)$ but with non-degenerate randomized ones: the law $\bar{u}(x) = (u_{j_1}, \ldots, u_{j_m})$ is replaced by the stochastic matrix $B_{\bar{u},\delta} = (b_{ri})$ having the rth row of the following form

$$\underbrace{\frac{\delta}{k-1}, \ldots, \frac{\delta}{k-1}}_{i_r - 1}, 1 - \delta, \frac{\delta}{k-1}, \ldots, \frac{\delta}{k-1}.$$

Such laws have been used in the previous section. The remaining automata $MD_{k^m,n,\delta}$ coincide completely with $MD_{k^m,n}$.

Lemma 2. *The automata $MD_{k^m,n,\delta}$ form an ε-optimal family with respect to the pair of parameters (n, δ) for each equiprofitable chain.*

Proof. We restrict attention to connected chains which are ergodic under the laws $B_{\bar{u},\delta}$. Then all arguments from Lemma 1 remain in force with the following supplement: the limiting average reward based on the law $B_{\bar{u},\delta}$ is equal to $W(\bar{u}) - \varepsilon(n, \delta)$. By increasing n ($\to \infty$) and decreasing δ ($\to 0$), we obtain the desired ε-optimality. $\qquad\square$

We can consider the automata $MD_{k^m,n}$ as families of ε-optimal adaptive algorithms for the classes of equiprofitable chains.

Theorem 1. *The automata $MD_{k^m,n}$ form an ε-optimal family for any compact class $\mathcal{K} \subset RM(X, U)$.*

Proof. By Lemma 1, the automata $MD_{k^m,n}$ serve as an ε-optimal family for every equiprofitable chain, i.e. for any $\varepsilon > 0$ there exists an n_ε such that $W(MD_{k^m,n}) > \sup_\sigma W(\sigma) - \varepsilon$ for all $n > n_\varepsilon$, where we choose a sequence of positive ε's converging to zero. It remains to show that in the compact class $\mathcal{K}$ the value n_ε can be chosen the same for all chains from this class. We define a system of neighborhoods in $\mathcal{K}$ as follows: fix $\varepsilon > 0$ and denote by $n_\varepsilon(C)$ an integer such that $W(MD_{k^m,n}) \geq \sup_\sigma W(\sigma) - \varepsilon$ for $C \in \mathcal{K}$ at $n \geq n_\varepsilon(C)$.

Let the set $U_{C',a}$ consist of all chains from $\mathcal{K}$ such that the condition

$$|n_\varepsilon(C) - n_\varepsilon(C')| \leq a n_\varepsilon(C'), \quad 0 < a < \frac{1}{2}$$

holds for the fixed C'. It is clear that the system of neighborhoods $U_{C',a}$ forms an open cover of $\mathcal{K}$. We can choose a finite subsystem $U_{C_1,a_1}, \ldots, U_{C_n,a_n}$ covering $\mathcal{K}$ as well. For all chains C from U_{C_j,a_j} we have $n_\varepsilon(C) \leq (1 + a_j) n_\varepsilon(C_j)$, and hence the inequality $n_\varepsilon(C) \leq 3/2 \max_j n_\varepsilon(C_j)$ holds for all chains from $\mathcal{K}$. $\qquad\square$

It is not difficult to make sure that the automata $\mathcal{MD}_{k^m,n,\delta}$ possess similar properties.

We now turn to techniques of construction of asymptotic-optimal automaton algorithms. For the automata $\mathcal{MD}_{k^m,n}$ one of them consists in increasing the memory n (as in Sec. 3, Chap. 2). Namely, we organize a cyclic change of branches, and after every complete cycle the depth of all branches increases by one with the appropriate reconstruction of the graph (in the case of encouragement we pass from any state into the deepest one). These modified automata are denoted by $\mathcal{MD}_{k^m}$.

Theorem 2. *The automata $\mathcal{MD}_{k^m}$ are asymptotic-optimal with respect to the class* $RM(X,U)$.

The proof of this theorem is almost obvious and hence omitted.

It is not difficult to use other known constructions of finite automata to synthesize optimal control strategies for Markov chains with rewards.

5.6. The Third Control Algorithm for a Class of Markov Chains (*stochastic approximation*)

In this section we study the asymptotic-optimal algorithm of controlling Markov chains. It is based on the stochastic approximation method considered and used in Sec. 4, Chap. 3 and Sec. 3, Chap. 4. This method consists of constructing a recurrent procedure which produces the randomized Markov laws B_t such that the objective function $W(\sigma, p)$ takes its extreme value in the limit (as $t \to \infty$). Without loss of generality we discuss below the minimization of this function. This means that the r.v. are interpreted as losses.

We begin with the simplest case when the regular chains forming the class $\rho(X, U)$ are under the control. In this situation the objective function $W(\sigma)$ does not depend on the initial distribution and has the following explicit form

$$W(\sigma) = \sum_{i=1}^{m} \sum_{l=1}^{k} r_i^l b_i^l \pi_i(\gamma)$$

where $\pi_i(\gamma) > 0$ is the limiting probability of the state x_i, $\gamma = (b_i^l)$ is the randomized law of the stationary strategy $\sigma(\gamma)$ used.

The weak form of the control aim being considered consists of minimizing the limiting average losses $W(\sigma)$. As we shall see later on the realization of this aim implies the fulfilment of the strong form of the control aim, i.e.

$$\lim_{T \to \infty} T^{-1} \sum_{t=1}^{T} \zeta_t = \inf_{\sigma} W(\sigma).$$

It is convenient to seek the optimal strategy in the new variables

$$z_i^l = b_i^l \pi_i(\gamma)$$

which satisfy the following evident restrictions

$$z_j^l > 0, \quad \sum_{l=1}^{k} z_j^l = \pi_j(\gamma), \quad \sum_{j,l=1}^{m,k} z_j^l = 1, \quad \sum_{l=1}^{k} z_j^l = \sum_{i,l=1}^{m,k} z_i^l p_{ij}^l,$$

$$j = 1, \ldots, m; \ l = 1, \ldots, k \quad (1)$$

and have been met before (under the notation x_j^l) in the linear programming problem for the regular chains in Sec. 2, Chap. 1. In these new variables the representation of the objective function is simplified

$$W(\sigma(\gamma)) = \sum_{i,l=1}^{m,k} r_i^l z_i^l. \quad (2)$$

From the variables z_i^l we can easily return to the former ones b_i^l

$$b_i^l = \frac{z_i^l}{\sum_{h=1}^{k} z_i^h}.$$

For the class $\rho(X, U)$ of regular Markov chains the solution of the adaptive problem consists of giving a minimization procedure of $W(\sigma(\gamma))$ with the restrictions (1), i.e. in solving the linear programming problem (1)–(2). We choose the gradient projective procedure of the form

$$z_{t+1} = \pi_t \left\{ z_t - a(t) \frac{\partial W(\gamma_t)}{\partial \gamma} \right\},$$

where

$$z_t = (z_i^l(t)), \quad \frac{\partial W(z_t)}{\partial \gamma} = \left(\frac{\partial W(z_t)}{\partial b_i^l} \right),$$

$a(t) > 0$ is some numerical sequence, and π_t is the appropriate projector. Only the variables observed up to the moment t can enter into description of this procedure: namely, the states of the chain $x(t)$ and reward $\zeta(t)$. Therefore this procedure should be rewritten in terms of the observations as it has been done in Chaps. 3, 4, i.e. it is necessary that the elements of the matrix $\partial W / \partial \gamma$ be represented as the mathematical expectations of some expressions containing only $x(t)$ and $\zeta(t)$, and that the estimates of the transition probabilities p_{ij}^l unknown beforehand and entering into restrictions (1) be obtained.

We begin by estimating the transition probabilities. The appropriate estimates are defined as follows (see Sec. 4)

$$\hat{p}_{ij}^l = \frac{N_{ijl}(t)}{N_{il}}.$$

The elements of the gradient (that the matrix is compiled of) can be represented in the form

$$\frac{\partial W(z_t)}{\partial z_i^l} = r_i^l = \mathbf{E} \left\{ \frac{\zeta_t \chi(x(t) = x_i, u(t) = u_l)}{b_i^l \pi_i(\gamma_t)} \right\}$$

$$= \mathbf{E} \left\{ \frac{\zeta_t \chi(x(t) = x_i, u(t) = u_l)}{z_i^l} \right\}, \quad i = 1, \ldots, m; \ l = 1, \ldots, k$$

where $\chi(A)$ is the indicator-function of the event A, i.e. it is equal either to one ("yes") or to zero ("no") depending on the fact whether this event occurred. Thus, the following r.v.

$$\Phi_l^t(t) = \frac{\zeta_t \chi(x(t) = x_i, u(t) = u_l)}{z_i^l}$$

serves as the statistical analog of the element of the matrix $\partial W / \partial \gamma$ at the moment t. Taking the above into account we can write this procedure in the form convenient in applications

$$z_{t+1} = \pi_t \{ z_t - a(t) A_{t_{n+1}} \} \tag{3}$$

where π_t is the projector on the simplex

$$\Sigma_{m,k}(\varepsilon_{t_{n+1}}) = \left\{ \|z_i^l\| : z_j^l \geq \varepsilon_{t_{n+1}}, \sum_{l=1}^{k} z_j^l = \sum_{i,l=1}^{m,k} z_i^l \hat{p}_{ij}^l (t_{n+1} - 1), \right.$$
$$\left. \sum_{i,l=1}^{m,k} z_i^l = 1, \; j = 1, \ldots, m; l = 1, \ldots, k \right\}.$$

Here the integers t_n, $n \geq 0$, characterizing the given simplex are the moments that the values z_i^l are calculated by using information obtained on the time interval $[t_{n-1}, t_n]$. These values characterize the laws forming the optimal adaptive strategy. The projector π_t is defined by the condition:

$$\|\pi_t(q) - q\| = \min_{p \in \Sigma_{m,k}(\varepsilon_t)} \|p - q\|$$

for any vector q. It remains to clarify the sense of the matrix A_{t_n}. It is the statistical estimate of the matrix $\partial W / \partial z$. Namely,[b]

$$A_{t_{n+1}} = \frac{1}{t_{n+1} - t_n} \sum_{t=t_n}^{t_{n+1}-1} \frac{\zeta_t e(x_t) e^T(u_t)}{e^T(x_t) z_t e(u_t)}.$$

The elements of this matrix are computed according to the following recurrent formula

$$A_{\mu\nu}(t+1) = A_{\mu\nu}(t) - \frac{1}{t - t_n + 1} \left(A_{\mu\nu}(t) - \frac{\zeta_t \chi(z(t) = z_\mu, u(t) = u_\nu)}{z_\mu^\nu(t)} \right).$$

[b]Recall that

$$e(x) = \sum_{i=1}^{m} e_i \chi(x = x_i), \quad x = (x_1, \ldots, x_m)$$

and the vector $e(u)$ differs from the vector $e(x)$ by the total number of summands. In the first case it is equal to k but in the second it is equal to m.

It remains to note that the initial values $z(1)$, $\hat{p}^l_{ij}(1)$ are supposed to be positive. This completes the description of the adaptive procedure for the Markov chains from the class $\rho(X, U)$.

To state the useful properties of procedure (3) it is necessary to impose some constraints on the numerical sequences entering into its description.

Condition D.

$$0 < \varepsilon_t \to 0, \quad a(t) > 0, \quad \frac{t_{n+1}}{t_n} \to 0, \quad \Delta t_n t_{n+1} - t_n \to \infty, \quad \text{as } t \to \infty, \ n \to \infty$$

$$\frac{t_{n+1} - t_n}{a(t+1)} \geq \frac{t_n - t_{n-1}}{a(t)}, \quad \sum_{n=1}^{\infty} \left[\frac{a(n)\Delta t_n}{\varepsilon_n t_n} + \left(\frac{\Delta t_n}{t_n} \right)^2 \right] < 0,$$

$$\lim_{t \to \infty} \frac{1}{a(t)} \left[\frac{\Delta t_n}{t_n} + |\varepsilon_{t+1} - \varepsilon_t| + \Delta t_n \left(\sum_{g=1}^{t} \varepsilon_g \Delta t_g \right)^{-1} \right] = 0$$

where $\Delta t_n = t_{n+1} - t_n$.

The main result is stated below.

Theorem 1. *Procedure (3) is asymptotically-optimal (under condition **D**), i.e. it minimizes the limit of the average losses:*

$$\lim_{T \to \infty} T^{-1} \sum_{t=1}^{T} \zeta_t = \inf_{\sigma} W(\sigma), \quad \text{a.s.}$$

for all chains from the class $\rho(X, U)$.

Proof. First, we observe that by regularity of the chain the estimates $\hat{p}^l_{ij}$ are consistent, i.e.

$$\mathbf{P}\left\{ \lim_{t \to \infty} \hat{p}^l_{ij}(t) = p^l_{ij}, \ \forall i, j, l \right\} = 1.$$

Due to this, the restrictions characterizing the set $\Sigma_{m,k}(\varepsilon_t)$ transform into the correlations to the limiting probabilities when passing to the limit. Let $z \in \Sigma_{m,k}(\varepsilon)$ and $\tilde{z}(t)$ be the projection of z on $\Sigma_{m,k}(\varepsilon_t)$. Making use of the properties of the operator π_t we have

$$\|z(n+1) - \tilde{z}(n+1)\|^2$$

$$= \sum_{i,l}(z^l_i(n+1) - \tilde{z}^l_i(n+1))^2 \leq \sum_{i,l} \left(z^l_i(n) - a(n)A_{il}(t_{n+1}) - \tilde{z}^l_i(n+1) \right)^2$$

$$= \|z(n) - \tilde{z}(n+1)\|^2 - 2a(n) \sum_{i,l} \left(z^l_i(n) - \tilde{z}^l_i(n+1) \right) A_{il}(t_{n+1}) + a^2(n)\|A(t_{n+1})\|^2$$

$$\leq \|z(n) - \tilde{z}(n)\|^2 - 2a(n) \sum_{i,l} \left(z^l_i(n) - \tilde{z}^l_i(n+1) \right) A_{il}(t_{n+1}) + a^2(n)\|a(t_{n+1})\|^2$$

$$+ K_1 \|\tilde{z}(n) - \tilde{z}(n+1)\|(1 + a(n)\|A(t_{n+1})\|), \quad n = 1, 2, \ldots.$$

Here and subsequently the quantities $K_1, K_2, \ldots, C_1, C_2, \ldots$ mean some positive constants (if they depend on some arguments, this will be shown in the notation of these constants). Multiplying these inequalities by $\Delta t_n / 2a(n)$ and summing over n, we obtain

$$\frac{1}{t_{n+1} - 1} \sum_{g=1}^{n} \Delta t_g \sum_{i,l} \left(z_i^l(g) - \tilde{z}_i^l(g) \right) A_{il}(t_{g+1})$$

$$\leq \frac{1}{2(t_{n+1} - 1)} \sum_{g=1}^{n} \frac{\Delta t_g}{a(g)} \left(\|z(g) - \tilde{z}(g)\| - \|z(g+1) - \tilde{z}(g+1)\| \right)$$

$$+ \frac{1}{2(t_{n+1} - 1)} \sum_{g=1}^{n} \frac{\Delta t(g)}{a(g)} \left[a^2(g) \|A_{t_{g+1}}\|^2 \right.$$

$$\left. + K_1 \|\tilde{z}(g) - \tilde{z}(g+1)\| (1 + a(g) \|A_{t_{g+1}}\|) \right]. \tag{4}$$

Substituting the representation of the matrix A_{t_n} on the left-hand side of this inequality, after transformations we obtain

$$\frac{1}{t_{n+1} - 1} \sum_{g=1}^{n} \sum_{i,l} (z_i^l(g) - \tilde{z}_i^l(g)) \sum_{h=t_g}^{t_{g+1}-1} \frac{\zeta_h \chi(x(h) = x_i, u(h) = u_l)}{e^T(x(h)) z_g e(u(h))}$$

$$= \frac{1}{t_{n+1} - 1} \sum_{h+1}^{t_{n+1}-1} \psi_h \overset{a.s.}{=} \sum_{h=1}^{t_{n+1}-1} \mathbf{E}(\Psi_h | \mathcal{F}_{h-1}) + o(1) \tag{5}$$

where the σ-algebra $\mathcal{F}_{h-1}$ is generated by the past history $\sigma(x_h; \zeta_s, x_s, u_s, s = 1, \ldots, h-1)$, and the r.v. Ψ_h is defined as follows

$$\Psi_h = \sum_{i,l} (z_i^l(t(h)) - \tilde{z}_i^l(t(h))) \frac{\zeta_h \chi(x(h) = x_i, u(h) = u_l)}{e^T(x(h)) z_g e(u(h))}.$$

Here the integer $t(h) = \{k : t_k \leq h < t_{k+1}\}$ denotes the number of intervals gathering data covered to the moment h. The last equality in (5) holds true a.s. due to the assumptions of Theorem 1, the equality $\sum_{l=1}^{k} z_i^l = \pi_i$ and a type of a strong law of large numbers.[c] According to the mentioned theorem, we first prove

[c]**Theorem.** *For $n \geq 0$, let a sequence of r.v. $\{\nu_n\}$ be measurable with respect to the flow of σ-algebras $\mathcal{F}_n$. If $\mathbf{E}(\nu_n | \mathcal{F}_{n-1}) < \infty$, $n = 1, 2, \ldots$, and*

$$\sum_{t=1}^{\infty} a_t^{-2} \mathbf{E} \left\{ (\nu_t - \mathbf{E}(\nu_t | \mathcal{F}_{t-1}))^2 | \mathcal{F}_{t-1} \right\} < \infty, \quad a.s.$$

where the positive numbers $a_n \uparrow \infty$ as $n \to \infty$ then

$$\lim_{t \to \infty} \left[a_t^{-1} \sum_{n=1}^{t} \nu_n - \frac{1}{a_t} \sum_{n=1}^{t} \mathbf{E}(\nu_n | \mathcal{F}_{n-1}) \right] = 0 \quad a.s.$$

This follows from the Kronecker theorem (see Sec. 3, Chap. 4) and the Martingale lemma.

the convergence of the appropriate series. Indeed

$$\sum_{\tau=1}^{\infty} \tau^{-2} \sum_{i,l} \mathbf{E}\left[(z_i^l(t(\tau)) - \tilde{z}_i^l(t(\tau)))^2 \frac{\zeta_\tau^2 \chi(x(t(\tau)) = x_i, u(t(\tau)) = u_i)}{(z_i^l(t(\tau)))^2}\Big|\mathcal{F}_{\tau-1}\right]$$

$$\leq K_2 \sum_{\tau=1}^{\infty} \tau^{-2}\left[\sum_{i,l} z_i^l(t(\tau)) \sum_{j=1}^{k} z_i^j(t(\tau))\right]^{-1} \leq C_1(\omega)\sum_{\tau=1}^{\infty} \tau^{-2}\varepsilon_{t(\tau)}^{-1}$$

$$\leq C_2(\omega)\sum_{t=1}^{\infty}\sum_{s=t_\tau}^{t_{\tau+1}-1} s^{-2}\varepsilon_{\tau(s)} \leq C_2(\omega)\sum_{\tau=1}^{\infty}\varepsilon_\tau^{-1}\int_{t_\tau}^{t_{\tau+1}-1} x^{-2}\,dx$$

$$= C_2(\omega)\sum_{\tau=1}^{\infty}\varepsilon_\tau^{-1}\left[\frac{1}{t_\tau - 1} - \frac{1}{t_{\tau+1} - 1}\right] < \infty.$$

The last inequality follows from Condition **D** since

$$\sum_{n=1}^{\infty} a(n)\Delta t_n \varepsilon_n^{-1} t_n^{-1} < \infty$$

and $\Delta t_{n+1} a(n) \to \infty$.

To obtain the desired result it remains to compute

$$\mathbf{E}(\Psi_\tau|\mathcal{F}_{\tau-1}) \overset{\text{a.s.}}{=} \sum_{i,l}(z_i^l(t(\tau)) - \tilde{z}_i^l(t(\tau)))\left(\sum_{s=1}^{k} z_i^s(t(z))\right)^{-1} \chi(x(\tau) = x_i)$$

$$\times \mathbf{E}(\zeta_\tau \mid \mathcal{F}_{t(\tau)}, u(\tau) = u_l)$$

where

$$\mathcal{F}_{t(\tau)} = \sigma\{x_{t(\tau)}; \zeta_s, x_s, u_s, s = 1, \ldots, t(\tau) - 1\}$$

or, finally, according to the properties of HPIV understood in the extended sense,

$$\mathbf{E}(\Psi_\tau|\mathcal{F}_{\tau-1}) \overset{\text{a.s.}}{=} \sum_{i,l}\left(z_i^l(t(\tau)) - \tilde{z}_i^l(t(\tau))\right)\left(\sum_{s=1}^{k} z_i^s(t(z))\right)^{-1} r_i^l\chi(x(\tau) = x_i).$$

Hence we obtain

$$\frac{1}{n}\sum_{s=1}^{n}\mathbf{E}(\Psi|\mathcal{F}_{s-1}) = \frac{1}{n}\sum_{s=1}^{n}\sum_{i,l} r_i^l\chi(x(\tau) = x_i)b_i^l(t(s)) - \sum_{i,l} r_i^l\tilde{z}_i^l - r_{i,n}^* \quad \text{a.s.}$$

where

$$r_{in}^* = \sum_{i,l} r_i^l\left[\frac{1}{n}\sum_{g=1}^{n} z_i^l(t(g))\chi(x(g) = x_i)\left(\sum_{h=1}^{k} z_i^h(t(g))\right)^{-1} - \tilde{z}_i^l\right].$$

The inequality (4) is equivalent to

$$\frac{1}{t_{n+1}-1}\sum_{s=1}^{t_{n+1}-1}\sum_{i,l}r_i^l\chi(x(s)=x_i)b_i^l t(s)-\sum_{i,l}r_i^l \tilde{z}_i^l$$

$$\le r_{1t_n}+r_{2t_n}+r_{3t_n}+o(1).$$

On the right-hand side the following notation has been used:

$$r_{1t_n}=r_{in}^*\big|_{n=t_{n+1}-1};$$

$$r_{2t_n}=\frac{1}{2(t_{n+1}-1)}\sum_{s=1}^{n}\frac{\Delta t_s}{a(s)}\big[\|z(s)-\tilde{z}(s)\|^2-\|z(s+1)-\tilde{z}(s+1)\|^2\big];$$

$$r_{3t_m}=\frac{1}{2(t_{n+1}-1)}\sum_{s=1}^{n}\frac{\Delta t_s}{a(s)}\big[\|A_{t_{s+1}}\|^2 a^2(s)+K_2\|\tilde{z}(s)-\tilde{z}(s+1)\|$$

$$\times\,(1+a(s)\|A_{t_{s+1}}\|)\big].$$

We show later on that these three sequences converge to 0 as $n\to\infty$. Having this in mind, we can complete the proof without difficulties. Indeed,

$$\overline{\lim_{n\to\infty}}\frac{1}{t_{n+1}-1}\sum_{s=1}^{t_{n+1}-1}\sum_{i,l}r_i^l\chi(x(s)=x_i)b_i^l(t(s)\le\sum_{i,l}r_i^l\tilde{z}_i^l,\quad\text{a.s.}$$

The left-hand side of this inequality defines the sequence

$$\bar{V}_t=t^{-1}\sum_{s=1}^{t}\mathbf{E}(\zeta_s|\mathcal{F}_{s-1})=t^{-1}\sum_{s=1}^{t}\sum_{i,l}r_i^l\chi(x(s)=x_i)b_i^l(t(s))$$

with respect to which we prove that it is equivalent to the current average losses at time t

$$V_t=t^{-1}\sum_{s=1}^{t}\zeta_s,$$

i.e. the equality

$$\bar{V}_t=V_t+o(t),\quad\text{a.s.}$$

holds. To do this, we use a type of strong law of large numbers besides the assumptions of theorem, i.e. we make sure that $\sum_{s=1}^{\infty}s^{-2}\mathbf{E}(\zeta_s^2|\mathcal{F}_{s-1})\le\infty$ a.s. We have

$$\sum_{s=1}^{\infty}s^{-2}\mathbf{E}(\zeta_s^2|\mathcal{F}_{s-1})\le C_3(\omega)\sum_{s=1}^{\infty}s^{-2}\sum_{i,l}^{\infty}(z_i^l(t(s))\sum_{j=1}^{k}z_i^j(t(s))^{-1}$$

$$\le C_4(\omega)\sum_{s=1}^{\infty}\varepsilon_{t(s)}^{-1}s^{-2}\le C_5(\omega)\sum_{s=1}^{\infty}\varepsilon_{t(s)}^{-1}\frac{\Delta t_s}{t_s t_{s+1}}<\infty.$$

Now, for any t and an appropriate n, namely $t_k=t'(n)$, we obtain

$$\bar{V}_t=\frac{t_n}{t}\bar{V}_{t_n}+\frac{1}{t}\sum_{s=t_n}^{t}\sum_{i,l}r_i^l\chi(x(s)=x_i)b_i^l(t_n),$$

and by assumption

$$|\bar{V}_t - \bar{V}_{t_n}| \le \frac{t - t_n}{t_n}\bar{V}_{t_n} + c_6\frac{\Delta t_n}{t} \le C_7\frac{\Delta t_n}{t} \xrightarrow[t\to\infty]{} 0.$$

The inequalities obtained above lead to the following

$$\overline{\lim_{t\to\infty}}\, V_t = \overline{\lim_{t\to\infty}}\, \bar{V}_{t_n} \le \sum_{i,l} r_i^l z_i^l = W(\tilde{z}).$$

From this and the arbitrariness of z the required assertion follows.

It remains to show that the quantities r_{jt_n}, $j = 1, 2, 3$, tend to 0 a.s. as $n \to \infty$. We begin by proving the equality

$$\mathbf{P}\left\{\lim_{n\to\infty} r_{1t_n} = 0\right\} = 1.$$

To achieve this, we write r_{1t_n} in the form $r_{1t_n} = r'_{1n} + r''_{1n}$, where

$$r'_{1n} = \sum_{i,l} n^{-1} r_i^l \sum_{s=1}^{n} \left(\tilde{z}_i^l(t(s)) - \tilde{z}_i^l\right)\chi(x(s) = x_i)\left(\sum_{g=1}^{k} z_i^g(t(s))\right)^{-1};$$

$$r''_{2n} = \sum_{i,l} n^{-1} r_i^l \tilde{z}_i^l \sum_{s=1}^{n}\left[\chi(x(s) = x_i)\left(\sum_{g=1}^{k} z_i^g(t(s))\right)^{-1} - 1\right].$$

The convergence of the estimates of transition probabilities and $\varepsilon_n \to 0$ imply that $\tilde{z}_i^l(n) \to \tilde{z}_i^l$. Since for regular chains the limiting probabilities are positive, the Cesaro averages converge to zero under these conditions and it follows that $r'_{1n} \to 0$. We now consider the second summand r''_{2n}. We have

$$\frac{1}{t_{n+1} - 1}\sum_{s=1}^{t_{n+1}-1}\chi(x(s) = x_i)\left(\sum_{g=1}^{k} z_i^g(t(s))\right)^{-1}$$

$$= \frac{1}{t_{n+1} - 1}\sum_{s=1}^{n}\left(\sum_{g=1}^{k} z_i^g(t(s))\right)^{-1}\sum_{h=t_s}^{t_{s+1}-1}\chi(x(s) = x_i)$$

$$\stackrel{\text{a.s.}}{=} \sum_{s=1}^{n}\left(\sum_{g=1}^{k} z_i^g(t(s))\right)^{-1}\sum_{h=t_s}^{t_{s+1}-1}\mathbf{P}\{x(h) = x_i|\mathcal{F}_{t_s}\} + o(1). \tag{6}$$

Here we have used again a type of strong law of large numbers because for any i we have the following inequalities

$$\sum_{n=1}^{\infty} t_n^{-2}\mathbf{E}\left\{\left(\sum_{j=1}^{k} z_i^j(t_n)\right)^{-1}\left[\sum_{g=t_n}^{t_{n+1}-1}\left(\chi(x(g)=x_i) - \mathbf{P}\{x(g)=x_i|\mathcal{F}_{t_n}\}\right)\right]^2\Bigg|\mathcal{F}_{t_n}\right\}$$

$$\le C_8(g)\sum_{n=1}^{\infty} t_n^{-2}(\Delta t_n)^2 < \infty, \quad \text{a.s.}$$

From (6) it follows that the quantities r''_{1n} can be written in the form

$$r''_{2n} = \sum_{i,l} \frac{r_i^l \tilde{z}_i^l}{t_{n+1} - 1} \sum_{s=1}^{n} \left[\left(\sum_{h=1}^{k} z_i^h(s) \right)^{-1} \sum_{g=t_s}^{t_{s+1}-1} \mathbf{P}(x(s) = x_i | \mathcal{F}_s) - \Delta t_s \right] + o(1).$$

As known from the theory of Markov chains, the vector of the state probabilities $\bar{p}_t$ converges to the vector of the limiting probabilities (as $t \to \infty$) with exponential rate. In our case

$$|\mathbf{E}\{\chi(x(\tau) = x_i) | \mathcal{F}_{t_n}\} - \sum_{g=1}^{k} z_i^g| \leq L \exp\{-\lambda(\tau - t_n)\}, \quad L, \ \lambda > 0$$

for all Markov laws and $\tau \in [t_n, t_{n+1} - 1]$. Our assumptions lead to the following estimate

$$r''_{1n} \leq \frac{C_9(\omega)}{t_{n+1} - 1} \sum_{s=1}^{n} \sum_{\tau=1}^{t_{s+1}-1} e^{-\lambda(\tau - t_s)} + o(1) \leq \frac{nC_{10}(\omega)}{t_{n+1} - 1} + o(1) \xrightarrow[n \to \infty]{\text{a.s.}} 0.$$

Hence $\mathbf{P}\{r_{1,n} \to 0\} = 1$.

Next, from the form of the sequence r_{2n} we see that it is sufficient to verify that as n goes to infinity, the sums below remain bounded:

$$\sum_{t=1}^{n} \left(\|z_t - \tilde{z}_t\|^2 - \|z_{t+1} - \tilde{z}_{t+1}\|^2 \right)^2 = \|z_1 - \tilde{z}_1\|^2 - \|z_{n+1} - \tilde{z}_{n+1}\|^2.$$

This is obvious. Our assumptions and an elementary lemma about the limit[d] imply the fulfilment of the required condition $\mathbf{P}\{r_{2n} \to 0\} = 1$.

Finally, it remains to consider the limiting behavior of r_{3n}. For this purpose, we note that procedure (3) implies

$$\|\tilde{z}_t - \tilde{z}_{t+1}\| \leq K_3 \left(|\varepsilon_t - \varepsilon_{t+1}| + \|\hat{P}(t_n) - \hat{P}(t_{n+1})\| \right)$$

where $\hat{P}(t)$ is the matrix composed of the estimates of the probabilities p_{ij}^l for all i, j, l. By this and the elementary inequalities: $2ab \leq a^2 + b^2$, $(\sum_{j=1}^{m} c_j)^2 \leq m \sum_{j=1}^{m} c_j^2$,

---

[d]**Lemma.** (*on the limit*) *For sequences* $\{\gamma_n\}, \{g_n\}$ *such that* $0 < \gamma_{n+1} < \gamma_n$, $n \geq 1$ *and* $G = \sup_n |\sum_{l=1}^{n} g_l| < \infty$ *the following estimate*

$$\left| \sum_{i=1}^{n} g_i \gamma_i^{-1} \right| \leq 2G\gamma_n^{-1}, \quad n \geq 0$$

holds. Moreover, if the numbers $h_n > 0$ *and* $\lim_{n \to \infty} h_n \gamma_n = \infty$ *then*

$$\lim_{n \to \infty} h_n^{-1} \sum_{i=1}^{n} \gamma_i^{-1} g_i = 0.$$

$m = 1, 2, \ldots$, we arrive at the following estimate

$$r_{3n} \leq \frac{k_4}{t_{n+1} - 1} \sum_{s=1}^{n} \frac{\Delta t_s}{a(s)} \left(a^2(s) \|A_{t_{s+1}}\|^2 + \|\tilde{z}_{(s)} - \tilde{z}_{(s+1)}\|^2 \right) \leq \frac{k_4}{t_{n+1} - 1}$$

$$\times \sum_{s=1}^{n} \left[a(s) \sum_{g=t_s}^{t_{s+1}-1} \frac{\zeta^2}{[e^T(x_g) z_g e(u_g)]^2} + C(\omega) \frac{\Delta t_s}{a_s} \left(|\varepsilon_s - \varepsilon_{s+1}| + \|\tilde{z}_{(s)} - \tilde{z}_{(s+1)}\| \right) \right]$$

$$= \frac{k_4}{t_{n+1} - 1} \sum_{\tau=1}^{t_{s+1}-1} \Phi'_\tau + \frac{k_4 C(\omega)}{t_{n+1} - 1} \sum_{s=1}^{n} \Phi''_s \tag{7}$$

where

$$\Phi'_\tau = a(t(\tau)) \zeta_t^2 [e^T(x_\tau) z_\tau e(u_\tau)]^{-2},$$

$$\Phi''_\tau = \frac{\Delta t_\tau}{a(\tau)} \left[|\varepsilon_\tau - \varepsilon_{\tau+1}| + \|\hat{P}(t_\tau) - \hat{P}(t_{\tau+1})\| \right].$$

We now have

$$\sum_{\tau=1}^{\infty} \tau^{-1} \mathbf{E}(\Phi'_\tau | \mathcal{F}_{t'(\tau)}) \leq C_{11}(\omega) \sum_{\tau=1}^{\infty} \tau^{-1} a(t(\tau)) \varepsilon_{t(\tau)}^{-1}$$

$$\leq C_{12} \sum_{k=1}^{\infty} a(k) \varepsilon_k^{-1} \sum_{\tau=t_k}^{t_{k+1}-1} \tau^{-1} \leq C_{13} \sum_{k=1}^{\infty} a(k) \varepsilon_k^{-1} \frac{t_{k+1} - t_k}{t_k} < \infty.$$

Therefore, in (7) the first summand converges to zero a.s. due to the assumptions and the Kronecker theorem (see Sec. 3, Chap. 3). The second one tends to zero by the Toeplitz theorem and condition **D**. Hence $r_{3n} \to 0$ as $n \to \infty$. $\quad\square$

From all possible ways in which the parameters of procedure (3) may be chosen the following is the simplest:

$$a(t) = at^{-\alpha}, \quad \varepsilon_t = \varepsilon t^{-\beta}, \quad t_n = [n^\nu]. \tag{8}$$

According to condition **D**

$$0 < \beta < \alpha < 1 - \alpha, \quad \nu > 1.$$

To obtain the rate of convergence of current losses to the minimal loss, more research should be done. But we restrict attention only to these results.

Theorem 2. *For regular chains the best convergence rate of procedure (3) has the order $o(n^{-1/5+\delta})$, $\delta > 0$, i.e.*

$$V_t = \sup_\sigma W(s) + o(n^{-1/5+\delta})$$

where $\delta > 0$ may be arbitrarily small. This rate is reached at

$$\alpha = 1/2, \quad \beta = 1/4, \quad \nu = 5/4.$$

Note that in the adaptive version the convergence rate of the current losses control is slower than that in the classical one (there its order was equal to $o(n^{-1/2+\delta})$). This may be interpreted as a pay for the actions undertaken under lack of *a priori* information.

We notice that Theorem 1 does not deal with convergence of the laws z_i^l or b_i^l that may not take place, but only with the convergence rate of the current losses to the minimum. However in adaptive control the attainability of the control aim serves as the decisive factor.

We now describe a situation when non-regular chains are controlled with the help of procedure (3). We start with connected chains. The assertion of Theorem 1 remains in force but we are forced to change the conditions imposed on the parameters of procedure (3).

Condition E.

$$0 < \varepsilon_n \to 0, \quad a(t) > 0, \quad nt_n^{-1} \to 0, \quad \varepsilon_n^{1+m^2/4}\Delta t_n \to \infty,$$

$$\lim_{n\to\infty} \frac{\Delta t_n}{a_n}\left[t_n^{-1} + \left(\sum_{s=1}^{n}\varepsilon_s^2\Delta t_s\right)^{-1}\right] = 0, \quad \frac{\Delta t_{n+1}}{a(t_{n+1})} \geq \frac{\Delta t_n}{a(t_n)}, \quad \text{as } n \to \infty,$$

$$\sum_{s=1}^{\infty}\left[\frac{a(s)\Delta t_s}{\varepsilon_s^2 t_s^2} + \left(\frac{\Delta t_s}{t_s}\right)^2\right] < \infty.$$

Choosing the parameters in accordance with (8) we assume that the following requirements

$$0 < 2\beta < \alpha < 1 - 2\alpha, \quad 1 + \beta(1 + 4^{-1}m^2) < \nu$$

are satisfied.

One can show that in the equality

$$V_t = \sup_\sigma W(\sigma) + o(n^{\delta-\lambda}), \quad \delta > 0$$

the order of convergence is at most equal to $\lambda = 4/(32 + m^2)$, where m is the number of chain states. This order is reached at

$$\alpha = 1/2, \quad \beta = 1/6, \quad \nu = 4/3 + m^2/24.$$

It is desirable to replace the two upper estimates obtained for regular and connected chains by exact ones. Such estimates being unknown as yet, we discuss here the available ones. The losses caused because of *a priori* uncertainty in the regular case have already been discussed above. In addition, we remark that in the case of connected chains the number of states m drastically influences the value of λ. For instance, for $m = 2$ $\lambda = 1/9$ but if $m = 10$, then $\lambda = 1/33$.

It remains to consider the stochastic approximation procedure (3) in the general case, i.e. to study its efficiency with respect to the class of all Markov chains $M(X,U)$.

The first (and main) step consists of proving that the chain leaves the inessential set N in finite time with probability one under the control algorithm (3). In that case the chain passes into one of the connective components (provided it was not there at the beginning). The assertion stated in Sec. 3, remains in force:

under adaptive control of chains from the class $M(X, U)$ the aim

$$\lim_{T \to \infty} T^{-1} \sum_{t=1}^{T} \zeta_t \geq \min_{p} \max_{\sigma} W(\sigma, p)$$

is attainable.

Theorem 3. *Let condition* **E** *hold and*

$$\lim_{n \to \infty} \left(\sum_{s=1}^{n} \Delta t_s \varepsilon_s^m \right) \ln n = 0. \tag{9}$$

Then the chain leaves the inessential set N in finite time with probability one if it was there at the initial moment.

Proof. On choosing an arbitrary non-degenerate law γ we consider the transition matrix $P(\gamma)$. In this matrix we distinguish m_0 rows corresponding to the states from N and obtain a square non-stochastic matrix $R_0(\gamma)$ describing the transitions into N. At least one row of the matrix $P(\gamma)$ is such that the sum of its elements is less then one for all non-degenerate laws γ, and not all matrices $R_1(\gamma), \ldots, R_L(\gamma)$ are trivial. Let us assume the convergence of the series

$$\sum_{n=1}^{\infty} \mathbf{P}\{x_{t_n} \in N\} \tag{10}$$

under condition (9). Then, by the Borel–Cantelli Lemma the current state of the chain remains in N for finite (a.s.) time.

To verify (10) we consider the transitions in the chain only at the ends of the time intervals $[t_n, t_{n+1}]$ which are defined by the probabilities $\mathbf{P}\{x(t_{n+1}) = x_j | x(t_n) = x_j, \gamma\}$, i.e. by the elements of the matrix $R_0^{\Delta t_n}(\gamma)$, $\Delta t_n = t_{n+1} - t_n$. We show that these probabilities decrease as $n \to \infty$. For the sake of simplicity we assume that $x_1, \ldots, x_{m_0}$ form the set of the inessential states N and introduce the notation $r_n = \Delta t_n / m_0$. For any $i = 1, \ldots, m_0$ we have

$$\sum_{j=1}^{m_0} \mathbf{P}\{x(t_{n+1}) = x_j | x(t_n) = x_i, \gamma\}$$

$$= \sum_{j=1}^{m_0} \sum_{j_1=1}^{m_0} \cdots \sum_{j_{r_n}=1}^{m_0} \left[R_0^{m_0}(\gamma)\right]_{ij} \cdots \left[R_0^{m_0}(\gamma)\right]_{i_{r_n-1} j_{r_n}} \left[R_0^{m_0}(\gamma)\right]_{i_{r_n} j}$$

$$\leq \sum_{j_1=1}^{m_0} \left[R_0^{m}(\gamma)\right]_{ij_1} \sum_{j_2=1}^{m_0} \left[R^{m_0}(\gamma)\right]_{j_1 j_2} \cdots \sum_{j_{r_n}=1}^{m_0} \left[R^{m_0}(\gamma)\right]_{j_{r_n-1} j_{r_n}}$$

$$\leq \left(\max_{1 \leq i \leq m_0} \max_{\gamma \in \Sigma_{mk}(\varepsilon_n)} \sum_{j=1}^{m_0} \left[R^{m_0}(\gamma)\right]_{ij} \right)^{r_n}$$

where

$$\Sigma_{mk}(\varepsilon) = \left\{ b_i^l : b_i^l \geq \varepsilon, \sum_{l=1}^{k} b_i^l = 1 \right\}.$$

From the linearity of the elements of the matrix $R(\gamma)$ with respect to b_i^l and from the form of the simplex $\Sigma_{mk}(\varepsilon)$ it follows that

$$\max_{\gamma \in \Sigma_{mk}(\varepsilon_n)} \sum_{j=1}^{m_0} \left[R^{m_0}(\gamma) \right]_{ij} \leq 1 - c\varepsilon^{m_0}, \quad C > 0$$

for any i and sufficiently small $\varepsilon > 0$. Hence

$$\sum_{j=1}^{m_0} \mathbf{P}\{x(t_{n+1}) = x_j | x(t_n) = x_i, \gamma\} \leq (1 - c\varepsilon^{m_0})^{r_n}.$$

This implies

$$\mathbf{P}\{x(t_{n+1}) \in N\} = \mathbf{E}\left[\sum_{j=1}^{m_0} \mathbf{P}\{x(t_{n+1}) = x_j | \mathcal{F}_{t_n}\} \right]$$

$$= \mathbf{E}\left[\sum_{i=1}^{m_0} \chi(x(t_n) = x_i) \sum_{j=1}^{m_0} \mathbf{P}\{x(t_{n+1}) = x_j | x(t_n) = x_i, \mathcal{F}_{t_n}\} \right]$$

$$\leq (1 - c\varepsilon_n^{m_0})^{r_n} \mathbf{P}\{x(t_n) \in N\}.$$

This and (9) imply the convergence of the series (10). $\square$

If the parameters of the procedure are chosen in the form (8), then condition (9) will be fulfilled, and hence it is not an additional restriction. For connected chains the estimate obtained of the highest order of convergence of the current losses to the limit remains in force along with the most profitable values of parameters of procedure (3).

5.7. Adaptive Optimization with Constraints on Markov Chains

The problems of adaptive control studied above can be regarded as unconditional optimization aims when an extremum of the objective function was found without any additional conditions. Apparently, this does not satisfy practical demands. Even in an elementary course of mathematical analysis one considers "Optimization with constraints" or "Problems on conditional extremum". Similar problems also arise in advanced studies of analysis. One can give some examples of real situations when conditional optimization of Markov chains is required. Below, we consider these problems both in classical and in adaptive versions.

Let the r.v. $\zeta(x, u)$, $\eta_1(x, u), \ldots, \eta_L(x, u)$ with finite mathematical expectations r_x^u, $g_x^u(1)$, $\ldots$, $g_x^u(L)$ respectively be given on $X \times U$. We determine the appropriate objective functions. Having in mind the applications of adaptive theory, we restrict attention to the limiting average (per step) rewards $W(\sigma, p)$ and $W_j(\sigma, p)$,

$j = 1, \ldots, L$. Assuming that the characteristics of the chain are known, and the values g_j are given, we put forward the following aim:

to find the conditional-optimal strategy σ^* such that

$$W(\sigma^*, p) \to \max; \quad W_j(\sigma^*, p) \geq g_j, \quad j = 1, \ldots, L \tag{1}$$

for all initial distributions p.

For the optimization problem with constraints to be solvable, it is necessary that there exist at least one strategy such that the inequalities $W_j(\sigma, p) \geq g_i$ are satisfied. We show that the stated condition (its necessity is obvious) is also sufficient. For the sake of simplicity we assume that $L = 1$.

Theorem 1. *If for a finite controlled Markov chain there exists a conditional-optimal strategy then this strategy will belong to the set Σ_{SRM}.*

To justify this assertion we can use the same approach as in the case of proving the optimality (without constraints) of the strategies from Σ_{SRM}. We now consider the techniques of calculating the conditional-optimal strategy. The control choice law will be sought by using linear programming. To achieve this we have to add some new constraints. We consider first the case of regular chains. The reduction of the optimization problem to linear programming has been done in Sec. 2, Chap. 1 but here the constraint (2b) is added:

$$\sum_{i,l} r_i^l z_i^l \to \max, \tag{2a}$$

$$\sum_{i,l} g_i^l z_i^l \geq g, \tag{2b}$$

$$\sum_l z_i^l - \sum_{i,l} p_{ij}^l z_i^l = 0, \quad j = 1, \ldots, m, \tag{2c}$$

$$\sum_{i,l} z_i^l = 1, \quad z_i^l \geq 0, \quad i = 1, \ldots, m, \ l = 1, \ldots, k. \tag{2d}$$

In the state x_i we use the $u_{l(i)}$ that the nonzero values of z_i^l correspond to. Let us analyze this problem. In (2c) one of the equations represent the linear combination of the rest and can be omitted. Let it be the mth equation. Then the basic solution contains no more than $m + 1$ nonzero elements. But the number of such elements is exactly equal to m in the linear programming problem for optimization without constraints. From this it follows that exactly one control $u_{l(i)} = u(x_i)$ is assigned to every one of the $m - 1$ states, say $x_1, \ldots, x_{m-1}$, but in the last "special" state x_m two controls u' and u'' are used with the probabilities $\alpha \ (\geq 1/2)$ and $1 - \alpha$ respectively. The stationary randomized Markov strategy with such a law serves as the required conditional-optimal strategy σ^*.

We notice that if the number of constraints is rather large ($L \geq 2$) then the law becomes more combersome.

If the discounted reward $W_\beta(\sigma, p)$ is used as the objective function then the previous arguments will remain in force except for the form of the linear programming problem (2a)–(2d).

Hence the existence of the conditional-optimal strategy is guaranteed by the constraint (2b). The regular Markov chain corresponds to such a strategy, and by the strong law of large numbers we have with probability one

$$\lim_{T \to \infty} T^{-1} \sum_{t=1}^{T} \zeta_t = W(\sigma^*) = \max_\sigma W(\sigma),$$

$$\lim_{T \to \infty} T^{-1} \sum_{t=1}^{T} \eta_t = W_1(\sigma^*) \geq g. \tag{3}$$

These arguments obviously hold for ergodic chains since the number $d \geq 2$ and the structure of the cyclic subclasses does not depend on the non-degenerate law γ used.

Let the controlled chain be connected. Any strategy from Σ_+ has only one ergodic class, and hence there is a unique stationary distribution, and if $W_1(\sigma, p) \geq 0$ at some p then this inequality will hold for all p. For the degenerate strategies this inequality may fail as the chain is decomposed into ergodic classes $E_1, \ldots, E_{h(\gamma)}$ and a class of the inessential states E_0. For some of them the constraints may fail.

The system (2a)–(2d) holds for the laws $\gamma = \Gamma(\gamma)$ with positive elements, i.e. with the constraints $z_i^l > 0, \forall i, l$ instead of $z_i^l \geq 0$. In the open polyhedron defined by these strict inequalities and by the conditions (2b)–(2d) the linear function (2a) does not reach its maximum. We examine in detail the behavior of the objective function on the boundary of this polyhedron. Let $W^{(i)}(\sigma)$, $i = 1, \ldots, h$, be the limiting average rewards associated with the ergodic classes E_i, $i = 1, \ldots, h$ (and there may exist a class of inessential states). Then the equality $W(\sigma, p) = q_1 W^{(1)}(\sigma) + \cdots + q_h W^{(h)}(\sigma)$ where $q_i > 0$ takes place. For a strategy σ^* let the inequalities

$$W^{(1)}(\sigma^*) \geq W^{(2)}(\sigma^*) \geq \cdots \geq W^{(h)}(\sigma^*)$$

hold (maybe we need to reindex the classes). Moreover, we assume that the constraints for the limiting rewards $W_1^{(j)}(\sigma^*)$ associated with the r.v. η are satisfied only for a part of the classes, for example, at $j \leq j_0 < h$. Then the Markov law generating the strategy σ^* can be "reconstructed" according to Theorem 2, Sec. 2, i.e. it can be replaced by a law that is deterministic on the set $X \backslash E_1$, and the controlled chain contains only one ergodic class E_1 with the maximum value of the limiting average reward, and the constraint is satisfied. The rest of the states prove inessential. If several classes turn out as the best we can retain either all or some of them. We have described the conditional-optimal strategy $\sigma^* \in \Sigma_{\text{SRM}}$ and for this strategy the structure of the chain has been found.

We have made sure again that keeping the constraints is sufficient for the conditional extremum problem to be solved.

In the general case when the controlled Markov chain has connective components and an inessential set N the common linear programming problem may be considered.

The law γ^* which generates the conditional-optimal strategy σ^* can be found by solving the corresponding general problem of linear programming that, in turn, can be applied to chains with any structure. It has the following form:

$$\sum_{i,l} r_i^l z_i^l \to \max, \quad \sum_{i,l} g_i^l z_i^l \geq g, \tag{4a}$$

$$\sum_{i,l} \left(\delta_{il} - p_{ij}^l\right) z_i^l = 0, \quad j = 1, \ldots, m, \tag{4b}$$

$$\sum_{l} z_j^l - \sum_{i,l} \left(\delta_{il} - p_{ij}^l\right) y_i^l = p_j^0, \quad j = 1, \ldots, m, \tag{4c}$$

$$z_i^l, \; y_i^l \geq 0, \quad i = 1, \ldots, m, \; l = 1, \ldots, k. \tag{4d}$$

The numbers p_j^0 are an initial distribution of the chain (it is convenient to let $p_j^0 = 1/m$). The basic solution defines the desired conditional-optimal law γ^*, the variables z_i^l, y_i^l refer to the states of the ergodic class and to the inessential states respectively. It is convenient to seek the conditional-optimal strategy separately for every connective component. Next, we need to check whether there are ergodic classes in the set N (under the degenerate or just deterministic strategies) and whether some of them are optimal. Under the proper initial condition the chain does not leave this class for ever. We now turn to the adaptive version of the problem.

We first consider a class of regular controlled chains $\rho(X, U)$ with $m = |X|$ states and $k = |U|$ controls. We study the conditional extremum problem from the view-point of the so-called identification approach.

At the beginning of the process of control the randomized law is assumed to be equiprobable, i.e. $b_i^l = 1/k$ for all i, j. Let $N_{ij}^l(t)$ denote the number of transitions from i to j at time t under the control l. If $N_i^l(t) = \sum_{j=1}^m N_{ij}^l(t) > 0$ then the frequencies

$$p_{ij}^l(t) = \frac{N_{ij}^l(t)}{N_i^l(t)}$$

serve, as usual, as the estimates of the transition probabilities. The estimates of the mathematical expectations of the rewards have the standard form

$$r_i^l = \frac{1}{N_i^l(t)} \sum_{s=1}^{N_i^l(t)} \zeta_s(x_s, u_s),$$

$$g_i^l(t) = \frac{1}{N_i^l(t)} \sum_{s=1}^{N_i^l(t)} \eta(x_s, u_s)$$

where $\zeta(\cdot)$ and $\eta(\cdot)$ are the observed values of the r.v. ζ and η at the time of transition from the state x_i under the control u_l. By using the empirical information obtained we can write down the linear programming problem. At the moment defined by the law $\mathcal{I}$ (at every time, in the fixed intervals, after an increase of all $N_{ij}^l(t)$ not smaller than $n_0 \geq 1$) it has the following form

$$\sum_{i,l} r_i^l z_i^l \to \max, \quad \sum_{i,l} g_i^l z_i^l > g - ct^{-\gamma}, \tag{5a}$$

$$\sum_l z_j^l - \sum_{il} p_{ij}^l(t) z_i^l = 0, \quad j = 1, \ldots, m, \tag{5b}$$

$$\sum_{i,l} z_i^l = 1, \quad z_i^l \geq 0, \quad i = 1, \ldots, m, \; l = 1, \ldots, k. \tag{5c}$$

Here t is the current time, $\gamma \in (0, 1/2)$. The regularizing term is added on the right-hand side of (5a) in order that in such exceptional cases when the hyperplane $\sum_i g_i^l z_i^l = g$ serves as one of the faces of the polyhedron (2b)–(2c) or passes through its vertex, the solution of the linear programming problem may be found.

It tourns out that the following randomized law $\gamma(t)$

$$\gamma(t) = \{u_{l(i)} = u(x_i), i = 1, \ldots, m - 1 : \text{ in the special state } x_m \text{ the controls}$$
$$u' \text{ and } u'' \text{ appear with the probabilities } \alpha' \text{ and } 1 - \alpha' \text{ respectively}\}$$

can be used as the conditional-optimal one γ^*. Using this we construct the non-degenerate δ-type law γ_δ. This law is represented by the matrix whose ith row $(i \leq m - 1)$ has the form

$$\left(\overbrace{\frac{\delta}{k-1}, \ldots, \frac{\delta}{k-1}}^{l(i)-1}, 1 - \delta, \frac{\delta}{k-1}, \ldots, \frac{\delta}{k-1} \right)$$

where δ is small and fixed. In the last mth row the numbers $\alpha' - \delta/2$ and $1 - \alpha' - \delta/2$ serve as the l'th coordinate and the l''th one respectively, and each of the remaining coordinates is equal to $\delta/(k - 2)$.

We now consider the properties of the constructed strategy. The regularity of the chain implies

$$\mathbf{P}\left\{ \lim_{t \to \infty} N_{ij}^l(t) = \infty, \forall i, j, l \right\} = 1$$

which implies that the estimates of the transition probabilities and the rewards of both types are consistent. Hence the problem (5) converges to the true problem (2), and its solution, in turn, to the solution of (2a)–(2d). In view of the solution structure we draw the conclusion that after the non-Markov moment τ (being finite a.s.) the required controls $u_{l(1)}, \ldots, u_{l(m-1)}$ will be assigned to the states $x_1, \ldots, x_{m-1}$ with probability one, and the controls u' and u'' correspond to the special state x_m but with the inaccurately computed probabilities α'' and $1 - \alpha''$. From this moment on the randomized laws γ_δ will ensure the ε-optimality[e] and, generally speaking, keeping the constraints.

Now, we slightly modify the procedure of constructing the control laws. Namely, the parameter δ is allowed to change, i.e. we put $\delta = \delta_n$ under the following constraints

$$\delta_n \downarrow 0, \quad \sum_{n=1}^{\infty} \delta_n = \infty.$$

The appropriate laws and the strategy to be made of them will be denoted by γ_n^* and $\sigma^* = \{\gamma_1^*, \gamma_2^*, \ldots\}$ respectively. By the Borel–Cantelli Lemma we can ensure that $\lim_{t\to\infty} N_{ij}^l(t) = \infty$ a.s., the estimates of the chain characteristics will be consistent, and for the time τ the conditional-optimal law will be found (with accuracy up to some probability α whose precision of calculating increases in time).

Theorem 2. *The strategy σ^* is conditional-optimal,* i.e. *it ensures the attainability of the control aim with respect to the class $\rho(X.U)$.*

Proof. For a regular chain we have

$$\lim_{t\to\infty} \frac{N_{ij}^l(t)}{t} = \pi_i^l, \quad \forall\, i, l, \ \ \text{a.s.}$$

where π_i^l is the limiting probability of the event that in the state x_i the control u_l is used. By the strong law of large numbers (Theorem 4., Sec. 2)

$$T^{-1} \sum_{t=1}^{T} \zeta_t = \sum_{i,l} \frac{N_i^l(T)}{T} \frac{1}{N_i^l(T)} \sum_{s=1}^{n_i^l(T)} \zeta_s \xrightarrow{\text{a.s.}} \sum_i \pi_i^{l(i)} r_i^{l(i)} = \max_{\sigma} W(\sigma)$$

since, according to the control choice laws, in the last sum only the probabilities $\pi_i^{l(i)}$ are positive where $l(i)$ are the numbers of controls which take part in forming the conditional-optimal law. Similar arguments can be applied to the average values η_t. Thus, the objective inequalities (3) hold. $\qquad\square$

The stated results can be extended to the case of ergodic chains without difficulties. We verify that they hold for connected chains as well. For this purpose, we use non-degenerate randomized δ-type laws for which the connected chains are ergodic. Then $N_{ij}^l(t) \to \infty$ a.s. and the estimates considered above are consistent. The choice laws of controls are sought by using the general form of the linear programming problem (4) with the regularizing correction. The solution of the last problem is transformed into the randomized law γ_δ. Further arguments coincide with the ones considered above.

---

[e]The value of the error ε depends on the choice of δ. The dependence $\varepsilon = \varepsilon(\delta)$ has no critical nature inside a *compact* class of regular chains with rewards (see Sec. 4).

We now turn to controlled Markov chains of the general form: $C_1 + \cdots + C_L + N$ where C_j, $j = 1, \ldots, L$, are the connected components, N is the set of inessential states. The aim of control is determined as follows

$$\lim_{T \to \infty} T^{-1} \sum_{t=1}^{T} \zeta_t = W^*, \quad \lim_{T \to \infty} T^{-1} \sum_{t=1}^{T} \eta_t \geq g, \quad \text{a.s.} \tag{6}$$

where $W^* = \sup_p \sup_\sigma W(\sigma, p)$.

In connection with this aim we introduce the following definition.

Definition 1. We call a controlled chain *conditional-equiprofitable* if $W_1(\sigma, p) \geq g$ for all p and the identity $\sup_\sigma W(\sigma, p) = W^{\mathrm{f}}$ holds.

It is clear that the connected chains are conditional-equiprofitable if the constraint $W_1 \geq g$ is satisfied.

Theorem 3. *The aim* (6) *is attainable with respect to a class of Markov chains if and only if this class is represented by the conditional-equiprofitable chains.*

The necessity is obvious. The sufficiency can be proved in a constructive way. The appropriate identificational type strategy has in fact been described above. The correlations (6) hold since the maximum of the limiting average (per step) reward is the same for all connective components (and for the optimal ergodic class of the set N provided it exists). Therefore, it does not matter to which component the chain passes to.

It remains to consider the class $M(X, U)$ of all Markov chains.

Theorem 4. *With respect to the chain subclass of* $M(X, U)$ *for which the constraint* $W(\sigma, p) \geq g$ *is satisfied, the conditional-optimal strategy* σ^* *secures the fulfilment a.s. of the following inequalities*

$$\lim_{T \to \infty} T^{-1} \sum_{t=1}^{T} \zeta_t \geq \min_p \max_\sigma W(\sigma, p), \quad \lim_{T \to \infty} T^{-1} \sum_{t=1}^{T} \eta_t \geq g.$$

Proof. The chain enters into one of the connective components and $\lim_{T \to \infty} \sum_{t=1}^{T} \zeta_t$ is equal to the maximum of the objective function in this component. This maximum cannot be less than $\min_p \max_\sigma W(\sigma, p)$. However this inequality can't be improved for the following reasons: (a) the initial state need not belong to the most profitable connective component, (b) if the optimal ergodic class l belongs to the inessential set N then under the strategy σ^* the chain will leave (a.s.) this class in finite time for ever. $\qquad\square$

---

[f] It is a class of states giving the maximum value of the objective function while keeping the constraint.

5.8. Minimax Adaptive Problems on Finite Markov Chains

In the optimization theory besides searching the unconditional or conditional extrema, the optimization problems with several variables have received a wide interest. For an objective function the variables are divided into two groups. The maximum of this function is sought in the variables of the first group while its minimum is sought in the variables belonging to the second group. The problems of this kind arise, for instance, in game theory where they are referred to as the minimax ones. Here we consider some problems of minimax control for finite Markov chains both in the classical statement and in the adaptive one. The dissimilarity of the previous discussion from the present one consists in forming the control which now contains two components, i.e. $u = (u', u'') \in U' \times U''$. The components u' and u'' are assumed to be placed at the disposal of player A and of player B respectively. In many problems it is reasonable to assume that in every state s of the chain each of the players has its own set of actions: the players A and B dispose of k' and k'' actions where $k' \neq k''$ respectively. For the sake of simplicity we assume that the sets U' and U'' do not vary in time. The states of the controlled chain belong to the set $S = \{s_1, \ldots, s_m\}$. In every state the rewards $\zeta(s, u', u'')$ depend on the actions of both players. The admissible strategies have the form $\sigma = (\sigma', \sigma'')$, i.e. there are two-components independent from each other. The latter means that the players do not join a coalition. The set of admissible strategies of the players is the product of the corresponding sets, i.e. $\Sigma = \Sigma' \times \Sigma''$. We agree upon the following course of the game. If at the moment t the chain is in the state $s(t) = s_i$ and both players choose the controls (or actions or moves) $u'(t)$ and $u''(t)$ then they will obtain the gains $\zeta(s(t), u'(t), u''(t))$ and $-\zeta(s(t), u'(t), u''(t))$ respectively. In other words, the game is antagonistic, i.e. it has zero-sum. At the next moment $t + 1$ the chain passes into the state s_j with probability $p_{ij}^{(u'(t), u''(t))}$, in accordance with the adopted strategies, the players make the moves $u'(t+1)$, $u''(t+1)$ and obtain their gains. This procedure is repeated sequentially.

The statements of the aims can be various. We consider here the so-called optimization aim:

player A wants to maximize the objective function while player B wants to minimize it.

The objective function has one of two forms. It represent either the discounted reward

$$W_\beta(\sigma', \sigma''; p) = \sum_{n=1}^{\infty} \beta^n \mathbf{E}_{\sigma', \sigma''; p} \zeta_n, \quad 0 < \beta < 1$$

or the limiting average (per step) reward

$$W(\sigma', \sigma''; p) = \lim_{t \to \infty} t^{-1} \sum_{n=1}^{t} \mathbf{E}_{\sigma', \sigma''; p} \zeta_n.$$

The Markov chain games defined above are called *non-terminating stochastic games* (for short NSG). They differ from common games by a random alternation, according to the Markov law, of the pay functions and the sets of rewards.

Games on Markov chains have the simplest form if every state of the chain has its own pay $k' \times k''$-matrix $M = (m_{j,l}^{(l)})$ where $k' = |U'|$, $k'' = |U''|$. If players A and B have chosen the controls $u'_{l'}$, $u''_{l''}$ and the chain was in a state s_i then they receive the gains $m_{l'l''}^{(i)}$ and $-m_{l'l''}^{(i)}$ respectively. The evolution of the chain with time implies the alternation of the pay matrices. If the pay matrix is fixed the optimal strategy of the player will be stationary. In this case it consists of using the randomized laws γ' and γ'' which are expressed in terms of the pay matrix. In a "dynamic" game the pay matrices alternate with each other according to a Markov law. We can expect the optimal strategy will be stationary. This is the case only for ergodic chains. We shall now consider this question in detail.

We assume that in the state s_i the reward is the r.v. $\zeta(s_i; u'_{l'}, u''_{l''})$ with $\mathbf{E}\zeta(s_i; u'_{l'}, u''_{l''}) = r_i^{l'l''} < \infty$, provided the players have used the actions $u'_{l'}$, $u''_{l''}$. We assume that the transition probability matrices $P^{(u',u'')}$ are irreducible for all stationary strategies from Σ. This means that we restrict attention to the ergodic chains. First, we study the structure of optimal strategies of the players. We have in mind that one player wants to maximize his own gain, but the other player wants to minimize his losses. In control theory of Markov chains optimal strategies are usually constructed in the following way: first, one finds them for the objective function W_β, i.e. for the discounted reward, and after that, one obtains the form of the optimal strategy for the limiting average (per step) reward by passing to the limit.

We deal with strategies from Σ_{SRM} generated by the laws $\gamma' = (b_i^{l'})$ and $\gamma'' = (c_i^{l''})$. Then

$$\mathbf{E}_{\sigma',\sigma'';p}\zeta_t = \bar{p}\big(P^{(\gamma',\gamma'')}\big)^{t-1}\bar{r}$$

where the vector of the average rewards has the following components

$$r_i(u', u'') = \sum_{l',l''} \mathbf{E}\zeta\big(s_i, u^{l'}, u^{l''}\big)b_i^{l'} c_i^{l''}$$

which represent some bi-linear functions of the arguments γ', γ''. It is clear that W_β is continuous in these arguments.

Theorem 1. *For the NSG on an ergodic Markov chain with discounted reward there exist randomized laws γ'_0, γ''_0 such that*

$$W_\beta(\gamma', \gamma''_0) \leq W_\beta(\gamma'_0, \gamma''_0) \leq W_\beta(\gamma'_0, \gamma''),$$

i.e. *the game has a saddle point on Σ_{SRM}.*

The proof of this assertion is beyond the scope of adaptive theory, and is for this reason omitted.

Theorem 2. *For the NSG on an ergodic Markov chain with limiting average reward there exist randomized laws γ_0', γ_0'' such that*

$$W(\gamma', \gamma_0'') \leq W(\gamma_0', \gamma_0'') \leq W(\gamma_0', \gamma''), \tag{1}$$

i.e. the game has a saddle point, and the optimal strategies of the players are stationary.

Proof. It is not difficult to verify that the function $W_\beta(\sigma'(\gamma'), \sigma''(\gamma''))$ is continuous in γ', γ''. In the following equality[g]

$$\lim_{\beta \to 1} (1 - \beta) W_\beta(\sigma'(\gamma', \sigma''(\gamma''))) = W(\sigma'(\gamma', \sigma''(\gamma'')))$$

the convergence is uniform with respect to the arguments γ', γ'' in the whole domain. Hence the function $W(\gamma', \gamma'')$ is continuous. The inequalities from Theorem 1 lead to the ones (1). From this the required assertion follows. $\qquad\square$

The stated results mean that under the assumptions of Theorem 2 the price g of the NSG is defined by the equality

$$g = \max_{\sigma'} \min_{\sigma''} W(\sigma', \sigma'') = \min_{\sigma''} \max_{\sigma'} W(\sigma', \sigma'').$$

We show by examples that the ergodicity condition is essential for Theorem 2 to be true. We first consider a connected controlled chain with three states, with the sets of controls $U' = \{u_1', u_2'\}$, $U'' = \{u_2', u_2''\}$ and the transition matrices

$$P^{(u_1', u_1'')} = P^{(u_2', u_1'')} = E, \quad P^{(u_1', u_2'')} = P^{(u_2', u_2'')} = \begin{pmatrix} 0 & 1/2 & 1/2 \\ 1/2 & 0 & 1/2 \\ 1/2 & 1/2 & 0 \end{pmatrix}.$$

The rewards given on the states of chain are represented by the matrices

$$R_1 = \begin{pmatrix} 1 & 0 \\ 0 & 0 \end{pmatrix}, \quad R_2 = \begin{pmatrix} 1 & 0 \\ 1 & 0 \end{pmatrix}, \quad R_3 = \begin{pmatrix} 0 & 0 \\ 0 & 0 \end{pmatrix}$$

where $R_i = \left(r_i^{l'l''}\right)$. The stationary strategies of players A and B are specified by the laws $\gamma' = (b_i^l) = (p(u_l'|s_i))$ and $\gamma'' = (c_i^l)$, respectively. We choose two laws for player B (the row corresponds to the number i of the state)

$$\gamma_1'' = \begin{pmatrix} 1 & 0 \\ 1 & 0 \\ 1 & 0 \end{pmatrix}, \quad \gamma_2'' = \begin{pmatrix} q & 1 - q \\ 1 & 0 \\ 1 & 0 \end{pmatrix},$$

---

[g]This equality written in conventional form

$$\lim_{x \to 1} (1 - x) \sum_{n=0}^{\infty} a_n x^n = \lim_{N \to \infty} \frac{1}{N} \sum_{n=0}^{N} a_n$$

is well-known from mathematical analysis in connection with the summation methods of numerical series.

while player A uses the strategy with the law

$$\gamma' = \begin{pmatrix} 1 & 0 \\ 1 & 0 \\ 1 & 0 \end{pmatrix}.$$

According to the formula $W(\sigma,p) = \bar{p}Q(\sigma)\bar{r}(\sigma)$ we compute the value of the limiting average reward under the stationary strategies $\sigma = (\sigma(\gamma'), \sigma(\gamma''_N))$ generated by the laws γ'_1, γ''_1. To do this, we need to calculate the transition probability matrices

$$\left(P^{(\gamma',\gamma''_1)}\right)^t = E, \quad \left(P^{(\gamma',\gamma''_2)}\right)^t = \begin{pmatrix} q^t & 2^{-1}(1-q^t) & 2^{-1}(1-q^t) \\ 0 & 1 & 0 \\ 0 & 0 & 1 \end{pmatrix}.$$

Then $Q(\gamma',\gamma''_1) = E$, and elementary computations lead to the following

$$Q(\gamma',\gamma''_2) = \begin{pmatrix} 0 & 1/2 & 1/2 \\ 0 & 1 & 0 \\ 0 & 0 & 1 \end{pmatrix}.$$

It remains to specify the initial distribution $\bar{p} = (1/3, 1/3, 1/3)$, and we can find the required values of the limiting average reward

$$W(\gamma',\gamma''_1,\bar{p}) = 2/3, \quad W(\gamma',\gamma''_2;\bar{p}) = 1/2, \quad \forall q \in (0,1).$$

We see from the above that the pay function has a discontinuity at the point (γ',γ''_1). The conclusion of Theorem 2 does not hold, the uniform convergence of the function $(1-\beta)W_\beta$ does not take place. For this reason the optimal stationary strategies may not exist.

 In the following example the controlled chain is not connected. Its sets U' and U'' are the same as in the previous example but the transition matrices are given as follows

$$P^{(u'_1,u''_1)} = P^{(u'_2,u''_1)} = \begin{pmatrix} 0 & 1 & 0 \\ 0 & 1 & 0 \\ 0 & 0 & 1 \end{pmatrix}, \quad P^{(u'_2,u''_2)} = \begin{pmatrix} 0 & 0 & 1 \\ 0 & 1 & 0 \\ 0 & 0 & 1 \end{pmatrix}.$$

The rewards are defined by the matrices

$$R_1 = \begin{pmatrix} 1 & 0 \\ 0 & 1 \end{pmatrix}, \quad R_2 = \begin{pmatrix} 1 & 1 \\ 1 & 1 \end{pmatrix}, \quad R_3 = \begin{pmatrix} 0 & 0 \\ 0 & 0 \end{pmatrix}.$$

The peculiarity of this chain is that in the second and third states the transitions do not depend on the controls. Hence the strategies of the players are only defined by the control choice probabilities in the first state. We let

$$b_1^1 = p, \quad b_1^2 = 1 - p, \quad c_1^1 = q, \quad c_1^2 = 1 - q.$$

Then it is natural to write $W(\sigma',\sigma'')$ in the form $W(p,q)$. We have

$$P(p,1) = P^t(p,1) = Q(p,1) = E,$$

at $q = 1$ and

$$
P(p,q) = \begin{pmatrix} q & p(1-q) & (1-p)(1-q) \\ 0 & 1 & 0 \\ 0 & 0 & 1 \end{pmatrix},
$$

$$
P^t(p,q) = \begin{pmatrix} q^t & p(1-q^t) & (1-p)(1-q^t) \\ 0 & 1 & 0 \\ 0 & 0 & 1 \end{pmatrix}
$$

at $q < 1$. Hence

$$
Q(p,q) = \begin{pmatrix} 0 & p & 1-p \\ 0 & 1 & 0 \\ 0 & 0 & 1 \end{pmatrix}.
$$

Finally, from the explicit representation of $W(p,q)$ it follows that

$$
W(p,q) = \begin{cases} p, & \text{if } q = 1, \\ 1 - p, & \text{if } q < 1. \end{cases}
$$

We notice that the upper price of the game and the lower price are equal to one and $1/2$ respectively. The convergence of the function $(1 - \beta)W_\beta$ is non-uniform again. In this game there are no stationary optimal strategies.

These examples show that beyond the class of ergodic controlled chains the stationary optimal strategies almost never exist. The word "almost" refers to the controlled chains with a fixed structure. These chains have several ergodic classes, their number and structure being the same for all possible stationary strategies of the players. Then the above-mentioned limiting reward is uniform with respect to (γ', γ''), and the function $W(\gamma', \gamma'', P)$ is continuous in (γ', γ'') but the conclusion of Theorem 2 holds.

We now turn to the problem of calculating the stationary optimal strategies by using the linear programming method. Denoting $z_i^l = b_i^l \pi_i$, $r_i^l = \sum_{s=1}^{k} r_i^{ls} c_i^s$, $p_{ij}^l = \sum_{s=1}^{k} p_{ij}^{ls} c_i^s$, where π_i is the limiting probability of the state i we write down once more (see Sec. 2, Chap. 1) the appropriate problem.

$$
\left.
\begin{aligned}
& \sum_{l=1}^{k} \sum_{i=1}^{m} r_i^l x_i^l \to \max; \\
& \sum_{l=1}^{k} \sum_{i=1}^{m} r_i^l z_i^l = 1; \\
& \sum_{l=1}^{k} z_j^l - \sum_{l=1}^{k} \sum_{i=1}^{m} r_i^l z_i^l p_{ij}^l = 0, \quad j = 1, \ldots, m; \\
& \quad z_i^l \geq 0, \quad i = 1, \ldots, m; \; l = 1, \ldots, k.
\end{aligned}
\right\} \quad (*)
$$

The law $\gamma' = (b_i^l)$ of player A serves as the required one against some arbitrary fixed strategy of the player B. Let $W(\gamma'')$ be the reward of player A provided player B

uses the law γ''. We treat the laws generating the stationary strategies $\sigma(\gamma'') = (c_i^l)$ as the points of an appropriate space $R^{mk''}$. The set of these laws is compact in $R^{mk''}$. In view of the continuous dependence of $r_i^{l'}, p_{ij}^{l'}$ and $c_i^{l''}$ on the strategies, we conclude that the function $W(\gamma'')$ is continuous, and hence it attains its minimum at least at one point of this compact set. We denote this point by γ_0''. It gives the optimal law of player B. Then by using $(*)$ on the previous page we can find the optimal law γ_0' of player A. The above can be interpreted as the description of the numerical algorithm of solving the game (i.e. the computation of its price and the optimal laws). Of course, such an algorithm is inefficient because it searches in the space of a large dimension defined as the m-multiple product of simplexes $\Sigma_{k''} \times \cdots \times \Sigma_{k''}$. For this reason we are interested in more constructive methods of solving the NSG. Great progress is possible here, but we make only one step sufficient for our purposes.

Let $\Gamma_i^{(v_1 \ldots v_m)}$ denote a $k' \times k''$-matrix game with the pay matrix $A = (a_{l'l''})$ where $a_{l'l''} = r_i^{l'l''} + \sum_{j=1}^m p_{ij}^{l'l''} v_j$. We denote the price (or the value) of this game by $\mathbf{Val}\,\Gamma_i^{(v_1 \ldots v_m)}$. Both for the price and for the optimal laws of the players explicit representations in terms of the elements of the matrix A are known (Shapley–Snow formula). The meaning for the notations introduced is to allow representing by them the price g and the optimal laws γ_0' and γ_0''. Indeed, let $(g^*, v_1^*, \ldots, g_m^*)$ be a solution of the problem dual to (5.8).

Theorem 3. *The following equations*

$$g^* + v_i^* = \mathbf{Val}\,\Gamma_i^{(v_1^* \ldots v_m^*)}, \quad i = 1, \ldots, m \tag{2}$$

take place. If two collections $(g^*, v_1^*, \ldots, v_m^*)$ *and* $(g, v_1, \ldots, v_m)$ *satisfy* (2) *then* $g^* = g$, $v_i^* - v_i = c$, $\forall i$.

We leave the proof of this simple assertion to the reader.

We notice that under the traditional condition $v_m = 0$ the system (2) has a unique solution. This solution of the dual linear programming problem defines uniquely the solution of the direct problem (5.8), and hence besides the price of the game it gives the optimal laws. The proposed algorithm can be improved. For example, it can be represented by a recurrent procedure which produces the price of the game and the optimal laws in the limit.

Thus, for non-terminating stochastic games on ergodic chains there are constructive solutions.

We turn now to the NSG in the adaptive version. Let a class $\mathcal{E}_{m,k',k''}$ of ergodic controlled chains with m states and with k' and k'' actions of the players be given. The players produce the laws of implementing the actions, based on information about their own moves and the obtained rewards. The aims of the players are the same as in the case of the NSG described above.

For the class $\mathcal{E}_{m,k',k''}$ the principle of designing the adaptive strategy consists of the construction of the identification. This is realized in the form of a strategy similar to AI_δ (Sec. 4). In the initial stage the players use the uniform laws and estimate the unknown numerical characteristics of the chain, i.e. the probabilities

$p_{ij}^{l'l''}$ and the average rewards $r_i^{l'l''}$. Afterwards, some numerical method of solving the NSG comes into action. If it leads to non-degenerate laws then we shall use them. Otherwise, we introduce, as shown above, an auxiliary parameter δ. Thus, each control is used an infinite number of times, and the chain enters all the states. Hence the estimates of the probabilities and average rewards are consistent. Next, we begin to decrease the values of the parameter δ, and the appropriate quasi-optimal laws begin to approach to the optimal one. The described strategy is denoted by $G\Gamma$.

Theorem 4. *The strategy $G\Gamma$ is asymptotic-optimal with respect to the class of chains $\mathcal{E}_{m,k',k''}$ under the minimax aim of control.*

In view of the results of Sec. 4 the proof of this assertion is obvious. We also notice that the stated theorem can immediately be extended to the class $\tilde{\mathcal{E}}_{m,k',k''}$ of controlled chains with fixed structure (with a fixed number of ergodic classes). But this is not the case for the wider class of Markov chains as even for connected ones optimal stationary strategies may not exist.

5.9. Controlled Graphs with Rewards

We have considered arbitrary controlled Markov chains with rewards and have obtained complete (or almost complete) results on their adaptive control. It is useful to discuss the particular case when the chains have simple transition probability matrices, i.e. matrices whose rows contain only one non-zero element (of course equal to one). It is reasonable to represent such chains in the form of automata with deterministic transition functions. We call them the *controlled graphs with rewards* and write

$$\Gamma = \{X, \Gamma^{(u)}, U; x^0; \zeta\}$$

where $X = (x_1, \ldots, x_m)$, $U = (u_1, \ldots, u_k)$ are the sets of vertices and controls respectively, $\Gamma^{(u)}$ is the set of edges which indicate the possible transitions from one vertex onto another one, x^0 is an initial state, $\zeta(x, u, \omega)$ is the reward in the state x under the control u. We assume that $\mathbf{E}|\zeta(x, u)| < \infty$ for all x, u. These objects are of independent interest and we shall meet them in the next chapter.

The optimization aim of control is *to maximize the limiting average reward*

$$W(\sigma) = \lim_{T \to \infty} T^{-1} \sum_{t=1}^{T} \mathbf{E}_\sigma \zeta_t.$$

As usual, two strong aims are considered: the ε-*optimality*

$$\lim_{T \to \infty} T^{-1} \sum_{t=1}^{T} \zeta_t > \sup_\sigma W(\sigma) - \varepsilon, \quad \varepsilon > 0, \quad \text{a.s.}$$

and the *asymptotic optimality*.

We first discuss the methods of their realization under complete information. According to the general theory we need to find the optimal Markov law $u(x)$.

The graph is supposed to be *strongly connected* (i.e. from each vertex one can pass into any another). The optimal control of the graph consists of moving along a closed path on the graph. This means that from an initial vertex x^0 one should pass into some vertex of the closed path, and afterwards this path will not be left for ever. If a path C is formed by the vertices $x_{j_1}, x_{j_2}, \ldots, x_{j_n}$ and the controls $u_{l_1}, u_{l_2}, \ldots, u_{l_n}$ the value

$$\xi(C) = \frac{1}{n} \sum_{i=1}^{n} \zeta(x_{j_i}, u_{l_i})$$

will be called the *reward on the path* C. Its mathematical expectation, i.e. the average reward on the path C

$$w(C) = \frac{1}{n} \sum_{i=1}^{n} \mathbf{E}\zeta(x_{j_i}, u_{l_i})$$

should be maximum. This represents the condition for the optimal path.

Definition 1. A closed path on the graph is called a *cycle* if it has no self-intersections, i.e. going rounds this path once no vertex is met twice.

It is easy to verify that in the optimization problem we can consider only cycles among all closed paths. The cycles' lengths are no greater than the number of vertices of the graph. Thus, for a strong connected graph the calculation of the optimal control consists of remaking the choice of the cycles C_j item-by-item in order to find a cycle C for which the average cyclic-reward $w(C)$ is maximal.

We turn now to the adaptive formulation of the graph control problem. The system of transitions $(\Gamma^{(u)}, u \in U)$ and the average values of rewards are assumed to be unknown. The sets X and U are known. The vertices of the graph and the rewards serve as the observable quantities. We deal with the class $\Gamma(X, U)$ of strong connected graphs with rewards and with fixed sets X and U.

Let us describe a version (the identification one) of adaptive control of graphs from the class $\Gamma(X, U)$.

In the first stage of control whose duration is greater than mk times, we reconstruct all edges. To do this, for any vertex we choose the control u_j at the moment of getting to this vertex for the jth time ($j \leq k$) and keep track of the path specified by this control. Thereafter the structure of the graph, i.e. the system $\Gamma^{(u)}$, becomes known but the average rewards $\mathbf{E}\zeta(x, u)$ are still unknown. In the second stage we seek the optimal cycle. We number all cycles $C_1, \ldots, C_M$ and all paths p_{ij} leading from the cycle C_i to the cycle C_j ($i \neq j$). We form two vectors $\bar{N}(t) = (N_1(t), \ldots, N_M(t))$ and $\bar{V}(t) = (V_1(t), \ldots, V_M(t))$ where $N_h(t)$ means the number of full rounds of the hth cycle up to the moment t, $V_h(t)$ being the average reward obtained as a result of these rounds (more exactly, the mean of the cyclic rewards $\xi(C)$). These quantities are re-computed in an obvious way. The change of cycles is regulated by the vector $\bar{p}(t) = (p_1(t), \ldots, p_m)$ whose ith component means the probability the cycle C_h is chosen with. On going the rounds of the cycle C_g the transition law from C_g to the cycle C_h (the path C_{gh}) is produced, a law of going the rounds of the cycle C_h being determinate. If the round of some cycle is made twice in secession then the path of the type C_{gh} will not be used. The vector $\bar{p}(t)$ is transformed in a similar manner as in the automaton of type G (in controlling the HPIV, see Sec. 3, Chap. 2), i.e. at the instants of time being specified by the law $\mathcal{I}$, we choose the maximum reward $V_{j_0}(t)$ among the empirical rewards $V_j(t)$ and the component $p_{j_0}(t)$ of the vector $\bar{p}(t)$ is taken equal to $1-\delta$, $(0 < \delta < 1/2)$, the other components being equal to $\delta(M-1)^{-1}$. Moreover, we form the leading group. It consists of such components of the vector $\bar{V}(t)$ which differ from the "leader" V_{j_0} no more than by the number $\varepsilon > 0$ fixed in advance. At the next transformation moment the leader and the leading group are formed again. If the new value of V_{j_0} proves to get to the new leading group the vector $\bar{p}(t)$ will remain the same. Otherwise, the probability $1 - \delta$ is assigned to the component corresponding to the new leader. This completes the description of the optimal strategy which we denote by $G\Gamma_\delta$.

We introduce the class $\Gamma(X, U; \rho) \subset \Gamma(X, U)$ of rewards which satisfy the condition $|\mathbf{E}\zeta| \leq \rho$. This class is compact in the topology we have used earlier.

Theorem 1. *The strategy $G\Gamma_\delta$ forms an ε-optimal family with respect to the class $\Gamma(X, U; \rho)$, i.e. for any $\varepsilon > 0$ there is a $\delta_\varepsilon > 0$ such that the strategies $G\Gamma_\delta$ with $\delta < \delta_\varepsilon$ ensure the fulfilment of the following inequality*

$$\lim_{T \to \infty} T^{-1} \sum_{t=1}^{T} \zeta_t > \sup_\sigma W(\sigma) - \varepsilon, \quad a.s.$$

for all processes from the class under consideration.

Proof. We verify that beginning from some non-Markov moment τ_δ such that $\mathbf{P}\{\tau_\delta < \infty\} = 1$ the choice probability of the optimal cycles becomes equal to $1 - \delta$. Indeed,

$$\mathbf{P}\left\{ \lim_{t \to \infty} N_h(t) = \infty, h = 1, \ldots, M \right\} = 1$$

since all choice probabilities of the cycles are bounded from below by some positive constant. Hence $\lim_{t \to \infty} V_h(t) = w(C_h)$ a.s. and there exists a moment the values $V_h(t)$ will be arranged in decreasing order and in the same order as $w(C_h)$ afterwards. In view of the transformation method of the vector $\bar{p}(t)$ the choice probability of the optimal cycle C_{h_0} will be equal to $1 - \delta$ and from the moment τ_δ on, this probability will remain the same. The corresponding distribution $\bar{p}_\delta$ is limiting on the set of cycles. Then the limiting average reward is equal to

$$W(\bar{p}_\delta) = (1 - \delta)W(C_{\text{opt}}) + \frac{\delta}{M - 1} \sum_{C \neq C_{\text{opt}}} W(C'CC''),$$

where C_{opt} is the optimal cycle, $W(C'CC'')$ is the total mathematical expectation of the rewards on the paths $C'CC''$ from the optimal cycle C_{opt} to the other cycles

C and back. The inequality $|W(C)| < \rho$ implies the existence of δ_ε such that $W(\bar{p}_\delta) > W(C_{\mathrm{opt}}) - \varepsilon$ for all $\delta < \delta_\varepsilon$ and for all processes from $\Gamma(X, U; \rho)$. $\qquad\square$

We now verify that for the strategies $G\Gamma_\delta$ there exists $\lim_{T\to\infty} T^{-1} \sum_{t=1}^T \zeta_t$. But this is obvious since on the vertex set X the probabilities of the transitions induced by the distribution $\bar{p}_\delta$ generate a regular Markov chain due to the strong connectedness of the graph.

We modify now the strategy $G\Gamma_\delta$ by the method used often before. For this purpose, we choose a sequence $\delta_n > 0$ decreasing monotonically to zero and such that $\sum_{n=1}^\infty \delta_n = \infty$. At the transformation moments of the vector $\bar{p}(t)$ defined by the law $\mathcal{I}$ we replace δ_n by δ_{n+1} and agree to denote this strategy by $G\Gamma_0$.

Theorem 2. *The strategy $G\Gamma_0$ is asymptotic-optimal with respect to the class $\Gamma(X, U)$, i.e. it ensures the fulfilment of the equality*

$$\lim_{T\to\infty} T^{-1} \sum_{t=1}^T \zeta_t = \sup_\sigma W(\sigma).$$

for all process of the class in question.

The proof can be done by the arguments used above. Therefore it is omitted.

The method of constructing adaptive strategies that was developed above can be extended to another class of controlled processes. They are described by automata with deterministic transition functions and with the output function $\mu(\cdot|u_{t-1}, \ldots, u_{t-r})$. We call them the *generalized processes with independent values in the narrow sense* (NGPIV).

We denote the class of scalar NGPIV with finite mathematical expectation by $\mathcal{O}_{k,r}$. We write them in the form of automata

$$\xi = (U, S, X; \lambda, \mu)$$

where $U(u_1, \ldots, u_r)$ and $X = R$ are an output alphabet and an input one respectively. $S = \{s\} = U^r$ is a set of states (i.e. the collection $u_{j_1} u_{j_2} \cdots u_{j_r}$), the deterministic transition function λ such that $\lambda : X \times S \to S$ is specified by the following equality $\lambda(u, u_{j_1} u_{j_2} \cdots u_{j_{r-1}} u_{j_r}) = u_{j_2} u_{j_3} \cdots u_{j_r} u$ and, finally, $\mu = \mathbf{P}\{\cdot|s\}$ is an output probabilistic function whose arguments are the collection of r preceding controls. These automata are strongly connected, i.e. from any state one can pass into any another in no more than r steps. The control methods described in this section can be applied to them.

In the situation under consideration the dependence of the optimal strategy on the current states implies the equality

$$u_t = g(u_{t-1}, \ldots, u_{t-r}),$$

i.e. the optimal strategy is of the program type.

Adding here the necessity to go round the states of the automaton cyclically, we conclude that for any process from the class $O_{k,r}$ the optimal strategy consists of repeating the controls $\bar{u}_{\mathrm{opt}} = u_{i_1} u_{i_2} \cdots u_{i_r}$ an infinite number of times. The closed path corresponding to it can be chosen without self-intersections and its maximum length does not exceed the number of states $|S| = k^r$. The cyclic reward

$$
\begin{aligned}
\chi_L(\bar{u}_L) = L^{-1}\big[& \zeta_t(u_{i_r} \cdots u_{i_2} u_{i_1}) + \zeta_{t+1}(u_{i_r+1} \cdots u_{i_2}) \\
& + \zeta_{t+L-r}(u_{i_L} u_{i_{L-1}} \cdots u_{i_{L-r+1}}) + \zeta_{t+L-r+1}(u_{i_1} u_{i_L} \cdots u_{i_{L-r+2}}) \\
& + \cdots + \zeta_{t+L-1}(u_{i_{r-1}} \cdots u_{i_1} u_{i_L})\big]
\end{aligned}
$$

is assigned to each collection $\bar{u}_L$. Its mathematical expectation $w(\bar{u})$ is called the *average cyclic reward*. It consists of the quantities

$$
w(u^{(1)} \cdots u^{(r)}) = \int_{-\infty}^{\infty} z\mu\big(dz|u^{(1)} \cdots u^{(r)}\big)
$$

which form the r-dimentional matrix $A(k,r) = (w(u^{(1)} \cdots u^{(r)}))$ of the average values. The element $w(u^{(i_1)} \cdots u^{(i_r)})$ is placed in the "intersection" of the rows $i_1, \ldots, i_r$. For this matrix the notion of "cycle" is similar to that on the set of states of the automaton, i.e. it is a sequence of average rewards on the chain of states which pass one into the other under the action of the input signals. Associating the cycle on the matrix $A(k,r)$ with the corresponding average cyclic reward we can find the "optimal cycle" by remaking the choice of the cycles item-by-item (their lengths do not exceed k^r). The maximum average cyclic reward corresponds to the optimal cycle found. The unlimited repetition of this cycle represents the optimal control in the classical version of the problem in question.

We turn to the adaptive variant of this problem for the class $O_{k,r}$ of the NGPIV, i.e. the distribution μ is unknown. *The adaptive strategy GO_δ produces collections of controls which generate, in turn, the cycles on the set of states: the one-component ones $(u_1, \ldots, u_r)$, the two-component ones $(u_1 u_2, u_2 u_3, \ldots, u_{k-1} u_k)$ and so on; up to the collections having the maximum length k^r. The cycle $\bar{u}_l$ having length l is reapeted h times where h satisfies the condition $(h-1)l \geq r$. The cyclic reward is defined by using l last values of the process. It is placed into the corresponding components V_i of the vector $\bar{V}(t)$ but the component N_i of the vector $\bar{N}(t)$ increases by one. If the component V_{i_0} proves to be the greatest, then the i_0th component p_{i_0} of the vector $\bar{p}(t)$ will be equal to $1 - \delta$ but the others will be identical. The subsequent description of the strategy GO_δ coincides with that of $G\Gamma_\delta$. The strategy CO_0 is defined by analogy with that $G\Gamma_0$.*

Let $\mathcal{O}_{k,r,\rho}$ be a class of NGPIV. This class represents the subset of the class $\mathcal{O}_{k,r}$ satisfying the additional condition $\mathbf{E}|\zeta| \leq \rho$.

Theorem 3. *The strategies GO_δ form an ε-optimal family with respect to the class $\mathcal{O}_{k,r,\rho}$. The strategy GO_0 is asymptotic-optimal with respect to the class $\mathcal{O}_{k,r}$.*

The proof of the stated theorem is similar to that of Theorems 1 and 2 and is therefore omitted.

CHAPTER 6

**CONTROL OF PARTIALLY OBSERVABLE MARKOV CHAINS
AND REGENERATIVE PROCESSES**

To some extent the present chapter is a continuation of Chap. 5. It deals with homogeneous finite Markov chains with partially observable states. This allows to narrow the set of the admissible strategies.

First, we study "conditional Markov chains" for which besides the rewards the "pseudostates" are observed. Next, control problems of Markov chains under a lack of information about their states are discussed, i.e. only the rewards are observed. We construct optimal strategies for the more general class of regenerative processes. For controlled graphs the optimal strategy takes a simpler form.

6.1. Preliminary Remarks

To describe most physical, chemical and many other phenomena of the surrounding world, Markov processes with discrete or continuous time are often used. Phenomena characterized by differential or difference equations can be regarded as such. But the practical use of such models has some difficulties: dynamics of the states are not accessible to the observer. This is the case when the available devices record the states of the object only partially. Then the accessible information on the phenomena is presented in the form of a functional given on the states and controls of the unobserved Markov process. Thus, the Markov process underlying the phenomenon is "hidden" but the investigator (or user) knows only the functional (scalar or vector) on this process.

Let us consider a class $M = M(X, U)$ of finite homogeneous controlled Markov chains with the functionals (rewards) $\zeta(x, u)$ that are the r.v. given on $X \times U$ and submitted to the condition $|\mathbf{E}\zeta(\cdot)| < \infty$.

Definition 1. A Markov chain is called **partially observable** if the control choice rules which form the admissible strategy $\sigma = \{\nu_t\}$ do not depend on the current and previous states.

The above-mentioned strategies form the "*set Σ of the admissible strategies*". The rewards $\zeta_t = \zeta(x_t, u_t)$ generated by the partially observable chains and by the strategies from Σ are, generally speaking, non-Markov random processes. This produces serious difficulties.

173

We now consider three types of partially observable Markov chains.

(a) The set $Z = \{z_1, \ldots, z_m\}$ of observations (pseudostates) is given. If at the moment t the state $x(t) = x_i$ is possible then with probability q_{ij} the pseudostate z_j appears. This can be interpreted as the observation of the true state $x \in X$ under the influence of noises. The numbers q_{ij} form the stochastic $n \times m$-matrix $Q = (q_{ij})$. In addition, it is assumed that the rewards ζ_t are observed;

(b) The rewards are only observed and there are no pseudostates;

(c) The state of the chain $x(t)$ has two components, i.e. $x = (\theta, \eta)$ and only one of them, say, η is observed. The evolution of the process η_t together with the control u_t produced are accessible to direct observation. The component θ_t is unobserved and it is not used in the course of control.

In the case (a) the rules have the form $F_t(\cdot | z^t, \zeta^t, u^{t-1})$ but in the cases (b) and (c) they have the form $F_t(\cdot | \zeta^t, u^{t-1})$, and $F_t(\cdot | \eta^t, u^{t-1})$ respectively. These rules characterize the partial-observability of the Markov chain. It is natural that narrowing of the class of admissible strategies and the lack of information about the states of the chain result in the decrease of the maximum of the objective function.

For the partially observable Markov chains an aim of control has the same form as in the case of total observability but the sets of the admissible strategies are different. In the course of investigation of the adaptive strategies for such chains we shall see that appropriate methods can be extended to a wider class of controlled processes called the regenerative processes and having great importance in applications, in particular, in queueing problems.

For the partially observable chains the optimal control problems were stated in the classical variants but only on a finite time interval. Its proper solution has not been found yet.

Apparently, this problem was not studied for an infinite time interval as well.

In the next section we discuss control of conditional Markov chains. Throughout the remainder of this chapter control algorithms of chains and regeneration processes with unobservable states will be studied.

6.2. Control of Conditional Markov Chains

We study finite homogeneous controlled Markov chains with rewards $C = \{X, P^{(u)}, U, \bar{p}; \zeta\}$. Their states x are assumed to be unobservable but we can observe their "pseudostates" z from some set Z where $|Z| = |X| = m$. These states appear according to following the rule: if at the moment t the chain C is in the state x_i then the pseudostate z_j will appear at the same moment with probability q_{ij} which does not depend on the past history. The numbers q_{ij} form a quadratic stochastic matrix $Q = (q_{ij})$.

Definition 1. The object

$$CC = \{X, Z, P^{(u)}, Q, U, \bar{p}; \zeta\}$$

is called a *controlled conditional Markov chain.*

We denote the set of admissible stationary strategies for controlled conditional Markov chains by $\Sigma^{(z)}$. The same set but for the original chain C is denoted by $\Sigma^{(x)}$. We introduce two subsets of $\Sigma^{(z)}$:

$\Sigma_+^{(z)}$ consists of non-singular strategies generated by the randomized non-degenerate rules $B = (b_i^l)$ where $b_i^l = \mathbf{P}\{u(t) = u_l | z(t) = z_i\} > 0$;

$\Sigma_d^{(z)}$ consists of deterministic strategies with k^m rules of the form $\bar{u}(z) = (u(z_1), \ldots, u(z_m))$ where $u(z_i) \in U$, $i = 1, \ldots, m$.

These strategies act on the Markov chain C and induce, in turn, the strategies that are naturally denoted by $\Sigma^{(x)}$. It is clear that the non-degenerate strategies for CC are such as for C.

The stationary strategies σ generate a homogeneous Markov chain in the usual sense with the transition probability matrix $p(\sigma) = (p_{ij}(\sigma))$ where

$$p_{ij}(\sigma) = \sum_{l=1}^{k} \sum_{h=1}^{m} p_{ij}^l q_{ih} b_h^l.$$

The triplet $(x_t, z_t, u_t) \in X \times Z \times U$, where the pair (z_t, u_t) is observable, is the result of applying the strategy σ to CC. There are two possible interpretations of the rewards:

(1) $\zeta_t = \zeta(x_{t-1}, x_t, u_{t-1}; \omega)$;
(2) $\zeta_t = \zeta(x_t, u_t; \omega)$.

Assuming, as usual, that mathematical expectations $r_i^l = \mathbf{E}\zeta(x_i, u_l)$ are finite, we shall use the second interpretation. An objective function is defined in the standard way, i.e. it is the limiting average reward (per step)

$$W(\sigma, p) = \lim_{T \to \infty} T^{-1} \sum_{t=1}^{T} \mathbf{E}_{\sigma,p} \zeta_t.$$

We can seek its maximum in various classes of strategies (but we restrict attention to the stationary ones)

$$W_+ = \sup_{\sigma \in \Sigma_+^{(z)}} W(\sigma, p), \quad W_d = \max_{\sigma \in \Sigma_d^{(z)}} W(\sigma, p).$$

In contrast with ordinary Markov chains we have $W_+ \geq W_d$, the inequality being strict in typical situations. We put forward the next optimization aim of the control *to achieve the equality*

$$\lim_{T \to \infty} T^{-1} \sum_{t=1}^{T} \zeta_t = W_d, \quad a.s. \tag{1}$$

with respect to some classes of controlled conditional Markov chains.

In the adaptive version we assume that at each moment t the quantities z_t, ζ_t are observed and the sets Z, U and the matrix Q are known. First, we briefly consider the classical version when the matrices $P^{(u)}$ and average rewards r_i^l are known as well. Here we give the method of attaining the aim (1). For this purpose, we introduce the notation

$$x_{ij}^l = \pi_i(\sigma) q_{ij} b_j^l \tag{2}$$

where $\pi_i(\sigma)$ is the limiting probability of the ith state of the original chain C when the strategy σ is used. We assume that the inverse matrix $\Lambda = Q^{-1} \overset{\text{def}}{=} (\lambda_{ij})$ is well defined and the chain C is connected.

Theorem 1. *The strategy σ is optimal (with respect to the aim (1)) if and only if the quantities x_{ij}^l represent the following linear programming problem*

$$\sum_{i=1}^{m}\sum_{j=1}^{m}\sum_{l=1}^{k} r_i^l x_{ij}^l \to \max, \tag{3}$$

$$\sum_{i=1}^{m}\sum_{j=1}^{m}\sum_{l=1}^{k} [\delta_{in} - p_{in}^l] x_{ij}^l = 0, \quad n = 1, \ldots, m, \tag{4}$$

$$\sum_{i=1}^{m}\sum_{j=1}^{m}\sum_{l=1}^{k} x_{ij}^l = 1, \tag{5}$$

$$\sum_{i=1}^{m}\sum_{j=1}^{m}\sum_{l=1}^{k} \lambda_{jn} x_{ij}^l \geq 0, \quad n = 1, \ldots, m, \tag{6}$$

$$x_{ij}^l \geq 0, \quad i, j = 1, \ldots, m, \quad l = 1, \ldots, k. \tag{7}$$

Proof. We introduce the matrix $\bar{X} = (x_{ij}^l)$ and the mapping

$$\varphi : \Sigma_+^{(z)} \to \{\bar{X}\}$$

of the non-degenerate strategies into the set of matrices $\bar{X}$ with elements of the form (2). Next, let

$$\Psi(\bar{X}) = \sum_{i=1}^{m}\sum_{j=1}^{m}\sum_{l=1}^{k} r_i^l x_{ij}^l$$

and $\mathfrak{M}$ be the family of matrices (x_{ij}^l) satisfying (4)–(7). Using these notations we can write the equalities (2) in the form $\bar{X} = \varphi(\sigma)$, $\sigma \in \Sigma_+^{(z)}$ and the linear programming problem (3)–(7) can be reformulated as follows:
    to find a matrix $\bar{X}_0$ such that

$$\sup_{\bar{X} \in \mathfrak{M}} \Psi(\bar{X}) = \Psi(\bar{X}_0).$$

In this interpretation we should prove the following equalities

$$W(\sigma) = \Psi(\varphi(\sigma)), \quad \sigma \in \Sigma_+^{(z)}; \quad \varphi(\Sigma_+^{(z)}) = \mathfrak{M}. \tag{8}$$

The first of them follows from the equalities

$$W(\sigma) = \sum_{i=1}^{m} \pi_i(\sigma)\mathbf{E}_\sigma(\zeta_t|x(t) = x_i)$$

$$= \sum_{i=1}^{m} \pi_i(\sigma) \sum_{j=1}^{m}\sum_{l=1}^{k} r_i^l \mathbf{P}_\sigma\{u(t) = u_l|z(t) = z_j, x(t) = x_i\}$$

$$\times \mathbf{P}\{z(t) = z_j|x(t) = x_i\}$$

$$= \sum_{i=1}^{m}\sum_{j=1}^{m}\sum_{l=1}^{k} r_i^l \pi_i(\sigma) q_{ij} b_j^l = \sum_{i=1}^{m}\sum_{j=1}^{m}\sum_{l=1}^{k} r_i^l x_{ij}^l = \Psi(\bar{X}).$$

To verify the second equality in (8) we define the inverse mapping φ^{-1} which assigns to the matrix X^{-1} the stochastic matrix $B = (b_j^l)$ with the elements

$$b_j^l = \frac{1}{m} \sum_{i=1}^{m} \frac{x_{ij}^l}{\sum_{h=1}^{k} x_{ij}^h},$$

where the external sum is taken over i such that $\sum_{h=1}^{k} x_{ij}^h > 0$. The equality $\varphi^{-1}(\varphi(\sigma)) = \sigma$ together with the stochasticity of B imply $\varphi^{-1}(\mathfrak{M}) \in \Sigma_+^{(z)}$, i.e. $\mathfrak{M} \subset \varphi(\Sigma_+^{(z)})$. It remains to verify the opposite inclusion. We choose $\sigma \in \Sigma_+^{(z)}$. Let $X = \varphi(\sigma)$. The relations (4), (5), (7) are obvious. We prove relation (6)

$$\sum_{i=1}^{m}\sum_{j=1}^{m}\sum_{l=1}^{k} \lambda_{jn} x_{ij}^l = \sum_{i=1}^{m}\sum_{j=1}^{m}\sum_{l=1}^{k} \pi_i(\sigma) q_{ij} b_j^l \lambda_{jn} = \sum_{i=1}^{m} \pi_i(\sigma) \sum_{j=1}^{m} q_{ij}\lambda_{jn} \sum_{l=1}^{k} b_j^l$$

$$= \sum_{i=1}^{m} \pi_i(\sigma)\delta_{in} = \pi_n(\sigma) \geq 0,$$

the equality holds for the strategies from $\Sigma_d^{(z)}$. $\qquad\square$

So, we have shown a calculation technique of the optimal rules which maximize (3). They generate the optimal strategy $u^*(= u^*(z))$. We focus attention on the class CC for which we construct the optimal adaptive strategy. We denote this class by $C\mathcal{K}(X, U; Q)$ and assume that:

1. The original Markov chain C is connected;
2. For any strategy $\sigma \in \Sigma_+^{(z)}$ the Markov chain C is regular.
3. $\det Q \neq 0$, matrix Q is known and the same for the whole class.

The construction of the adaptive strategy is based on the identificational approach: first, we estimate the elements of the matrices $P^{(u)}$; next, making use the linear programming problems (3)–(7) and the current observations z_t, ζ_t, we find the quasi-optimal rules of control which will approach the optimal rules in the class of strategies $\Sigma_d^{(z)}$.

We start with the construction of estimates of the transition probabilities of the chain. Let $N_{rs}^l(t)$ denote the number of transitions for time t from the pseudostate z_r into z_s under the control u_l. The frequency of this event is equal to $\varkappa_{rs}^l(t) = t^{-1}N_{rs}^l(t)$. The number $N_r^l = \sum_{s=1}^m N_{rs}^l(t)$ will characterize the number of visits in the state z_r for time t provided that the control u_l is only used in this state. We put

$$p_t'(i,j,l) = \frac{\sum_{r,s=1}^m \lambda_{ri}\lambda_{sj}N_{rs}^l(t)}{\sum_{r,s,j=1}^m \lambda_{ri}\lambda_{sj}N_{rs}^l(t)} = \frac{\sum_{r,s=1}^m \lambda_{ri}\lambda_{sj}\varkappa_{rs}^l(t)}{\sum_{r,s,j=1}^m \lambda_{ri}\lambda_{sj}\varkappa_{rs}^l(t)}.$$

If the denominator is equal to zero we shall put $p_t'(\cdot) = 0$. Here λ_{ij} are the elements of the matrix Q^{-1}. In the special case when the chain is observable, i.e. $Q = \Lambda = E$ this expression takes the ordinary form

$$p_t'(i,j,l) = \frac{N_{ij}^l(t)}{N_i^l(t)}.$$

Theorem 2. *Let $\sigma(z) \in \Sigma_+^{(z)}$. For any CC from the class $CK(X, U; Q)$ the equality*

$$\mathbf{P}\left\{ \lim_{t\to\infty} p_t'(i,j,l) = p_{ij}^l, \ \forall(i,j,l) \right\} = 1$$

holds.

Proof. We consider the original controlled Markov chain C under the strategy $\sigma(x)$ induced by the primary non-degenerate strategy $\sigma(z)$. According to the strategy $\sigma(x)$, the probability to choose u_l in the unobservable state x_i is equal to $\mathbf{P}\{u(t) = u_l | x(t) = x_i\} = \sum_{h=1}^m q_{ih}b_h^l$. We define the following function on the chain

$$\chi_t(i,j,l) = \delta_{z_{t-1},i}\delta_{z_t,j}\delta_{u_{t-1},l}.$$

Recalling that the non-degenerate strategy $\sigma(x)$ turns the connected chain C into a regular chain, we use the strong law of large numbers:

$$\lim_{t\to\infty} \varkappa_{ij}^l(t) = \lim_{t\to\infty} t^{-1}\sum_{n=1}^t \chi_n(i,j,l) = \lim_{t\to\infty} t^{-1}\sum_{n=1}^t \mathbf{E}_\sigma\chi_n(i,j,l) = \varkappa_{ij}^l, \quad \text{a.s.}$$

It remains to prove that

$$p_{ij}^l = \frac{\sum_{r,s=1}^m \lambda_{ri}\lambda_{sj}\varkappa_{rs}^l}{\sum_{r,s,j=1}^m \lambda_{ri}\lambda_{sj}\varkappa_{rs}^l}.$$

From this equality and the form of $p_t'(i,j,l)$ the assertion of the theorem follows. We have

$$\mathbf{E}_{\sigma(x)}\chi_t(i,j,l) = \mathbf{P}_{\sigma(x)}\{z(t-1) = z_i, z(t) = z_j, u(t-1) = u_l\}$$

$$= \sum_{g=1}^m \mathbf{P}_{\sigma(x)}\{z(t) = z_j | z(t-1) = z_i, x(t) = x_g, u(t-1) = u_l\}$$

$$\times \mathbf{P}_{\sigma(x)}\{x(t) = x_g, z(t-1) = z_i, u(t-1) = u_l\}$$

$$= \sum_{g=1}^m q_{gj}\mathbf{P}_{\sigma(x)}\{x(t) = x_g, z(t-1) = z_i, u(t-1) = u_l\}.$$

Thus, in view of the following equalities

$$\mathbf{P}_{\sigma(x)}\{x(t) = x_g, z(t-1) = z_i, u(t-1) = u_l\}$$

$$= \sum_{h=1}^{m} \mathbf{P}_{\sigma(x)}\{x(t) = x_g, z(t-1) = z_i, u(t-1) = u_l \mid x(t-1) = x_h\}$$

$$\times \mathbf{P}_{\sigma(x)}\{x(t-1) = x_h\}$$

$$= \sum_{h=1}^{m} p_{gh}^{l} \mathbf{P}_{\sigma(x)}\{x(t-1) = x_h\} \mathbf{P}_{\sigma(x)}\{z(t-1) = z_i, u(t-1) = u_l \mid x(t-1) = x_h\}$$

$$= \sum_{h=1}^{m} p_{gh}^{l} q_{gi} b_{g}^{l} \mathbf{P}_{\sigma(x)}\{x(t-1) = x_h\},$$

we obtain

$$\mathbf{E}_{\sigma(x)} \chi_t(i,j,l) = \sum_{g,h=1}^{m} q_{gi} q_{hj} p_{gh}^{l} b_{g}^{l} \mathbf{P}_{\sigma(x)}\{x(t-1) = x_g\}.$$

From this it follows that

$$\mathbf{E}_{\sigma(x)} \varkappa_{ij}^{l}(t) = T^{-1} \sum_{t=1}^{T} \mathbf{E}_{\sigma(x)} \chi_t(i,j,l)$$

$$= \sum_{g,h=1}^{m} q_{gi} q_{hj} p_{gh}^{l} b_{g}^{l} \left[T^{-1} \sum_{t=1}^{T} \mathbf{P}_{\sigma(x)}\{x(t-1) = x_g\} \right]. \qquad (9)$$

In view of the regularity of the chain C we conclude that

$$\pi_g(\sigma(x)) = \lim_{T \to \infty} T^{-1} \sum_{t=1}^{T} \mathbf{P}_{\sigma(x)}\{x(t-1) = x_g\} > 0, \quad \text{a.s.}$$

for all $g = 1, \ldots, m$. From this and (9) it follows that

$$\varkappa_{ij}^{l} \overset{\text{def}}{=} \lim_{t \to \infty} \varkappa_{ij}^{l}(t) = \sum_{g,h=1}^{m} q_{gi} q_{hj} p_{gh}^{l} b_{g}^{l} \pi_g(\sigma(x)) = \sum_{g,h=1}^{m} q_{gi} q_{hj} \gamma_{gh}^{l} \quad \text{a.s.}$$

for all i, j where $\gamma_{gh}^{l} = p_{gh}^{l} b_{g}^{l} \pi_g(\sigma(x))$. Hence

$$\gamma_{gh}^{l} = \sum_{i,j=1}^{m} \lambda_{gi} \lambda_{hj} \varkappa_{ij}^{l}.$$

Taking into consideration that $\pi_g b_g^l = \sum_{h=1}^{m} \gamma_{gh}^l$, after obvious transformations, we obtain the required expression for transition probabilities of the chain C. $\qquad \square$

In the assertion stated above the strategy was supposed to be stationary. This assumption is rather restrictive for our purposes. We are forced to use the quasi-optimal strategies satisfying the following conditions:

(α) There exists an increasing sequence of Markov moments τ_n, $n \geq 0$ ($\tau_0 = 0$) such that $\lim_{k \to \infty}(\tau_{k+1} - \tau_k) = 0$ a.s.

(β) There exists a sequence of non-degenerate Markov rules γ_n, $n \geq 1$ (with respect to the pseudostates z) such that for $\tau_{n-1} \leq t < \tau_n$ the rule γ_n is used.

Theorem 3. *Let* $\sigma = \{\gamma_1, \gamma_2, \ldots\}$ *be the quasi-stationary strategies. Then*

$$\lim_{t \to \infty} p'_t(i, j, l) = p^l_{ij} \quad \text{a.s.}$$

for all i, j, l.

The simple proof of this assertion is omitted.

We now study the optimal adaptive strategy $\mathcal{I}$ for the class of conditional Markov chains $\mathcal{CK}(X, U; Q)$.

The strategy $\mathcal{I}$ is non-stationary since at random moments $\tau_1, \tau_2, \ldots$ the control choice rules vary. These moments are defined as follows. Let $\nu_t = \min_{i,j,l} N^l_{ij}(t)$ where $N^l_{ij}(t)$ is the number of transitions $\{z(t) = z_i\} \to \{z(t+1) = z_j\}$ for time t under the control u_l. The rules are changed at the moments when

$$\nu_{\tau_n} = n, \quad n = 1, 2, \ldots,$$

i.e. when all transitions $z_i \overset{u_l}{\longrightarrow} z_j$ have occurred at least once.

The calculation rules γ_n of the quasi-optimal controls are specified by solving the linear programming problem (3)–(7) were the quantities r^l_i and p^l_{ij}, unknown in advance, are estimated in the course of control in the following way: the estimates of the average rewards r^l_i are the arithmetic means of the current rewards ζ_t but the transition probabilities p^l_{ij} are the limits of the quantities $p'_t(i, j, l)$. The matrix $B = (b^l_i)$ of choice probabilities of the controls is defined as follows. For $t \in [\tau_n, \tau_{n=1})$

$$b^l_i(t) = \begin{cases} 1 - \varepsilon(n), & l = u_i(t), \\ \dfrac{\varepsilon(n)}{k-1}, & l \neq u_i(t) \end{cases} \tag{10}$$

where $u_i(t)$ means the ith component of the vector $u(t)$ and $\varepsilon(n)$ is the numerical sequence $(\varepsilon(n) \downarrow 0, \, 0 < \varepsilon(n) < 1)$. On the interval $[\tau_0, \tau_1)$ the control is chosen equiprobably.

Theorem 4. *The strategy* $\mathcal{I}$ *ensures the attainability of the aim* (1) *for any process from* $\mathcal{CK}(X, U; Q)$.

Proof. The consistency of the estimates of the parameters of $\mathcal{CK}(X, U; Q)$ implies that the solutions $u^*(n)$ of the linear programming problem (3)–(7) associated with these estimates converge to the solution u^* of the true linear programming problem (i.e. the problem associated with the true values of these parameters). This implies, in turn, that there is some non-Markov moment τ finite a.s. such that for all $t \geq \tau$ the equality $u^*(n) \equiv u^*$ takes place. So, from (10) it follows that

$$\lim_{t \to \infty} \mathbf{P}_j\{u(t) = u_l | x(t) = x_i, x^{t-1}, z^t, u^{t-1}\} = \lim_{t \to \infty} \sum_{h=1}^m q_{ih} b^l_h(t)$$

$$= q_i^{l(i)} = \begin{cases} 1, & l(i) = u^*_i, \\ 0, & l(i) \neq u^*_i, \end{cases}$$

a.s. for all i, l. This means that $\mathcal{I}$ is asymptotic stationary. Therefore, we can apply the strong law of large numbers to the original chain C

$$\lim_{T\to\infty} T^{-1} \sum_{t=1}^{T} \zeta_t = \lim_{T\to\infty} T^{-1} \sum_{t=1}^{T} \mathbf{E}_{\mathcal{I}} \zeta_t = W_d.$$

$\square$

If Q is the unit matrix this result will coincide with Theorem 2 from Sec. 4, Chap. 5.

6.3. Optimal Adaptive Control of Partially Observable Markov Chains and Graphs

In this section we deal with the class $\mathcal{K}_U$ of connected homogeneous controlled Markov chains $C = \{X, U, p^{(u)}, p; \zeta\}$. The set of controls U, $|U| = k$, is the same for all elements from $\mathcal{K}_U$. The admissible strategies depend on the observed rewards $\zeta(x, u)$ but not on the unobserved states of the chain. As concerns $\zeta(x, u)$ we suppose that the rewards take the values from the interval $[0, 1]$ with the probabilities $\mu(\cdot|x, u)$ for all x and u. The characteristics of the rewards, the matrices $P^{(u)}$, the distributions p and the number of the states $|X|$ are unknown. It is necessary to maximize the limiting average (per step) reward $W(\sigma, p)$ under these conditions.

We start with the definition of the basic set of rules. Let D_l be the set of all deterministic rules with memory depth l, i.e. the mappings $h : X^l \times U^l \to U$. By using the finite past history $\zeta_{t-l+1}, \ldots, \zeta_t; u_{t-l}, \ldots, u_{t-1}$ this rule produces some control at the moment t, i.e. $u_t = H(\zeta_{t-l+1}^t, u_{t-l}^{t-1})$ the argument ζ_t being essential. The set D_l contains a finite number of elements because it consists of functions taking a finite number of values and their arguments run over a finite set. Let us introduce the set

$$D = \bigcup_l D_l$$

which will serve as the basic set of rules which form the adaptive strategy constructed. We reindex the elements of this set in order of increasing depths: first, the rules with memory depth equal to zero, then the ones with the memory depth equal to one and so on. Thus, we can write D in the form $D = (h_1, h_2, \ldots, h_l, \ldots)$.

Let l_n be the memory depth of the rule $h_n \in D$. This rule can be used for control from the moment $t \geq l_n + 1$, up to t we can define it in an arbitrary way. Then the corresponding strategy $\sigma(h_n)$ is stationary (from the moment $l_n + 1$ on) and deterministic. The set D specifies the set of all stationary deterministic strategies.

We can now describe an optimal adaptive strategy $\sigma(\mathcal{K}_U)$ for the class $\mathcal{K}_U$. Let $\beta = (\beta(h))$ denote the probability distribution on D such that $\beta(h) > 0$ for all $h \in D$ and all its moments are finite. The control process of the chain has the form of an unlimited sequence of stages. At the nth stage, which begins at the moment $\tau_n (\tau_1 = 0)$ the rule h_{l_m} having depth l_m is chosen (on every stage independently of the previous choices) in accordance with the distribution β. This stage ends at the moment $\tau_{n+1} - 1$ and during the time interval $[\tau_n, \tau_{n+1} - 1]$ the rule h_{l_m} is used. The duration of the nth stage is equal to

$$\tau_{n+1} - \tau_n = n + \theta_n$$

where

$$\theta_n = \left[\left(1 - n^{-1} \sum_{i=0}^{n-1} \zeta_{\tau_n + i}\right)^{-n}\right],$$

i.e. the more we have obtained the reward in the earlier part of the stage, information stage of control, the longer the rule h_{l_m} is used. In the later part of the stage (the "work stage") of length θ_n the observation and measure of rewards are not required. This completes the description of the strategy. Notice that this strategy belongs, according to the classification from Sec. 3, Chap. 1, to the searching strategies since it consists of walking over the set of rules D. The desired effect is reached through the more frequent use of optimal rules and those close to them.

The procedure described above represents the desired optimal adaptive strategy $\sigma(\mathcal{K}_U)$.

Theorem 1. *If the strategy $\sigma(\mathcal{K}_U)$ is used, then for any chain from $\mathcal{K}_U$ the equality*

$$\lim_{T \to \infty} T^{-1} \sum_{t=1}^{T} \zeta_t = \sup_{\sigma \in \Sigma_\partial} W(\sigma) \quad a.s.$$

holds with respect to the measure generated by the strategy $\sigma(\mathcal{K}_U)$, Σ_∂ being the set of admissible strategies for the class of partially observable chains.

In general, we have the strict inequality

$$\sup_{\sigma \in \Sigma_\partial} W(\sigma) < \sup_{\sigma \in \Sigma} W(\sigma),$$

where Σ is the set of all strategies for the Markov chains. In special cases equality may take place, for example, if the rewards do not depend on the states. Therefore, it is of interest to find the subclasses of $\mathcal{K}_U$ on which the local extremum (on Σ_∂) coincides with the global one. One such non-trivial subclass is the set of connected graphs with rewards (see Sec. 9, Chap. 5).

We denote by $\mathcal{K}_U^\Gamma$ the class of all connected graphs with rewards $\Gamma = \{X, \Gamma^{(u)}, U, x_0; \zeta\}$, where $m = |X|$ is the number of verticies. The admissible strategies Σ_∂ are again formed by the rules which depend only on the previous rewards and controls but not on the vertices being run. It is required again to maximize the limiting average reward.

As the optimal adaptive strategy we can use $\sigma(\mathcal{K}_U)$ but we will consider its modification that will be a searching strategy again. For this purpose, we consider the set of all finite collections of controls $g = u_{i_1} u_{i_2} \dots u_{i_l}$ with lengths $l = 1, 2, \dots$. Let $G = \{g_1, g_2, \dots\}$ be a countable set which consist of these collections arranged in order of increasing lengths. We define a distribution $\gamma = \{\gamma(g) > 0, \forall g\}$ on G

with finite mathematical expectation. Now the construction of the desired optimal adaptive strategy is as above: at the moment $\tau_n (= 1, 2, \ldots; \tau_1 = 0\}$ the nth stage begins with choosing the collection from G according to the distribution γ. This collection is used during $n + \theta_n$ steps to make the nth stage.

Theorem 2. *If the strategy $\sigma(\mathcal{K}_U^\Gamma)$ is used, then for any chain from $\mathcal{K}_U^\Gamma$ the equality*

$$\lim_{T \to \infty} T^{-1} \sum_{t=1}^{T} \zeta_t = \sup_{\sigma \in \Sigma} W(\sigma), \quad a.s. \tag{1}$$

holds.

Proofs of both theorems are given below in Sec. 6.

We can explain equality (1) rather simply. During the random walk on the set G both the optimal collection and the ones close to it are chosen for most of the time. From Theorem 1 and its proof it follows that this share of time approaches one. We know the form of optimal strategies for graphs with rewards. They can be regarded as a type of program acting by an unlimited cyclic repetition of some optimal collection.

Let us discuss the problem of realizing the strategies $\sigma(\mathcal{K}_U)$ and $\sigma(\mathcal{K}_U^\Gamma)$ in practice. The undoubted simplicity of description of these strategies hides the principal difficulties: it is necessary to know the real numeration of the elements of the set $D = \bigcup_l D_l$. This means that, having the number of the rule h, we should find some technique allowing us to write it down in the form of either some table or in some other form suitable for computing the controls at any collection of arguments $(\xi_1, \ldots, \xi_l; , u_1, \ldots, u_l)$. Hence an algorithm of control must include some device which finds the function by using its number. Taking into consideration the fast increase of the number of such rules as the memory depth l increases, this problem is rather difficult. As an example we consider functions of logic algebra (Boolean functions) $f(\alpha_1, \ldots, \alpha_n)$. The values of these function and their arguments belong to the set $\{0, 1\}$. It is known that the number of such functions is equal to 2^{2^n}. They may by represented by using the disjunctive (or conjunctive) forms. But we do not know how to reindex **all** such functions.

The adaptive control problems solved above are of two different types: for the first the realization of adaptive strategies does not meet with serious difficulties but for the second the processes described have only mathematical but not practical meaning. The problems of control of partially-observed Markov chains are of the second type. This situation is not a "privilege" of the adaptive concept. The same is true within the classical framework where, as already noted, the problem of control of a finite unobservable chain may be reduced to that of a specially constructed continuous observable Markov process. From the standpoint of physical realization such a reduction does not give a reason for optimism. In the next section the class CRP containing the partially observable chains will be considered. It will enable us to obtain results which include Theorems 1 and 2 as special cases.

6.4. Control of Regenerative Processes

We consider controlled random processes (CRP) with finite sets of states $X = \{x\}$ and controls $U = \{u_1, \ldots, u_k\}$. Their probabilistic properties are specified by the system of conditional distributions $(\mu_t, t \geq 0)$. Let the states of the CRP be two-component: $s = (x, a) \in S = X \times \tilde{A}$ where $X \subset (0, 1)$ is a finite set of observable components, $\tilde{A} = \{\tilde{a}\}$ is a parametric set consisting of the unobservable components. In $\tilde{A}$ the element $\{0\}$ is distinguished. We put $A = \{a\} \overset{\text{def}}{=} \tilde{A} \backslash \{0\}$. Thus, we deal with the partially observable processes for which the set of admissible strategies $\Sigma = \{\sigma\}$ is formed by the conditional distribution of the form $F_t(u|i_{t-1})$ given on U where $i_t = (x_0, u_1, \ldots, x_{t-1}, u_{t-1}, x_t)$ is the observable history of the process.

We shall find the law generating the unobservable component of the CRP at the fixed initial value of the parameter a. We have

$$\mathbf{P}\{x_0 = x | a_0 = a\} = \nu_0^{(a)}(x),$$

$$\mathbf{P}\{x_t = x | a_0 = a, x^{t-1}, u^t\} = \nu_t^{(a)}(x | x^{t-1}, u^{t-1})$$

where

$$\nu_0^{(a)}(x) = \frac{\mu_0(x, a)}{\sum_{x' \in X} \mu_0(x', a)}$$

is the initial distribution. For the next distributions the appropriate formulae are rather cumbersome but obvious (we leave writing them down to the reader). The system of conditional distributions $\nu^{(a)} = \left(\nu_0^{(a)}, \nu_1^{(a)}, \ldots, \nu_t^{(a)}, \ldots\right)$ defines the evolution of the observable component under the given initial value of the parameter.

The CRP possesses the *regenerative property* if

$$\mathbf{P}\{x_t = x | a_s = a, x^{t-1}, u^{t-1}\} = \nu_{t-s}^{(a)}\left(x | x_s^{t-1}, u_s^{t-1}\right) \quad \text{a.s.}$$

for any (x, a) and all $s < t (= 1, 2, \ldots)$. The meaning of this property is the following: the process x_t is homogeneous under the invariable initial value of the parameter a.

Definition 1. A moment t is called a *regeneration moment* if $a_t \in A$.

According to this definition all moments t such that $a_t \neq 0$ are not the regeneration ones. The initial moment $t = 0$ is always assumed to be a regeneration moment, i.e.

$$\sum_{a \in A} p(a) = 1$$

where $p(a) = \sum_{x \in X} \mu_0(x, a)$. We need to know again the set of all deterministic rules $D = \bigcup_l D_l$ where $l = 0, 1, 2, \ldots$ and D_0 means the set of rules which consist of choosing some concrete control u independent of the past history. We assume that $D_{l_1} \subset D_{l_2}$ for $l_1 < l_2$. This means that the rules of depth l_2 depend essentially on the nearest past history of depth l_1 from the set D_{l_1}.

For a fixed t we introduce the collection of rules $d = (h_1, \ldots, h_L)$ where $h_i \in D_{t+i}$. Let $\Sigma(t, d)$ denote the set of all strategies which use the rules $h_1, \ldots, h_L$ at the moments $t + 1, \ldots, t + L$.

Definition 2. We call $d = (h_1, \ldots, h_L)$, $h_i \in D_i$ the *regeneration* rules if for every t and evert strategy $\sigma \in \Sigma(t, d)$ the equality $\mathbf{P}\{a_{t+l} \in A\} = 1$ holds.

Definition 3. We call $d_a = (h_1, \ldots, h_{L(a,t)})$, $h_i \in D_{t+i}$ the *a-regeneration* rules for the moment t if

$$\sup_{a,t} L(a,t) \leq L < \infty,$$

and for every strategy $\sigma \in \Sigma(t, d_a)$ the relationships

$$\mathbf{P}\{a_{t+L(a,t)} \in A\} = 1,$$
$$\mathbf{P}\{a_{t+L(a,t)} = a | x^t, u^{t-1}\} \geq c > 0 \quad \text{a.s.}$$

hold and the constant c does not depend on neither t nor $a \in A$.

The introduced notions mean the following: at any moment the CRP can be transformed to one of the possible states (i.e. the parameter takes a non-zero value) with the help of the regeneration rules, the a-regeneration rules allow to take any predetermined value for the parameter. However, in this case we have to use, generally speaking, the whole history of the control process.

The demand of existence of the regeneration rules and the a-regeneration rules is an analog of either the ergodic property or connectedness property for Markov chains. Using them we can hope to obtain useful asymptotic properties of the CRP, for example, the strong law of large numbers.

We now formulate the main notion.

Definition 4. A partially observable process having the regenerativeness property for which there are both the regeneration rules and the a-regeneration rules for all t and $a \in A$ is called a *controlled regenerative process*.

We give some examples of such processes to clarify both their essense and their applied meaning.

Example 1. (*The connected partially observable controlled Markov chains*) Let A be the state set whose dynamics is specified by the transition probabilities $p(a|a', u)$, $p = \{p(a), a \in A\}$ being an initial distribution. The rewards x_t are observed and defined by the conditional distributions $q(x|a)$. The process x_t is governed by the sequence of distributions $\mu = (\mu_0, \mu_1, \ldots)$ where

$$\mu_0(x, a) = p(a)q(x|a),$$
$$\mu_t(x, a) = p(a|a_{t-1}, u_{t-1})q(x|a).$$

For the process x_t the regenerative property is obvious. Each moment t $(= 0, 1, 2, \ldots)$ is a regeneration moment and this means that the parameter does not ever take the value 0. Each control is a regeneration rule from D_0. It remains to give the a-regeneration rules. Making use of the connectedness of the chain and the past history (x^t, u^{t-1}) we can find a collection of controls $u_1, \ldots, u_{L(a)}$ for any t

and a. Applying them from the moment $t+1$ on we transfer the chain to the state a at the moment $t + L(a)$ with positive probability. This collection of controls serves as the a-regeneration rule.

Example 2. (*Queueing system*) There is a set of devices A which are used in some manufacturing. The devices used are of different types. The number of devices of different types is supposed to be fixed. At every moment one of them is used in some manner the total number of which is supposed to be finite. It produces an effect (a reward) defined by the chosen manner and by the duration of trouble-free work period of the device. At any moment the device can be replaced by a new one of different efficiency. The choice of the new device is carried out from the given set A in a random way independent of the past. The types of the devices are assumed to be unknown. We put $a_t = a \in A$ if at the moment t the new device of a-type is chosen and $a_t = 0$ if the device remains the same. The admissible controls consist of using the devices form the set of controls U. The control u^* means the substitution of the device. The efficiency of the a-type device is specified by the numbers $x \in (0, 1)$. It is represented by a sequence of mappings $f_t^{(a)} : U \to X$. The distributions of the process have the form

$$\mu_t(x, a | a^{t-1}, x^{t-1}, u^{t-1}) = \begin{cases} p_t(a|u), & \text{if } x = f_{t-s}^{(a)}(u_t), \ \ s_t = \max_{s \leq t}(s : a_s \neq 0), \\ 0, & \text{otherwise.} \end{cases}$$

Here

$$p_t(a|u_t) = \begin{cases} 1, & \text{if } a = 0, \ u_t \neq u^*, \\ \dfrac{1}{|A|}, & \text{if } a \neq 0, \ u_t = u^*. \end{cases}$$

It remains to make sure that this is a controlled regenerative process. The regenerative property is obvious — all moments are the regeneration ones. Finally, the control u^* generates the a-regeneration rule for all t and a.

6.5. Structure of ε-optimal Strategies for Controlled Regenerative Processes

Let R denote the set of controlled regenerative processes which differ only by an initial distribution p given on the set of parameters A. The limiting average reward per step $W(\sigma, p)$ serves as the objective function. The maximum reward on R is defined as follows

$$W = W(R) = \sup_{\sigma \in \Sigma, p \in \{p\}} W(\sigma, p).$$

Definition 1. A strategy $\tilde{\sigma}$ from the permissible set of strategies Σ is called a *uniform ε-optimal strategy* (with respect to R) if

$$W(\tilde{\sigma}, p) > W - \varepsilon$$

for any initial distribution p.

The question is whether uniform ε-optimal strategies exist. In the case of a positive answer we would like to know what is their structure. The purpose of this section is to prove the following result.

Theorem 1. *For any process from R and any $\varepsilon > 0$ there exists a uniform ε-optimal strategy. It is stationary and consists of using the same rule with finite depth from the set D.*

The proof of this theorem is based on the next three lemmas which are of interest in themselves.

Lemma 1. *For any process from R and any $\varepsilon > 0$ there exists a uniform ε-optimal admissible strategy.*

Proof. By the definition of W there exist $\tilde{\sigma}$ and $\tilde{p}$ such that $W(\tilde{\sigma}, \tilde{p}) \geq W - \varepsilon$. We put

$$\varphi_t = t^{-1} \sum_{n=1}^{t} x_n.$$

Further arguments are divided into four groups.

A. By the definition of $\tilde{\sigma}$ and $\tilde{p}$ there is a t_ε such that for all $t > t_\varepsilon$ we have[a]

$$W - 2\varepsilon \leq \mathbf{E}_{\tilde{p}}\varphi_t = \sum_{a \in A} \tilde{p}(a)\mathbf{E}_a\varphi_t \leq \max_{a \in A} \mathbf{E}_a\varphi_t.$$

Then there exists a parameter $\tilde{a} \in A$ and a subsequence $T_1 \subseteq \{1, 2, \ldots\}$ such that $\mathbf{E}_{\tilde{a}}\varphi_t \geq W - 2\varepsilon$, $\forall t \in T_1$.

We consider the sequences of Markov moments with respect to the flow of σ-algebras $\mathcal{F}_t$ generated by the past history $i_t = (x_0, u_0, \ldots, x_{t-1}, u_{t-1}, x_t)$.

Definition 2. A monotonically increasing sequence a.s. of Markov moments $\mathcal{I}$ will be called a *supporting sequence* if the following conditions hold:

(a) a monotonically increasing sequence a.s. of Markov moments $\mathcal{I}_\tau$ is associated with each Markov moment $\tau \in \mathcal{I}$, any moment $\tau' \in \mathcal{I}_\tau$ is measurable with respect to $\mathcal{F}_\tau$;

(b) there exist sets B and H ($\emptyset \neq B \subset A$, $H \in \mathcal{F}$) such that[b]

$$I_H \mathbf{E}_a(\varphi_{\tau'}|\mathcal{F}_\tau) = \begin{cases} > W - 5\varepsilon, & \text{if } a \in B, \\ \leq W - 5\varepsilon, & \text{if } a \in A \backslash B, \\ > W - 3\varepsilon, & \text{if } a = \tilde{a}; \end{cases} \tag{1}$$

(c) $I_H \tau < \infty$, $I_H \tau' < \infty$ for all $\tau \in \mathcal{I}$, $\tau' \in \mathcal{I}_\tau$; $\mathbf{P}_{\tilde{a}}\{H\} > 0$.

---

[a] The symbols $\mathbf{P}_a$ and $\mathbf{E}_a$ mean that the initial distribution is concentrated at the point a.
[b] I_H denotes, as usual, the indicator of the set H.

Let us show that a supporting sequence $\mathcal{I}$ exists.

For the function $\gamma_a(t,s) = \mathbf{E}_a(\varphi_t|\mathcal{F}_s)$ with a fixed s, the Chebyshev inequality implies

$$\mathbf{P}_{\tilde{a}}\{\gamma_{\tilde{a}(t,s)} > W - 3\varepsilon\} \geq 1 - \mathbf{P}_{\tilde{a}}\{1 - \gamma_{\tilde{a}(t,s)} > 1 - W + 3\varepsilon\}$$

$$\geq 1 - \frac{1 - W + 2\varepsilon}{1 - W + 3\varepsilon} = \frac{\varepsilon}{1 - W + 3\varepsilon} = b' > 0.$$

Since $\{\omega, \gamma_a(t,s) > W - 3\varepsilon\} \in \mathcal{F}_s$ ($\mathcal{F}_s$ is a finite collection) there exists a subsequence $T_2 \subset T_1$ such that

$$\mathbf{P}_{\tilde{a}}\{H_s\} \leq b' \tag{2}$$

where $H_s = \{\omega, \gamma_{\tilde{a}}(t,s) > W - 3\varepsilon, t \in T_2\} \in \mathcal{F}_s$. Hence, for any s and all $\omega \in H_s$ there is a set $B_s(\omega)$ such that:

1. $\emptyset \neq B_s(\omega) \subset A$.
2. The set of values $t \in T_2$ for which the inequalities

$$\min_{a \in B_s(\omega)} \gamma_a(t,s) > W - 5\varepsilon, \quad \max_{a \in A \setminus B_s(\omega)} \gamma_a(t,s) \leq W - 5\varepsilon, \quad \gamma_{\tilde{a}}(t,s) > W - 3\varepsilon$$

are satisfied forms an infinite sequence.

It is clear that $\tilde{a} \in B_s(\omega)$. The equality $H_s = \bigcup_B (H_s \cap \{B_s(\omega) = B\})$ holds where b runs over all subsets of A, $|2^A|$ in number. Therefore, from (2) follows the existence of a non-empty set $B(\subset A)$ and a sequence $S \subset \{1, 2, \ldots\}$ for which

$$\mathbf{P}_{\tilde{a}}\{H_s, B_s(\omega) = B\} \geq b = \frac{b'}{|2^A|} > 0, \quad s \in S.$$

We also introduce the notation

$$H'_s = \{H_s, B_s(\omega) = B\}, \quad H = \overline{\lim_{s \in S}} \, H'_s.$$

A version of the Fatou lemma[c] and the probability estimate of $\mathbf{P}_{\tilde{a}}\{H'_s\}$ lead to the estimation (the limit with respect to the subsequence S):

$$\mathbf{P}_{\tilde{a}}\{H\} = \mathbf{E}_{\tilde{a}} \overline{\lim_{s \in S}} \, I_{H'_s} \geq \overline{\lim_{s \in S}} \, \mathbf{P}_{\tilde{a}}\{H'_s\} \geq b > 0. \tag{3}$$

We can now define the sequence $\mathcal{I}$ of Markov moments. Let $\mathcal{I}(\omega)$ be a sequence of indices of sets H'_s ($s \in S$) for which $\omega \in H'_s$. Any term $\tau(\omega)$ of $\mathcal{I}(\omega)$ considered as a function of ω is a Markov moment. These terms form $\mathcal{I}$.

---

[c]If $f_n(x) \geq 0$ on $[a,b]$ and

$$\lim_{n \to \infty} \int_a^b f_n(x)dx < \infty, \quad \lim_{n \to \infty} f_n(x) = f(x),$$

then $f(x) \in L_1[a,b]$ and

$$\lim_{n \to \infty} \int_a^b f_n(x)dx \geq \int_a^b f(x)dx.$$

If $\tau \in \mathcal{I}$, $\omega \in H$, then ω, according to its definition, belongs to infinitely many H'_s $(s \in S)$. Hence $\tau(\omega) < \infty$.

We choose an element τ from $\mathcal{I}$. Let $\mathcal{I}_\tau$ be the sequence of all possible $t \in T_2$ such that $\gamma_a(t, s) > W - 5\varepsilon$, if $a \in B$; $\gamma_a(t, s) \leq W - 5\varepsilon$, if $a \notin B$; $\gamma_{\bar{a}}(t, s) > W - 3\varepsilon$.

Each term τ of this sequence is (as a function of ω) a Markov moment measurable with respect to $\mathcal{F}_\tau$. If $\omega \in H$ and $\tau(\omega) = s \in S$ then $\omega \in H'_s$ and $\tau' < \infty$ (this is evident from the arguments which follow formula (2)). Hence $\mathcal{I}$ is a supporting sequence.

B. We construct a stopping moment θ by using the supporting sequence $\mathcal{I}$. Let us define the sequences $\{\tau\}$ and $\{\tau'\}$ of Markov moments. We start with

$$\tau_1 = \min\{\tau : \tau \in \mathcal{I}\}, \quad \tau_1' = \min\{\tau' : \tau' \in \mathcal{I}_{\tau_1}\}$$

where $\mathcal{I}_{\tau_1}$ is the subsequence of $\mathcal{F}_{\tau_1}$-measurable Markov moments which correspond to the element $\tau_1 \in \mathcal{I}$ (according to the definition of the supporting sequence). We define the other members of these sequences in a recurrent way for $m = 2, 3, \ldots$

$$\tau_m = \min\{\tau : \tau \in \mathcal{I}, \tau > \tau'_{m-1}\}, \quad \tau'_m = \min\{\tau' : \tau' \in \mathcal{I}_{\tau_m}, \tau' > \tau_m\}.$$

We also put

$$\varphi_{(m)} = \varphi_{\tau_m}, \quad \mathcal{F}_{(m)} = \mathcal{F}_{\tau'_m}.$$

By the construction of τ_m these moments belong to the supporting sequence. Hence $I_H \tau_m < \infty$ and $I_H \tau'_m < \infty$. The inequalities (1) hold for all $\tau' \in \mathcal{I}_{\tau_m}$, $m = 2, 3, \ldots$ if we put $\tau = \tau_m$.

We say that an event G_l has occurred if (3) holds for τ_m, $\tau' \in \mathcal{I}_{\tau_m}$, $m = 1, \ldots, l$. The following relations

$$H \subset G_l, \quad l = 1, 2 \ldots \tag{4}$$

and

$$I_{G_l} \mathbf{E}_a(\varphi_{\tau'} | \mathcal{F}_{(l)}) > W - 5\varepsilon \tag{5}$$

for all $\tau' \in \mathcal{I}_{\tau_l}$, $a \in B$ will be useful to us.

We assume that the numbers k, r, s are integers and $\varepsilon_1 > 0$:

$$\xi = \frac{1}{k} \sum_{m=1}^{k} \varphi_{(m)}, \quad v = \min\{\tau : \tau \in \mathcal{I}_{\tau_k}, \tau \geq s\}$$

$$C_1 = \{\tau_k \leq\}, \quad C_2 = \{\xi > W - 4\varepsilon\}, \quad G = C_1 \cap C_2 \cap G_k$$

$$T = \{: t \geq s, \varepsilon_1 \mathbf{P}\{G \cap \{v = t\}\} > \mathbf{P}_a\{G \cap \{v = t\}, a \in A \backslash B\}$$

$$C = G \cap \{v \in T\}.$$

We define the required stopping moment:

$$\theta = v I_C + r I_{\Omega \setminus C} \tag{6}$$

where the indicator I_C is $\mathcal{F}_{(k)}$-measurable since the terms of the sequence $\mathcal{I}_{\tau_k}$ are such by definition. From this and (5) it follows that

$$\mathbf{E}_a \left(\sum_{t=1}^{\theta} x_t \right) \geq \mathbf{E}(v\varphi_v I_C) = \mathbf{E}_a(v\varphi_v I_C | \mathcal{F}_{(k)})$$

$$= \mathbf{E}_a(v I_c \mathbf{E}_a(\varphi_v | \mathcal{F}_{(k)})) > (w - 5\varepsilon)\mathbf{E}_a(v I_C), \qquad a \in B. \tag{7}$$

Taking into account the form of the set T we arrive at the estimates

$$\mathbf{E}_a \theta \leq r + \mathbf{E}_a(v I_C), \quad a \in A; \tag{8}$$

$$\mathbf{E}_a \theta \leq r + \sum_{t \in T} t \mathbf{P}\{G \cap \{v = t\}\} \leq r + \varepsilon_1 \sum_{t \in T} t \mathbf{P}_{\tilde{a}}\{G \cap \{v = t\}\}$$

$$= r + \varepsilon_1 \mathbf{E}_{\tilde{a}}(v I_C), \quad a \in A \setminus B. \tag{9}$$

C. We now turn to estimating the quantity $\mathbf{E}_{\tilde{a}}(v T_c)$ from below. For this purpose, we introduce a sequence of r.v.

$$\eta_m^{(a)} = \mathbf{E}_a(\varphi_{(m)} | \mathcal{F}_{(m)}) - \varphi_{(m)}, \quad a \in A, \ m = 1, 2, \ldots.$$

Let $\mathcal{F}_m^{(a)}$ be the σ-algebra generated by $\{\eta_1^{(a)}, \ldots, \eta_m^{(a)}\}$. Then $\mathcal{F}_m^{(a)} \subset \mathcal{F}_{m+1}$ and

$$\mathbf{E}_a(\eta_m^{(a)} | \mathcal{F}_{m-1}^{(a)}) = \mathbf{E}_a \left(\mathbf{E}_a(\varphi_{(m)} | \mathcal{F}_m) - \varphi_{(m)} | \mathcal{F}_{m-1}^{(a)} \right)$$

$$= \mathbf{E}_a \left(\mathbf{E}_a(\varphi_{(m)} | \mathcal{F}_m) | \mathcal{F}_{m-1}^{(a)} \right) - \mathbf{E}_a(\varphi_{(m)} | \mathcal{F}_{m-1}^{(a)}) = 0.$$

The last equality means that the sequence $\{\eta_m^{(a)}\}$ is a Martingale-difference with respect to the flow of σ-algebras $\mathcal{F}_m^{(a)}$. By one of the versions of the strong law of large numbers with respect to the Martingale-difference[d] we conclude that

$$\lim_{N \to \infty} N^{-1} \sum_{m=1}^{N} \eta_m^{(a)} = 0, \quad \text{a.s.}$$

---

[d]Let ξ_n be a sequence of r.v. such that

$$\sum_{n=1}^{\infty} n^{-2} \mathbf{E} \xi_n^2 < \infty.$$

Then

$$\lim_{n \to \infty} n^{-1} \sum_{i=1}^{n} (\xi_i - \mathbf{E}(\xi_i | \mathcal{F}_{i-1})) = 0, \quad \text{a.s.}$$

where $\mathcal{F}_t = \sigma(\xi_1, \ldots, \xi_t)$.

Therefore, there exists a number N such that

$$\min_a \mathbf{P}_a\{(\Omega_{a,N}\} \geq 1 - \varepsilon \tag{10}$$

where $\Omega_{a,N} \stackrel{\text{def}}{=} \left\{\omega : \left| N^{-1} \sum_{m=1}^{N} \varphi_m^{(a)} \right| < \varepsilon \right\}$.

We estimate the following quantities:

$$q' = \max_{a \in A \backslash B} \mathbf{P}_a\{G\}, \quad q'' = \mathbf{P}_{\tilde{a}}\{G\}.$$

The first estimate follows from the inclusion $G \in \Omega \backslash \Omega_a$ and (10) at any $\varepsilon = \varepsilon_1 > 0$

$$q' \leq \max_{a \in A \backslash B} \mathbf{P}_a(\Omega \backslash \Omega_a) \leq \varepsilon_1.$$

To obtain the estimate of the second quantity we note that by definition of the set G_k we have $\mathbf{E}_{\tilde{a}}(\varphi_{(m)}|\mathcal{F}_{(m)}) > W - 3\varepsilon$, $m = 1, \ldots, N$. Hence $C_2 \supset \Omega_{\tilde{a}} \cap G_N$, and according to (4) and (10) we arrive at the following

$$q'' \geq \mathbf{P}_{\tilde{a}}\{C_1 \cap \Omega_{\tilde{a}} \cap G_n\} \geq \mathbf{P}_{\tilde{a}}\{C_1 \cup H\} - \varepsilon_1$$
$$= \mathbf{P}_{\tilde{a}}\{H\} - \mathbf{P}_{\tilde{a}}\{H \backslash (H \cap C_1)\} - \varepsilon_1.$$

From $I_H \tau_m < \infty$ it follows that the sequence of sets $C_1 = C_1(\tau)$ possesses the property $\lim_{r \to \infty} C_1(r) \cap H = H$. Therefore, there exists r_0 such that

$$\mathbf{P}_{\tilde{a}}\{H \backslash (H \cap C_1(r))\} \leq \varepsilon_1$$

for all $r \geq r_0$. For such r with the help of (3) we find that

$$q'' \geq b - 2\varepsilon_1.$$

The relation of the probabilities q' and q'' is specified by using the sets T and C introduced. Indeed

$$q'' = \sum_{t \in T} \mathbf{P}_{\tilde{a}}\{G \cap \{v = t\}\} + \sum_{t \notin T} \mathbf{P}_{\tilde{a}}\{G \cap \{v = t\}\}$$
$$\leq \mathbf{P}_{\tilde{a}}\{G \cap \{v \in T\}\} + \varepsilon_1 \sum_{t \notin T} \max_{a \in A \backslash B} \mathbf{P}_a\{G \cap \{v = t\}\}$$
$$\leq \mathbf{P}_{\tilde{a}}\{G\} + \varepsilon_1 q''.$$

From this, together with the estimates of q' and q'' obtained, the required inequality follows

$$\mathbf{E}_{\tilde{a}}(v I_C) \geq s \mathbf{P}_{\tilde{a}}\{C\} \geq s(q'' - \varepsilon_1 q') \leq s(b - 3\varepsilon_1). \tag{11}$$

D. We can now complete the proof of the lemma. We define a sequence $\theta^{(n)}$ of Markov moments in a recurrent way

$$\theta^{(0)} = 0, \quad \theta^{(n)} = \theta^{(n-1)} + v_n + \theta_n, \qquad n = 1, 2, \dots.$$

Here v_n is the length of the collection of a-regeneration rules for the moment $\theta^{(n_1)}$ with $v_n \leq L$. The members of the sequence $\{\theta_n\}$ constructed in accordance with (6) are the stopping moments with respect to the part of the path from the moment $\theta^{(n-1)} + v_n$ on.

We can now define the strategy $\sigma^{(\varepsilon)}$ whose existence has been declared in the lemma. At the moments $\theta^{(n-1)} + 1, \dots, \theta^{(n-1)} + v_n$ its rules coincide with the $\tilde{a}$-regeneration rules for the moments $\theta^{(n-1)}$. At the next moments $\theta^{(n-1)} + v_n + 1, \dots, \theta^{(n)}$ they coincide with the initial rules of the strategy $\tilde{\sigma}$, i.e.

$$\sigma^{(\varepsilon)}_{\tau_n + i}\left(\cdot \, | x^{\tau_n + i - 1}, u^{\tau_n + i - 2} \right) = \tilde{\sigma}\left(\cdot \, | x_{\tau_n}^{\tau_n + i - 1}, u_{\tau_n}^{\tau_n + i - 2} \right)$$

where $\tau_n = \theta^{(n-1)} + v_n$, $i = 1, \dots, \theta_n$. We estimate the reward φ_t under this strategy provided the initial distribution p belongs to the class R. Let the measure $\mathbf{P}$ be generated by the pair $(\sigma^{(\varepsilon)}, p)$. The moments τ_n are the regeneration ones. According to the definition of a controlled regenerative process we have

$$\mathbf{P}\{\cdot | \mathcal{F}_n\} = \mathbf{P}_{\tilde{\sigma}, a(n)}\{(\cdot)\}, \quad \mathbf{P}\{a_{(n)} = a | \mathcal{F}_{\theta^{(n-1)}}\} \geq q > 0, \quad \text{a.s.} \tag{12}$$

where $a_{(n)} = a_{\tau_n}$. The stochastic situation on every interval $[\theta^{(n-1)} + v_n + 1, \theta^{(n)}]$ has been studied in **A** and **B** above. The distribution $a_{(n)}$ serves as the initial one. We put

$$\eta_n = \max\{n : \theta^{(n)} \leq t\}, \quad S_n = \sum_{t=\tau_n}^{\theta^{(n)}} x_t.$$

We already know that quantities $\mathbf{E}_{\tilde{\sigma}, a}(S_n)$ and $\mathbf{E}_{\tilde{\sigma}, a}(\theta_n)$ do not depend on n and the estimates (7)–(9) hold for them. Then, from the inclusions $\{\eta \geq n\} = \{\eta > n - 1\} \in \mathcal{F}_{\theta^{(n-1)}} \subset \mathcal{F}_{\tau_n}$, it follows that

$$\mathbf{E}\left(\sum_{i=1}^{t} x_i \right) \geq \mathbf{E}\left(\sum_{n=1}^{\eta} S_n \right) = \sum_{l=1}^{\infty}\left(\sum_{n=1}^{l} S_n; \eta = l \right) = \sum_{m,l=1}^{\infty} \mathbf{E}(S_n; \eta = l)$$

$$= \sum_{n=1}^{\infty} \sum_{a \in A} \mathbf{E}(S_n; \eta \geq n, a_{(n)} = a)$$

$$= \sum_{n=1}^{\infty} \sum_{a \in A} \mathbf{E}_{\tilde{\sigma}, a}(S_n) \mathbf{P}\{\eta \geq n; a_{(n)} = a\}$$

$$\geq \sum_{a \in B} (w - 5\varepsilon) \mathbf{E}_{\tilde{\sigma}, a}(v I_C) \sum_{n=1}^{\infty} \mathbf{P}\{\eta \geq n; a_{(n)} = a\}.$$

By analogy with the above we have

$$t \geq \mathbf{E}\left(\sum_{n=1}^{\eta+1}(v_n + \theta_n)\right) = L\mathbf{E}(\eta + 1) + \sum_{n=1}^{\infty}\mathbf{E}(\theta_n; \eta \geq n + 1)$$

$$= L\mathbf{E}(\eta + 1) + \sum_{n=1}^{\infty}\sum_{a \in A}\mathbf{E}_{\tilde{\sigma},a}(\theta_n)\mathbf{P}\{\eta_t \geq n - 1, a_{(n)} = a\}$$

$$\leq L\mathbf{E}(\eta + 1) + \sum_{a \in B}\mathbf{E}_{\tilde{\sigma},a}(\theta_n)\mathbf{P}\{\eta_t \geq n - 1, a_{(n)} = a\}$$

$$\leq L\mathbf{E}(\eta + 1) + \sum_{n=1}^{\infty}\sum_{a \in B}\left(r + \mathbf{E}_{\tilde{\sigma},a}(vI_C)\right)\mathbf{P}\{\eta_t \geq n - 1, a_{(n)} = a\}$$

$$+ \sum_{a \in A\backslash B}\left(r + \varepsilon_1\mathbf{E}_{\tilde{\sigma},\tilde{a}}(vI_C)\right)\sum_{n=1}^{\infty}\mathbf{P}\{\eta_t \geq n - 1, a_{(n)} \in A\backslash B\}.$$

From the estimates obtained and the fact that $\tilde{a} \in B$ it follows that (for the sake of simplicity, we denote $m_a = \mathbf{E}_{\tilde{\sigma},a}(vI_C)$)

$$t\left(\mathbf{E}\sum_{i=1}^{t}x_i\right)^{-1} \leq \frac{(L + r)\sum(\eta_t + 1)}{(w - 5\varepsilon)m_{\tilde{a}}\sum_{n=1}^{\infty}\mathbf{P}\{\eta_t \geq n, a \in B\}}$$

$$+ \frac{\sum_{a \in B}m_a\sum_{n=1}^{\infty}\mathbf{P}\{\eta_t \geq n - 1, a_{(n)} = a\}}{(w - 5\varepsilon)\sum_{a \in B}\sum_{n=1}^{\infty}\mathbf{P}\{\eta_t \geq n, a \in B\}}$$

$$+ \frac{\varepsilon_1 m_{\tilde{a}}\sum_{n=1}^{\infty}\mathbf{P}\{\eta_t \geq n - 1, a_{(n)} \in A\backslash B\}}{(w - 5\varepsilon)m_{\tilde{a}}\sum_{n=1}^{\infty}\mathbf{P}\{\eta_t \geq n, a_{(n)} = \tilde{a}\}}$$

$$= r_t^{(1)} + r_t^{(2)} + r_t^{(3)}.$$

We now find $\lim_{t \to \infty} r_t^{(i)}$ for $i = 1, 2, 3$. According to (11) and (12) we have

$$r_t^{(1)} \leq \frac{(L + r)\mathbf{E}(\eta_t + 1)}{(w - 5\varepsilon)s(b - 3\varepsilon_1)q\mathbf{E}\eta_t} \leq \frac{2(L + r)}{(w - 5\varepsilon)s(b - 3\varepsilon_1)q} \to 0$$

as $s \to \infty$. Therefore, one can choose s so that $\lim_{t \to \infty} r_t^{(1)} < \varepsilon$. For $r_t^{(i)}, i = 2, 3$ the correlations

$$\lim_{t \to \infty} r_t^{(2)} = \frac{1}{w - 5\varepsilon}, \qquad \overline{\lim_{t \to \infty}} \, r_t^{(3)} \leq C\varepsilon_1,$$

where a constant C does not depend on ε_1 hold.

It follows that for any initial distribution p and any $\varepsilon > 0$ we have the required inequality

$$W(\sigma^{(\varepsilon)}, p) > w - \varepsilon.$$

This completes the proof. $\qquad\square$

Definition 3. A strategy σ is called a **periodic strategy with period** T if all control choice rules $\sigma_t(\cdot|\cdot)$ forming it satisfy the relations

$$\sigma_t(\cdot|x^t, u^{t-1}) = \sigma_s(\cdot|x^t_{t-s-1}, u^{t-1}_{t-s}) \tag{13}$$

for $t = kT + s$, $k = 0, 1, 2, \ldots$, $s = 1, \ldots, T$.

Lemma 2. *For any controlled regenerative process from R there exists a periodic uniform ε-optimal strategy.*

Proof. We consider the uniform ε-optimal strategy $\sigma^{(\varepsilon)}$ constructed by Lemma 1. For any initial distribution p there exists a sequence $r_t(p)$ $(r_t(p) \downarrow 0$) such that

$$W_t(\sigma^{(\varepsilon)}, p) = t^{-1} \sum_{i=1}^{t} \mathbf{E}_{\sigma(\varepsilon), p} x_i \geq w - \varepsilon/2 - r_t(p).$$

Consequently,

$$W_t(\sigma^{(\varepsilon)}, p) = \sum_{a \in A} p(a) W_t(\sigma^{(\varepsilon)}, a) \geq \min_{a \in A} W(\sigma^{(\varepsilon)}, a)$$

$$\geq W_t(\sigma^{(\varepsilon)}, a(t)) \geq w - \varepsilon/2 - r_t(a(t))$$

where $a(t) = \operatorname{argmin}_a W_t(\sigma^{(\varepsilon)}, a)$. This means that there exists a numerical sequence $r(t)$ not depending on p such that $W_t(\sigma^{(\varepsilon)}, p) \geq w - \varepsilon/2 - r(t)$ and $r(t) = r(a(t)) \to 0$ as $t \to \infty$ for any p. Therefore, for any $\delta > 0$ there exists an integer N such that

$$W_N(\sigma^{(\varepsilon)}, p) \geq w - \delta. \tag{14}$$

We can now give the required strategy $\sigma^{(\varepsilon, T)}$. Its first T rules coincide with the first T' rules of the strategy $\sigma^{(\varepsilon)}$. The next L rules are the regeneration ones. Thereafter, we have cyclic repetitions of these $T = T' + L$ rules as in the formula (13). It is clear that $t = kT$, $k = 0, 1, \ldots$, are the regeneration moments. Hence, by (14), for sufficiently large T' we have constructed the periodic uniformly ε-optimal strategy. $\qquad\square$

Lemma 3. *For any controlled regenerative process from R there exists a deterministic periodic uniform ε-optimal strategy.*

Proof. It is sufficient to verify that for any p there exists some deterministic ε-optimal strategy. In other words, for any p there exists a strategy $\sigma^{(D)}(p) = (h_1, h_2, \ldots)$, $h_l \in D_l$, such that $W(\sigma^{(D)}, p) > w - \varepsilon$. If we prove this, then using arguments from the proofs of Lemmas 1 and 2, we obtain the periodic uniform ε-optimal strategy where $\sigma^{(D)}(p)$ is chosen as the primary strategy $\tilde{\sigma}$ which is deterministic like $\sigma^{(D)}(p)$.

So, let p be fixed. We prove the existence of the strategy $\sigma^{(D)}$. By Lemma 2, there exists a periodic uniform ε-optimal strategy $\sigma^{(\varepsilon, T)}$ representing the cyclic repetition of the rules $\sigma_1, \ldots, \sigma_T$; L of them are the regenerations rules which will be deterministic according to the definition, and inequality (14) holds. We replace, step by step, the randomized rules of the strategy $\sigma^{(\varepsilon,T)}$ by the deterministic ones. Let σ_1 be the randomized rule. We have

$$S_{T'}(\sigma^{(\varepsilon,T)}, p) = T' w_{T'}(\sigma^{(\varepsilon,T)}, p)$$

$$= \sum_{x \in X} p(x_0 = x) \mathbf{E}\left(\sum_{t=1}^{T'} x_t | x_0 = x, u_0 = h \right) = S_{T'}(\sigma(1), p).$$

Here $\sigma(1) = (h_1, \sigma_2, \ldots)$ differs from the initial strategy $\sigma^{(\varepsilon,T)}$ only by the first rule h_1 that is expressed as follows

$$h_1 = \operatorname{argmin}_{u \in U} \sum_{x \in X} p(x_0 = x) \mathbf{E}\left(\sum_{t=1}^{T'} x_t | x_0 = x, u_0 = h \right).$$

Suppose the strategy

$$\sigma(l) = (h_1, \ldots, h_l, \sigma_{l+1}, \ldots), \quad h_i \in D_i, l \leq T'$$

has already been constructed. It differs from $\sigma^{(\varepsilon,T)}$ by the first l rules. For this strategy we have

$$S_{T'}(\sigma(l), p) \geq S_{T'}(\sigma^{(\varepsilon,T)}, p).$$

If the next rule σ_{l+1} is the randomized one we shall replace it with the deterministic rule $h_{l+1} \in D_{l+1}$ by using the following relations (where $\mathbf{P}_l$, $\mathbf{E}_l$ mean $\mathbf{P}_{\sigma(l),p}$, $\mathbf{E}_{\sigma(l),p}$ respectively)

$$S_{T'}(\sigma(l), p)$$

$$= \sum_{j \in X^{l+1} \times U^l} \mathbf{P}_l\{(x^l, u^{l-1}) \in j\} \mathbf{E}_l\left(\sum_{t=1}^{T} x_t | (x^l, u^{l-1}) \in j, u_{l+1} = u \right)$$

$$= \sum_{j \in X^{l+1} \times U^l} \mathbf{P}_l\{(x^l, u^{l-1}) \in j\} \mathbf{E}_l\left(\sum_{t=1}^{T} x_t | (x^l, u^{l-1}) \in j, u_{l+1} = h_{l+1}(j) \right)$$

$$= S_T(\sigma(l+1), p)$$

where

$$\sigma(l+1) = (h_1, \ldots, h_l, h_{l+1}, \sigma_{l+2}, \ldots), \quad h_{l+1} = \{h_{l+1}(j), j \in X^{l+1} \times U^l\}.$$

For j such that $\mathbf{P}_l\{(x^l, u^{l-1}) = j\} > 0$ we put

$$h_{l+1}(j) = \operatorname{argmax}_{u \in U} \mathbf{E}_l\left(\sum_{t=1}^{T} x_t | (x^l, u^{l-1} = j, u_{l+1} = u \right)$$

but for the other j we define $h_{l+1}(j)$ in an arbitrary way. So, we obtain T deterministic rules. Because $T, 2T, \ldots$ are the regeneration moments we can replace the rules $h_{kT+1}, \ldots, h_{(k+1)T}$ in a similar manner and the rewards on the intervals $[kT + 1, (k+1)T]$ don't decrease. This completes the proof of the existence of the required strategy $\sigma^{(d)}(p)$. $\qquad\square$

Now we can return to the main theorem declared at the beginning of this section.

Proof of Theorem 1. This is based on the existence of a periodic deterministic uniform ε-optimal strategy σ which consists of repeating the collection of rules $h_1, \ldots, h_T$ with $h_i \in D_i \in D_T$.

We put

$$I_t(k) = \left\{ i : u_{t-k} = h_{C^l(i)}\left(x^{t-l}_{t-l-C^l(i)}, u^{t-l-1}_{t-l-C^l(i)+1}\right), \ l = 0, 1, \ldots, k\right\}$$

for $k \leq t - T$, where $C^l(i)$ means the cyclic permutation of the index-set $\{1, \ldots, T\}$, i.e.

$$C^0(i) = i, \quad C^1(i) = \begin{cases} i - 1, & \text{if } 1 < i \leq T, \\ T, & \text{if } i = 1, \end{cases}$$

$$C^{k+1}(i) = C^1(C^k(i)), \quad k = 1, 2, \ldots .$$

These sets are defined on the paths of the random process (x_t, u_t). Therefore they are random sets. They point out those rules from the collection $h_1, \ldots, h_T$ which could form the sequence of controls $u_{t-k}, \ldots, u_{t-1}, u_t$ at the moments $t - k, \ldots,$ $t - 1, t$ under the given past history $\left(x^t_{t-k}, u^{t-1}_{t-k}\right)$. Obviously, for the strategy σ and sufficiently large k the set $I_t(k)$ contains only one element, namely, the number of the rule h_i specified by the strategy σ at the moment t. Making use of these sets we want to define a single rule $\bar{h}$ with a larger depth, i.e. $\bar{h} \in D_k$. This can be done as follows: for $t \geq k+t$ we put $i(t) = \min\{i : i \in I_t(k)\}$, the rule $h_{i(t)} \in D_k$ is chosen as $\bar{h}$.

Finally, we note that for all $t \geq k + T$ and all pairs $k_1 < k_2$ the relation

$$\emptyset \neq I_t(k_1) \subset I_t(k_2), \quad |I_t(k_1)| \leq |I_t(k_2)|$$

are implemented. To form the strategy $\sigma^{(0)}$ given in the theorem we take preliminarily some sufficiently large integer k. The first $k + T - 1$ rules of the desired strategy are formed by cyclic repetition of the rules $h_1, \ldots, h_t$. Then, beginning from the moment $t = k + T$ on, we use the rule $\bar{h} = h_{i(t)}$. The required properties of the strategy $\sigma^{(0)}$ follow from the ε-optimality of the initial strategy σ. This completes the proof of Theorem 1. $\qquad\square$

6.6. Adaptive Strategies for Controlled Regenerative Processes

In this section we are concerned with optimal adaptive strategies for the class of controlled regenerative processes satisfying the following conditions:

(i) A set X of observable components of states, a set U of controls and a collection of regeneration rules of length L are assumed to be known;

(ii) For any stationary strategy $\sigma(h)$ generated by a rule $h \in D$ with finite depth and for any initial value of the parameter a there exists

$$w_{h,a} = \lim_{T \to \infty} T^{-1} \sum_{t=1}^{T} \mathbf{E}_{h,a} x_t$$

which depends on the distributions $\{\mu_t\}$ of the process;

(iii) For each stationary strategy $\sigma(h)$

$$\lim_{T \to \infty} \mathbf{P}_{h,a}\left\{\left| T^{-1} \sum_{t=1}^{T} x_t - w_{h,a} \right| > \varepsilon \right\} = 0$$

for every a and $\varepsilon > 0$.

The next condition imposed on the class of CRP requires existence of the distribution β defined on a countable set D of all finite-depth rules. We suppose that these rules are indexed in order of increasing memory depth. The distribution $\beta = \{\beta(h), h \in D\}$ is non-degenerate, i.e. $\beta(h) > 0$ for all h, and its moments are finite: $\sum_{n=1}^{\infty} n^l \beta(h_n) < \infty$, $l = 1, 2, \ldots$;
(iv)

$$\sum_{h \in D} \sum_{n=1}^{\infty} \sum_{t=n}^{\infty} \beta(h_t) \mathbf{P}_{h,a}\left(\left| t^{-1} \sum_{i=1}^{t} x_i - w_{h,a} \right| > \varepsilon \right) < \infty.$$

We discuss below the fulfilment of these conditions for concrete controlled regenerative processes, especially, for connected Markov chains.

Let $\mathcal{K} = \mathcal{K}(X, U; L; \beta)$ denote the class of controlled regenerative processes satisfying the conditions mentioned above. As the aim of control we choose asymptotic optimality in the strong sense, i.e.

to find an optimal adaptive strategy σ_0 which ensures the fulfilment of the equality

$$\lim_{T \to \infty} T^{-1} \sum_{t=1}^{T} x_t = w_0, \qquad \mathbf{P}_{\sigma_0} - \text{a.s.}$$

where $w_0 = w_0(\xi, \sigma_0) = \sup_{\sigma} W_\xi(\sigma)$ for any process $\xi \in \mathcal{K}$.

We now focus attention on the construction of the adaptive strategy σ_0. We put

$$\tau_0 = 0, \qquad \tau_n = \tau_{n-1} + L + n + \theta_n, \qquad n = 1, 2, \ldots$$

where $\theta_n = [(1 - n^{-1} \sum_{\tau=1}^{n} x_{\tau_{n-1}+L+\tau})^{-n}]$, and $n^{-1} \sum_{\tau=1}^{n} x_{\tau_{n-1}+L+\tau}$ means the empirical reward on the interval $[\tau_{n-1} + L + 1, \tau_{n-1} + L + n]$. At the moments $\tau_{n+1} + 1, \ldots, \tau_{n_1} + L$ the rules of the strategy σ_0 are the regeneration rules, which are supposed to be known. On the next time-interval of length n, we use the rule $h^{(n)}$ chosen from D according to the distribution β independent of the past history. Finally, on the last time-interval of the random length θ_n we use the rule $h^{(n)}$ without observing the course of the controlled process. This completes the description of the strategy σ_0.

We would like to emphasize that while describing the class $\mathcal{K}(X, U; L; \beta)$ we have not used the following sets: the family of conditional distributions $\{\mu_t\}$ (including the initial distribution of the parameter) and the a-regeneration rules.

The main result about efficiency of the strategy σ_0 for the processes from $\mathcal{K}(X, U; L; \beta)$ is the following.

Theorem 1. *For any process from $\mathcal{K}(X, U; L; \beta)$ the strategy σ_0 leads to the equality*

$$\mathbf{P}_{\sigma_0}\left\{\lim_{T\to\infty} T^{-1}\sum_{t=1}^{T} x_t = w_0\right\} = 1.$$

Proof. We choose some process $\xi \in \mathcal{K}$ and put

$$\varphi_n(m) = m^{-1}\sum_{t=\tau_{n-1}+L+1}^{\tau_{n-1}+L+m} x_t, \quad \varphi_n^{(1)} = \varphi_n(n), \quad \varphi_n^{(2)} = \varphi_n(n+\theta_n)$$

for $m, n = 1, 2, \ldots$. For a fixed $\varepsilon > 0$ we introduce the sets

$$A_n^{(l)} = \left\{\omega : \varphi_n^{(l)} \geq w - l\varepsilon\right\}, \quad l = 1, 2,$$

$$\Sigma_1 = \sum_{k=1}^{n} I_{A_k^{(1)} \cap A_k^{(2)}} \Delta_k, \quad \Sigma_2 = \sum_{k=1}^{n} I_{A_k^{(1)} \cap \bar{A}_k^{(2)}} \Delta_k, \quad \Sigma_3 = \sum_{k=1}^{n} I_{\bar{A}_k^{(1)}} \Delta_k.$$

The average reward up to the moment τ_n can be estimated from below as follows

$$\Psi_n = \frac{1}{\tau_n}\sum_{t=1}^{\tau_n} x_t \geq \frac{\sum_{k=1}^{n}\varphi_k^{(2)}\Delta_k}{\sum_{k=1}^{n}\Delta_k} \geq (w - 2\varepsilon)\frac{\Sigma_1}{\Sigma_1 + \Sigma_2 + \Sigma_3}. \tag{1}$$

We put

$$\alpha_n = a_{\tau_{n-1}+L}, \quad w_n = w_{h^{(n)},\alpha_n}, \quad \mathbf{P}\{\cdot\} = \mathbf{P}_{\sigma_0}\{\cdot\}$$

and $\mathcal{F}_{\tau_{n-1}+L}$ denotes the σ-algebra generated by the history of the process up to the moment $\tau_{n-1} + L$. It is important that the moments $\tau_{n-1} + L$ are the regeneration ones. Hence

$$\mathbf{P}\{\cdot|\mathcal{F}_{\tau_{n-1}+L}\} = \mathbf{P}_{h^{(n)},\alpha_n}\{\cdot\}.$$

Let us estimate the sums Σ_i, $i = 1, 2, 3$. We have

$$\Sigma_1 \geq \Delta_{\nu_n}, \quad \nu_n = \max\{k : k \leq n, \varphi_k^{(1)} \geq w - \varepsilon/2, \varphi_k^{(2)} \geq w - 2\varepsilon\}.$$

We consider now the event

$$A_n = \{\nu_n \leq n - \ln n\} \subset \bigcap_{n-\ln n < k \leq n}\{\varphi_k^{(1)} - \varepsilon/2\} \cap A_k^{(2)}.$$

Denoting $\mathbf{P}_{(k)}\{\cdot\} = \mathbf{P}\{\cdot|\mathcal{F}_{\tau_{k-1}}\}$, we have

$$\mathbf{P}_{(k)}\left\{\left\{\varphi_k^{(1)} \geq w - \varepsilon/2\right\} \cup A_k^{(2)}\right\} \leq \sum_{h:w_n \leq w-\varepsilon/2}\mathbf{P}_{(k)}\{h^{(n)} = h, \alpha_n = a\}$$

$$+ \sum_{a\in A}\sum_{h:w_n > w-\varepsilon/2}\mathbf{P}_{(k)}\{h^{(n)} = h, \alpha_n = a\}\mathbf{P}_{h,a}\left\{\bigcup_{m=k}^{\infty}\{\varphi_k(m) < w - \varepsilon/2\}\right\}$$

$$\leq \mathbf{P}\{w_n \leq w - \varepsilon/2\} + \sum_{h\in D}\sum_{m=k}^{\infty}\beta(h)\mathbf{P}_{h,a}\{\varphi_k(m) \leq w - \varepsilon/2\}.$$

By Theorem 1 from the previous section, the first term on the right-hand side of these relations is less than one. By condition (iv), the second term can be made however small by the appropriate choice of k. Hence $\mathbf{P}\{A_n\} \leq q^{n-\ln n}$, $(q < 1)$ for all $n > n_0$, and by the Borel–Cantelli Lemma we see that for the set $A = \overline{\lim}_{t\to\infty} A_n$ we have

$$\mathbf{P}\{A\} = 0. \tag{2}$$

From this it follows that

$$\Sigma_1 \geq \Delta_{\nu_n} \leq (1 - w + 2^{-1}\varepsilon)^{-n\ln n} \xrightarrow[n\to\infty]{} \infty \quad \mathbf{P} - \text{a.s.}$$

We now turn to the sum Σ_2. First, we estimate the probability of the event

$$B_n = A_n^{(1)} \cup \left\{ \inf_{n < m \leq n+\theta_n} \varphi_n(m) < w - 2\varepsilon \right\}.$$

We have

$$
\begin{aligned}
\mathbf{P}\{B_n\} &= \mathbf{P}\{B_n, w_n < w - 3\varepsilon/2\} + \mathbf{P}\{B_n, w_n \geq w - 3\varepsilon/2\} \\
&\leq \mathbf{P}\{\varphi^{(1)} \geq w - \varepsilon, w_n < w - 3\varepsilon/2\} \\
&\quad + \mathbf{P}\left\{ \inf_{n < m \leq n+\theta_n} \varphi_n(m) < w - \varepsilon, w_n \geq w - 3\varepsilon/2 \right\} \\
&\geq \sum_{a \in A} \left[\sum_{h:w_{h,a} < w-3\varepsilon/2} \mathbf{P}\{h^{(n)} = h, \alpha_n = a\}\mathbf{P}_{h,a}\left\{ n^{-1}\sum_{i=1}^{n} x_i \geq w - \varepsilon \right\} \right. \\
&\quad \left. + \sum_{h:w_{h,a} \geq w-2\varepsilon/2} \mathbf{P}\{h^{(n)} = h, \alpha_n = a\}\mathbf{P}_{h,a}\left\{ t^{-1}\sum_{i=1}^{t} x_i < w - \varepsilon, t \geq n \right\} \right] \\
&\geq \sum_{h \in D} \beta(h)\mathbf{P}_{h,a}\left\{ \left| t^{-1}\sum_{i=1}^{t} x_i - w_{h,a} \right| \geq \varepsilon/2, t \geq n \right\}.
\end{aligned}
$$

From condition (iv) it follows that $\sum_{n=1}^{\infty} \mathbf{P}\{B_n\} < \infty$ and, consequently, for $B = \overline{\lim}_{t\to\infty} B_n$ we have

$$\mathbf{P}\{B\} = 0.$$

Hence the sum Σ_2 is bounded a.s. as n increases. The estimate for Σ_3 is obvious. Indeed

$$\Sigma_3 \leq \sum_{k=1}^{n}[L + k + (1 - w + \varepsilon)^{-k}] \leq n(L + n) + n(1 - w + \varepsilon)^{-n}.$$

Substituting the estimates obtained in (1), we obtain

$$\Psi_n \geq (w - 2\varepsilon)\left(1 + \frac{\Sigma_2}{\Sigma_1} + \frac{\Sigma_3}{\Sigma_1} \right)^{-1}$$

$$\geq (w - 2\varepsilon)\left[1 + \frac{C_1}{\Sigma_1} + \frac{n(L + n) + n(1 - w + \varepsilon)^{-n}}{(1 - w + \varepsilon/2)^{-n}} \right]^{-1} \geq w - 3\varepsilon$$

for all $n \geq n_0$. Hence

$$\mathbf{P}\left\{ \lim_{n \to \infty} \Psi_n = w \right\} = 1. \tag{3}$$

Let us find the estimate of the average reward $\varphi_t = t^{-1} \sum_{i=1}^{t} x_i$. To achieve this we consider the set

$$\Omega' = \left\{ \omega : \lim_{n \to \infty} \Psi_n = w \right\} \cap \{\Omega \backslash A\} \cap \{\Omega \backslash B\}.$$

According to (2)–(3) we have $\mathbf{P}\{\Omega'\} = 1$. We introduce the following events

$$C_{n,t}^{(1)} = \{\omega : \tau_{n-1} < t \leq \tau_{n-1} + L + n\} \cap \Omega',$$
$$C_{n,t}^{(2)} = \{\omega : \tau_{n-1} + L + n < t \leq \tau_n\} \cap \Omega' \cap A_n^{(1)},$$
$$C_{n,t}^{(3)} = \{\omega : \tau_{n-1} + L + n < t \leq \tau_n\} \cap \Omega' \cap \bar{A}_n^{(1)}.$$

On the set $C_{n,t}^{(1)}$ we estimate the average reward by making use of (3)

$$\varphi_t \geq \frac{\tau_{n-1}}{\tau_{n-1} + L + n} \Psi_n \geq w - a_n^{(1)}$$

where $\lim_{n \to \infty} a_n^{(1)} = 0$.

On the set $C_{n,t}^{(2)}$ we have

$$\theta_n \leq (1 - w + \varepsilon)^{-n}, \quad \nu_n > n - \ln n, \quad n \geq n_0.$$

Here the second inequality follows from the definition of A, A_n and $C_{n,t}^{(2)}$. Thus, on the set $C_{n,t}^{(2)}$ we obtain

$$\varphi_t \geq \frac{\tau_{n-1}}{\tau_{n-1} + L + n} \Psi_n \geq w - a_n^{(2)}$$

where $\lim_{n \to \infty} a_n^{(2)} = 0$.

By the inclusion $C_{n,t}^{(3)} \subset \{\omega : \inf_{n < m \leq n + \theta_n} \varphi_n(m) \geq w - 2\varepsilon\}$ on the set $C_{n,t}^{(3)}$ we have

$$\varphi_t \geq \frac{\tau_{n-1}}{\tau} \Psi_n + \left(1 - \frac{\tau_{n-1} + L}{t}\right) (t - \tau_n - L)^{-1} \sum_{i=\tau_{n-1}+L+1}^{t} x_i$$

$$\geq \frac{\tau_{n-1}}{\tau} \Psi_n + \left(1 - \frac{\tau_{n-1} + L}{t}\right) (w - 2\varepsilon) \geq w - 2\varepsilon - a_n^{(3)}$$

where $a_n^{(3)} \downarrow 0$ as $n \to \infty$. Thus, on the union of the sets $C_{n,t} = \bigcup_{l=1}^{3} C_{n,t}^{(l)} = \{\omega : \tau_{n-1} < t \leq \tau_n\} \cap \Omega'$ we have

$$\varphi_t \geq w - 2\varepsilon - a_n$$

where $a_n \downarrow 0$ as $n \to \infty$. Now, from the obvious relationships

$$\Omega = \bigcup_{n=0}^{\infty} \{\tau_{n-1} < t \tau_n\}, \quad \lim_{t \to \infty} I_{C_{n,t}} = 0, \quad \forall n$$

the theorem follows. $\qquad \square$

The application sphere of the above theorem is rather wide and includes, in particular, partially observable Markov chains and graphs, and queueing systems. We consider in short their applications to connected chains and graphs, that is to the results of Sec. 3.

It is obvious that conditions (i)–(iii) hold for $\mathcal{K}(X, U; L; \beta)$. However, it may appear that condition (iv) is not, but it holds since we have the exponential estimate of the probabilities $\mathbf{P}\{|\cdot| \geq \varepsilon\}$, i.e. more precisely,

$$\mathbf{P}\left\{\left|t^{-1}\sum_{i=1}^{t} x_i - w_{h,a}\right| \geq \varepsilon\right\} \leq \beta e^{-\alpha t^{\gamma}}, \quad \alpha, \ \beta, \ \gamma > 0.$$

Hence the series in condition (iv) converges.

Thus, Theorem 1 from Sec. 3 is valid but we make one supplement to it. Let a class of all partially observable Markov chains (but not only the connected ones) be considered. Again $W(\sigma, p)$ is the average reward and $\sigma \in \Sigma$ is the set of admissible strategies (they depend on past controls and rewards). Then $\sup_{\sigma \in \Sigma} W(\sigma, p) = W(p)$ that depends on the distribution of the states. Like the result on observable chains stated in Sec. 3, Chap. 5 in the case of partially observable chains the strategy σ_0 leads to the inequality

$$\lim_{T \to \infty} T^{-1} \sum_{t=1}^{T} x_t \geq \inf_{p} \sup_{\sigma \in \Sigma} W(\sigma, p), \quad \text{a.s.}$$

for every (not connected) chain. It cannot be improved.

It is easy to understand that Theorem 2 from Sec. 3 about controlling the class of connected graphs with rewards follows immediately from the theorem proved above. The achievement of the global maximum by the objective function $W(\sigma)$ under the strategy $\sigma(\mathcal{K}'_u)$ is explained by the fact that the set of controls contains the optimal rules for connected controlled graphs with rewards.

CHAPTER 7

**CONTROL OF MARKOV PROCESSES WITH DISCRETE TIME
AND SEMI-MARKOV PROCESSES**

Controlled Markov processes are the main tool in modelling applied problems. In this chapter we consider the methods of adaptive control for Markov processes with continuous state space and discrete time and countable semi-Markov processes. These are both interesting and often used in various applications. We study both identification and serching strategies for these processes. Some strategies can be used in applications without difficulty, while others have only a theoretical meaning.

7.1. Preliminary Results

The controlled objects studied in this chapter differ from those considered in Chaps. 5 and 6 by their phase spaces. Earlier these spaces were finite but now they are assumed to be continuous. We give below some definitions and results concerned with Markov processes with discrete time and continuous state space.

Let $(X, \mathcal{F})$ be a topological measurable space which is assumed to be separable with metric[a] ρ. The symbol $\mathfrak{F}$ denotes the Borel σ-algebra, i.e. the least σ-algebra containing all open sets from X. The $\mathfrak{F}$-measurable functions are called the *Borel functions*. We define a metric on the set of all compact subsets of the space X in the following way

$$h(A, B) = \max \left(\max_{a \in A} \rho(a, B), \max_{b \in B} \rho(A, b) \right)$$

where the sets A, B are non-empty and compact. This metric is called the *Hausdorff metric*. Let $C(X)$ be the Banach space of all bounded, continuous, real-valued functions with the norm $\|f\| = \sup_x |f(x)|$ which gives the uniform convergence of a sequence of functions in this space. Sometimes it is useful to consider the wider Banach space $B(X)$, namely, the space of bounded Borel functions with the norm $\|f\| = \operatorname{esssup} |f(x)|$.

Let $\mathcal{P}_X = \{\mu\}$ be the family of all probability measures which are given on the Borelean σ- algebra $\mathcal{F}$ of X. If in $\mathcal{P}_X$ the convergence is understood as weak convergence of measures, i.e. $\mu_n \xrightarrow[n \to \infty]{w} \mu$ means

$$\int_X f \, d\mu_n \xrightarrow[n \to \infty]{} \int_X f \, d\mu$$

[a]A metric space X is called *separable* if it contains a countable dense subset. A metric $\rho(x, y)$ is a non-negative function such that: I. $\rho(x, y) = 0$ if and only if $x = y$; II. $\rho(x, y) = \rho(y, x)$, $\forall x$, $y \in X$; III $\rho(x, y) \leq \rho(x, z) + \rho(z, y)$, $\forall x, y, z \in X$.

203

for any $f \in C(X)$, then $\mathcal{P}_X$ becomes a topological space and the topology generated by this convergence is called the "weak topology". This space inherits many properties of the initial space X. Compactness and metrizability are the main of them. This means that the space $\mathcal{P}_X$ has at least one metric generating the weak topology. We give two examples of such metrics. The first of them is the Dudley metric

$$d(\mu, \nu) = \sup_{f \in F} \left| \int_X f(x)\, d\mu - \int_X f(x)\, d\nu \right|$$

where $F = \{f \in C(X) : \|f\| \leq 1, |f(x) - f(y)| \leq \rho(x, y)\}$. The second metrics is the Lévy-Prohorov one

$$\delta(\mu, \nu) = \inf\{\varepsilon : \mu(A) \leq \nu(A_\varepsilon) + \varepsilon, \nu(A) \leq \mu(A_\varepsilon) + \varepsilon\}$$

where A is any open set from X and $A_\varepsilon = \{x : \rho(x, A) < \varepsilon\}$ is the ε-neighborhood of A.

So, the space $\mathcal{P}_X$ inherits the following properties of X: compactness, completeness and being Borelean.[b]

Let us introduce one more metric (and, in fact, one more topology) in $\mathcal{P}_X$ as follows

$$V(\mu, \nu) = \sup_{f \in B(X), \|f\| \leq 1} \left| \int_X f(x)\, d\mu - \int_X f(x)\, d\nu \right|,$$

known as the metric (or distance) *in variation*. For any μ, ν we have $d(\mu, \nu) \leq V(\mu, \nu)$. Convergence with respect to this metric in $\mathcal{P}_X$ is stronger than weak convergence. One concrete result about the convergence of probability measures will be stated below in connection with limiting properties of Markov processes. We now define that notion. Let $(X, \mathcal{F})$ be a separable metric space with the Borel σ-algebra $\mathcal{F}$ and $(\Omega, \mathcal{A}, \mathbf{P})$ be some probability space. Let $x_t(\omega)$, $t = 0, 1, 2 \ldots$ be a random process given on $(\Omega, \mathcal{A}, \mathbf{P})$ that takes the values from X. This process can be defined by a family of conditional distributions $\{\mu_t, \ t = 0, 1, 2, \ldots\}$, $\mu_0(\cdot) = \mathbf{P}\{x_0 \in \cdot\}$, $\mu_{t+1}(\cdot) = \mathbf{P}\{x_{t+1} \in \cdot | x^t\}$ where $x^t = (x_0, x_1, \ldots, x_t)$.

Definition 1. A random process $x_t(\omega), \omega \in \Omega$, is called a *Markov process with discrete time* if $\mathbf{P}\{x_{t+1} \in \cdot | x^t\} = \mathbf{P}\{x_{t+1} \in \cdot | x_t\}$. If the function $\mathbf{P}\{x_{t+1} \in \cdot | x_t\}$ does not depend on t, i.e. $\mathbf{P}\{x_{t+1} \in \cdot | x_t\} = \mathbf{P}\{x_1 \in \cdot | x_0\}$ this is called a *homogeneous* process.

In other words, properties of a Markov process at the moment $t + 1$ depend solely on its previous value x_t but not on its more remote past history.

We consider only homogeneous Markov processes and assume that the conditional measure $\mu(A|x) \overset{\text{def}}{=} \mathbf{P}\{x_1 \in A | x_0 = x\}$ is a probability measure in $A \in \mathcal{F}$ for all fixed $x \in X$ and a Borel function in x for all fixed A. In this case $\mu(\cdot|\cdot)$

---

[b]A topological space is called *borelian* if it is homeomorphic to a Borel subset of a complete separable metric space.

is called a *transition function*. This function together with the initial distribution $\mu_0(\cdot)$ defines the process x_t completely. Indeed, if we define

$$\mu^{(k)}(\cdot|x) \stackrel{\text{def}}{=} \int_X \mu^{(k-1)}(\cdot|y)\mu(dy|x), \quad \mu^{(1)}(\cdot|x) \equiv \mu(\cdot|x)$$

then, for example,

$$\mathbf{P}\{x_2 \in A|x_0 = x\} = \mu^{(2)}(A|x) = \int_X \mu(A|y)\mu(dy|x).$$

Obviously, the unconditional distribution at the moment t is defined as follows

$$\mathbf{P}\{x_t \in A\} = \int_X \mu^{(t)}(A|y)\mu_0(dy).$$

Definition 2. A measure μ is called *invariant* if $\mu^{(t)} \equiv \mu$ for all $t \geq 1$, i.e. if we take μ as the initial distribution then the unconditional distributions of the process x_t coincide with μ

$$\mu(A) = \int_X \mu^{(t)}(A|y)\mu(dy), \quad \forall A \in \mathcal{F}.$$

An initial distribution μ_0 and a transition function μ allow constructing a probability measure $\mathbf{P}$ on the path space of the given Markov process. The appropriate formulas have been given in Sec. 2, Chap. 1. Hence we may consider probabilities of events depending on the paths of the process — for example, $\mathbf{P}\{\sup_t |x_t| > a\}$.

Any measure μ can be decomposed into an absolutely continuous and a singular component. These notions are defined as follows:

The *absolute continuity* of μ with respect to some basic measure λ (on the real axis it is usually the Lebesgue measure) means that $\mu(A) = 0$ whenever $\lambda(A) = 0$.

The *singularity* of μ with respect to λ means that there exists a set M such that $\mu(M) = 1$ and $\lambda(X \setminus M) = 1$.

Any measure μ which is absolutely continuous with respect to λ has a density $p(x)$, i.e.

$$\mu(A) = \int_A p(x)\lambda(dx), \quad A \in \mathcal{F}.$$

Any non-atomic measure μ has a unique representation in the form

$$\mu(A) = a \int_A p(x)\lambda(dx) + (1-a)s(A), \quad 0 \leq a \leq 1$$

where $s(A)$ means the singular component of μ with respect to λ. For the transition functions this result means that

$$\mu(A|x) = a \int_A p(y,x)\lambda(dy) + (1-a)s(A|x), \quad \forall x \in X, A \in \mathcal{F}$$

where $\lambda(\cdot)$ is some probability measure.

If $\mu(\cdot|x)$ has density $p(y,x)$ then the measure $\mu^{(t)}(\cdot|x)$ has the density

$$p^{(t)}(y,x) \stackrel{\text{def}}{=} \int_X p^{(t-1)}(y,u)p(u,x)du, \quad p^{(1)}(y,x) \equiv p(y,x).$$

We define a measurable function $q(x)$ on X which will be interpreted as a *reward* under the state x. The mathematical expectation $\mathbf{E}q(x_t) = \int_X q(x)\mu^{(t)}(dx)$ can be interpreted as the *average reward at the moment t*.

We shall try to find the limiting average reward (as $t \to \infty$), whose existence depends on ergodic properties of the process x_t. We impose a number of sufficient conditions on the process to guarantee the existence of the limiting average reward. We state one such condition.

Assume that the absolutely continuous component of the transition function has a density $p(y,x)$ with respect to a probability measure λ. We impose on this density the following condition.

D*. There exists a set $A \in \mathcal{F}$ and an integer $l \geq 1$ such that

(1) $\lambda(A) > 0$;
(2) $p^{(l)}(y,x) \geq c > 0, \quad \forall x \in X, \ y \in A.$

By a version of the ergodic theorem for the compact space X there exists a unique invariant measure μ_∞ such that

$$V\big(\mu^{(t)}, \mu_\infty\big) \leq 2(1-c)^{t/l-1}$$

under the condition **D***. The assertion means, in particular, that for $l = 1$ and any $A \in \mathcal{F}$

$$|\mu^{(t)}(A|x) - \mu_\infty(A)| \leq 2(1-c)^{t-1}.$$

For the average rewards on the paths of a Markov process under the condition **D*** the limiting average reward exists and does not depend on the initial state. It is equal to

$$W = \lim_{t \to \infty} \mathbf{E}q(x_t) = \int_X q(x)\mu_\infty(dx).$$

If condition **D*** fails this limit may not exist. Instead, we can use the limit in the Cesaro average sense which can depend on the initial state. In other words, for any $A \in \mathcal{F}$ we have

$$\lim_{T \to \infty} T^{-1} \sum_{t=1}^{T} \mu^{(t)}(A|x) = \mu_\infty(A|x), \quad \forall x.$$

In the case of a unique ergodic class the limiting measure $\mu_\infty(\cdot|x)$ does not depend on x.

It is useful to know whether the limit of the average rewards exists. To answer this question we introduce the following principal condition.

Condition D. (Doeblin) There exists a probability measure λ on $\mathcal{F}_X$, an integer $l \geq 1$ and a number $\varepsilon > 0$ such that

$$\mu^{(l)}(A|x) \leq 1 - \varepsilon, \quad \text{if } \lambda(A) \leq \varepsilon.$$

Under this condition we can define the ergodic classes $\mathcal{E}_\alpha$ and the set of non-essential states N and show that such Markov processes decompose into these ergodic classes $\mathcal{E}_\alpha$ and the set N. The ergodic class $\mathcal{E}_\alpha$ has the limiting distribution μ_α that is understood as the limit of the Casaro averages in variation in presence of the cyclic subclasses. Otherwise, it is interpreted similarly to the ergodic theorem stated above.

Let the reward $q(x)$ be a Borel function satisfying $\int_{\mathcal{E}_\alpha} |q(x)| \mu_\alpha(dx) < \infty$ for all α. Then we have

Proposition. *Strong law of large numbers: if condition* **D** *holds then for any initial distribution there exists*

$$\lim_{T \to \infty} T^{-1} \sum_{t=1}^{T} q(x_t) = \int_{\mathcal{E}_\alpha} |q(x)| \mu_\alpha(dx) < \infty \quad \text{a.s.}$$

at the initial condition $x_0 \in \mathcal{E}_\alpha$.

Now we define the notions used in the theory of controlled Markov processes.

Let $(X, \mathcal{F})$ and $(U, \mathcal{G})$ be separable metric spaces with Borel σ-algebras. They are called the state and control spaces respectively. Let some model $\{\mu_0(\cdot), \mu_{t+1}(\cdot | x^t, u^t), \ t \geq 0\}$ be defined on these spaces. We say that this model is *Markov* if $\mu_{t+1}(\cdot | x^t, u^t) = \mu(\cdot | x_t, u_t)$. The measure $\mu(\cdot | x, u)$ is called the *controlled transition function*. It is assumed to be homogeneous and borelean in (x, u). A Markov model can be represented in the form

$$x_{t+1} = \Phi(x_t, u_t, \omega),$$

and it is convenient to consider it as a solution of the difference equation with a random disturbance

$$x_{t+1} = F(x_t, u_t, \xi_t(\omega))$$

where $\xi_t(\omega)$, $t \geq 0$, are independent, identically distributed random variables. A special case of great significance is the linear equation

$$x_{t+1} = Ax_t + Bu_t + \xi_t$$

where A and B are linear operators (the first of them is given on the space $X = \mathrm{R}^n$, the second one is a mapping of $U = \mathrm{R}^m$ into R^n). Chapter 10 will be completely devoted to the objects of this type.

We assume the admissible controls to be taken from some given set $U_x \subseteq U$ for each state x whenever a model path enters this state. In many cases $U_x \equiv U$, $\forall\, x \in X$.

As a set of admissible strategies Σ we take all unanticipated strategies under which the rewards $q(x^t, u^{t-1})$ have finite mathematical expectations $\mathbf{E}_{\sigma,\mu_0} q$. Here $q(\cdot)$ is a Borel function of appropriate arguments. Instead, we can consider the r.v. ζ_t whose distribution is a Borel function depending on the history (x^t, u^{t-1}).

Definition 3. A pair $\zeta = [\mu_0, \mu; \Sigma]$ consisting of a Markov model (μ_0, μ) and a set Σ of admissible strategies is called a **controlled Markov process**.

Under a fixed strategy $\sigma \in \Sigma$ we have a random process in the usual sense whose paths belong to $(X \times U)^\infty$.

In what follows we assume that the following condition holds:

$\tilde{\mathbf{D}}$. There exist a probability measure λ on $\mathcal{F}_X$, an integer $l \geq 1$ and a number $\varepsilon > 0$ such that

$$\mu^{(l)}(A|x, u) \leq 1 - \varepsilon, \quad \text{if } \lambda(A) \leq \varepsilon.$$

This condition allows considering ergodic classes of controlled Markov processes.

In connection with the optimization aims of control we use two objective functions. The first of them is called the *discounted reward* and has the form

$$W_\beta(\sigma, \mu_0) = \sum_{n=0}^{\infty} \beta^n \mathbf{E}_{\sigma, \mu_0} q_n, \quad 0 < \beta < 1.$$

In Chap. 5 we have explained why in the adaptive statement of the optimal control problem one should not consider searching of an extremum of this function as the aim of control. For this reason we exclude this function from consideration. The second objective function is the limiting average reward per step or *average one-step reward* for short. It has the form

$$W(\sigma, \mu_0) = \varlimsup_{T \to \infty} T^{-1} \sum_{t=1}^{T} \mathbf{E}_{\sigma, \mu_0} q_t.$$

In fact, we have used this aim in all optimization problems of adaptive control. Now we briefly consider such problems within the framework of the classical control theory always supposing the Markov model (μ_0, μ) to be known.

The search of an optimal strategy for a controlled Markov process with the objective function $W(\sigma, \mu_0)$ is usually based on the dynamic programming method which uses the Bellman "optimality principle"[c] leading to the Bellman optimality equation. This equation serves as a sufficient condition for the optimal control to exist and as a tool for computing the optimal control and the maximum of the objective function.

In the optimization problem for the average one-step reward the Bellman equation has the form

$$W + V(x) = \sup_{u \in U_x} \left\{ q(x, u) + \int_X V(y)\mu(dy|x, u) \right\}. \tag{1}$$

Note the similarity of this equation to equation (1) in Sec. 2, Chap. 1 for finite Markov chains. We now consider properties of its solution due to which it serves as a tool for designing an optimal strategy.

---

[c] If $x^0(t)$ and $u^0(t)$ are the optimal path and optimal control respectively over the time-interval $[0, T]$ then they will be optimal over any subinterval $[t_0, T]$, $0 < t_0 < T$.

Lemma 1. *If there exists a constant W and a bounded Borel function $V(x)$ satisfying the Bellman equation then $\sup_{\sigma \in \Sigma} W(\sigma, x) \leq W$ for any $V(x)$.*

Proof. Let $h_t = \{x^t, u^{t-1}\}$ be the history of the controlled process. For any strategy $\sigma \in \Sigma$ we have

$$\mathbf{E}_{\sigma, x} \left\{ \sum_{l=1}^{t} \left[V(x_l) - \mathbf{E}_\sigma (V(x_l)|h_{l-1}) \right] \right\} = 0. \tag{2}$$

The Bellman equation and Markov property of the process imply the following

$$\mathbf{E}_\sigma \{ V(x_l)|h_{l-1} \} = \int V(y) \mu(dy|x_{l-1}, u_{l-1})$$

$$= q(x_{l-1}, u_{l-1}) + \int V(y) \mu(dy|x_{l-1}, u_{l-1}) - q(x_{l-1}, u_{l-1})$$

$$\leq W + V(x_{l-1}) - q(x_{l-1}, u_{l-1}). \tag{3}$$

Using the relationship (3) and (2) we obtain

$$\mathbf{E}_{\sigma, x} \left\{ \sum_{l=1}^{t} q(x_{l-1}, u_{l-1}) + V(x_t) - V(x_0) \right\} \leq tW. \tag{4}$$

This and the boundedness of $V(x)$ imply the assertion of Lemma 1. $\qquad\square$

Theorem 1. *If the assumptions of Lemma 1 hold and*

$$W + V(x) = q(x, d(x)) + \int_X V(y) \mu(dy|x, d(x))$$

for some Borel function $q(x)$ taking the values from U, then the simple strategy $\sigma_0 = \{d^\infty\}$ is optimal and $W(\sigma_0, x) = W$.

Proof. The required assertion follows immediately from the fact that the inequalities (3), (4) and $W(\sigma, x) \leq W$ turn, in fact, into equalities. $\qquad\square$

Thus, if the Bellman equation has a solution then there exists an optimal simple strategy, i.e. a stationary, deterministic, Markov strategy. We give some general conditions for the existence of the solution of the Bellman equation and hence for the existence of a simple optimal strategy.

Definition 4. We call a Markov model *semi-continuous* if:

(1) the sets U_x are closed[d];
(2) if $x_n \to x$ and $u_n \in U_{x_n}$, then the sequence u_n has a limiting point belonging to U_x;

---

[d]In a compact space X they are compact.

(3) the function $g(x) = \int_X f(y)\mu(dy|x,u)$ is continuous if f is continuous and bounded;

(4) the reward $q(x,u)$ is a semi-continuous[e] function bounded from above.

Definition 5. We call a measure ν on $\mathcal{F}$ such that $0 < \nu(X) < 1$ a *minorant* of a transition function $\mu(\cdot/x,u)$ if

$$\nu(A) \le \mu(A|x,y), \quad \forall A \in \mathcal{F},\ x \in X,\ y \in U_x.$$

The main result is

Theorem 2. *If the transition function of a semi-continuous Markov model has a minorant then the corresponding Bellman equation has a solution (i.e. a simple optimal strategy exists).*

We note that if a function μ has a minorant then the operator

$$Tf(x) = \sup_{u \in U_x} \left\{ q(x,u) + \int_X V(y)\mu(dy|x,u) \right\}$$

is a contracting operator on $B(X)$. This implies the convergence in variation of the transition probabilities of the process to the limiting distribution with exponential rate for any stationary strategy.

It remains to clear up some details concerning the calculation procedure of the optimal strategy. This is connected with the maximization of the right-hand side of the Bellman equation: for every x we have to find a $u(x) \in U_x$ at which that side is maximum. If the Bellman equation has a unique solution then the unique solution $u(x)$ will be obtained. Otherwise, we deal with the multi-valued mapping

$$x \to M_x = \left\{ x^* \in U_x : q(x,u^*) + \int_X v(y)\mu(dy|x,u^*) \right\}$$
$$= \min_{u \in U_x} \left\{ q(x,u) + \int_X v(y)\mu(dy|x,u) \right\},$$

assigning to each x the set $M_x \subset U$.

It is rather difficult to use such mappings. One may consider whether it is possible to select from M_x a "convenient" function: either continuous or semi-continuous or, at least, measurable.

It turns out that one may speak only about measurable functions (they are often called "selectors"). Assertions about their existence are called "measurable choice theorems". Before proceed to considering them we introduce some definitions and notation. We agree to denote:

(a) the set of real-valued, semi-continuous and bounded from above functions by $\mathcal{L}(X)$;

---

[e] A real-valued function $f(x)$ is called *upper semi-continuous* if the sets $\{x : f(x) \le c\}$ are closed for any c or if for any sequence $x_n \to x$ we have $\overline{\lim}_{n \to \infty} f(x_n) \le f(x)$.

A function $f(x)$ will be upper semi-continuous and bounded from above if there exists a sequence $f_n \in C(X)$ such that $f_n(x) \downarrow f(x)$.

(b) the set of numerical functions $f(x, u)$ which are the limits of non-increasing sequences of bounded functions $f_n(x, u)$, measurable in x and u and continuous in u at any x by $L(X \times U)$;

(c) the set of non-empty compacts which belong to the space U by $F(U)$.

Definition 6. A set-valued function $g(x)$ (representing a mapping of X to $F(U)$) is called *quasi-continuous* if for every sequence $x_n \to x$ the sequence $u_n \in g(x_n)$ has a limiting point belonging to the set $g(x)$.

Measurable Choice Theorems

Theorem A. *Let X be a measurable space and U be a separable metric space (with metric ρ). Then if the following conditions hold:*

(1) *to any $x \in X$ a non-empty compact set $M_x \subset U$ is assigned,*

(2) *at any $u \in U$ the function $\rho(M_x, u)$ is measurable in x, then the mapping $x \to M_x$ will admit of a measurable choice.*

Theorem B. *Let X and U be separable metric spaces and $g(x)$ be a quasi-continuous mapping from X to $F(U)$. If $f \in \mathcal{L}(U)$ then:*

(1) *the function $q(x) = \sup_{u \in g(x)} f(u) \in \mathcal{L}(X)$;*

(2) *the sets $\{u \in g(x) : f(u) = g(x)\}$, $x \in X$ are non-empty;*

(3) *the mapping $g(x)$ admits a measurable choice.*

Theorem C. *If $W \in L(X \times U)$ and the mapping $\varphi : X \to F(U)$ is measurable then there exists a measurable mapping $h : X \to U$ such that*

$$h(x) \in \varphi(x), \quad W(x, h(x)) \equiv \max_{u \in \varphi(x)} W(x, u).$$

Theorem B is a corollary of Theorem A.

We note that these theorems assert only that "measurable choices" exist but say nothing about how to find them. For this reason they are only used to prove the existence of optimal control but not for practical computation.

7.2. Optimal Automaton Control for Markov Processes with A Compact State Space and A Finite Control Set

We consider controlled Markov processes x_t given on a compact space X with a finite set of the controls $U = \{u_1, \ldots, u_k\}$. The transition distributions $\mu(\cdot|x, u)$ of such processes are the collections consisting of k transition functions $\mu_j(\cdot|x) = \mu(\cdot|x, u_j)$, $j = 1, \ldots, k$, which are $\mathcal{F}$-measurable in x. At any moment the evolution of the process x_t is governed by one of these measures in accordance with the strategy being applied. For the sake of simplicity we assume that the sets of admissible controls are the same for all states x, i.e. $U_x \equiv U$. The deterministic control laws

$h(x)$ are characterized by a measurable decomposition of the space X, i.e. there is a collection $[M] = \{M_1, \ldots, M_k\}$ of k disjoint measurable sets such that

$$X = \bigcup_{i=1}^{k} M_i$$

and

$$h(x) = u_i, \quad \forall\, x \in M_i, \quad i = 1, \ldots, k.$$

Let two laws h and h' defined by the measurable decompositions $[M]$ and $[M']$, respectively, be given. These laws coincide on the set $\tilde{M} = \bigcup_i (M_i \cap M_i')$ but not on the set[f] $\tilde{N} = X \backslash \tilde{M} = \bigcup_i (M_i \triangle M_i')$. We introduce one more decomposition of the space X into a family of subsets called elementary. Any measurable decomposition will be approximated by these subsets. For this purpose, fixing some $\delta > 0$ we decompose X into "small" measurable sets with diameters less or equal to δ. We denote by $m(\delta)$ the number of such "small" sets. For example, in the case of the real axis the required decomposition of the interval $X = [a, b]$ consists of dividing this interval into subintervals having the same length δ (the last interval can be, of course, shorter). In n-dimensional Euclidean space the hypercubes with sides not exceeding δ can be taken as the elements of such a decomposition.

Making use of elementary decompositions we can approximate the laws of general form h defined by an arbitrary measurable decomposition $[M]$. This can be done in the following manner. Let the set $M_1^{(\delta)}$ be formed by all elements of the elementary decomposition $[M^\delta]$ which have a non-empty intersection with M_1. From the remaining elements of the decomposition $[M^\delta]$ we choose the elements having common points with M_2. They form the set $M_2^{(\delta)}$ and so on. This procedure is obvious in the case of an interval $X = [a, b]$ and other Euclidian spaces. If δ is fixed the approximation possibilities of the decomposition $[M^\delta]$ will be rather limited. For this reason we take a sequence δ_l, $l = 1, 2, \ldots$, such that $\delta_l \downarrow 0$ and consider the decompositions $[M_l] \stackrel{\text{def}}{=} [M^{\delta_l}]$. Then $M_i^{(\delta_k)} \to M_i$ as $k \to \infty$ for every i and the appropriate sequence of the laws h_l

$$h_l(x) = u_i, \quad \forall\, x \in M_i^{\delta_l}$$

converges in the measure sense to the given law h.

For a given diameter δ let the decomposition $[M^\delta]$ contain $m = m(\delta)$ elements. Then 2^m subsets of X can be formed from them. We denote the number of different piecewise constant laws of the described kind by $\varkappa = \varkappa(\delta)$ (or $\varkappa(l)$ if $\delta = \delta_l$).

We shall return to the controlled transition functions when the simple strategies based on the law h are used. Two functions

$$\mu = \mu(\cdot\,|x, h(x)), \quad \mu_l = \mu(\cdot\,|x, h_l(x)),$$

---

[f] Here the symbol $\triangle$ means the symmetrical difference of sets, i.e., $A \triangle B = (A\backslash B) \cup (B\backslash A)$.

correspond to $\mu(\cdot|x,u)$. The first of them is based on the decomposition $[M_l]$ and both coincide on the set $\tilde{M}_l \in X$ where $\tilde{M}_l = \bigcup_l \left(M_i \cap M_l^{(\delta_l)} \right)$ but they differ on the set $\tilde{N}_l = X \backslash \tilde{M}_l$. We estimate the distance in variation between these functions

$$\text{Var}(\mu, \mu_l)(x) = \sup_{f \in F} \left| \int_X f(y)\mu(dy|x) - \int_X f(y)\mu_l(dy|x) \right|$$

$$\leq \sup_{f \in F} \left| \int_X f(y)(\mu(dy|x) - \mu_l(dy|x)) \right| \leq r_l.$$

For the distance between the iterations $\mu^{(t)}$ and $\mu_l^{(t)}$ the same inequality holds, i.e. $\text{Var}\left(\mu^{(t)}, \mu_l^{(t)}\right) \leq r_l$.

If there exist μ_∞, $\mu_{l,\infty}$, then, of course,

$$\text{Var}(\mu_\infty, \mu_{l,\infty}) \leq r_l.$$

Thus, if the sequence of the subdividing partitions admits an approximation of any measurable decomposition and $\lim_{l\to\infty} r_l = 0$, then the functions $\mu_l^{(t)}$ will converge to $\mu_{l,\infty}$ in variation

$$\lim_{t,l\to\infty} \text{Var}\left(\mu^{(t)}, \mu_l^{(t)}\right) = 0.$$

Let us return to our subject. We wish to construct either an optimal or an ε-optimal strategy to maximize the limiting average reward $W(\sigma)$. In the classical version when a given measurable decomposition and a controlled transition function are specified we have to solve the Bellman equation. In the adaptive situation we should use a different approach. The measurable decomposition associated with the optimal law is unknown. Let us represent the optimal adaptive strategy in the form of an automaton. One of the constructions discussed Chap. 2 will be used as a basis. For the sake of simplicity we use automata of type $\mathcal{D}$ (Sec. 2, Chap. 2) with modifications similar to that which have been made for Markov chains in Sec. 5, Chap. 5.

The input signal at the moment t is a pair (x_t, ζ_t) where x_t is the state of the process and ζ_t is the reward, which is supposed to be bounded and whose value, without restricting generality, belongs to the interval $[0,1]$ for all x and u. If some known function $q(x_t, u_{t-1})$ serves as the reward then it will be sufficient to know only the current state at the input. The output signals are the controls $u \in U$. They are formed in the following way. The automaton denoted by $\mathcal{MD}_{\varkappa,n}$ has $\varkappa = \varkappa(l)$ branches each of which has n states. The number $\varkappa$ is the number of piecewise constant laws. These laws assign the law h_j to the branch with the number j. In each branch the nth state (deep) serves as the input state but the output state is the first one (of unit depth). The change of states is realized in the same way as for ordinary automata $\mathcal{D}_{k,n}$ from Chap. 2. The rewards coming into the automaton are transformed into either "encouragement" or "punishment" and the empirical rewards on a branch are given by

$$\eta_t = t^{-1} \sum_{s=1}^{t} \zeta_s.$$

We define a sequence of *independent* r.v. ρ_1, ρ_2, ... such that

$$\rho_t = \begin{cases} 1, & \text{with probability } \eta_t, \\ 0, & \text{with probability } 1 - \eta_t. \end{cases}$$

The value $\rho_t = 1$ is interpreted as "encouragement". In this case the automaton passes from state i into state n. If $\rho_t = 0$, which is interpreted as "punishment," the automaton will pass from state i ($i \neq 1$) into state $i - 1$, but if $i = 1$ it will pass into the other branches in cyclic or equiprobable manner. In the jth branch the law h_j is used. Therefore, in response to the input signal x_t this branch produces the answer $u_t = h_j(x_t)$ which will define the next value of the controlled Markov process. This completes the description of the automaton $\mathcal{MD}_{\varkappa,n}$ as the strategy of control.

It remains to define the class of controlled Markov processes to be studied. The symbol $\mathcal{M}_k$ denotes a class of Markov processes with a compact state space X and a finite set of controls $U = \{u_1, \ldots, u_k\}$. Every process is assumed to be ergodic and have a simple optimal strategy. For all x and u the rewards $\zeta_t \in [0, 1]$. A sufficient condition for a Markov process to belong to this class is the existence of a density of the transition function $\mu(\cdot|x, u)$ such that $p(y|x, u) \geq c > 0$. There are other sufficient conditions for the process to belong to $\mathcal{M}_k$.

We state below the main result about the control of the processes from $\mathcal{M}_k$ using the automata $\mathcal{MD}_{\varkappa,n}$.

Theorem 1. *The automata $\mathcal{M}_{\varkappa(l),n}$ form an ε-optimal family for the compact subclass $\mathcal{M}_k$. This means that for every $\varepsilon > 0$ there exist natural numbers n_ε and l_k such that for all $n \geq n_\varepsilon, l \geq l_\varepsilon$ the automata $\mathcal{MD}_{\varkappa(l),n}$ satisfy the inequality $W(\mathcal{MD}_{\varkappa(l),n}) > \sup_\sigma W(\sigma) - \varepsilon$ for every process from the compact subclass considered.*

The proof of this theorem is similar to that of Theorem 1, Sec. 4, Chap. 5. However, the role of the additional parameter δ_l which defines the size of the elements of the partition $[M^{\delta_l}]$ should be taken into consideration. The reader will not meet difficulties here. When δ_l decreases as $l \to \infty$ the accuracy of approximating the sets M_i by the sets $M_i^{\delta_l}$ increases. Hence, the appropriate maximum limiting average reward W_{δ_l} approaches $\sup W(\delta)$. This can easily be proved.

The ergodicity of the process, the strong law of large numbers and the properties of the algorithm above show that ε-optimality in the weak sense implies the ε-optimality in the strong sense:

$$\lim_{T \to \infty} T^{-1} \sum_{t=1}^{T} \zeta_t > \sup_\sigma W(\sigma) - \varepsilon, \quad \text{a.s.}$$

By analogy with results on finite Markov chains, we now discuss attaining accurately the maximum of the limiting average reward $W(\sigma)$ by automata with increasing memory (Sec. 4, Chap. 2). The class of Markov processes under consideration is $\mathcal{M}_k$ and the aim is to maximize $W(\sigma)$. As the control algorithm we take the automaton with increasing memory $\mathcal{MD}_\varkappa$. The control laws (or the branches of the automata) change cyclically. On making the round of all branches the memory increases linearly together with l, i.e. finer partitions of the space X into subsets are used. The joint growth of n and l leads to increasing the number of branches and the memory depth. This complicated construction ensures the required optimality with respect to the whole class $\mathcal{M}_k$ of ergodic Markov processes.

7.3. Searching Optimal Strategies for Ergodic Markov Processes with Compact Spaces of States and Controls

We denote by MP the class of controlled Markov processes with compact state space X and compact control space U (with metrics ρ and r respectively) which satisfies the following conditions:

(i) The reward at the moment T is defined by a function $q(x_t, u_{t-1})$ which is continuous on U for any $x \in X$ and satisfies the Lipschitz condition

$$|q(x, u) - q(x', u)| \le L_1 \rho(x, x');$$

(ii) For the transition function $\mu(\cdot|x, u)$, the mapping $u \to \mu(\cdot|x, u)$ is continuous in variation and

$$\mathrm{Var}(\mu(\cdot|x, u), \mu(\cdot|x', u)) \le L_2 \rho(x, x');$$

(iii) There exist positive numbers $L_3 < \infty$, $\lambda \in (0, 1)$ and a probability distribution ν_σ on X such that

$$\mathrm{Var}\left(\nu_{x,\sigma}^{(t)}, \nu_\sigma\right) \le L_3 \lambda^t,$$

for every simple strategy σ, where $\nu_{x,\sigma}^{(t)}$ denotes the distribution of the process at the moment t for the initial state x and the strategy σ.

The constants L_1, L_2, L_3, λ are the same for all processes from the class MP but $MP(X, U; L_1, L_2, L_3, \lambda)$ is the detailed notation for this class.

We note that the finite sets of controls are, of course, compact.

The aim of control is optimality: for a given value $\varepsilon > 0$ it is required to ensure the fulfilment of the inequality

$$\lim_{T \to \infty} T^{-1} \sum_{t=1}^{T} q(x_t, u_{t-1}) > \sup_\sigma W(\sigma) - \varepsilon, \quad \text{a.s.}$$

In the classical version of this problem, when the transition function of the process from MP is known, this aim can be reached by using a simple strategy. Moreover, it is enough to use a stationary strategy with Markov piecewise constant rules. We prove this now by using the next three lemmas which follow from assumptions (i)–(iii). The simple proofs of the first two lemmas are omitted.

Lemma 1. *There exists a bounded function $g(x)$ ($|g(x)| \leq c = 2q_0(1-\lambda)^{-1}$, $q_0 = \max_{x,u} q(x,u)$) satisfying the Lipschitz condition with the constant $L = L_1 + cL_2$ such that*

$$W^* + g(x) = \max_{u \in U} \left\{ q(x,u) + \int_X g(z)\mu(dz|x,u) \right\}$$

where $W^* = \sup_\sigma W(\sigma)$.

We put

$$H(x,u) = q(x,u) + \int_X g(z)\mu(dz|x,u).$$

Lemma 2. *If*

$$H(x, h_\varepsilon(x)) \geq \max_u H(x,u) - \varepsilon$$

for every x, then the stationary Markov strategy $\sigma_\varepsilon = \{h_\varepsilon^\infty\}$ which consists of using the law h_ε has the following property

$$W(\sigma_\varepsilon) \geq W^* - \varepsilon.$$

Let the sets $M_1, \ldots, M_k$ with the maximum diameter $\delta = \varepsilon(3(L_1 + cL_2))^{-1}$ be a decomposition of the compact space X. We introduce the collection H of functions on X taking values from U and constant on every set M_i, $i = 1, \ldots, k$. We denote the set of simple strategies which use the functions from H by Σ_H.

Lemma 3. *For any process from MP the inequality*

$$\sup_{\sigma \in \Sigma_H} W(\sigma) \geq W^* - 2\varepsilon/3$$

holds.

Proof. According to conditions (i), (ii) and Lemma 1, there exists an optimal simple strategy σ_0 consisting of repetition of the law h_0 at which the equality

$$H(x, h_0(x)) = \max_u H(x,u)$$

takes place. Choose points $x_j \in M_j$ and define the law $h(x) = h_0(x_j)$ provided $x \in M_j$, $j = 1, 2, \ldots, k$, i.e. $h(x) \in H$. To prove the lemma we show that $H(x, h(x)) \geq \max_u H(x,u) - \varepsilon/3$ and the required assertion will follow from Lemma 2. It is obvious (Lemma 1 and condition (ii)) that $h(x)$ is a Lipschitz function with the constant $L = L_1 + cL_2$. Hence, $|H(x,u) - H(x_j,u)| \leq \delta L$ for $x \in M_j$, $u \in U$ and

$$\left| \sup_u H(x,u) - \sup_u H(x_j,u) \right| \leq \delta L.$$

From this it follows that

$$H(x, h(x)) = H(x_j, h_0(x_j)) \geq H(x_j, h_0(x_j)) - \delta L = \max_u H(x_j, u) - 2\delta L.$$

for $x \in M_j$. Therefore $H(x, h(x)) \geq \sup_u H(x,u) - 2\varepsilon/3$. $\qquad\square$

Our hope to restrict attention to piecewise constant laws in the case of adaptive control problem from MP is based on these results. We now turn to the description of the appropriate adaptive strategy. We introduce the following time intervals: $[t_i, \tau_i - 1]$ is called an informational intervals and $[\tau_i, t_{i+1} - 1]$ is called a work interval. Concerning the sequence $\{t_i, \tau_i\}$ we assume that:

1. $t_1 = 0,\ t_{i+1} - \tau_i > \tau_i - t_i,\quad i = 1, 2, \ldots;$

2. $\displaystyle \lim_{n \to \infty} \frac{\sum_{i=1}^{n}(\tau_i - t_i)}{\sum_{i=1}^{n}(t_{i+1} - \tau_i)} = 0;$

3. $\tau_n - t_n \sim n^\gamma, \qquad\qquad \gamma > 1/2.$

We have already used such partitions of the real axis and shall use them below.

The required strategy is formed by the piecewise constant laws $h_i \in H$ on the sets $M_1, \ldots, M_k$ and taking the values from U. It is convenient to represent such laws as the vectors $u = (u^{(1)}, \ldots, u^{(k)})$, where $u^{(j)} = h(x)$ and $x \in M_j$.

Moreover, we choose a sequence of probability distributions $\varkappa_l(\cdot)$ on the Borel σ-algebra of the space U^k, i.e. the set of values of the laws h_j. They are supposed to satisfy the condition $\inf_l \varkappa_l(B) > 0$ for every open set $U^{(k)}$.

For the class MP we define the optimal adaptive strategy as follows:

a. On an information interval $[t_i, \tau_i - 1]$ the law h_i is defined by the distribution $\varkappa_i$ independently of the past history;
b. On a work interval $[\tau_i, t_{i+1} - 1]$ it is defined recurrently by using the results of the previous controls. We denote this law by g_i;
c. We assign the empirical rewards

$$\tilde{S}_i(g_i) = \frac{1}{t_{i+1} - \tau_i} \sum_{t=\tau_i}^{t_{i+1}-1} q(x_i, g_i(x_{t-1})),$$

$$S_i(h_{i+1}) = \frac{1}{\tau_{i+1} - t_{i+1}} \sum_{t=t_{i+1}}^{\tau_{i+1}-1} q(x_i, h_i(x_{t-1}))$$

when the laws h_{i+1} and g_i have been used on the intervals $[\tau_i, t_{i+1} - 1]$ and $[t_{i+1}, \tau_{i+1} - 1]$ respectively;
d. We put $g_1 = \lambda_1$ and

$$g_{i+1} = \begin{cases} h_{i+1}, & \text{if } S_i(h_{i+1}) > \tilde{S}_i(g_i) + \varepsilon/12, \\ g_i, & \text{otherwise.} \end{cases} \tag{1}$$

This completes the description of the desired strategy σ_ε. For a fixed initial state it generates the distribution $\bar{\mathbf{P}}$ on the path space $(X \times U)^\infty$ of the process (x_t, u_t).

Theorem 1. *If the strategy σ_ε, $\varepsilon > 0$, is used then*

$$\mathbf{P}\left\{ \lim_{T \to \infty} T^{-1} \sum_{t=1}^{T} q(x_t, u_{t-1}) \geq \sup_\sigma W(\sigma) - \varepsilon \right\} = 1$$

for every process from MP and every initial state.

Proof. For the simple strategy $\sigma = h^\infty$ we denote the limiting distribution (according to condition (ii)) of the process from MP by $\nu_h(x)$. We put also

$$R(h) = \int_X q(x, h(x))\, d\nu_h(x).$$

We have $W^* = \sup_h R(h)$ (h belongs to the set of all measurable laws). This fact follows from the definition of W^*, Lemma 1 and the ergodicity of the process under consideration. Lemma 3 and the equality $\bar{W} \overset{\text{def}}{=} \sup_{h \in H} R(h) = \sup_{\sigma \in \Sigma_H} W(\sigma)$ imply

$$\bar{W} \geq W^* - 2\varepsilon/3. \tag{2}$$

Let $d(h, h') = \max_j r(u_j, u'_j)$, where $h = (u_1, \ldots, u_k)$, $h' = (u'_1, \ldots, u'_k)$, be a metric on U^k. We show that the function $R(h)$ is continuous in this metric on the set $H = U^k$.

The function $q(x, u)$ is uniformly continuous since it is continuous on the compact set $X \times U$ by conditions (i) and (iii). Hence

$$\sup_x |q(x, h(x)) - q(x, h'(x))| \leq \varphi(d(h, h'))$$

where φ is some monotonic function and $\lim_{s \to 0} \varphi(s) = 0$. We have

$$|R(h) - R(h')| \leq \int_X |q(x, h(x)) - q(x, h'(x))|\, d\nu_h(x)$$
$$+ \left| \int_X q(x, h'(x))\, d\nu_h(x) - \int_X q(x, h'(x))\, d\nu_{h'}(x) \right|$$
$$\leq \varphi(d(h, h')) + a\mathrm{Var}(\nu_h, \nu_{h'}).$$

It remains to verify that $\lim_{h' \to h} \mathrm{Var}(\nu_h, \nu_{h'}) = 0$. To achieve this, we define a family of linear operators T_h, $h \in H$, given on the space of probability measures on the Borel σ-algebra $\mathcal{F}_X$ in X in the following way

$$T_h \mu(\cdot) = \int_X \mu(\cdot | x, h(x)) \mu(dx).$$

Then for any measure μ we have

$$\mathrm{Var}\big(T_h \mu, T_{h'} \mu\big) = \sup_{M \in \mathcal{F}_X} \left| \int_X \mu(dx) \big[\mu(M | x, h(x)) - \mu(M | x, h') \big] \right|$$
$$\leq \sup_{M \in \mathcal{F}_X} \sup_x |\mu(M | x, h(x)) - \mu(M | x, h')|$$
$$\leq \sup_x \mathrm{Var}\big(\mu(\cdot | x, h(x)), \mu(\cdot | x, h')\big).$$

By condition (ii), the mapping $(x, u) \to \mu(\cdot|x, u)$ is continuous on the compact space $X \times U$. Hence

$$\sup_x \mathrm{Var}[\mu(\cdot|x, h(x)), \mu(\cdot|x, h')] \xrightarrow[h' \to h]{} 0. \tag{3}$$

For every n we have

$$\mathrm{Var}(\nu_h, \nu_{h'}) \leq \mathrm{Var}(\nu_h, T_h^n \delta_x) + \mathrm{Var}(T_h^n \delta_x, T_{h'}^n \delta_x) + \mathrm{Var}(\nu_{h'}, T_{h'}^n \delta_x) \tag{4}$$

where δ_x stands for the probability measure concentrated at the point x, and $T_h^t \delta_x = \nu_{x,\sigma}^{(t)}$ means the distribution of the process at the moment t. By induction from (3) it follows that $\lim_{h' \to h} \mathrm{Var}\left(T_h^n \delta_x, T_{h'}^n \delta_x\right) = 0$ for a fixed n. Thus, $\lim_{h' \to h} \mathrm{Var}(\nu_h, \nu_{h'}) = 0$. Therefore the function $R(h)$ is continuous on H. We define random variables ξ_i, ζ_i as

$$\xi_i = S_i(h_{i+1}) - R(h_{i+1}), \qquad \zeta_i = \tilde{S}_i(g_i) - R(g_i), \quad i = 1, 2, \ldots.$$

Then the rules from (1) take the form

$$g_{i+1} = \begin{cases} h_{i+1}, & \text{if } R(h_{i+1}) + \xi_i > R(g_i) + \varepsilon/12, \\ g_i, & \text{otherwise.} \end{cases} \tag{5}$$

We are going to prove that $\lim_{i \to \infty} \xi_i = 0$ and $\lim_{i \to \infty} \zeta_i = 0$ a.s. For both equalities the arguments are the same. Therefore, we prove only the first of them. We put

$$\gamma_1^{(i)} = \frac{1}{\tau_i - t_i} \left| \sum_{t=t_i}^{\tau_i - t_i} \left[q(x_t, h_i(x_{t-1})) - \mathbf{E}q(x_i, h'(x_{t+1}))\right] \right|,$$

$$\gamma_2^{(i)} = \frac{1}{\tau_i - t_i} \left| \sum_{t=t_i}^{\tau_i - 1} \mathbf{E}q(x_t, h_i(x_{t-1})) - R(h_i) \right|,$$

$$\Delta_i > 0, \quad \lim_{i \to \infty} \Delta_i = 0.$$

We have

$$\mathbf{P}\{|\xi_{i-1}| > 2\Delta_i\} \leq \mathbf{P}\{\gamma_1^{(i)} + \gamma_2^{(i)} > 2\Delta_i\} \leq \mathbf{P}\{\gamma_1^{(i)} > \Delta_i\} + \mathbf{P}\{\gamma_2^{(i)} > \Delta_i\},$$

$$\mathbf{P}\{\gamma_1^{(i)} > \Delta_i\} \leq \Delta_i^{-4}\mathbf{E}(\gamma_1^{(i)})^4.$$

We can verify, using condition (iii), that there exists a constant a depending only on L_3 and λ such that

$$\mathbf{E}\left(\gamma_1^{(i)}\right)^4 \leq (\tau_i - t_i)^{-4} a(\tau_i - t_i)^2 = a(\tau_i - t_i)^{-2}.$$

In view of the fact that $\tau_i - t_i \sim i^\gamma$ ($\gamma > 1/2$) we obtain $\sum_{i=1}^{\infty} \mathbf{P}\{\gamma_1^{(i)} > \Delta_i\} < \infty$ provided the sequence $\{\Delta_i\}$ tends to zero sufficiently slowly. By condition (iii), $\lim_{i \to \infty} \gamma_2^{(i)} = 0$. Hence $\gamma_2^{(i)} \leq \Delta_i$ for all sufficiently large i (this may require changing the sequence $\{\Delta_i\}$). This implies the convergence of the series $\sum_{i=1}^{\infty} \mathbf{P}\{|\xi_i| > 2\Delta_i\}$ and $\mathbf{P}\{\lim_{i \to \infty} \xi_i = 0\} = 1$ according to the Borel–Cantelli Lemma. Similar arguments can be used to prove $\mathbf{P}\{\lim_{i \to \infty} \zeta_i = 0\} = 1$.

Using Lemma 4 (see below) we obtain

$$\lim_{i\to\infty} R(h_i) \geq W^* - 3\varepsilon/4. \quad \text{a.s.}$$

This fact together with continuity of the function R and (2), (5) leads to the following

$$\lim_{i\to\infty} \tilde{S}_i(g_i) = \lim_{i\to\infty} R(g_i) \geq W^* - 3\varepsilon/4. \quad \text{a.s.}$$

Let ω be an elementary event for which the last relationship holds. Then for all $i \geq t_i(\omega)$ we have $\tilde{S}_i(g_i) \geq W^* - \varepsilon$. We introduce the index sets

$$v = \{i : t_i \geq t_{i(\omega)}, \ \tau_i \leq T\}, \quad w = \{i : t_i \geq \tau_{i(\omega)}, \ t_i \leq T\}.$$

The arithmetic mean that we are interested in can be written in the form

$$T^{-1}\sum_{t=1}^{T} q(x_t, u_{t-1}) = T^{-1}\sum_{t=1}^{t(\omega)} q(x_t, u_{t-1}) + T^{-1}\sum_{i\in v}(\tau_i - t_i)S_{i-1}(h_i)$$

$$+ T^{-1}\sum_{i\in w}(t_{i+1} - \tau_i)\tilde{S}_i(g_i) + T^{-1}S^* \tag{6}$$

Let us estimate the terms on the right-hand side of this equality. The number of members of the sum S^* is not greater than $t_{i(T)+1} - \tau_{i(T)}$ where $i(T) = \min\{i : t_{i+1} \geq T\}$. From conditions 1–3 on the sequences $\{\tau_i\}$, $\{t_i\}$ it follows that $t_{j+1} - \tau_i = o\left(\sum_{i=1}^{j+1}(t_{i+1} - \tau_i)\right)$. Hence $\lim_{T\to\infty} T^{-1}S^* = 0$. Since the sums $S_{i_1}(h_i)$ are bounded and $\sum_{i\in v}(\tau_i - t_i) = o(T)$ the second terms on the right-hand side of (6) will tend to 0 as $T \to \infty$. This means that

$$\lim_{T\to\infty} T^{-1}\sum_{t=1}^{T} q(x_t, u_{t-1}) = \lim_{T\to\infty} T^{-1}\sum_{i\in w}(t_{i+1} - \tau_i)\tilde{S}_i(g_i)$$

$$\geq (W^* - \varepsilon)\lim_{T\to\infty} T^{-1}\sum_{i\in w}(t_{i+1} - \tau_i) = W^* - \varepsilon.$$

This completes the proof. $\qquad\square$

Now we consider a result (Lemma 4 below) from stochastic approximation theory. Consider the following procedure for searching the supremum f_0 ($f_0 = \sup f(z)$) of the real-valued function $f(z)$ upper semi-continuous and bounded from below on a metric space Z. Let η_i, $i = 1, 2, \ldots$, be the realizations of random vectors with distributions and z_i, $i = 1, 2, \ldots$ be points of Z to be specified later on. We assume that the function $f(z)$ is measured at the points η_i, z_i with some errors ξ_i, ζ_i respectively which are, in general, mutually dependent. We put

$$X_i(\eta_i) = f(\eta_i) + \xi_i, \qquad Y_i(z_i) = f(z_i) + \zeta_i, \quad i = 1, 2, \ldots.$$

We can now define the sequence $\{z_i, \ i = 1, 2, \ldots\}$. Let $z_1 = \eta_1$ and

$$z_{i+1} = \begin{cases} \eta_{i+1}, & \text{if } X_{i+1}(\eta_{i+1}) > Y_i(z_i) + \delta, \\ z_i, & \text{otherwise} \end{cases}$$

where $\delta > 0$, $i = 1, 2, \ldots$.

Lemma 4. *If* $\lim_{i\to\infty} \xi_i = \lim_{i\to\infty} \zeta_i = 0$ a.s. *then*

$$\varlimsup_{i\to\infty} f(z_i) \geq f_0 - \delta, \quad \text{a.s.}$$

It is interesting to remark that even for $\delta = 0$ it may happen that the sequence $f(z_i)$ does not converge to f_0. We omit the proof of Lemma 4.

To improve the convergence rate when the strategy σ_ε is used it is reasonable "to control" the distributions μ_i in the course of control but we will not consider the practical realization of this idea.

We consider a simplified procedure for practical applications of the strategy σ_ε. Using the compactness of the space X this technique discretizes the set of the controls used and can be described as follows. For a given small number $\Delta > 0$ we construct the Δ-net $u_1, \ldots, u_{N(\Delta)}$ in U. Thereafter we act in a way described above, i.e. for a given partition $[N^\delta]$ of the space X we consider the set $H_{\delta,\Delta}$ of all piecewise constant functions. These functions take the values $u_l \in \{u_1, \ldots, u_{N(\Delta)}\}$ from the set M_j. Then, the set $H_{\delta,\Delta}$ turns out to be finite, the distributions μ_i being discrete. It is clear that such an approach allows to simplify the practical realization of the discussed strategy. However, to have the desired accuracy in the relationship

$$\varliminf_{T\to\infty} T^{-1} \sum_{t=1}^{T} q_t > W^* - \varepsilon(\delta, \Delta)$$

it is necessary that the partition $[M^\delta]$ be rather small but the number of elements of the Δ-net be rather large.

7.4. Control of Finite Semi-Markov Processes

For the first time, we consider processes with continuous time. The time t starts at the moment $t_0 = 0$ and runs over the positive axis.

We deal with a Markov processes having finite state space $X = \{x_1, \ldots, x_m\}$ and control space $U = \{u_1, \ldots, u_k\}$. These are called *jump processes*. They are defined by (1) the matrices $F^{(u)} = \left(F_i^{(u)}(t)\right)$, where $F_i^{(u)}$ is the distribution of the sojourn time in the state x_i after the control u was used, (2) the transition probability matrices $\mathcal{P}^{(u)} = \|p_{ij}^u\|$ where p_{ij}^u is the probability of the transition $x_i \to x_j$ under the control u.

In the case of a Markov process, $F_i^{(u)}(t) = 1 - \exp\{-t/m_i(u)\}, \forall j$, where $m_i(u)$ denotes the mathematical expectation of the sojourn time in the state x_i provided the control u was used.

For the *semi-Markov processes*, in contrast to the jump Markov processes, the distributions of their sojourn time $F_{ij}^{(n)}(t)$ depend not only on the initial state x_i and the control u but on the next state x_i appearing with the probability p_{ij}^u. The distributions $F_{ij}^{(u)}$ can be arbitrary but with finite mathematical expectations.

The interest in semi-Markov processes was motivated by inadequacy of Markov processes when describing problems of controlled queueing systems where the distributions $F_i^{(u)}(t)$ were not exponential. This led to wide use of semi-Markov processes in applications. Now their theory is well developed. In particular, it has been discovered that the optimal strategy (to maximize the limiting average reward) is simple. It can be calculated by using the reduction of the original problem to linear programming. We now focus our attention on adaptive problems of optimal control for semi-Markov processes. But first, we give a detailed description of such a process.

Let x_t and u_t denote, respectively, the state and control of a controlled semi-Markov process (CSMP for short) at the moment t. This process is supposed to be continuous on the right and piecewise constant. Let $\tau_0 = 0$, $\tau_n = \sup\{t > \tau_{n-1} : x_{t-0} \neq x_t\}$, $n = 1, 2, \ldots$, be the sequence of the skips of the process x_t. Then $x_t = x(n)$, $\tau_n \leq t < \tau_{n+1}$, and $u_t = u(n)$, $\tau_n \leq t < \tau_{n+1}$, are the state and control on the interval $[\tau_n, \tau_{n+1})$. The average sojourn time in the state x_i of the process under the control u is equal to

$$m_i(u) = \sum_{j=1}^m p_{ij}^u \mathbf{E}\tau_{ij}^u = \sum_{j=1}^m p_{ij}^u \int_0^\infty t F_{ij}^{(u)}(dt).$$

It is finite according to the assumptions made.

The evolution of the CSMP is defined by the chosen strategy. Its laws are supposed to be deterministic and depend on the past history $h(u(n)|x^n, \tau^{n-1}, u^{n-1})$, $n \geq 0$. The stationary Markov laws have the form $\bar{u} = (u(x_1), \ldots, u(x_n))$ and their applications turn the CSMP into a usual homogeneous semi-Markov process with a transition matrix $\mathbf{P}^{(\bar{u})} = (p_{ij}^{u(x_i)})$.

Definition 1. The pair $(X, \mathbf{P}^{(\bar{u})})$ is called the *embedded Markov chain* corresponding to the simple strategy $\sigma = \bar{u}^\infty$.

The evolution of the semi-Markov process is readily illustrated by that of the enclosed chain with random sojourn times in the states.

We define the rewards on the paths of the CSMP by the matrix of the one-step rewards $R = (\zeta(x_i, u_l))$, where $\zeta(x, u)$ are some r.v. such that $|\mathbf{E}\zeta(x, u)| < \infty$. These rewards are random piecewise linear process described as follows. If the CSMP enters the state x_i when we choose the control u_l, and the process has spent time τ in this state, then the reward will be equal to $\zeta(x_i, u_l)\tau$. Sometimes, it is useful to operate with the numeral matrix $R = (r_{ij})$ which generates the rewards in the same way as above.

Under the given initial distribution $\bar{p}$ and strategy σ the limiting average reward is defined as follows

$$W(\sigma, \bar{p}) = \lim_{T \to \infty} T^{-1} \int_0^T \mathbf{E}_{\sigma, \bar{p}} \zeta(x_s, u_s) \, ds.$$

We put

$$V_t(\sigma, \bar{p}) = t^{-1} \int_0^t \zeta(x_s, u_s)\, ds.$$

The aim of control consists of strong asymptotic optimality, i.e.

$$\lim_{t \to \infty} V_t(\sigma, \bar{p}) = W = \sup_{\sigma, \bar{p}} W(\sigma, \bar{p}), \quad \text{a.s.} \tag{1}$$

The adaptive formulation of this problem means that the matrices $\mathbf{P}^{(u)}$ and the distributions $F^{(u)}$ are unknown but belong to some given class.

The CSMP is *ergodic* if any two of its states comunicate with each other at any simple strategy. It is *connected* if for any law $\tilde{u}(x)$ the enclosed chain $(X, \mathbf{P}^{u(x)})$ is connected. Any CSMP can be uniquely decomposed into the sum of its connected components $C_1, \ldots, C_L$ and the set of non-essential states C_0. For the initial distribution $\bar{p}$ concentrated on C_j we put $W_j(\bar{p}) = \sup_\sigma W(\sigma, \bar{p})$. Then $W_j(\bar{p}) \equiv W_j$ and it does not depend on the choice of $\bar{p}$. We say that the CSMP is *equiprofitable* if $W_1 = \cdots = W_L$. Then, for the supremum of the reward we write just W.

Theorem 1. *If an optimal strategy ensuring the achievement of aim* (1) *with respect to the class of* CSMP *exists then all chains from this class are equiprofitable.*

The proof is obvious. The inverse assertion is more substantial.

Theorem 2. *There exist optimal adaptive strategies with respect to the class of equiprofitable* CSMP *with* $|X| = m$, $|U| = k$.

Proof. We give one such strategy of the identification type, by evaluating the following quantities: the matrices $\mathbf{P}^{(u)}$ and average times $M_i(u)$. To estimate p_{ij}^u we introduce the numbers $N_{ijl}(n)$, i.e. the number of transitions from x_i to x_j during n transitions under the control u_l. Let $N_{il}(n) = \sum_{j=1}^m N_{ijl}(n)$. The required estimates have the following form

$$p_{ij}^l(n) = \begin{cases} \dfrac{N_{ijl}(n)}{N_{il}(n)}, & \text{if } N_{il}(n) > 0, \\ 0, & \text{otherwise.} \end{cases}$$

For the average times we choose the estimates in the form of arithmetic means

$$\mu_{ij}^l(n) = \begin{cases} \dfrac{1}{N_{il}(n)} \sum_{s=0}^{n-1} \tau(x(s), x(s+1)) \delta_{x(s),j} \delta_{x(s+1),j}, & \text{if } N_{il}(n) > 0, \\ 0, & \text{otherwise.} \end{cases}$$

Consider the sequence of matrices $\mathbf{P}(n) = (p_{ij}^l(n))$, $i, j, = 1, \ldots, m$, $l = 1, \ldots, k$, and the collection $M(n) = \{\mu_i^l(n)\}$. We use them to construct the "optimal" simple strategy, always supposing the empirical data are correct. This is called a *quasi-optimal strategy*. It uses either reduction of the original problem to linear programming or a recurrent procedure. Without going into particulars about these methods we note that the required Markov law u^* can be found by the following algorithm that is convenient in description but cumbersome in practical application.

For any simple strategy defined by the law $\bar{u}$ we can write down the explicit representation of the limiting average reward

$$W(\bar{u}^\infty) = \frac{\sum_{i=1}^m r_i^{u(i)} m_i(\bar{u}) \pi_i(\bar{u})}{\sum_{i=1}^m m_i(\bar{u}) \pi_i(\bar{u})}$$

where $\pi_i(\bar{u})$ is an ergodic distribution for the CSMP, and $r_i^{u(i)}$ are the mathematical expectation of the reward corresponding to the state x_i and the ith component of the law $\bar{u}$. If we calculate $W(\bar{u}^\infty)$ for every law, the total number of which is equal to k^m, we find the optimal law $\bar{u}^*$.

We choose a numerical sequence ε_n such that:

$$\varepsilon_n \in (0,1), \quad \varepsilon_n \downarrow 0, \quad \sum_{n=1}^\infty \varepsilon_n = \infty.$$

Let $\nu_n = \min_{i,l} N_{il}(n)$.

First, we describe the adaptive strategy that we are interested in for the connected CSMP. This strategy is formed by the randomized laws defined using the stochastic matrices $B = (b_i^l(n))$ where $b_i^l(n) = \mathbf{P}\{u_t = u_l | x(n) = x_i, x^{n-1}, \tau^{n-1}, u^{n-1}\}$ is the probability of selecting the control u_l in the state x_i at the nth changing of this state. At the initial moment we take $b_i^l = 1/k$ for all i, l. If at the moment t we have $\nu_{n+1} = \nu_n$, then the probability b_i^l will remain the same. But if $\nu_{n+1} > \nu_n$ and $x_t = x_i$ we put

$$b_i^l(n+1) = \begin{cases} 1 - \varepsilon_{\nu_n+1}, & \text{if } u_l = u^*(i), \\ \dfrac{\varepsilon_{\nu_n+1}}{k-1}, & \text{if } u_l \neq u^*(i), \end{cases}$$

where u_l stands for the ith component of the quasi-optimal law u_n^*. The other rows of the matrix B remain the same.

We denote this strategy by $\mathcal{I}$ and prove that it guarantees asymptotic optimality with respect to the class of connected CSMP. We show that $\mathbf{P}\{\lim_{n\to\infty} N_{ijl}(n) = \infty\} = 1$ for all i, j, l. Suppose the contrary holds. This means that beginning from some n_0, the probabilities $b_i^l(n)$ are fixed (being positive by construction). But from the inequality $b_i^l(n) > 0$ at $n > n_0$ it follows that $N_{ijl}(n) \to \infty$, i.e. the transitions $x_i \to x_j$ (for each u) occur infinitely often. The above and the estimates of the probabilities and average times imply the following

$$\mathbf{P}\left\{ \lim_{n\to\infty} p_{ij}^l(n) = p_{ij}^l, \ \lim_{n\to\infty} \mu_i^l(n) = m_i(u_l), \ i,j = 1,\ldots,m, \ l = 1,\ldots,k \right\} = 1.$$

It follows that beginning from some non-Markov moment finite a.s. the quasi-optimal law u^* is the optimal u_{opt}. Hence $W(\mathcal{I}) = W$. According to the strong law of large numbers the desired aim is attainable independently of the initial distribution. So, the assertion is proved for the connected CSMP. It remains to consider an arbitrary equiprofitable process. If its initial state belongs to the connected component the previous arguments remain in force.

Let the initial state x_0 belong to the set of non-essential states C_0. Visiting, maybe many times, the states $\bar{x}_1, \ldots, \bar{x}_{N-1}, \bar{x}_N \in C_0$ the process remains in C_0 during N time-intervals, and thereafter it enters some connected component C_i with probability one. It will never visit any of the states $\bar{x}_0, \bar{x}_1, \ldots, \bar{x}_{N-1}, \bar{x}_N$ and, hence, the quantity $\min_{i=0,1,\ldots,N} N_{il}(n)$ will not increase. So, the recomputation rule of the matrix $B(n)$ described above stops to work, the choice probabilities of the controls remain unchanged. Therefore the control aim is unattainable. For this reason the choice rule of the recomputation moments of the probabilities $b_i^l(n)$ must be modified in the case of equiprofitable CSMP of general form. There are different methods to do this but we shall use the rule $\mathcal{I}$ from Sec. 4, Chap. 5: either at every moment, or for every $s > 1$ times with the probability p.

The positivity of the matrix $B(n)$ guarantees that the process leaves the set C_0 for a finite time with probability one. The arguments used for the connected CSMP demonstrate that in the connected component the maximum limiting average reward (which is the same for all components) will be achieved. $\qquad\square$

As concerns the efficiency of the strategy $\mathcal{I}$ with respect to the class of all CSMP, not only of the equiprofitable ones, we state the final result: for any CSMP with $|X| = m$, $|U| = k$ the strategy $\mathcal{I}$ guarantees the fulfilment of the inequality (here p is the initial distribution)

$$\lim_{T \to \infty} T^{-1} \int_0^T \zeta_t \, dt \geq \min_p \max_\sigma W(\sigma, p).$$

For a jump Markov process with finite sets X and U these results remain in force.

For the CSMP it is also of interest to consider the other aims of control such as optimization with constraints and the minimax problem. The situations with either unobserved states (of course, the changing moments of the states must be known) or with observable "pseudostates" which are similar to the case of conditional Markov chains are likely to be important. But here we will not consider these problems.

7.5. Control of Countably Valued Semi-Markov Processes

Until now, we studied controlled random processes with either a finite state space or a continuous one. Here we consider the countably valued processes.

Let $X = \{x_1, x_2, \ldots\}$ be the state space. For the sake of convenience we identify the state x_i with the index i, $i = 1, 2, \ldots$. The set U is a measurable topological space of controls with a Borel σ-algebra of measurable sets, the set of admissible controls being a compact set $U(i)$ for every i. Any state i is supposed to be observed and, in addition, there are some "additional" observations belonging to a topological measurable space Z (with a Borel σ-algebra $\mathcal{F}_Z$).

A semi-Markov model is defined by a transition function given on the space $X \times Z$

$$\mu(M|i, u) = \mathbf{P}\{x(t + \tau) = j, \tau \in B | x(t) = i, u(t) = u\}, \quad \tau > 0$$

where $M = \{j\} \times B$, $B \in \mathcal{F}_Z$, $u \in U(i)$, which means the conditional probability to enter the state j from i at the random time τ under the control u. This function is assumed to be independent of t and measurable in u. We define the "sojourn function" $\tau(i, u, z, j)$ giving the time that the process $x(t)$ will be in the state i before it enters the state j under the control u and the value of additional information z used. In problems of queueing theory the latter means either the waiting time or the service time or their sum. The function $\tau(i, u, z, j)$ is supposed to be measurable with respect to (u, z). Often it is convenient to assume that $\tau(\cdot) \equiv z$. Finally, a function $q(i, u, z, j)$ serves as the reward per unit time which is also assumed to be measurable with respect to (u, z).

In what follows we deal with a class of the processes described which differ by the values of some parameter θ. Let us assume that all permissible values of this parameter form a measurable topological space Θ with a Borel σ-algebra. To distinguish the considered models we supply all functions specifying the model with the index θ, i.e. instead of $\mu(\cdot)$, $\tau(\cdot)$, $q(\cdot)$ we write $\mu_\theta(\cdot)$, $\tau_\theta(\cdot)$, $q_\theta(\cdot)$. This dependence on θ is assumed to be measurable. Assuming the following quantities are well defined, we put

$$q_\theta(i, u) = \sum_{j \in X} \int_X q_\theta(i, u, z, j) \mathbf{P}_\theta\{j, dz | i, u\},$$

$$\tau_\theta(i, u) = \sum_{j \in X} \int_X \tau_\theta(i, u, z, j) \mathbf{P}_\theta\{j, dz | i, u\}.$$

We denote the set of all strategies, including the randomized ones, by $\Sigma = \{\sigma\}$. Let Σ_s be the subset consisting of the simple strategies. It is generated by the laws $h : X \to U$ with $h(i) \in U(i)$. Let $H = \{h\}$ be the set of such laws. Under the stated assumptions this set is compact. Every simple strategy from Σ_s can be written in the form $\sigma = h^\infty$.

For fixed σ, θ, i (an initial state) there exists a probability measure $Q^i_{\sigma,\theta}(\cdot) = Q_{\sigma,\theta}(\cdot | x_0 = i)$ on the path space $(X \times U \times Z)^\infty$. Let $\mathbf{E}_{\sigma,\theta}(\zeta | x_0 = i)$ denote the conditional mathematical expectation under this measure. We introduce the objective function

$$W(\sigma, i, \theta) = \lim_{T \to \infty} \frac{\sum_{t=1}^T \mathbf{E}_{\sigma,\theta}(q_\theta(x_t, u_t, z_t, x_{t+\tau} | x_0 = i)}{\sum_{t=1}^T \mathbf{E}_{\sigma,\theta}(\tau_\theta(x_t, u_t, z_t, x_{t+\tau} | x_0 = i)}$$

for all $(\theta, i) \in \Theta \times X$ provided this expression is well defined. We assume that the initial observation z_0 is fixed. In particular, if $\tau_\theta(i, u, z, j) \equiv z$ we have the limiting average reward in the usual form.

Definition 1. A strategy σ^0 is called *θ-optimal in the weak sense* if $W(\sigma^0, i, \theta) = \sup_{\sigma \in \Sigma} W(\sigma, i, \theta)$. The strategy is called *optimal* if it is θ-optimal for all $\theta \in \Theta$.

Our aim is to construct an optimal adaptive strategy for a class of CSMP. To define this class we introduce some notation. Let $p_\theta(j|i, u) = \mathbf{P}_\theta(j, Z|i, u)$ be the marginal distribution on the set X. Then the pair (X, p_θ) is a controlled Markov chain with the state space X. For a fixed $h \in H$ we define the matrix $P_{h,\theta}$ with the elements $p_{h,\theta}(j|i, h(i)) = p_{h,\theta}(j, i)$ and obtain a usual Markov chain referred to as the enclosed chain. We define the conditional mathematical expectation of a real-valued function v given on X by the formula

$$\mathbf{E}_{h,\theta}v(i) = \sum_{j \in X} v(j)p_{h,\theta}(i, j),$$

provided the series on the right-hand side of the equality converges.

Next, we state the assumptions concerning the class of the CSMP under consideration. They consist of three groups.

I. Conditions C.

$\mathbf{C_1}$ the functions $q_\theta(i, u)$, $\tau_\theta(i, u)$, $p_\theta(j|i, u)$ are continuous with respect to (θ, u) on $\Theta \times U(i)$ for every i, j;

$\mathbf{C_2}$ there exists a constant $\eta > 0$ such that $\tau_\theta(i, u) \geq \eta$ for all (θ, i, u).

For a fixed set $I \subset X$ we put $\bar{p}_{h,\theta} = (\bar{p}_{h,\theta}(i, j)) = (p_{h,\theta}(i, j)(1 - J_I(j)))$, where J_I stands for the indicator of the set I. This matrix is obtained from $p_{h,\theta}$ by replacing all rows corresponding to the indices $j \in I$ with zeros. The following conditions $\mathbf{L(v)}$ refer to some positive-valued functions $v(\theta, h, i)$.

II. Conditions L(v).

There exists a finite non-empty set $I \subset H$ and a function $y_\theta(i) > 0$ on the set $\Theta \times X$ such that

$\mathbf{L_1(v)}$ $v(\theta, h, i) + 1 + \bar{p}_{h,\theta}y_\theta(i) \leq y_\theta(i)$, $\quad \forall\, \theta, h, i$;

$\mathbf{L_2(v)}$ $\lim_{n \to \infty} p_{h,\theta}^n y_\theta(i) = 0$, $\quad \forall\, \theta, h, i$;

$\mathbf{L_3(v)}$ the numerical sequence $p_{h,\theta}y_\theta(i)$ depends continuously (in the obvious topology) on the law h.

III. Conditions R.

For all $\theta \in \Theta$ and $h \in H$ the enclosed chain $(X, p_{h,\theta})$ is ergodic (it contains only the communicating states).

Here is one more version of the conditions $\mathbf{L(v)}$.

II$^+$. Conditions $\tilde{\mathbf{L}}(\mathbf{v})$.

$\tilde{\mathbf{L}}_1(\mathbf{v})$ and $\tilde{\mathbf{L}}_2(\mathbf{v})$ are the same as $\mathbf{L_1(v)}$, $\mathbf{L_2(v)}$ but $I \subset X$;

$\tilde{\mathbf{L}}_3(\mathbf{v})$ the function $y_\theta(i)$ is continuous with respect to θ for all i;

$\tilde{\mathbf{L}}_4(\mathbf{v})$ the function $p_{h,\theta}y_\theta(i)$ depends continuously on (h, θ), $h \in H$.

In the last group of conditions the function y is the same as in $\mathbf{L(v)}$. The enclosed chain $(X, p_{h,\theta})$ is supposed to be either ergodic and positive recurrent[g] or the set I contains a single element.

---

[g]It means that the average returning time is finite for every state.

Let $\mathcal{K}$ be a class of controlled semi-Markov processes with countable state space X, topological measurable space of controls U and space of additional observations Z. In this situation, the conditions $\mathbf{C}$, $\mathbf{L}$, $\tilde{\mathbf{L}}(|\mathbf{q}| + \tau)$, $\mathbf{R}$ hold. The set $U(i)$ of admissible controls is assumed to be compact for all states i. It remains to define a strategy. For the controlled semi-Markov processes with parameter θ this is the identificational strategy based on the knowledge of the optimal law $h_\theta = h(\cdot|\theta)$.

Instead of the true parameter θ entering into this law we use an estimate $\hat{\theta}_t = \theta(i_0, z_0, u_0 :, \ldots; i_{t-1}, z_{t-1}, u_{t-1}; i_t, z_t)$. The value $h(i_t, \hat{\theta}_t) \in U_{i_t}$ serves as the control at the moment t. For calculating h_θ there is a special procedure described below. The sequence of such laws $h_t = h(\cdot|\hat{\theta}_t)$ forms the desired strategy. As concerns $\hat{\theta}_t$ we suppose that

$$\lim_{t \to \infty} \mathbf{P}_{Q^i_{\sigma,\theta}}(|\hat{\theta}_t - \theta| > \varepsilon) = 0, \quad \forall \varepsilon > 0, \ \theta \in \Theta,$$

i.e. these estimates are assumed to be weakly consistent with respect to the measure $Q^i_{\sigma,\theta}$ defined on the path space of the process. We denote this strategy by $\sigma_\mathcal{K}$. The calculation procedure of h_θ will be specified latter on. It is obvious that it has the features of an identificational strategy.

We now give a short survey (without proof) of the main results on optimal control of controlled semi-Markov processes under consideration. The Bellman Equation

$$v_\theta(i) = \max_{u \in U(i)} \left\{ q_\theta(i, u) - g_\theta \tau_\theta(i, u) + \sum_{j \in X} v_\theta(j) p_\theta(j|i, u) \right\}, \quad \forall \theta, \ i$$

where the functions g_θ, $v_\theta(i)$ are unknown, is the basis of these results. If there exists a solution (g, v) of this equation, then the law h maximizing the function on the right-hand side of the Bellman Equation will be called a *maximizor of the optimality equation*. We introduce the following notation

$$q_{h,\theta}(i) = q_\theta(i, h(i)), \quad \tau_{h,\theta}(i) = \tau_\theta(i, h(i)).$$

Let $\Pi_{h,\theta} = (\pi_{h,\theta}(i))$ denote the unique stationary distribution of the enclosed Markov chain which corresponds to the transition matrix $P_{h,\theta}$. Moreover, let $\Pi_{h,\theta}\zeta$ denote the mathematical expectation of the random sequence $\zeta = \zeta(i)$ with respect to this distribution. It remains to introduce the function

$$g(h, \theta) = \frac{\Pi_{h,\theta}\, q_{h,\theta}}{\Pi_{h,\theta}\, \tau_{h,\theta}}$$

which will be rather important later on.

Lemma 1. *If the conditions* $\mathbf{C}$, $\mathbf{L}(|\mathbf{q}| + \tau)$ *and* $\mathbf{R}$ *hold, then:*

(1)

$$\sup_h g(h, \theta) \le \eta^{-1} \max_h y_\theta(h),$$

$$W(h, i, \theta) = g(h, \theta), \quad \forall h, \theta, i;$$

(2) *there exists a solution (g, v) of the Bellman equation such that*

$$g_\theta = \max_h g(h, \theta), \quad |v_\theta(i)| \leq \alpha(\theta)y_\theta(i), \quad \forall\, \theta, i;$$

(3) *the function $g(h, \theta)$ is continuous with respect to h and θ provided the function $y_\theta(i)$ is continuous with respect to θ and the function $p_{h,\theta}y_\theta(i)$ with respect to h and θ;*

(4) *if we can choose the set I in the condition $\mathbf{L}(|\mathbf{q}| + \tau)$ to consist of one-element then the function g_θ has the same form as in (2) and*

$$v_\theta = \sup_h \sum_{n=0}^{\infty} p_{h,\theta}^n (q_{h,\theta} - g_\theta \tau_{h,\theta}),$$

moreover, the pair (g, v) satisfies the Bellman equation with

$$|v_\theta(i)| \leq (1 + |g_\theta|)y_\theta(i), \quad \forall\, \theta, i.$$

Under the above conditions the quantity $g_\theta = \max_h g(h, \theta)$ is defined in a unique way but v_θ may not. So, we suppose that v_θ satisfies the inequality from (2) but if the set $I = \{i_*\}$ is one-element then this function will have the form in (4).

We define the following function

$$\Phi_\theta(i, u) = q_\theta(i, u) - g_\theta \tau_\theta(i, u) + \sum_{j \in X} v_\theta(j)p_\theta(j|i, u) - v_\theta(i)$$

whose role is shown by the lemma below.

Lemma 2. *If the conditions $\mathbf{C}$, $\mathbf{L}(|\mathbf{q}| + \tau)$ and $\mathbf{R}$ hold then:*

(1) *$g(h, \theta) = g_\theta$ if and only if $\Phi_\theta(i, h) = 0$ with the chain (X, p) positive recurrent for all h and θ;*

(2) *for $\sigma \in \Sigma$ the inequality*

$$\lim_{T \to \infty} T^{-1} \sum_{t=1}^{T} \mathbf{E}_{\sigma,\theta}\big(\Phi_\theta(x_t, u_t)|x_0 = x\big) \geq -\varepsilon$$

with $\varepsilon > 0$ implies the inequality $W(\sigma, i, \theta) \geq g_\theta - \varepsilon/\eta$;

(3) *$W(\sigma, i, \theta) \leq g_\theta$.*

If for the strategy σ the inequality in (2) holds with $\varepsilon = 0$ then this strategy is θ-optimal.

We can now explain the definition of the strategy $\sigma_{\mathcal{K}}$: the laws forming it are the solutions of the equation $\Phi_\theta(i, u) = 0$.

We now turn to the problem of adaptive control of semi-Markov processes from the class $\mathcal{K}$ above.

Theorem 1. *The strategy $\sigma_{\mathcal{K}}$ is optimal with respect to the class $\mathcal{K}$.*

First, we find specific sufficient conditions for the introduced identificational strategy to be optimal, and establish these conditions for the class $\mathcal{K}$ and the strategy $\sigma_{\mathcal{K}}$. We need the next four lemmas.

Lemma 3. *Let the conditions* **C**, **L**$(|\mathbf{q}| + \tau)$ *and* **R** *hold and the laws* $h_\theta = h(\cdot|\theta)$ *be such that* $\Phi_\theta(i, h(i, \theta)) = 0$ *for all* i, θ. *If, in addition, the following conditions hold:*

(α) *the function* $h(\cdot, \theta)$ *is measurable with respect to* θ;
(β) *the function* $\Phi_{\theta_0}(i, h(i, \theta))$ *is continuous at any point* θ_0 *for all* i;
(γ)

$$\lim_{l \to \infty} \overline{\lim_{T \to \infty}} \, T^{-1} \sum_{t=1}^{T} \mathbf{E}_{\sigma(\hat{\theta}), \theta} \big\{ y_{\theta_0}(x_t) J_{[x_t \geq l]} | x_0 = i \big\} = 0$$

for all i *and* θ;
(δ) *the estimates* $\hat{\theta}_t$ *converge to the true value* θ *in the distribution sense* $Q^i_{\sigma(\hat{\theta}), \theta}$
for all i, θ_0, *then the strategy* $\sigma_{\mathcal{K}}$ *will be optimal, i.e.* $W(\sigma_k, i, \theta_0) = g_{\theta_0}, \forall i, \theta_0$.

Proof. By assertion (2) of Lemma 1, there exists a law h_θ such that $\Phi_\theta(i, h_\theta(i)) = 0$. Condition (α) implies the measurability of the laws $h_\theta(\cdot|\hat{\theta}_t)$ which belong to the strategy $\sigma_{\mathcal{K}}$. Next, from the conditions (β), (γ) it follows that $\lim_{t \to \infty} \Phi_{\theta_0}(j, h_{\hat{\theta}_t}(j)) = \Phi_{\theta_0}(j, h_{\theta_0}(j)) = 0$ in the measure $Q^i_{\sigma(\theta), \theta_0}$. Hence

$$\lim_{t \to \infty} \Phi_{\theta_0}(x_t, h_{\hat{\theta}_t}(x_t)) J_{[x_t < l]} = 0$$

in probability.

It is easy to verify that the quantity $\Phi_{\theta_0}(x_t, h_{\hat{\theta}_t}(x_t)) J_{[x_t < l]}$ is bounded for $t \geq 0$. Thus

$$\lim_{t \to \infty} \mathbf{E}_{\sigma_k, \theta_0} \big\{ |\Phi_{\theta_0}(x_t, u_t)| J_{[x_t < l]} | x_0 = i \big\} = 0.$$

This means that for every $\varepsilon > 0$ there exist an integer l_ε such that for $l > l_\varepsilon$

$$\overline{\lim_{T \to \infty}} \, T^{-1} \sum_{t=1}^{T} \big| \mathbf{E}_{\sigma_{\mathcal{K}}, \theta_0} \big\{ \Phi_{\theta_0}(x_t, u_t) | x_0 = i \big\} \big|$$

$$\leq \overline{\lim_{T \to \infty}} \, T^{-1} \sum_{t=1}^{T} \mathbf{E}_{\sigma_{\mathcal{K}}, \theta_0} \big\{ |\Phi_{\theta_0}(x_t, u_t)| J_{[x_t < l]} | x_0 = i \big\}$$

$$+ \overline{\lim_{T \to \infty}} \, T^{-1} \sum_{t=1}^{T} \mathbf{E}_{\sigma_{\mathcal{K}}, \theta_0} \big\{ y_{\theta_0}(x_t) J_{[x_t \geq l]} | x_0 = i \big\} \leq \varepsilon.$$

Since ε is arbitrary, the required optimality of the strategy ε_k follows from Lemma 2 (ii. (2), (3)). $\qquad\square$

We now verify that the conditions imposed on the strategy $\sigma_{\mathcal{K}}$ and class $\mathcal{K}$ imply the fulfilment of the conditions of Lemma 3. Here we need some topological notions about set-valued mappings.

Let V and W be topological spaces and Γ be a function given on V, whose values are non-empty subsets of W, i.e. $\Gamma : V \to 2^W$.

Definition 2. We say that a set-valued function Γ is upper semi-continuous (S-USC for short) if for any open set $G \subset W$ the set $\{v : \Gamma(v) \subset G\}$ is open in V.

A necessary and sufficient condition for a set-valued function to be upper semi-continuous at a point $v_0 \in V$ is the following: for any open set $G \subset W$ such that $\Gamma(v_0) \subset G$ there exists an open set $U \subset V$ containing v_0 such that $\Gamma(v) \subset G$ for all $v \in U$. We remind (see Sec. 1) that the numerical function $f : V \to \mathrm{R}^1$ is upper semi-continuous (USC for short) if the set $\{v \in V : f(v) < a\}$ is open for all $a \in \mathrm{R}^1$.

Let Γ be an S-USC mapping at $v_0 \in V$ and $\gamma : V \to W$ be a mapping such that $\gamma(v) \in \Gamma(v)$. If the set $\Gamma(v_0)$ is one-element then γ will be continuous at the point v_0.

Lemma 4. *Let Γ be a S-USC mapping of a topological space V into the set of non-empty subsets of a topological space W and let real-valued functions f on W and u on $V \times W$ be continuous. Then:*

(1) *the mapping $v \to \sup\{f(w) \mid w \in \Gamma(v)\}$ is USC;*
(2) *the mapping $v \to f(\Gamma(v)) = \{f(w) \mid w \in \Gamma(v)\}$ is S-USC.*
   Under the additional condition that W is compact, we have
(3) *the mapping $v \to \max\{u(v, w) \mid w \in W\}$ is continuous;*
(4) *the mapping $v \to \Gamma_\varepsilon = \{w \in W \mid u(v, w) \geq \max_{w' \in W} u(v, w') - \varepsilon\}$ is S-USC, compact and non-empty for all $\varepsilon > 0$.*

The proof is rather simple. We illustrate this by proving the second claim. We choose $v_0 \in V$ and an open $G \subset \mathrm{R}^1$ such that $f(\Gamma(v)) \subset G$. It is clear that $f^{-1}(G)$ is open and $\Gamma(v) \subset f^{-1}(G)$. Hence, there exists an open $U \subset V$ ($v_0 \in U$) such that $\Gamma(v) \subset f^{-1}(G)$ for all $v \in U$. Therefore $f(\Gamma(v)) \subset G$. These arguments remain in force for all $v_0 \in V$.

In accordance with the program of proof of Theorem 1 we deduce the conditions of Lemma 3 from Theorem 1.

Lemma 5. *Let besides the conditions $\mathbf{C}$ and $\mathbf{L}(|\mathbf{q}| + \mathbf{i})$ at least one of the following conditions hold:*

(1) *the chain $(X, p_{h,\theta})$ is irreducible and positive recurrent for all h and θ;*
(2) *the set $I = \{i_*\}$ contains a single element.*
   Then
(α) *there exists a law h_θ such that $\Phi_\theta(i, h_\theta(i)) = 0$ for all θ and i and the function h_θ is measurable with respect to θ;*
(β) *the function $\Phi_{\theta_0}(i, h_\theta(i))$ is continuous at any point θ_0 for any law h_θ such that $\Phi_\theta(i, h_\theta(i)) = 0$.*

Proof. We establish both assertions by using a real-valued continuous function $\Psi(h, \theta)$ possessing the following property: its maximizor h_θ satisfies the equation

$\Phi_\theta(i, h_\theta(i)) = 0$. Assertion (α) follows immediately from Theorem C on a measurable choice (Sec. 3). Indeed, the function Ψ is continuous and the set of its values is compact. We define the required function Ψ as follows: $\Psi(h, \theta) = g(h, \theta)$. By (3) of Lemma 1 and (1) of Lemma 2 this function possesses all the necessary properties.

To prove assertion (β) we introduce the set $I(\theta) = \{h \in H : \Psi(h, \theta) = \max_{h'} \Psi(h', \theta)\}$ and choose an element h belonging to it. By (3) of Lemma 4 the set-valued mapping $\theta \to I(\theta)$ is upper semi-continuous. This together with (3) of Lemma 4 and the continuity of $\Phi_\theta(i, u)$ with respect to u imply that the mapping

$$\theta \to G(\theta) = \{\Phi_{\theta_0}(i, h(i))|h \in I(\theta)\}$$

is SUC and $\Phi_{\theta_0}(i, h(i)) \in G(\theta)$. From the remark before Lemma 4 assertion (β) follows. It remains to define the function Ψ. We do this in the following manner: $\Psi(h, \theta) = \Phi_\theta(i, h)$. This function possesses all the necessary properties. $\qquad\square$

We consider now condition (γ) of Lemma 3 which can be interpreted as a restriction imposed on the limiting average reward generated by the function $y_\theta(i)J_{[x_t \geq l]}$. We suppose that this function satisfies the condition $\mathbf{L}(\mathbf{y})$ with $y_\theta(i)$ instead of $v(h, \theta, i)$. Let $Z_\theta(i)$ denote the appropriate bounded function.

Lemma 6. *Let conditions* $\mathbf{C}$, $\mathbf{L}(\mathbf{y})$ *and* $\mathbf{R}$ *take place. Then for all* θ

$$\varlimsup_{l \to \infty} \varlimsup_{T \to \infty} T^{-1} \sum_{t=1}^{T} \mathbf{E}_{\sigma,\theta}\big\{y_\theta(x_t)J_{[x_t \geq l]}|x_0 = i\big\} = 0$$

uniformly with respect to i *and* $\sigma \in \Sigma$.

Proof. We divide the proof into two stages. First, making use of condition $\mathbf{L_1}(\mathbf{y})$, we remark that it is possible to choose a bounded function Z so that the following inequality

$$\tilde{r}_l(i) + 1 + p_{h,\theta}Z_\theta(i) \leq Z_\theta(i)$$

holds for $\tilde{r}_l(i) = y_\theta(i)J_{[i \geq l]}$. Hence conditions $\mathbf{L}(\tilde{\mathbf{r}}_1 + \mathbf{e})$, where $e = (1, \ldots, 1)$, and (3) of Lemma 2 imply that

$$0 \leq \sup_i \sup_{\sigma \in \Sigma} \varlimsup_{T \to \infty} T^{-1} \sum_{t=1}^{T} \mathbf{E}_{\sigma,\theta}\big\{\tilde{r}_l(x_t)|x_0 = i\big\} \leq \sup_{h \in H} \Pi_{h,\theta}\tilde{r}_l, \quad \forall l, \theta.$$

Next, denoting $g_l(h, \theta) = \Pi_{h,\theta}\tilde{r}_l$ we can show that $\lim_{l \to \infty} \sup_{h \in H} g_l(h, \theta) = 0$ for all θ. Indeed, from the continuity of $g(h, \theta)$ with respect to h for all θ follows the continuity of $g_l(h, \theta)$ with respect to h for all l and θ. In view of Lemma 1, (3) with $\eta = 1$ we have

$$\Pi_{h,\theta}\tilde{r}_l \leq \Pi_{h,\theta}y_\theta \leq \max_{h \in H} Z_\theta(h) < \infty, \quad \forall l$$

and by the monotone convergence theorem we obtain

$$\lim_{l\to\infty} g_l(h,\theta) = \sum_{j\in X} \pi_{h,\theta}(j) \lim_{l\to\infty} \tilde{r}_l(j) = 0.$$

Since the space H is compact, by the Dini Theorem we have

$$\lim_{l\to\infty} \sup_{h\in H} g_l(h,\theta) = \sup_{h\in H} \lim_{l\to\infty} g_l(h,\theta) = 0,$$

which leads to the required assertion. Thus, under our assumptions about the class $\mathcal{K}$ and strategy $\sigma_\mathcal{K}$ all conditions of Lemma 3 hold. This completes the proof of the lemma. $\qquad\square$

We remark that condition **R** implies the validity of at least one of the assumptions of Lemma 5.

Theorem 1 is proved.

The result stated ensures optimality in the weak sense only. But if the estimates $\hat{\theta}_t$ are strongly consistent then the strategy $\sigma_\mathcal{K}$ will be optimal in the strong sense. The objective function has the following form

$$V(\sigma,\theta) = \lim_{T\to\infty} \frac{\sum_{t=1}^{T} q_\theta(x_t, u_t, z_{t+\tau}, x_{t+\tau})}{\sum_{t=1}^{T} \tau_\theta(x_t, u_t, z_{t+\tau}, x_{t+\tau})}$$

and means the empirical average reward (per unit time) for the controlled Markov process with parameter θ under the strategy σ.

It is worth to discuss here the design methods of good estimates of unknown parameters of the CSMP. In absence of a general approach we consider a special situation from queueing theory. More precisely, the queueing system of the $M/G/1$ type is considered. This means that we deal with a single server and a Poisson input stream with the parameter λ, the service time has the distribution P of a general form and the service discipline used is "first in–first out". The controllability of such a system means that both the parameter λ and the distribution P depend on a control u, i.e. $\lambda = \lambda(u)$, $P(\cdot|u) = \mathbf{P}\{\cdot|u\}$. The controls are set at the beginning of the next service interval by using the current information about the queue length, the waiting and service times.

The set $X = \{0,1,2,\ldots\}$ is the state space, the sets of controls U and additional information $Z = \mathrm{R}_+ \times \mathrm{R}_+$ (where $\mathrm{R}_+ = [0,\infty]$) remain the same. So does the parameter space Θ whose elements serve as the indices of the mathematical model: $\lambda_\theta(u)$ and $P_\theta(\cdot|u)$. The transition probabilities have the following form (for $j \geq i - 1 \geq 0$)

$$Q_\theta(j, [0,t_1], [0,t_2]|i, u) = \int_0^{t_1} e^{\lambda_\theta(u)s} \frac{(\lambda_\theta(u)s)^{j-i+1}}{(j-i+1)!} P_\theta(ds|u)$$

and are equal to 0 if $j < i - 1 \geq 0$ (for all θ), $i, j, u \in U(i)$, $t_1, t_2 \in \mathrm{R}_+$. This expression signifies the probability of the following event: the arrival of new calls during servicing the previous one and that the service time is no more than t_1. If $i = 0$ (system is empty, i.e. while waiting for the first customer) we put

$$Q_\theta\big(j, [0, t_1], [0, t_2] | 0, u\big) = \int_0^{\max(t_1, t_2)} \lambda_\theta(u) e^{\lambda_\theta(u)s} Q_\theta\big(j, [0, t_1 - s], \mathrm{R}_+ | 1, u\big)\, ds.$$

Clearly, both the function $\lambda_\theta(u)$ and the distribution $p_\theta(\cdot | u)$ have to depend on the pair (θ, u) in a measurable way. As usual, the reward function $q_\theta(i, u, z, j)$ is assumed to be measurable for any θ. Hence, there exists $q_\theta(i, u) = \sum_j \int_0^\infty Q_0(j, dz | i, u) q_\theta(i, u, z, j)$. Let $\tau_\theta(i, u, (t_1, t_2), j) \equiv t_1$ and T be service times of a customer. Then

$$\tau_\theta(i, u) = \mathbf{E}_{\sigma, \theta}\{T | x_0 = i, u_0 = u\} = \begin{cases} m_1(\theta, u) + \lambda_\theta^{-1}(u,) & i = 0, \\ m_1(\theta, u), & i > 0 \end{cases} \tag{1}$$

for any strategy $\sigma \in \Sigma$ where

$$m_l = \int_0^\infty s^l P_\theta(ds | u), \quad l \geq 0.$$

The average one-step reward is as follows

$$W(\sigma, \theta, i) = \lim_{T \to \infty} \frac{\mathbf{E}_{\sigma, \theta}\left\{\sum_{t=1}^T q_\theta(x_t, u_t) | x_0 = i\right\}}{\mathbf{E}_{\sigma, \theta}\left\{\sum_{t=1}^T \tau_t | x_0 = i\right\}}.$$

The control aim consists of minimizing this function with respect to σ.

For this model of queueing theory it is necessary to find the conditions which guarantee application of the theory developed above. First, we introduce some restrictions.

Condition K.

(1) The spaces U and $U(i)$ are compact for all i;
(2) The functions $q_\theta(i, u)$ and $m_1(\theta, u)$ are continuous with respect to (θ, u) for all i. The mapping $(\theta, u) \to \mathbf{P}_\theta(\cdot | u)$ is continuous in the weak topology of the set $P(\mathrm{R}_+)$ of all probability measures given on the positive real axis R_+;
(3) $m = \inf_{\theta, u} m_1(\theta, u) > 0$.

The next two restrictions differ by the growth order of the function $q_\theta(u)$ which is either polynomial or exponential.

Condition KP.

(1) There exists a continuous positive function $G(\theta)$ and an integer N such that

$$|q_\theta(i, u)| \leq G(\theta)(1 + i^N);$$

(2) $\sup_{\theta, u} m_{N+4}(\theta, u) < \infty$;
(3) $\delta(\theta) = \sup_u \lambda_\theta(u) m_1(\theta, u) < 1$.

Condition KE.

(1) There exist continuous functions $R(\theta) > 0$ and $\varkappa(\theta) > 0$ such that

$$|q_\theta(i, u)| \le R(\theta)\varkappa^i(\theta);$$

(2) The mapping

$$(\theta, u) \to \Psi(\theta, u) = \int_0^\infty \exp\{\lambda_\theta(u)(\varkappa(\theta) - 1)s\}\mathbf{P}_\theta\{ds|u\}$$

is finite and continuous;

(3) $\sup_u \Psi(\theta, u) < \varkappa(\theta), \quad \forall\,\theta.$

Theorem 2. *If either conditions* **K** *and* **KP** *or* **K** *and* **KE** *hold then the conditions* **C**, **R**, $\tilde{\mathbf{L}}(|\mathbf{q}| + \tau)$ *and* **L(y)** *take place at* $I = \{0\}$.

We omit the cumbersome proof and focus on constructing consistent estimates of parameters of the controlled processes in two important special cases. We consider a model whose parameter $\lambda = \lambda(u)$ is known but the distribution $P_\theta(\cdot)$ is unknown and does not depend on the control. We need to estimate the mathematical expectation of the service time $m_1(\theta)$ by using the observations of the process. We use the observable r.v. T_n, i.e. the random time intervals between the neighbouring services each of which is the sum of the waiting and service times S_n. If the times S_n are observed as well then the estimate

$$\hat{\theta}_N = N^{-1} \sum_{l=1}^{N} (T_l - S_l)$$

will be strongly consistent (under the condition $m_1(\theta) < \infty$)

$$\mathbf{P}\left\{ \lim_{n \to \infty} \hat{\theta}_n = m_1(\theta) \right\} = 1.$$

The situation when the times S_n are not observable is more complicated:

Theorem 3. *If* $\inf_u \lambda(u) > 0$, *then for all* $\sigma \in \Sigma$, $i \in X$ *and for all* θ *such that* $m_1(\theta) < \infty$ *we have*

$$\lim_{N \to \infty} N^{-1} \sum_{l=0}^{N-1} \left(T_l - \lambda^{-1}(u_l)J_{\{x_l=0\}}\right) = m_1(\theta), \quad Q_{\sigma,\theta}^i - \text{a.s.}$$

Proof. We start by estimating the conditional second moments

$$\mathbf{E}_{\sigma,\theta}\{T_{l+1}^2|x_l, u_l\} = \mathbf{E}_{\sigma,\theta}\{(T_{l+1} - S_{l+1} + S_{l+1})^2|x_l, u_l\}$$

$$\le 3\mathbf{E}_{\sigma,\theta}\{(T_{l+1} - S_{l+1})^2|x_l, u_l\} + 3\mathbf{E}_{\sigma,\theta}\{S_{l+1}^2|x_l, u_l\}$$

$$\le 3\left(m_2(\theta) + 2\sup_u \lambda^{-2}(u)\right).$$

This enables us to estimate another second moment (i.e. variation)

$$
\begin{aligned}
\mathrm{Var}_{\sigma,\theta}\big(T_{l+1}-\tau_\theta(x_l,u_l)|x_0{=}i\big) &= \mathbf{E}_{\sigma,\theta}\big\{(T_{l+1}-\tau_\theta(x_l,u_l))^2|x_0{=}i\big\}\\
&= 3\mathbf{E}_{\sigma,\theta}\big\{\mathbf{E}_{\sigma,\theta}(T_{l+1}^2|x_l,u_l)+q_\theta^2(x_l,u_l)|x_0=i\big\}\\
&= 6\mathbf{E}_{\sigma,\theta}\big\{\mathbf{E}_{\sigma,\theta}\{T_{l+1}^2|x_l,u_l\}|x_0=i\big\}\\
&\le 18\Big(m_2(\theta)+2\sup_u \lambda^{-2}(u)\Big)
\end{aligned}
$$

for all l, θ, σ, i where we have used the Jensen inequality.[h] Consequently

$$
\sum_{l=1}^{\infty} l^{-2}\mathrm{Var}_{\sigma,\theta}\big(T_{l+1}-\tau_\theta(x_l,u_l)|x_0=i\big) < \infty
$$

and we can use the strong law of large numbers for the martingale difference: recalling the expression (1) for $\tau_\theta(x,u)$ we have ($Q^i_{\sigma,\theta}$-a.s.)

$$
\begin{aligned}
0 &= \lim_{N\to\infty} N^{-1}\sum_{l=0}^{N-1}(T_{l+1}-\tau_\theta(x_l,u_l))\\
&= \lim_{N\to\infty} N^{-1}\sum_{l=0}^{N-1}(T_{l+1}-\lambda^{-1}(u_l)J_{[x_l=0]}-\tau_\theta(x_l,u_l)+\lambda^{-1}(u_l)J_{[x_l=0]}\\
&= \lim_{N\to\infty} N^{-1}\sum_{l=0}^{N-1}(T_{l+1}-\lambda^{-1}(u_l)J_{[x_l=0]}-m_1(\theta))
\end{aligned}
$$

which leads to the required result. $\qquad\qquad\qquad\qquad\qquad\qquad\qquad\square$

We now consider another model. Let the parameter λ_θ be unknown but the distribution of the service time $p(\cdot|u)$ be known and depend only on the control u. We need to construct estimates of the unknown parameter λ_θ which are strongly consistent at any admissible strategy.

Theorem 4. *Let $m_2(u) \ge m > 0$ for all u. Then:*

(a) *if $\sup_u m_2(u) < \infty$ then for all θ, σ and i the estimates*

$$
\hat{\theta}'_n = \frac{x_n+n}{\sum_{l=1}^{n} T_l}
$$

will converge (as $n\to\infty$) to λ_θ $G^i_{\sigma,\theta}$-a.s.;

(b) *if $\sup_u m_3(u) < \infty$ and $\sup_u m_1(u)\lambda_\theta < 1$ then for all θ, i and σ the estimates*

$$
\hat{\theta}''_n = \Big(n^{-1}\sum_{l=1}^{n} T_l\Big)^{-1}
$$

will converge (as $n\to\infty$) to λ_θ $G^i_{\sigma,\theta}$-a.s.

---

[h] Let $F(x)$ be a convex function and ξ be a r.v. such that $\mathbf{E}|\xi| < \infty$. Then

$$
\mathbf{E}F(\xi) \ge F(\mathbf{E}\xi).
$$

We omit the cumbersome proof of this theorem.

There are more general algorithms of constructing strongly consistent estimates of parameters of controlled semi-Markov processes.

7.6. Optimal Control of Special Classes of Markov Processes with Discrete Time

Here we discuss methods of control for some special types of Markov processes. These processes can be represented, as usual in the discrete case, by recurrent equations

$$x_{t+1} = \Phi(x_t, u_{t-1}, \xi_{t+1}), \quad x_0 = x, \ t \geq 0,$$

where ξ_t is a sequence of independent, identically distributed r.v. with a common distribution function F. The controls u_t belong to a subset $U_{(x_t)} \subset U$ which depends on the current state x_t. We need to extremize a functional. A peculiarity of these problems is that the function Φ is known but the distribution F of the "noise" ξ is unknown. The special form of such objects requires some motivation. We consider briefly two examples of real processes which are described by these models. One of them is the control problem of supplies, namely, the problem of stocking goods. In the storehouse there are l types of goods. Let $x = \left(x^{(1)}, \ldots, x^{(l)}\right)$ be a vector whose component $x^{(i)}$ denotes the supply of the ith type of goods (if this component is positive) or the size of the unsatisfied demand (if it is negative). The capacity V of the storehouse is limited. Defining the norm of the vector x as $\|x\| = \sum_{i=1}^{l} |x^{(i)}|$ we can write this condition as follows: $\|x\| \leq V$. In the space R^l, we define the subset $R^l_+ = \{x : x^{(i)} \geq 0, \ i = 1, 2, \ldots, l\}$. For every $x \in R^l$ the vector $x^+ = \left(\max(0, x^{(1)}), \ldots, \max(0, x^{(l)})\right)$, which gives the amount of goods in stock, belongs to R^l_+. The initial supply x_0 is such that $\|x_0\| \leq V$. The evolution of supplies is modeled by the equation

$$x_{t+1} = x_t^+ + u_t - \xi_{t+1},$$

where the control u_t is the size of the order of goods at the moment t, $\xi_t \in R^l_+$ is a random vector of demand at the moment t. A reasonable assumption about the "noise" ξ_t is the absence of any information about its distribution. The operation of the supplyhouse involves expenses associated with safekeeping of goods and ordering of new parties as well as fines either for unsatisfied demand or for delay of deliveries. The control aim is to minimize average losses per unit time. The controls satisfy the following constraint

$$\|x_t^+ + u_t\| \leq V.$$

Another problem is concerned with regulating the water-supply. Here we have the following equation

$$x_{t+1} = \min(x_t - u_t + \xi_t, V),$$

where x_t means the amount of water in the reservoir of capacity V, u_t is the control (the planned consumption of water) and ξ_t the inflow of water, i.e. the r.v. defined by the amount of rains, the melting of snow and so on. It is reasonable to expect that the distribution of the r.v. ξ_t is unknown. Under the obvious condition $0 \le u \le V$ we search the most profitable water-use.

There are other situations when a model of a Markov process is defined up to the distribution of "noise". Thus, there are reasons to study adaptive control for such models. Let us turn now to some concrete results.

Let a state space X be a locally compact separable metric space with metric ρ and the space of controls U be separable with metric r. The controlled Markov process is specified by the relationship

$$x_{t+1} = \Phi(x_t, u_t; \xi_{t+1}), \qquad x_0 = x, \quad t \ge 0 \tag{1}$$

where $\Phi(x, u; \xi)$ is a known function which is continuous with respect to x and u and measurable with respect to ξ. If at the moment t the process x_t is in the state x then the control u will belong to $U(x) \subset U$. The states of the process x_t are observed but the values ξ_t are either observed directly or can be calculated by using Eq. (1) and belong to a measurable space $(S, \mathcal{F}_s)$. The distributions of the r.v. ξ_t are unknown. Initially, we assume they do not vary in time.

As the admissible strategies Σ we take all deterministic strategies with the laws of the form $h_t = h$.

The control aim consists in finding the extremum (for the sake of concreteness let it be the maximum) of the objective function $W(\sigma, x)$ (the average "one-step" reward)

$$W(\sigma, x) = \varlimsup_{T \to \infty} T^{-1} \sum_{t=1}^{T} \mathbf{E}_{\sigma,x} q(x_t, u_t)$$

where x is the initial state of the process and $q(\cdot)$ is some bounded measurable function. We introduce the notation:

$$W^* \overset{\text{def}}{=} W(\sigma, x), \quad Q(x) = \{x\} \times U(x),$$
$$Q = \bigcup_x Q(x), \qquad v = (x, u) \in Q.$$

$d \overset{\text{def}}{=} \max\{\rho, r\}$ is the metric in Q; F_v denotes the distribution of the r.v. $\Phi(x, u; \xi)$ on the Borel σ-algebra in the space X at $(x, u) = v$. We impose the following constraints on the controlled Markov processes:

(i) $|q(v) - q(v')| \le L_0 d(v, v')$, $|q(v)| \le q_0 < \infty$;
(ii) $h\big(Q(x), Q(x')\big) \le L_1 \rho(x, x')$ where h stands for the Hausdorff metric in the space Q;
(iii) $\mathrm{Var}\,(F_v, F_{v'}) \le L_2 d(v, v')$;
(iv) Let $\nu_t(x, \sigma)$ be the distribution of the Markov process at the moment t with the initial state x and some deterministic stationary strategy σ. Then on the

Borel σ-algebra of the space X there exists a distribution $\nu(\sigma)$ such that

$$\mathrm{Var}(\nu_t(x,\sigma),\nu(\sigma)) \leq \lambda_t, \quad \sum_{t=1}^{\infty} \lambda_t < \infty;$$

(v) The space X is compact and

$$\rho\big(\Phi(v,s),\Phi(v',s)\big) \leq L_3(s)d(v,v')$$

where $\mathbf{E}L_3(\xi) = K < \infty$.

These constraints are satisfied, for instance, for the above-mentioned equations of queueing theory and water-supply regulating. Condition (ii) holds if the sets $U(x)$ do not depend on x. Condition (iv) holds if the distribution F_v has an absolutely continuous component F_v^0 with respect to a bounded measure μ on a σ-algebra of the space X and there exists a set M $(\mu(M) > 0)$ such that $dF_v^0(x)/d\mu > 0$ for all $x \in M$, $u \in U$. Then, as is well known, we can take $\lambda_t = a\alpha^t$, $a > 0$, $0 < \alpha < 1$. If the function q specifying the rewards does not depend on the controls then in condition (iv) the distance in variation can be replaced by the Lévy–Prohorov metric.

We denote the class of controlled Markov processes which satisfy Eq. (1) and conditions (i)–(v) by $\mathbf{G}$.

We now describe an adaptive strategy which is optimal in the weak sense above.

Let β_t be a non-decreasing numerical sequence such that $0 < \beta_t < 1$, $\lim_{t\to\infty} \beta_t = 1$ and $\lim_{n\to\infty} \varkappa(n)/n = 0$ where $\varkappa(n)$ denotes the number of alterations of values of this sequence on the interval $[0, n]$. Later on we impose on β_t some additional restrictions.

We consider the space $C(X)$ of bounded and continuous functions on X and a family of operators T_t, $t \geq 1$, on this space and depending on r.v. $\xi_1, \ldots, \xi_t$

$$T_t\varphi(x) = \max_{u \in U(x)} H(x,u;\varphi), \quad \varphi \in C(X)$$

where

$$H(x,u;\varphi) = q(x,u) + \beta_t t^{-1} \sum_{k=1}^{t} \varphi\big(\Phi(x,u;\xi_k)\big).$$

It is easy to show that T_t are contracting operators with respect to the norm of the space $C(X)$, i.e. $\|T_t\varphi\| \leq \beta_t\|\varphi\|$. Let φ_t be the fixed point of the operator T_t, i.e. $T_t\varphi_t = \varphi_t$. We form the set of controls corresponding to the state x at the moment t in the following way

$$M_t(x) = \left\{ u^* \in U(x_t) : H(x,u^*;\varphi_t) = \max_{u \in U(x)} H(x,u;\varphi_t) \right\}.$$

If for every t and x this set contains a single element $u_t = h_t(x_t)$ then h_t will be the required choice law (selector) of the control at the moment t. Otherwise we have a mapping of X into a collection of compact subsets of the space U. In view of Theorem A, Sec. 1 whose conditions are fulfilled in this case there exists some measurable function h_t taking the values from the set $M_t(x)$. The set of such selectors forms the required optimal adaptive strategy $\sigma^0 = \prod_{t=0}^{\infty} h_t$. We will not prove this assertion because it involves in fact some serious difficulties. The first of them is finding the fixed point φ_t. We can do this by using the well-known iterative procedure: having taken an arbitrary f_0 as an initial approximation we compute successively $f_n \overset{\text{def}}{=} T_t f_{n-1}$, $n \geq 1$ and obtain $\varphi_t = \lim_{n\to\infty} T_t f_n$ according to the contractibility property of the operator T_t. So, we have to calculate the corresponding approximation times out of number. For this reason we will proceed in another manner: having taken some initial function f_0 we define $f_t = T_t^{n(t)} f_{t-1}$, $t \geq 1$ where $n(t)$ is some unbounded, increasing integer-valued sequence. One can make sure that $\|\varphi_t - f_t\| \leq c(1 - \beta_t)^{-1}\beta_t^{n(t)}$. The sequence $n(t)$ can be chosen in such a way that $\lim_{t\to\infty} \beta_t^{n(t)}(1 - \beta_t)^{-1} = 0$.

The second difficulty is the necessity to find accurately the extremum of the function H. Instead of this we shall consider the less difficult problem: that is to find the approximate value of the desired extremum. For this purpose, we take positive numbers γ_t such that $\lim_{t\to\infty} \gamma_t = 0$ and define the set of controls $M_t(x)$ as follows

$$M_t(x) = \left\{ u^* \in U(x) : H(x, u^*; f_t) \geq \max_{u \in U(x)} H(x, u; f_t) - \gamma_t \right\}.$$

From now on, the construction is similar to that described above: by Theorem B on a measurable choice there exists a function (selector) $\Psi_t : X \to \bigcup_x U(x)$, $t \geq 0$, Ψ_0 being arbitrary. The sequence $\Psi_0, \Psi_1, \ldots$ forms the required strategy σ_G.

It remains to impose the necessary constraint on the sequence β_t. We introduce the constants $c = 2q_0 \sum_{t=1}^{\infty} \lambda_t$, $L = L_1(L_0 + cL_2)$ and a sequence of numbers ε_t, $(\varepsilon_t > 0, \lim_{t\to\infty} \varepsilon_t = 0)$. Putting

$$\delta_t = \delta_t(h) = \frac{\varepsilon_t(1 - \beta_t)}{2L(K + h)}, \quad h > 0,$$

we require that there exist a sequence ε_t such that

$$\lim_{t\to\infty} (1 - \beta_t)^{-1} \inf_{h>0} \left[tKh^{-1} + 2N_{\delta_t(h)} \exp\left\{ -\frac{\varepsilon_t^2 t(1 - \beta_t)^2}{8c^2} \right\} \right] = 0 \qquad (2)$$

where N_ε is the number of elements of the minimal ε-net for the compact set G. This technical condition will be used below.

Theorem 1. *For the class* **G** *of controlled Markov processes the following equality*

$$W(\sigma_G, x) = W^*$$

holds.

Proof. First, we consider the problem of maximizing a discounted reward[i]

$$W_\beta(\sigma, x) = \sum_{n=0}^{\infty} \beta^n \mathbf{E}_{\sigma,x} q_n, \quad 0 < \beta < 1$$

where x is the initial state and β is taken according to (2). The restrictions (i) and (ii) in the definition of the class $\mathbf{G}$ imply the semi-continuity of the Markov model. This and the boundedness of $q(\cdot)$ imply, in turn, the existence of an optimal stationary simple strategy σ_β for the objective function $W_\beta(\sigma, x)$. The function $W_\beta(x) = \sup_\sigma W_\beta(\sigma, x)$ represents the unique solution of the appropriate Bellman Equation

$$W_\beta(x) = \sup_{u \in U(x)} \{q(x, u) + \beta \mathbf{E} W_\beta(\Phi(x, u; \xi))\}.$$

Let $x_0 \in X$ be any fixed point. We put

$$g_\beta = (1 - \beta) W_\beta(x_0), \quad \lambda_\beta(x) = W_\beta(x) - W_\beta(x_0).$$

From the above equation it follows that $\lambda_\beta(x)$ and g_β are the solutions of the Bellman Equation that have the form

$$g_\beta + \lambda_\beta(x) = \sup_{u \in U(x)} \{q(x, u) + \mathbf{E}\lambda_\beta(\Phi(x, u; \xi))\}. \tag{3}$$

We now discuss some asymptotic properties (as $\beta \to 1$) of the solution.

Lemma 1. *If* $|\varphi(v) - \varphi(v')| \leq K d(v, v')$, *for all* v, $v' \in Q$ *and* $f(x) = \sup_{u \in U(x)} \varphi(x, u)$, *then*

$$|f(x) - f(x')| \leq L_1 \rho(x, x'), \quad x, x' \in X.$$

Proof. The compactness of $U(x)$ and the continuity of φ imply that $f(x) = \varphi(x, u_*)$. Let $|f(x) - f(x')| = \delta \geq 0$. Then for any $u \in U(x')$ we have $0 \leq \delta \leq \varphi(x, u_*) - \varphi(x', u)$. For $\varepsilon > 0$ we can find $u' \in U(x')$ such that

$$\begin{aligned}
d((x, u_*), (x', u')) &= \max\{\rho(x, x'), r(u_*, u')\} \\
&\leq \max\{\rho(x, x'), h(U(x), U(x')) + \varepsilon\} \\
&\leq h(U(x), u(x')) + \varepsilon \leq L_1 \rho(x, x') + \varepsilon.
\end{aligned}$$

From this the required inequality $\delta \leq L_1 \rho(x, x') + \varepsilon$ follows. $\square$

Lemma 2. *For all* $\varepsilon \in (0, 1)$ *and* $x, y \in X$ *the following inequalities:*

$$|\lambda_\beta(x)| \leq C, \quad |\lambda_\beta(x) - \lambda_\beta(y)| \leq L\rho(x, y), \quad \beta \in (0, 1),$$

are satisfied.

---

[i]We need to do this since the problem under consideration is the limiting problem for $\beta \to 1$. This method is often used in optimal control theory of Markov chains and processes.

Proof. According to the definitions of $\lambda_\beta(x)$ and W_β we have

$$\lambda_\beta(x) = \sum_{t=1}^{\infty} \beta^{t-1}\left[\mathbf{E}_{\sigma_\beta,x}q(x_{t-1},\Psi_\beta(x_{t-1})) - \mathbf{E}_{\sigma_\beta,x_0}q(x'_{t'-1},\Psi_\beta(x'_{t-1}))\right] = \sum_{t=1}^{\infty}\beta^t S_t,$$

where $\sigma_\beta = \prod_0^\infty \Psi_\beta$ is the optimal strategy for the β-discounted reward. By condition (iv)

$$|S_t| \leq \left|\int_X q(y,\Psi_\beta(y))[dF_{x_{t-1}}(y) - dF_{x'_{t-1}}(y)]\right|$$

$$\leq 2q_0\mathrm{Var}(\nu_{t-1}(\sigma_\beta,x)), \quad \nu_{t-1}(\sigma_\beta,x_0)) \leq 2q_0\lambda_{t-1}.$$

From this it follows that the set of functions $\{\lambda_\beta(x)\}$ is bounded. The Lipshitz condition can be verified by using Eq. (3), conditions (i), (iii) and Lemma 1. $\quad\square$

Lemma 3.

$$\lim_{\beta\to 1} g_\beta = W^*.$$

Proof. We consider the family of functions $\{\lambda_\beta\}$ from the space $C(X)$. It is relatively compact (i.e. its closure is compact) with respect to the uniform convergence topology on compact subsets of X. This follows immediately from Theorem 2 and the Ascoli theorem.[j] Hence in $C(X)$ there exists a function λ and a sequence β_k such that $\lim_{k\to\infty}\lambda_{\beta_k} = \lambda$. From the inequality

$$|W_\beta| = (1-\beta)|W_\beta(x_0)| \leq (1-\beta)\sum_{t=0}^{\infty}\beta^t\mathbf{E}_{\sigma_\beta,x}|q(x_t,u_t)| \leq q_0$$

it follows that the sequence $\{\beta_k\}$ is such that W_{β_k} converges to some number g. We show that $g = W^*$. First, we prove that for any given x the sequence $\varphi_k(u) = \mathbf{E}\lambda_{\beta_k}\big(\Phi(x,u;\xi)\big)$ converges uniformly with respect to $u \in U(x)$ to the function $\varphi(u) = \mathbf{E}\lambda\big(\Phi(x,u;\xi)\big)$.

Let $\varepsilon > 0$ and $S_\varepsilon = \{u_j\}$ be some finite ε-net for the compact $U(x)$. According to condition (iii) and Lemma 2 we have

$$|\varphi_k(u) - \varphi(u)| \leq |\varphi_k(u) - \varphi_k(u_j)| + |\varphi_k(u_j) - \varphi(u_j)| + |\varphi(u_j) - \varphi(u)|$$

$$\leq 2cL_2\varepsilon + \max_{u_j\in U(x)}|\varphi_k(u_j) - \varphi(u_j)|.$$

We now use the well-known fact: any probability measure μ on a locally-compact separable space is tight.[k] Hence for any $u_j \in S_\varepsilon$ there is a compact K_j such that

---

[j]**Theorem.** (Ascoli) *If a family of functions from $C(X)$ is uniformly bounded and equicontinuous then it is relatively compact with respect to the uniform convergence topology on compact subsets of X.*

[k]This means that for every $\varepsilon > 0$ there is a compact K such that $\mu(X\backslash K) \leq \varepsilon$.

$F_{x,u_j}(X\backslash K_j) \leq \varepsilon$. Then for the compact $K = \bigcup_j K_j$, $F_{x,u_j}(X\backslash K) \leq \varepsilon$ for all $u_j \in S_\varepsilon$. By uniform convergence of λ_{β_k} to λ with respect to $x \in K$ and the boundedness of λ_{β_k} we conclude that

$$\max_{u_j \in U(x)} |\varphi_k(u_j) - \varphi(u_j)| \leq 2c\varepsilon.$$

Hence $\lim_{k\to\infty} \sup_{u\in U(x)} |\varphi_k(u) - \varphi(u)| = 0$ and the required uniform convergence is proved. So, we can pass to the limit in (3) as $\beta_k \to 1$. Then

$$W + \lambda(x) = \max_{u\in U(x)} \{q(x,u) + \mathbf{E}\lambda(\Phi(x,u;\xi))\}.$$

Instead of *sup* we use *max* because of the continuity of the function in the brackets. According to Theorem B on measurable choice from Sec. 1 the mapping

$$x \to \left\{ u_* \in U(x) : q(x,u_*) + \varphi(u_*) = \max_{u\in U(x)} [q(x,u) + \varphi(u)] \right\}$$

admits a measurable choice. So, there exists a function (selector) Ψ such that

$$g + \lambda(x) = q(x, \Psi(x)) + \mathbf{E}\lambda(\Phi(x, \Psi(x); \xi))$$

for all $x \in X$. By Theorem 1, Sec. 1, we conclude that $g = W^*$. If there exists another sequence β_n converging to one and $W_{\beta_n} \to g' \neq g$ then for some subsequence $\{\beta_m\} \subset \{\beta_n\}$ the previous arguments would lead to $g' = W^* = g$. $\qquad\square$

We now prove two auxiliary results.

Lemma 4. $W_{\beta_t}(x) \in C(X)$ *for* $t \geq 1$ *and*

$$\|\varphi_t - W_{\beta_t}\| \geq \frac{\beta_t}{1 - \beta_t} \sup_{v\in G} \left| \mathbf{E}\lambda_{\beta_t}(\Phi(v;\xi)) - t^{-1} \sum_{m=1}^{t} \lambda_{\beta_t}(\Phi(v;\xi_m)) \right|.$$

Proof. The inclusion follows from Lemma 2 in view of the equality $W_{\beta_t}(x) = \lambda_{\beta_t}(x) + W_{\beta_t}(x_0)$. To prove the estimate we consider the following operator

$$T\varphi(x) \overset{\text{def}}{=} \sup_{u\in U(x)} \{q(x,u) + \beta_t \mathbf{E}W_{\beta_t}(\Phi(x,u;\xi))\}.$$

By the definition of φ_t and the Bellman equation for the discounted reward we have

$$\|\varphi_t - W_{\beta_t}\| = \|T_t\varphi_t - TW_{\beta_t}\|$$
$$\leq \|T_tW_{\beta_t} - TW_{\beta_t}\| + \|T_tW_{\beta_t} - TW_{\beta_t}\|,$$
$$\|T_t\varphi_t - T_tW_{\beta_t}\| \leq \beta_t\|\varphi_t - W_{\beta_t}\|,$$
$$T_t(\varphi + c) = T_t\varphi + a, \quad T(\varphi + c) = T\varphi + a, \quad a = \text{const}.$$

Hence

$$(1 - \beta_t)\|\varphi_t - W_{\beta_t}\| \le \|T_t \lambda_{\beta_t} - T\lambda_{\beta_t}\|$$

$$= \sup_x \sup_{u \in U(x)} \left| \{q(x, u) + \beta_t \mathbf{E}\lambda_{\beta_t}(\Phi(x, u; \xi))\} - \sup_{u \in U(x)} \Phi(x, u; \lambda_{\beta_t}) \right|$$

$$\le \sup_x \sup_{u \in U(x)} \beta_t \left| \mathbf{E}\lambda_{\beta_t}(\Phi(x, u; \xi))\} - t^{-1} \sum_{m=1}^{t} \lambda_{\beta_t} \Phi(x, u; \xi_m) \right|.$$

The inequality obtained is equivalent to the required inequality. $\qquad\square$

Let Z be a metric space with a metric τ, V be a compact subset of Z, N_ε be the number of elements of the minimal ε-net for the compact V and η be a real-valued function given on $Z \times S$ and measurable with respect to $s \in S$ such that

$$|\eta(z, s)| \le C, \quad |\eta(z, s) - \eta(z', s)| \le L_*(s)\tau(z, z'), \quad \forall\, s, z, z'. \tag{4}$$

Let $\xi_1, \xi_2, \ldots$ be independent r.v. taking values from S and having the same distribution as ξ. We put

$$I_t(z) = \mathbf{E}\eta(z, \xi) - t^{-1} \sum_{k=1}^{t} \eta(z, \xi_k).$$

Lemma 5. *If* $\mathbf{E}L_*(\xi) = K < \infty$ *then for every* $\varepsilon > 0$ *we have*

$$\mathbf{P}\left\{ \sup_{z \in V} |\zeta_t(z)| > \varepsilon \right\} \le \inf_{h > 0}\left\{ tKh^{-1} + 2N_\delta \exp\left(-\frac{\varepsilon^2}{8c^2} t \right) \right\},$$

where $\delta = \varepsilon(2L(K + h))^{-1}$.

Proof. From the continuity with probability one of the random function $\eta(z, \cdot)$ and the separability of V it follows that the function $Y_t = \sup_{z \in V} |\zeta_t(z)|$ is a r.v. We introduce the set

$$A = \{\omega : L_*(\xi_k) \le h, \ k = 1, \ldots, t\}, \quad h > 0.$$

Then we have

$$\mathbf{P}\{Y_t > \varepsilon\} \le \sum_{k=1}^{t} \mathbf{P}\{L_*(\xi_k) > h\} + \mathbf{P}\{Y_t > \varepsilon; A\}. \tag{5}$$

Let S_δ be a finite δ-net for the compact subset V and H_j be a sphere with diameter δ and center at the point $z_j \in S_\delta$ in Z. Then

$$\{Y_t > \varepsilon; A\} \subset \bigcup_j \left\{ \sup_{z \in H_j} |\zeta_t(z)| > \varepsilon; A \right\}. \tag{6}$$

From (4) and Hölder inequality, on the set A the function $\zeta_t(z)$ satisfies the Lipschitz condition with a constant $L(K + h)$. Hence

$$\left\{ \sup_{z \in H_j} |\zeta_t(z)| > \varepsilon; A \right\} \subset \left\{ \sup_{z \in H_j} |\zeta_t(z)| > \varepsilon; \sup_{z \in H_j} |\zeta_t(z) - \zeta_t(z_j)| \le L(K + h)\delta \right\}$$

$$\subset \{|\zeta_t(z_j)| > \varepsilon/2\}.$$

Since $\zeta_t(z_j)$ is the sum of independent, bounded (by C) r.v. with zero means we can use the Hoeffding theorem (see, p. 59). Finally, we obtain

$$\mathbf{P}\{|\zeta_t(z_j)| > \varepsilon/2\} \le 2\exp\left\{-\frac{\varepsilon^2}{8c^2}t\right\}.$$

This, the Chebyshev inequality and (5) and (6) imply the assertion of the lemma. $\qquad\Box$

Let us return to the proof of Theorem 1. We introduce the following notation:

$$W_t = W_{\beta_t}, \quad g_t = g_{\beta_t}, \quad \lambda_t = \lambda_{\beta_t};$$
$$\hat{\varphi}_t(x) = \varphi_t(x) - W_t(x_0), \quad \hat{f}_t = f_t(x) - W_t(x_0)$$

we put

$$R(x,u) = q(x,u) + \beta_t \mathbf{E}\lambda_t\big(\Phi(x,u;\xi)\big), \quad u \in U(x).$$

Let $v_t = (x_t, u_t)$ be a pair consisting of state and control provided the strategy σ_G with the initial state x was used at the moment t.

We put $R(x_t, u_t^*) = \max_{u \in U(x_t)} R(x_t, u)$, where the mapping $x \to u_t^* \in U(x)$ is measurable. According to (3)

$$R(x_t, u_t^*) = g_t + \lambda_t(x_t). \tag{7}$$

For the difference $L_t = R(x_t, u_t) - R(x_t, u_t^*)$ the following estimate takes place

$$|L_t| \le |R(x_t, u_t) - H(x_t, u_t; \hat{\varphi}_t)| + |H(x_t, u_t; \hat{\varphi}_t) - H(x_t, u_t; \hat{f}_t)|$$
$$+ |H(x_t, u_t; \hat{f}_t) - R(x_t, u_t^*)|.$$

We estimate the second term on the right-hand side of the above inequality

$$|H(x_t, u_t; \hat{\varphi}_t) - H(x_t, u_t; \hat{f}_t)| \le \|\varphi_t - f_t\|.$$

Since the first and third terms can be estimated in the same manner we consider only the third one. So,

$$|H(x_t, u_t; \hat{f}_t) - R(x_t, u_t^*)| \le \gamma_t + \left| \max_{u \in U(x_t)} H(x_t, u; \hat{f}_t) - \max_{u \in U(x_t)} R(x_t, u) \right|$$

$$\le \gamma_t + \max_{u \in U(x_t)} \left| t^{-1} \sum_{m=1}^{t} \{\hat{f}_t\big(\Phi(x_t, u; \xi_m)\big) - \hat{\varphi}_t\big(\Phi(x_t, u; \xi_m)\big)\} \right|$$

$$+ \max_{u \in U(x_t)} \left| t^{-1} \sum_{m=1}^{t} |\hat{\varphi}_t\big(\Phi(x_t, u; \xi_m)\big) - \lambda_t\big(\Phi(x_t, u; \xi_m)\big) \right|$$

$$+ \max_{u \in U(x_t)} \left| t^{-1} \sum_{m=1}^{t} |\lambda_t\big(\Phi(x_t, u; \xi_m)\big) - \mathbf{E}\lambda\big(\Phi(x_t, u; \xi_m)\big) \right|$$

$$\le \gamma_t + \|\varphi_t - f_t\| + \|\varphi_t - W_t\| + (1 - \beta_t)\Lambda_t$$

where $\lambda_t = (1 - \beta_t)^{-1} \sup_{v \in Q} |t^{-1} \sum_{k=1}^{t} \lambda_t\big(\Phi(v; \xi_k)\big) - \mathbf{E}\lambda_y\big(\varphi(v; \xi)\big)|$. Making use of Lemma 4, we arrive at the estimate

$$|L_t| \leq \gamma_t + 2\|\varphi_t - f_t\| + 4\Lambda_t. \tag{8}$$

If $I_t = (x_0, u_0; x_1, u_1; \cdots; x_t, u_t)$ denotes the history of the process up to the moment t, then by the Markov property and (7) we see that

$$\beta_t \mathbf{E}_{\sigma_G}\big\{\lambda_t(x_{t+1}) | I_t\big\} = q(v_{t+1}) + \beta_t \mathbf{E}\lambda_t\big(\Phi(v_t; \xi)\big) + q(v_t)$$
$$= g_t + \lambda_t(x_t) - q(v_t) + L_t.$$

Thus, for $n \geq k > 1$ we conclude that

$$n^{-1}\mathbf{E}_{\sigma_G, x} \sum_{t=k}^{n} [q(v_t) - g_t] = n^{-1}\mathbf{E}_{\sigma_G, x} \sum_{t=k}^{n} [\lambda_t(x_t) - \beta_t\lambda_t(x_{t+1})]$$
$$+ n^{-1} \sum_{t=k}^{n} \mathbf{E}_{\sigma_G, x} L_t. \tag{9}$$

We put

$$I_n \stackrel{\text{def}}{=} \sum_{t=k}^{n} [\lambda_t(x_t) - \beta_t\lambda_t(x_{t+1})]$$
$$= \sum_{t=k}^{n} [\lambda_t(x_t) - \beta_t\lambda_t(x_t)] + \sum_{t=k}^{n} \beta_t[\lambda_t(x_t) - \beta_t\lambda_t(x_{t+1})].$$

According to Lemma 2

$$n^{-1}|I_n| \leq (1 - \beta_t)c + c_1 n^{-1}\varkappa(n), \quad c_1 = const$$

where $\varkappa(n)$ is the number of alternations of the values of the sequence $\{\beta_t\}$ on the interval $[0, n]$. We show that the second term on the right-hand side of (9) tends to zero. In view of (8) and the convergence of the quantities γ_t, $\|\varphi_t - f_t\|$ to zero as $t \to \infty$ it is sufficient to prove that

$$\lim_{t \to \infty} \mathbf{E}_{\sigma_G, x}\Lambda_t = 0. \tag{10}$$

We choose the numbers ε_t according to the choice of $\{\beta_t\}$ in condition (2). By Lemma 2, $|\Lambda_t| \leq 2c(1 - \beta_t)^{-1}$. Hence

$$\mathbf{E}_{\sigma_G, x}\Lambda_t \leq \varepsilon_t + \frac{2c}{1 - \beta_t}\mathbf{P}(\Lambda_t > \varepsilon_t), \quad \mathbf{P}\{\Lambda_t > \varepsilon\} = \mathbf{P}\{\zeta_t > (1 - \beta_t)\varepsilon_t\}$$

where $\zeta_t = (1 - \beta_t)\Lambda_t$. In Lemma 5 we put that $Z = V = Q$, $\eta(v, s) = \lambda_t\big(\Phi(v, s)\big)$. According to the assumptions

$$|\eta(v, s)| \leq c, \quad |\eta(v, s) - \eta(v', s)| \leq L_3(s)d(v, v'), \quad \mathbf{E}L_3(\xi) = k < \infty,$$

and we can use Lemma 5 which together with condition (2) implies (10). Therefore, we have proved that

$$n^{-1}\mathbf{E}_{\sigma_G, x} \sum_{t=m}^{n} [q(v_t) - g_t] = (1 - \beta_n)c + o(1)$$

for every $m \geq 1$ as $n \to \infty$. From Lemma 3 and $\lim_{t \to \infty} \beta_t = 1$ it follows that

$$W(\sigma_G, x) = \lim_{n \to \infty} n^{-1} \mathbf{E}_{\sigma_G, x} \sum_{t=1}^{n} q(v_t) = W^*.$$

The theorem is proved completely. $\qquad \square$

Corollary 1. *Let X and V be subsets of finite-dimensional Euclidean spaces and $\mathbf{G}$ be a class of controlled Markov processes. Let $\beta \sim 1 - at^{-\gamma}$ $(0 < \gamma < 1/2)$. Then*

$$W(\sigma_G, x) = W^*.$$

Proof. It follows from the assumptions that the space Q is compact in the finite-dimensional Euclidian space R^m. Hence we can use the well-known estimate $N_\varepsilon \leq b\varepsilon^{-m}$ (for $\varepsilon \leq \varepsilon_0$) for the number of elements of the minimal ε-net for the compact space Q. Putting $\varepsilon = t^{-f}$, $h = t^{-g}$ we choose f, g such that $\lim_{t \to \infty} \beta_t = 1$. From this the assertion follows. $\qquad \square$

The stated result can be extended to non-compact state spaces X. In this connection we state a result related to separable metric spaces S of values of the r.v. $\xi_1, \xi_2, \ldots$.

Theorem 2. *If the set of functions $\{\Phi(v, \cdot), v \in Q\}$ is equicontinuous at every point $s \in S$ then there is a sequence β_t for which the strategy σ_G leads to the equality*

$$W(\sigma_G, x) = W^*.$$

We now turn to another aim of control: to obtain the maximum (with respect to σ) of the quantity $\lim_{t \to \infty} \mathbf{E}_{\sigma_G, x} q(x_t, u_t)$. To achieve this we construct a randomized strategy by using the selectors Ψ_t applied already in designing the strategy σ_G. This strategy is represented as follows: $\sigma_{GR} = \prod_{t=1}^{\infty} \bar{\Psi}_t$, where the selectors $\bar{\Psi}_t$ are constructed in a recurrent way. We put $\bar{\Psi}_0 = \Psi_0$ and $\bar{\Psi}_t = \Psi_{t-1}$ with the probability $1 - p_t$ and $\bar{\Psi}_t = \Psi$ with the probability p_t, for $t \geq 1$. The choice of the selector at the moment t is performed with probability $p_t = t^{-\alpha}$, $0 < \alpha < 1$ independent of the previous choices.

Theorem 3. *Under the conditions of Corollary 1 the equality*

$$\lim_{t \to \infty} \mathbf{E}_{\sigma_{GR}, x} q(x_t, u_t) = W^*.$$

takes place for all x.

The proof is omitted.

Up to now we have dealt with homogeneous controlled Markov processes of form (1) where the noise ξ_t had the constant distribution F. Of course, the case of non-homogeneous processes with nonconstantly distributed noises is more important. However the optimal control problem of such processes is much more difficult. It is clear in advance that for the optimal adaptive control to exist the law of varying the distributions cannot be arbitrary. For this reason, studying the control of nonhomogeneous processes, we first consider admissible laws of varying the distributions (probability measures) F_t in the space S of values of the r.v. ξ.

Let d denote a semi-metric (the condition $d(\mu, \nu) = 0 \Rightarrow \mu = \nu$ may fail) in the space $\mathcal{P}(S)$ of all probability measures on S.

Definition 1. A sequence of measures $\{\mu_n\} \subset \mathcal{P}(S)$ is called *slowly varying with respect to d* if

$$\lim_{N \to \infty} N^{-1} \sum_{n=1}^{N} d(\mu_n, \mu) = 0.$$

We give some examples of slowly varying sequences of measures (SVSM for short):

1. If $\{\mu_n\}$ converges to μ with respect to the semi-measure d then it will be SVSM.
2. Here and in (ii) 3–4 below we choose the real axis as the set S. Let μ_t be the uniform distribution on $[0, \ln t]$. It is easy to verify that μ_t form SVSM with respect to convergence in variation and it has no limit points in $\mathcal{P}(\mathbb{R}^1)$.
3. Let μ_t be the exponential distribution on $\mathbb{R}^1$ with the parameter $\lambda_t = t^{1/(\ln t)^\gamma}$, $\gamma \in (0, 1)$, $t \geq 2$. The sequence μ_t is SVSM with respect to convergence in variation.
4. Let $F(s)$ denote an arbitrary distribution function on $\mathbb{R}^1$ and $F_t(s) = F(s_{v_t})$, $v_t = \ln L(t)$ where $L(t)$ is a slowly varying (in the Karamata sense) function.[1] Then F_t is SVSM with respect to the Dudley metrics and we can take $v_t = b(\ln t)^\gamma$, $0 < \gamma < 1$. But the sequence $F_t(s) = F(s + \ln t)$ may not be slowly varying with respect to the Dudley metrics.

We now turn to the description of a class of nonhomogeneous Markov processes $\tilde{G}$ represented by correlation (1). Its state space X is separable, metric (with metric ρ) and with the Borel σ-algebra $\mathcal{F}_X$. A space of controls U has the same properties (with metric r). A compact set $U(x) \subset U$ corresponds to each $x \in X$. The noises $\xi_t \in \mathbb{R}^k$ are unknown and have the distributions F_t. The choice law h_t of the control at the moment t can be randomized. It depends in a measurable way on the history $\mathcal{I}_t = \{x_0, \ldots, x_t; u_0, \ldots, u_{t-1}; \xi_1, \ldots, \xi_t\}$, i.e. the states x_t and the noises ξ_t are observed. The control u_t selected at the moment t belongs to $U(x_t)$. The

---

[1]A positive function $L(x)$, $x \in [0, \infty]$ is called *slowly varying in the Karamata sense* if

$$\lim_{x \to \infty} L(tx)/L(x) = 1, \quad \forall t > 0.$$

The function $L(x)$ is slowly varying if and only if

$$L(x) = a(x) \exp\left\{ \int_1^x z^{-1}\varepsilon(z)\,dz \right\},$$

where $\varepsilon(x) \to 0$, $a(x) \to a < \infty$ as $x \to \infty$.

admissible strategies σ form a set Σ. The notation $F_{v,t}$ signifies the distribution of the r.v. $\Phi(v, \xi_t)$.

The objective function means, as before, the average reward

$$W(\sigma, x) = \lim_{N \to \infty} N^{-1} \sum_{t=0}^{N} \mathbf{E}_{\sigma,x} q(x_t, u_t).$$

The aim of control is to maximize this function.

Let a class of processes Γ be specified by the following conditions:

(i) The function $q(v)$ is uniformly continuous and bounded on G, i.e. $|q(v)| \leq q_0$;

(ii) The mapping $x \to U(x)$ is uniformly continuous in the Hausdorff metric h;

(iii) $\lim_{s' \to s} \sup_{v \in Q} \rho(\Phi(v, s'), \Phi(v, s)) = 0$, $\forall s$;

(iv) $\rho(\Phi(v, s), \Phi(v', s)) \leq L(s)\gamma(u, u')$ where a function $L(s) > 0$ is bounded on the bounded sets in R^k, $\gamma(y)$ is defined for $y \geq 0$ and $\gamma(y) \to 0$ as $y \to \infty$;

(v) There exists a measure ν (minorant) on $\mathcal{F}_X$ such that $0 < \nu(X) < 1$ and $F_{v,t}(M) \geq \nu(M)$ for all $M \in \mathcal{F}_X$, $v \in Q$, $t \geq 1$;

(vi) $\lim_{n \to \infty} \sup_{t \geq 1} F_t\{s : \|s\| > n\} = 0$;

(vii) The sequence of distributions F_t of the r.v. ξ_t is slowly varying with respect to a semi-metric d_D.

We now have to specify the semi-metric d_D. For every t we consider the following equation

$$\varphi(x) = \sup_{u \in U(x)} \{q(x, u) + \mathbf{E}\varphi(\Phi(x, u; \xi_t))\} - \int_X \varphi(y)\nu(dy).$$

Under conditions (i)–(v) it has a solution $\varphi_t \in C(X)$. Under the conditions (i)–(vi) the set of functions $\Pi_0 = \{\varphi_t, t \geq 1\}$ is bounded and equicontinuous. We omit the proof of these facts. As the set Π we choose a bounded set of equicontinuous functions[m] such that $\Pi_0 \subset \Pi$. We put

$$D = \{\varphi(\Phi(v, \cdot)); v \in Q, \varphi \in \Pi\}.$$

Then

$$d_D(\mu, \mu') \stackrel{\text{def}}{=} \sup_{f \in \Pi} \left| \int_{\mathrm{R}^k} f(y)\, d\mu(y) - \int_{\mathrm{R}^k} f(y)\, d\mu'(y) \right|$$

for all measures from $\mathcal{P}(\mathrm{R}^k)$. From the definitions in Sec. 1, both Dudley metrics and metrics in variation are special cases of d_D for proper sets D. In constructing the semi-metric d_D, one of the difficulties consists of choosing the set of functions Π_0.

Comparing the sets of conditions which specify the classes G and Γ, we note that the restrictions on Γ are weaker than those on G. Condition (v) is special as it requires existence of a uniform minorant for the class Γ.

---

[m]In the definition of continuity of a function f at point x one says: for any $\varepsilon > 0$ there is a number $\delta(\varepsilon, x, f)$ depending, generally speaking, on ε, x and f such that $|f(x) - f(x')| < \varepsilon$ if $\rho(x, x') < \delta(\varepsilon, x, f)$. For equicontinuity of a family of functions the number δ depends only on ε but neither on x nor on f.

We now construct an optimal adaptive strategy σ_Γ with respect to the class Γ. As usual, we introduce the family of operators T_t on the space $C(X)$

$$T_t\varphi(x) = \sup_u H_t(x, u; \varphi) - \int_X \varphi(y)\nu(dy)$$

where

$$H_t(x, u; \varphi) = q(x, u) + t^{-1} \sum_{n=1}^{t} \varphi\big(\Phi(x, u; \xi_n)\big).$$

Let $\varepsilon_t > 0$, $\chi_t > 0$, $t \geq 1$ be some numerical sequences such that $\lim_{t\to\infty} \varepsilon_t = 0$, $\lim_{t\to\infty} \beta_t = 0$ and $f_t(x)$ be functions from Π satisfying the inequalities

$$\|T_t f_t - f_t\| \leq \inf_{\varphi\in\Pi} \|T_t\varphi - \varphi\| + \varepsilon_t, \quad t \geq 1.$$

We introduce the following subset of the control space U

$$U_t(x) = \left\{ u^* \in U(x) : H_t(x, u^*; f_t) \geq \sup_{u\in U(x)} H_t(x, u; f_t) - \chi_t \right\}.$$

By Theorem A, Sec. 1 and conditions (i)–(iv) the set-valued mapping $x \to U_t(x)$ admits a measurable choice. Let Ψ_t be such a measurable function (selector), i.e. $\Psi_t(x) \in U_t(x)$. Then $\sigma_\Gamma = \prod_{t=0}^{\infty} \psi_t$ where ψ_0 is an arbitrary measurable function is the required strategy. As usual, this strategy consists of using the control $u_t \in \Psi_t(x_t)$ at the moment t. The following property of the maximum reward $W^*(x) = \sup_\sigma W(\sigma, x)$ is a feature of this strategy.

Theorem 4. *For the class of processes Γ, the strategy σ_Γ leads to the equality*

$$W^*(x) = W^* = \varlimsup_{T\to\infty} T^{-1} \sum_{t=1}^{T} g_t, \quad g_t = \int_X \varphi_t(y)\nu(dy)$$

for every $x \in X$.

Corollary 2. *If the sequence of distributions F_t converges to a distribution F in the semi-metric d_D then the strategy σ_Γ implies the equality*

$$\lim_{T\to\infty} T^{-1} \sum_{t=1}^{T} \mathbf{E}_{\sigma_\Gamma, x} q(x_t, u_t) = W^*$$

for every initial state x.

The proof of Theorem 4 is similar to that of Theorem 1. For this reason it is omitted. We could give the estimates for the quantity $\sup_x |W_t(\sigma_\Gamma, x) - W^*|$ but because they are cumbersome we will not do it.

The adaptive strategies considered in this section may be regarded as direct. There is no identification. In the homogeneous case identification would mean finding proper estimates of an unknown distribution of the noise. In the non-homogeneous case even the statement of the identification problem is not a simple task.

CHAPTER 8

CONTROL OF STATIONARY PROCESSES

We consider optimal adaptive strategies for discrete, stationary in a wide sense, processes. Time t runs over the set of integers, i.e. $t \in \{\ldots, -1, 0, 1, \ldots\}$. The processes considered are characterized by measures which are invariant with respect to shifts in time. The optimization aims differ slightly from the ones studied earlier.

8.1. Formulation of the Problem

In the previous chapters a wide range of adaptive control problems has been studied. For various combinations of a "class of controlled processes" and an "aim of control" appropriate strategies have been constructed. Most problems considered have an applied significance. In this connection the following questions arise. What are the possible extensions of the concept of adaptive control? Should one consider "complex" processes which cannot be reduced to "simple" ones, for example, to Markov processes? The description of general classes of CRP can be done by means of a set of controlled conditional probabilities $\mu_{t+1}(\cdot \,|x^t, u^t)$ which depend essentially on the whole past history of the process. Of course, the set $\{\mu_t\}$ must satisfy some restrictions to allow existence of the adaptive strategy. We now define a class of stationary processes in terms of μ_t.

For the first time in this book we deal with processes defined on the set of all integers $\{\ldots, -2, -1, 0, 1, 2, \ldots\}$. So, we now put $z^t = (\ldots, z_{t-2}, z_{t-1}, z_t)$, in particular, $z^0 = (\ldots, z_{-2}, z_{-1}, z_0)$.

Definition 1. A stationary random process ξ_t is called a *controlled stationary process* (CSP) if its conditional distributions are invariant with respect to shifts in time, i.e.

$$\mu_{t+1}\big(M|x^t, u^t\big) \equiv \mu_1\big(M|x^0, u^0\big), \quad \forall t, \ M \in \mathfrak{M}.$$

The stationary processes have well-known practical significance, however we are interested in studying them as mathematical objects. They appear in problems of data processing and in control of communication systems. These processes also play a crucial role in regulation theory, selecting useful signals against a background noise, and in prediction and filtration problems. Among numerous examples we mention functionals defined on the paths of a partially observable homogeneous Markov chain.

As usual, a concrete random process is obtained from a CSP by choosing the appropriate strategy. Let $\xi_t(\sigma)$ denote a process controlled by the strategy σ. We

251

are especially concerned with the stationary strategies generated by the rules h with finite memory depth. Such rules can be described by a measurable mapping $h : X^m \to U$ such that $u_t = h(x_t, \ldots, x_{t-m+1})$. The set of all rules having memory depth m will be denoted by D_m. We put $D = \bigcup_m D_m$. The strategies $\sigma(h)$ generated by such rules will be identified with these rules. The following assertion takes place.

Proposition. *For any strategy $\sigma \in D$ and any CSP ξ_t the process $\xi_t(\sigma)$ is stationary in the wide sense.*

We now turn to the aims of control considered. Let f be an arbitrary bounded measurable function defined on X. One and the same symbol $W(\sigma)$ denotes the following limits[a] (if they exist)

$$\lim_{t \to \infty} \mathbf{E}(f(x_t)|x^0, u^0), \quad \underset{T \to \infty}{\mathrm{l.i.m}} \; T^{-1} \sum_{t=1}^{T} f(x_t),$$

where the conditions (x^0, u^0) are assumed to be fixed and arbitrary. We also put $\bar{W} = \sup_{\sigma \in D} W(\sigma)$.

The aim of control is to *construct the strategies such that the following limits exist and fulfill one of the following equalities*

1. $\mathrm{l.i.m.}_{T \to \infty} T^{-1} \sum_{t=1}^{T} f(x_t) = \bar{W}$,
2. $\lim_{t \to \infty} \mathbf{E}\big(f(x_t)|x^0, u^0\big) = \bar{W}$.

To attain the given aims the classes of CSP must satisfy some restrictions. As concerns the spaces X and U the assumptions can be rather general. For example, we can assume that these spaces are compact subsets of a separable metric space. However, for the sake of simplicity, X and U are assumed to be finite intervals on the real axis. In this case *mes* $X = \mu$ denotes the length of the interval X. In addition, we suppose that the conditional distributions of the process have densities $p\big(x|x^t, u^t\big)$ which satisfy the following condition

$$0 < \delta_1 \leq p\big(x|x^t, u^t\big) \leq \delta_2 < \infty$$

for all x, x^t, u^t. We assume also that $\delta_2 \mu > 1$. This condition implies the ergodicity property.

8.2. Some Properties of Stationary Processes

Here we give some results on dependence and ergodicity of ordinary (without control) stationary processes specified by conditional densities $p(x|x^t)$ which are invariant with respect to shifts in time. These densities are assumed to satisfy the inequality stated at the end of the previous section.

---

[a]The symbol l.i.m. denotes the limit in the mean square sense, i.e.

$$\lim_{T \to \infty} \mathbf{E}\left(T^{-1} \sum_{t=1}^{T} f(x_t) - W(\sigma)\right)^2 = 0.$$

First, we study the dependence of probabilistic properties of the process on past history for the last $k + 1$ steps. We introduce the following notation:

$$\varepsilon_k = \sup_{x^0, \tilde{x}^{-k-1}} \left| p\left(x_1 | x^0_{-k}, x^{-k-1}\right) - p\left(x_1 | x^0_{-k}, \tilde{x}^{-k-1}\right) \right|,$$

$$A_k = \sup_{x_t, x^0, \tilde{x}^{-k-1}} \left| p\left(x_t | x^0_{-k}, x^{-k-1}\right) - p\left(x_t | x^0_{-k}, \tilde{x}^{-k-1}\right) \right|$$

where $x^n_m \stackrel{\text{def}}{=} (x_m, x_{m+1}, \ldots, x_n)$, $m < n$.

Lemma 1. *If* $\sum_{k=1}^{\infty} \varepsilon_k < \infty$, *then* $A_k \to 0$ *as* $k \to \infty$.

Proof. Let

$$a_{t,k} = \sup_{x_t, x^0, \tilde{x}^{-k-1}} \left| p\left(x_t | x^0_{-k}, x^{-k-1}\right) - p\left(x_t | x^0_{-k}, \tilde{x}^{-k-1}\right) \right|.$$

Then $A_k = \sup_t a_{t,k}$ and from the estimate

$$p(x_t | x^1) = \int_{X^{t-2}} p(x_t | x^{t-1}) p(x_{t-1} | x^{t-2}) \cdots p(x_2 | x^1) \, dx_{t-1} \cdots dx_2$$

$$\leq \delta_2 \int_{X^{t-2}} p(x_{t-1} | x^{t-2}) p(x_{t-2} | x^{t-3}) \cdots p(x_2 | x^1) \, dx_{t-2} \cdots dx_2 = \delta_2,$$

we obtain

$$a_{t,k} = \sup_{x_t, x^0, \tilde{x}^{-k-1}} \left| \int_X p(x_t | x^1_{-k}, x^{-k-1}) p(x_1 | x^0_{-k}, \tilde{x}^{-k-1}) \, dx_1 \right.$$

$$\left. - \int_X p(x_t | x^1_k, \tilde{x}^{-k-1}) p(x_1 | x^0_{-k}, \tilde{x}^{-k-1}) \, dx_1 \right|$$

$$\leq \sup_{x^0, \tilde{x}^{-k-1}} \left[\int_X p(x_t | x^1_{-k}, x^{-k-1}) | p(x_1 | x^0) - p(x_1 | x^0_{-k}, \tilde{x}^{-k-1}) | \, dx_1 \right.$$

$$\left. + \int_X p(x_1 | x^0_{-k}, \tilde{x}^{-k-1}) | p(x_t | x^1) - p(x_t | x^0_{-k}, \tilde{x}^{-k-1}) | \, dx_1 \right]$$

$$\leq \sup_{x^0, \tilde{x}^{-k-1}} \int_X [p(x_t | x^1) \varepsilon_k + p(x_1 | x^0_{-k}, \tilde{x}_{-k-1} a_{t-1,k-1}] \, dx_1$$

$$\leq \delta_2 \mu \varepsilon_k + a_{t-1,k-1}.$$

Iterating the inequality obtained we get $(t < k)$

$$a_{t,k} \leq \delta_2 \mu (\varepsilon_k + \varepsilon_{k+1} + \cdots + \varepsilon_{k+t-2}) + a_{1,k+t-1}$$

$$= \delta_2 \mu (\varepsilon_k + \varepsilon_{k+1} + \cdots + \varepsilon_{k+t-2}) + \varepsilon_{k+t-1} \leq \delta_2 \mu \sum_{i=k}^{k+t-1} \varepsilon_i.$$

Thus $A_k = \sup_t a_{t,k} \leq \delta \mu \sum_{i=k}^{\infty} \varepsilon_i \to 0$ as $k \to \infty$. $\qquad \square$

Lemma 2. *If* $\lim_{k\to\infty} \varepsilon_k(\delta_1\mu)^{-k} = 0$ *then for every* x^0 *we have*

$$\lim_{t\to\infty} p(x_t = x|x^0) = p(x),$$

where the limiting density does not depend on x^0.

The proof is carried out by estimating the difference

$$M_t(x) - m_t(x)$$

in a standard way, where $M_t(x) = \sup_{x_0} p(x_t = x|x_0)$ and $m_t(x) = \inf_{x^0} p(x_t = x|x^0)$ are non-increasing and non-decreasing functions of the argument t at every x respectively.

Corollary 1. *There exists a function* $\varphi(t)$ *vanishing as* $t \to \infty$ *such that*

$$|p(x_t = x|x^0) - p(x)| \le \varphi(t).$$

Well-known methods of random processes theory enable us to obtain the following two results.

Lemma 3. *Under the condition of Lemma 2 there exists*

$$\lim_{T\to\infty} \mathbf{E}(f(x_T)|x^0) = \int_X f(x)p(x)\,dx$$

which does not depend on x^0. *The convergence rate is evaluated as follows*

$$\left|\mathbf{E}(f(x_T)|x^0) - \int_X f(x)p(x)\,dx\right| \le \varphi(T)\int_X |f(x)|\,dx.$$

Lemma 4. *Under the condition of Lemma 2*

$$\operatorname*{l.i.m.}_{T\to\infty} T^{-1}\sum_{t=1}^{T} f(x_t) = \int_X f(x)p(x)\,dx$$

with the following estimate of the convergence rate

$$\mathbf{E}\left[T^{-1}\sum_{t=1}^{t} f(x_n) - \int_X f(x)p(x)\,dx\right]^2 \le \Psi(T) = 4FT^{-1}\left(1 + \mu\sum_{i=1}^{T}\varphi(i)\right)$$

where $F = \sup|f(x)|$, $\lim_{T\to\infty}\Psi(T) = 0$.

8.3. Auxiliary Results for CSP

The main goal of this section is to extend the results of the previous section to controlled stationary processes. Because the calculations are lengthy we shall omit the proofs of some simple lemmas below.

Here the parameters ε_k and A_k are defined as follows

$$\varepsilon_k = \sup \left| p\big(x_1|x^0, u^0\big) - p\big(x_1|x^0_{-k}, \tilde{x}^{-k-1}, u^0_{-k}, \tilde{u}^{-k-1}\big) \right|,$$
$$A_k = \sup \left| p\big(x_t|x^0, u^0\big) - p\big(x_t|x^0_{-k}, \tilde{x}^{-k-1}, u^0_{-k}, \tilde{u}^{-k-1}\big) \right|,$$

where both suprema are taken over all variables involved in these notations. For the process $\zeta_t(\sigma)$ with the stationary strategy generated by a rule h having memory depth m, i.e. $h \in D_m$, the parameters $\varepsilon_{k,m}, A_{k,m}$ and the lower and upper estimates $\delta_{1,m}, \delta_{2,m}$ for the densities are defined by analogy with the above.

Lemma 1. *For all integers k and m*

$$\varepsilon_{k,m} \leq \varepsilon_{k-m}, \quad \delta_{1,m} \geq \delta_1, \quad \delta_{2,m} \leq \delta_2.$$

Lemma 2. *If for a CSP ζ_t the relation $\lim_{k\to\infty} \varepsilon_k(\delta_1\mu)^{-k} = 0$ holds, then for $\zeta_t(\sigma)$, $\sigma \in D_m$, we have the following*

$$\lim_{k\to\infty} \varepsilon_{k,m}(\delta_{1,m}\mu)^{-k} = 0.$$

From this assertion and Lemmas 2–4 of the previous section three lemmas given below follow. In their statements p_σ means the density corresponding to the stationary strategy σ.

Lemma 3. *For every integer m there exists a monotonically decreasing positive function $\varphi_m(t)$ vanishing as $t \to \infty$ such that*

$$|p_\sigma(x_t = x|x^0, u^0) - p_\sigma(x)| \leq \varphi_m(t), \quad \sigma \in D_m.$$

The function $\varphi_m(t)$ is constructed using the known parameters μ, ε_k, δ_1, δ_2. It will be the same for all CSP provided the parameters mentioned are fixed.

Lemma 4. *If $\lim_{k\to\infty} \varepsilon_k(\delta_1\mu)^{-k} = 0$, then for $\sigma \in D_m$ the limit*

$$\lim_{t\to\infty} \mathbf{E}_\sigma(f(x_t|x^0, u^0)) = \int_X f(x)p_\sigma(x)\,dx = W(x)$$

exists and does not depend on (x^0, u^0). Moreover,

$$\left| \mathbf{E}_\sigma(f(x_T|x^0, u^0)) - \int_X f(x)p_\sigma(x)\,dx \right| \leq \varphi_m(T) \int_X |f(x)|\,dx.$$

Lemma 5. *If $\lim_{k\to\infty} \varepsilon_k(\delta_1\mu)^{-k} = 0$, then for $\sigma \in D_m$ the limit*

$$\mathrm{l.i.m.}_{T\to\infty}\, T^{-1} \sum_{t=1}^{T} f(\zeta_t(\sigma)) = \int_X f(x)p_\sigma(x)\,dx$$

exists and does not depend on (x^0, u^0). Moreover,

$$\mathbf{E}\left[T^{-1} \sum_{t=1}^{T} f(\zeta_t(\sigma)) - \int_X f(x)p_\sigma(x)\,dx \right]^2 \leq \Psi_m(T) = 4F^2 T^{-1}\left(1 + \mu \sum_{i=1}^{T} \varphi_m(i) \right).$$

The form of the function Ψ_m shows that it is the same for all CSP with fixed values μ, δ_1, δ_2.

So, we have obtained sufficient conditions for a CSP to be ergodic.

Let D'_m be a finite subset of D_m. For a stationary strategy $\sigma \in D'_m$ we put $z_\sigma^{(n)} = n^{-1} \sum_{t=1}^{n} f(\zeta_t(\sigma))$.

Definition 1. A strategy $\sigma_m^0 \in D'_m$ is called the *best strategy* if

$$W(\sigma_m^0) = \sup_{\sigma \in D'_m} W(\sigma).$$

Let $\tilde{\sigma}$ be a (random!) strategy to which the maximum value of $z_\sigma^{(n)}$ corresponds. The following lemma is concerned with the "distance" between σ_m^0 and $\tilde{\sigma}$.

Lemma 6. *For every past history* (x^0, u^0) *the estimate*

$$\mathbf{E}([W(\tilde{\sigma}) - W(\sigma_m^0)]^2 | x^0, u^0) \leq 4|D'_m|\Psi(n)$$

holds.

We consider the non-decreasing numerical sequence $W_m = \sup_{\sigma \in D_m} W(\sigma)$. Hence

$$\bar{W} = \sup_{D} W(\sigma) = \lim_{m \to \infty} W_m.$$

We introduce auxiliary sets of rules (corresponding stationary strategies) D_m^l, $l = 1, 2, \ldots$ with finite depths. We partition the sets X and U into intervals of equal length. Let $x^{(0)}$, $x^{(1)}, \ldots, x^{(l)}$ and $u^{(0)}$, $u^{(1)}, \ldots, u^{(l)}$ be the points of these partitions. The set D_m^l is formed by picewise-constant functions $f(x_1, \ldots, x_m)$ on the hypercube X^m which on every element of the partition with the boundaries $\left[x_1^{(i_1)}, x_1^{(i_1+1)}\right]$, $\left[x_2^{(i_2)}, x_2^{(i_2+1)}\right], \ldots, \left[x_m^{(i_m)}, x_m^{(i_m+1)}\right]$ take one of the values u^i belonging to the partition of the interval U into l parts. Any measurable function $h(x_1, \ldots, x_m)$ can be approximated by the picewice-constant functions $\tilde{h}(x_1, \ldots, x_m)$. From this it follows that for arbitrary positive numbers η_1, η_2 and any rule $h \in D_m$ we can find a rule $\tilde{h} \in D_m^l$ (at some l) such that $|h - \tilde{h}_l| < \eta_1$ for all collections of $x_1, \ldots, x_m$ where $x_i \in X \backslash X_{\eta_2}$, and the subset $X_{\eta_2} \subset X$ has measure not greater than η_2.

Let the transition function density $p\left(x_1 | x^0_{-k+1}\right)$ of a Markov process of order k[b] satisfying the condition $0 < \delta_1 \leq p(\cdot) \leq \delta_2 < \infty$. The same inequalities are assumed to hold for the densities $p'\left(x_1 | x^0_{-k+1}\right)$, $\tilde{p}\left(x_1 | x^0_{-k+1}\right)$ and, moreover,

$$|p(x_1 | x^0_{-k+1}) - p'(x_1 | x^0_{-k+1})| \leq \eta_1$$

for all $x_1, \ldots, x_m$ and

$$p'(x_1 | x^0_{-k+1}) = \tilde{p}(x_1 | x^0_{-k+1})$$

for all $x_1 \in X$, $x_i \in X \backslash X_{\eta_2}$, $i = -k+1, \ldots, -1, 0$.

---

[b]This means that $\mathbf{P}\{x_{n+1} \in \cdot | x_0, \ldots, x_n\} = \mathbf{P}\{x_{n+1} \in \cdot | x_{n-k}, \ldots, x_n\}$, $n \geq k$.

The Markov processes (of order k) with such densities have the densities of the limiting distributions $p(x)$, $p'(x)$ and $\tilde{p}(x)$. We estimate the distance between p and $\tilde{p}$ in metric L_1. We need the three lemmas formulated below.

Lemma 7.

$$\|p - \tilde{p}\|_{L_1} \leq k\eta_1 \frac{\delta_2^{k-1}}{\delta_1^k}.$$

Proof. Let us consider the limiting distributions for the collections x_1^k. We introduce the following two integral operators

$$Tf = \int_X f(x_1, \ldots, x_k) p\big(x_{k+1}^{2k} | x_1^k\big) \, dx_1^k,$$

$$T'f = \int_X f(x_1, \ldots, x_k) p'\big(x_{k+1}^{2k} | x_1^k\big) \, dx_1^k$$

with the kernels $p\big(x_{k+1}^{2k} | x_1^k\big)$ and $p'\big(x_{k+1}^{2k} | x_1^k\big)$. The eigenfunctions $p\big(x_{k+1}^{2k}\big)$ and $p'\big(x_{k+1}^{2k}\big)$ corresponding to their eigenvalues are equal to one. According to the triangle inequality we have $\big($here $L_1^{(k)} = L_1(\mathrm{R}^k)\big)$

$$I \overset{\text{def}}{=} \|Tp - T'p'\|_{L_1^{(k)}} \leq \|Tp - T'p\|_{L_1^{(k)}} + \|T'p - T'p'\|_{L_1^{(k)}}.$$

First, we estimate the first summand on the right-hand side of the last inequality. Indeed

$$\|Tp - T'p\|_{L_1^{(k)}} \leq \int_X^k \int_X^k |p\big(x_{k+1}^{2k} | x_1^k\big) - p'\big(x_{k+1}^{2k} | x_1^k\big)| p\big(x_1^k\big) \, dx_1^k \, dx_k^{2k}.$$

Taking into account the equality

$$p\big(x_{k+1}^{2k} | x_1^k\big) = p\big(x_{2k} | x_1^{2k-1}\big) p\big(x_{2k-1} | x_1^{2k-2}\big) \cdots p\big(x_{k+1} | x_1^k\big),$$

similar equalities for p' and the elementary inequality

$$\left| \prod_{i=1}^k a_i - \prod_{i=1}^k b_i \right| \leq |a_1 - b_1| \prod_{i=2}^k a_i + b_1 |a_2 - b_2| \prod_{i=3}^k a_i + \prod_{i=1}^{k-1} b_i |a_k - b_k|,$$

we obtain

$$|p\big(x_{k+1}^{2k} | x_1^k\big) - p'\big(x_{k+1}^{2k} | x_1^k\big)| \leq k\eta_1 \delta_2^{k-1}.$$

Finally we have

$$\|Tp - T'p\|_{L_1^{(k)}} \leq k\eta_1 \delta_2^{k-1} \int_{X^k} \int_{X^k} p\big(x_1^k\big) \, dx_1^k \, dx_k^{2k} = k\eta_1 \delta_2^{k-1} \mu^k.$$

We now consider the second summand. Since

$$p'_{\delta_1}\big(x_{k+1}^{2k} | x_1^k\big) = p'\big(x_{k+1}^{2k} | x_1^k\big) - \delta_1^k \geq 0,$$

we obtain

$$\|T'p - T'p'\|_{L_1^{(k)}} \le \int_{X^k} \left| \int_{X^k} \left(p(x_1^k) - p'(x_1^k) \right) p'\left(x_{k+1}^{2k} \big| x_1^k \right) dx_1^k \right| dx_{k+1}^{2k}$$

$$= \int_{X^k} \left| \int_{X^k} p'_{\delta_1}\left(x_{k+1}^{2k} \big| x_1^k \right) \left(p(x_1^k) - p'(x_1^k) \right) dx_1^k \right| dx_{k+1}^{2k}$$

$$\le \int_{X^k} \left[\int_{X^k} p'_{\delta_1}\left(x_{k+1}^{2k} \big| x_1^k \right) dx_{k+1}^{2k} \right] \left| p(x_1^k) - p'(x_1^k) \right| dx_1^k$$

$$= \left(1 - (\mu \delta_1)^k \right) \|p - p'\|_{L_1^{(k)}}.$$

It follows immediately that

$$I = \|p - p'\|_{L_1^{(k)}} \le k \eta_1 \mu^k \delta_1^{k-1} + \left(1 - (\mu \delta_1)^k \right) \|p - p'\|_{L_1^{(k)}},$$

and, hence,

$$\|p - p'\|_{L_1^{(k)}} \le k \eta_1 \delta_1^{-k} \delta_2^{k-1}.$$

The one-dimensional limiting density that we are interested in is equal to $p(x_k) = \int_X^{k-1} p(x_1^k) \, dx_1^{k-1}$. From the inequality

$$\|p - p'\|_{L_1} = \int_X \left| \int_{X^{k-1}} [p(x_1^k) - p'(x_1^k)] \, dx_1^{k-1} \right| dx_k$$

$$\le \int_{X^k} \left| p(x_1^k) - p'(x_1^k) \right| dx_1^k = \|p - p'\|_{L_1^{(k)}}$$

the assertion follows. $\square$

Lemma 8.

$$\|p' - \tilde{p}\|_{L_1} \le (2k - 1)\eta_2 \mu^{k-1} \delta_1^{-k} \delta_2^{2k}.$$

Proof. For the densities $p'(x_1^k)$ and $\tilde{p}(x_1^k)$ we consider the following equalities

$$\tilde{T}\tilde{p} = \tilde{p}, \quad T'p' = p',$$

where $\tilde{T}$ and T' denote the integral operators with the kernels $\tilde{p}\left(x_{k+1}^{2k} \big| x_1^k \right)$ and $p'\left(x_{k+1}^{2k} \big| x_1^k \right)$ respectively. We have

$$I \overset{\text{def}}{=} \|p' - \tilde{p}\|_{L_1^{(k)}} = \|T'p' - \tilde{T}\tilde{p}\|_{L_1^{(k)}} \le \|T'p' - \tilde{T}\tilde{p}\|_{L_1^{(k)}}$$

$$\le \|T'p' - \tilde{T}p'\|_{L_1^{(k)}} + \|\tilde{T}p' - \tilde{T}\tilde{p}\|_{L_1^{(k)}}. \tag{1}$$

For the first summand on the right-hand side we have

$$\|T'p' - \tilde{T}p'\|_{L_1^{(k)}} \le \int_{X^k} \int_{X^k} \left| p'\left(x_{k+1}^{2k} \big| x_1^k \right) - \tilde{p}\left(x_{k+1}^{2k} \big| x_1^k \right) \right| p'(x_1^k) \, dx_1^k \, dx_{k+1}^{2k}.$$

According to the definitions of the densities p' and $\tilde{p}$

$$p'\left(x_{k+1}^{2k}\big|x_1^k\right) = \tilde{p}\left(x_{k+1}^{2k}\big|x_1^k\right), \quad i = 1, \ldots, 2k - 1 \tag{2}$$

if $x_i \in X \setminus X_{\eta_2}$ for $i = 1, \ldots, 2k - 1$. The measure of the set of all collections x_1^{2k} for which equality (2) fails is equal to

$$1 - (1 - \mu^{-1}\eta_2)^{2k-1}\mu^{2k} \le (2k - 1)\eta_2\mu^{2k-1}.$$

Therefore

$$\|T'p' - \tilde{T}\tilde{p}\|_{L_1^{(k)}} \le (2k - 1)\eta_2\mu^{2k-1}\delta_2^{2k}.$$

The second summand on the right-hand side of (1) can be estimated as in the previous lemma, and hence

$$\|\tilde{T}p' - \tilde{T}\tilde{p}\|_{L_1^{(k)}} \le (1 - (\mu\delta_1)^k)\|p' - \tilde{p}\|_{L_1^{(k)}}.$$

This leads to the inequality

$$I = \|p' - \tilde{p}\|_{L_1^{(k)}} \le (2k - 1)\eta_2\mu^{2k-1}\delta_2^{2k} + (1 - (\mu\delta_1)^k)\|p' - \tilde{p}\|_{L_1^{(k)}}$$

that, in turn, implies the following

$$\|p' - \tilde{p}\|_{L_1^{(k)}} \le (2k - 1)\eta_2\frac{\mu^{k-1}\delta_2^{2k}}{\delta_1^k}.$$

The desired estimate of $\|p' - \tilde{p}\|_{L_1}$ is obtained as at the end of Lemma 2. $\qquad\square$

Using the triangle inequality, from Lemmas 7, 8 we have

Lemma 9.

$$\|p - \tilde{p}\|_{L_1} \le (2k - 1)k\eta_1\frac{\delta_2^{k-1}}{\delta_1^k} + (2k - 1)\eta_2\frac{\delta_2^{2k}\mu^{k-1}}{\delta_1^k}.$$

We now compare the density of the conditional distribution $p(x_1|x^0)$ of a stationary process with that of $p\left(x_1|x_{-k+1}^0\right)$ of a Markov process of order k satisfying the condition

$$\inf_{x^{-k}} p(x_1|x_{-k+1}^0, x^{-k}) \le p(x_1|x_{-k+1}^0) \le \sup_{x^{-k}} p(x_1|x_{-k+1}^0, x^{-k}).$$

Lemma 10. *For every past history* x^0 *the introduced densities are related by the inequalities*

$$|p(x_t|x_{-k+1}^0) - p(x_t|x^0)| \le \mu\delta_2\varepsilon_{k-1}t.$$

Lemma 11. *Let* $p(x)$ *be the density of the limiting distribution of a Markov process of order k with the transition probability density* $p(x_1|x_{-k+1}^0)$. *Then*

$$|p(x_t = x|x_{-k+1}^0) - p(x)| \le c[1 - (\delta_1\mu)^k]^{t/k}$$

where c is some constant not depending on t.

We impose the constraint G on the conditional distribution densities of a CSP ζ_t.

G. For every two sequences of controls u^0 and $\tilde{u}^0$ we have

$$|p(x_1|x^0, u^0) - p(x_1|x^0, \tilde{u}^0)| \le \sum_{n=1}^{\infty} d_n|u_{-n} - \tilde{u}_{-n}|,$$

where $d_n > 0$ and $d = \sum_{n=1}^{\infty} d_n < \infty$.

Theorem 1. *Let $\sigma_m^{l,0} \subset D_m^l$ be the best strategy. If conditions **G** holds and $\lim_{k\to\infty} k\varepsilon_k(\delta_2/\delta_1)^k = 0$, then for every increasing integer-valued sequence $l(m)$ the following equalities are satisfied*

$$\lim_{l\to\infty} W\big(\sigma_m^{l,0}\big) = \lim_{l\to\infty} \max_{\sigma \in D_m^{l(m)}} W(\sigma) = W_m.$$

Proof. For every strategy $\sigma \in D_m$ and numbers η_1, η_2 we find an integer $l = l(m)$ and a strategy $\tilde{\sigma} \in D_m^l$ such that

$$|\sigma(x_{-m}, \ldots, x_{-1}) - \tilde{\sigma}(x_{-m}, \ldots, x_{-1})| \le \eta_1 \quad \forall\, x_i \in X\backslash X_{\eta_2}, \;\; i = 1, \ldots, m.$$

Then for the same $\{x_{-i}\}$

$$|p_\sigma(x_1|x^0, u^0) - p_{\tilde{\sigma}}(x_1|x^0, \tilde{u}^0)| \le \sum_{n=1}^{\infty} d_n|u_{-n} - \tilde{u}_{-n}| \le d\eta_1.$$

We associate with the stationary processes $\zeta_t(\sigma)$ and $\zeta_t(\tilde{\sigma})$ the Markov processes of order k having the densities $p_\sigma^{(k)}(x_1|x_{-k+1}^0)$ and $p_{\tilde{\sigma}}^{(k)}(x_1|x_{-k+1}^0)$ satisfying the conditions

$$\inf_{x_1, x_{-k+1}^0} p_\sigma(x_1|x_{-k+1}^0) \le p_\sigma^{(k)}(x_1|x_{-k+1}^0) \le \sup_{x_1, x_{-k+1}^0} p_\sigma(x_1|x_{-k+1}^0),$$

$$\inf_{x_1, x_{-k+1}^0} p_{\tilde{\sigma}}(x_1|x_{-k+1}^0) \le p_{\tilde{\sigma}}^{(k)}(x_1|x_{-k+1}^0) \le \sup_{x_1, x_{-k+1}^0} p_{\tilde{\sigma}}(x_1|x_{-k+1}^0),$$

and to the obvious inequalities

$$|p_\sigma^{(k)}(x_1|x_{-k+1}^0) - p_\sigma(x_1|x^0)| \le \varepsilon_{k-m-1},$$

$$|p_{\tilde{\sigma}}^{(k)}(x_1|x_{-k+1}^0) - p_{\tilde{\sigma}}(x_1|x^0)| \le \varepsilon_{k-m-1}.$$

Under these assumptions the limiting densities exist. The following inequality holds

$$\|p_\sigma(x_\infty|x^0) - p_{\tilde{\sigma}}(x_\infty|x^0)\|_{L_1}$$

$$\le \|p_\sigma(x_\infty|x^0) - p_\sigma^{(k)}(x_t|x^0)\| + \|p_\sigma(x_t|x^0) - p_\sigma^{(k)}(x_t|x_{-k+1}^0)\|$$

$$+ \|p_\sigma^{(k)}(x_t|x_{-k+1}^0) - p_\sigma^{(k)}(x_\infty|x_{-k+1}^0)\| + \|p_\sigma^{(k)}(x_\infty|x_{-k+1}^0) - p_{\tilde{\sigma}}^{(k)}(x_\infty|x_{-k+1}^0)\|$$

$$+ \|p_{\tilde{\sigma}}^{(k)}(x_\infty|x_{-k+1}^0) - p_{\tilde{\sigma}}^{(k)}(x_t|x_{-k+1}^0)\| + \|p_{\tilde{\sigma}}^{(k)}(x_t|x_{-k+1}^0) - p_{\tilde{\sigma}}(x_t|x^0)\|$$

$$+ \|p_{\tilde{\sigma}}(x_t|x^0) - p_\sigma(x_\infty|x^0)\|. \tag{3}$$

We estimate each term on the right-hand side of this inequality. In view of condition **G** we have

$$|p_\sigma^{(k)}(x_1|x_{-k+1}^0) - p_{\tilde\sigma}^{(k)}(x_1|x_{-k+1}^0)|$$
$$\leq |p_\sigma^{(k)}(x_1|x_{-k+1}^0) - p_{\tilde\sigma}^{(k)}(x_1|x^0)| + |p_\sigma^{(k)}(x_1|x^0) - p_\sigma(x_1|x_{-k+1}^0)|$$
$$+ |p_{\tilde\sigma}(x_1|x^0) - p_\sigma^{(k)}(x_1|x_{-k+1}^0)| \leq 2\varepsilon_{k-m-1} + d\eta_1.$$

Taking into account Lemma 9 we estimate the forth term on the right-hand side of (3) as follows

$$\|p_{\tilde\sigma}^{(k)}(x_\infty|x_{-k+1}^0) - p_{\tilde\sigma}^{(k)}(x_\infty|x_{-k+1}^0)\|$$
$$\leq k\delta_1^{1-k}(d\eta_1 + \varepsilon_{k-m-1})\delta_2^{k-1} + (2k-1)\eta_2\delta_1^{-k}\mu^{k-1}\delta_2^{2k}.$$

With the help of Lemmas 3, 10, 11 we can estimate the other terms on the right-hand side of (3). Indeed,

$$\|p_\sigma(x_\infty|x^0) - p_{\tilde\sigma}(x_\infty|x^0)\|_{L_1}$$
$$\leq 2\mu\varphi_m(t) + 2\delta_2\mu^2\varepsilon_{k-m} + 2\mu c(1-(\mu\delta_1)^k)^{t/k} + k\delta_2^{k-1}(d\eta_1 + \varepsilon_{k-m-1})\delta_1^{1-k}$$
$$+ (2k-1)\eta_2\mu^{k-1}\delta_2^{2k}\delta_1^{-k}. \tag{4}$$

We show that the right-hand side of the last inequality can take an arbitrarily small value by choosing t, k, η_1, η_2 appropriately. The equality $\lim_{k\to\infty} k\varepsilon_k\delta^{-1}\delta_2^k = 0$ is a sufficient condition for this. Indeed, the function

$$g(k,t) = 2\delta_2\mu\varepsilon_{k-m} + 2\mu c(1-(\mu\delta_1)^k)^{t/k},$$

being the sum of the second and third summands from (3) takes the minimal value in t at the point $t_{\min}^{(k)}$ and under the above condition we have

$$\lim_{k\to\infty} g(k, t_{\min}^{(k)}) = 0, \quad \lim_{t\to\infty} t_{\min}^{(k)} = \infty.$$

Thus, the sum $g(k,t)$ can be made however small by putting $t = t_{\min}^{(k)}$ and $k \to \infty$. The first term on the right-hand side of (4) can be made however small with $g(k,t)$ simultaneously. Finally, the remaining terms can be made small by choosing η_1 and η_2 properly.

So, we have proved that

$$\|p_\sigma(x_0|x^0) - p_{\tilde\sigma}(x_\infty|x^0)\| \to 0$$

as $l = l(m) \to 0$. The rest of the proof is rather simple. First, we note that

$$|W(\sigma) - W(\tilde\sigma)| = \left|\int_X f(x_\infty)(p_\sigma(x_\infty|x^0) - p_{\tilde\sigma}(x_\infty|x^0))\,dx_\infty\right|$$
$$\leq \int_X \|f(x_\infty)\|\|p_\sigma(x_\infty|x^0) - p_{\tilde\sigma}(x_\infty|x^0)\|\,dx_\infty \leq F\|p_\sigma - p_{\tilde\sigma}\|_{L_1}.$$

Hence $\lim_{l\to\infty} W(\tilde\sigma) = W(\sigma)$. Using the notation $W_m^l = \max_{\sigma\in D_m^{l(m)}} W(\sigma)$, from the inequality

$$|W_m - W_m^l| \leq |W_m - W(\sigma)| + |W(\sigma) - W(\tilde\sigma)| \quad \text{for all } \sigma \in D_m$$

we finally obtain

$$\lim_{l \to \infty} |W_m - W_m^l| \leq |W_m - W(\sigma)|, \quad \sigma \in D_m.$$

Therefore $\lim_{l \to \infty} W_m^l = W_m$. $\qquad\qquad\qquad\qquad\qquad\qquad\qquad\square$

8.4. Adaptive Strategies for CSP

First, the control aim is the fulfilment of the objective equality

$$\underset{T \to \infty}{\text{l.i.m}} \, T^{-1} \sum_{t=1}^{T} f(x_t) = \bar{W}$$

for all processes from a class of CSP. The achievement of this aim is realized by a strategy $A(n, N)$ specified as follows. The process of control is divided into successive stages. In the mth stage, the rules having the same memory depth from the set D_m are used. Each of them is applied to the process n_m times in succession. We consider the r.v.

$$Z = n_m^{-1} \sum_{t=t'+1}^{t'+n_m} f(x_t),$$

where $t' + 1$ is the first moment of using the chosen rule by the reactions of the process in response to the applied rules from D_m. On finishing using all rules from D_m, the rule that produces the maximum value of Z is chosen. The rule found is applied N_m times. Thereafter, stage $m + 1$ begins. This completes the description of the strategy $A(n, N)$ that is defined by the two sequences $\{n_m\}$, $\{N_m\}$ given beforehand.

We define the classes of CSP denoted by $\mathcal{S}(\varepsilon, \delta_1, \delta_2)$ as the class of all CSP such that: the parameters ε_1', ε_2', ... and the conditional probability densities satisfy the inequalities:

(i) $\varepsilon_k' \leq \varepsilon_k$, $k \geq 1$,
(ii) $0 < \delta_1 \leq p(x_1|x^0) \leq \delta_2 < \infty$.

Moreover,

(iii) $\lim_{k \to \infty} k \varepsilon_k \delta_1^{-k} \delta_2^k = 0$

and condition **G** from Sec. 3 holds as well.

All processes from $\mathcal{S}(\varepsilon, \delta_1, \delta_2)$ have the same collection of functions Ψ_m.

Theorem 1. *With respect to the class $\mathcal{S}(\varepsilon, \delta_1, \delta_2)$ the strategy $A(n, N)$ secures the attainment of the stated aim*

$$\underset{T \to \infty}{\text{l.i.m}} \, T^{-1} \sum_{t=1}^{T} f(x_t) = \bar{W}$$

if the following conditions hold

$$\lim_{m\to\infty} \left|D_m^{l(m)}\right| \Psi_m(n_m) = 0, \quad \lim_{m\to\infty} \frac{\left|D_m^{l(m)}\right| n_m}{N_{m-1}} = 0,$$

where $l(m)$ is a subsequence of the positive integers.

Proof. For r.v. η we introduce the norm $\|\eta\| = (\mathbf{E}(\eta^2|x^0, u^0))^{1/2}$. We also introduce τ_m' and τ_m'' which are the moments of finishing the mth stage of using the strategy and comparison period of "usefulness" of the applied rules with the same depth m, respectively, i.e.

$$\tau_m' = \sum_{i=1}^{m}\left(\left|D_i^{l(i)}\right|n_i + N_i\right), \quad \tau_m'' = \tau_{m-1}' + \left|D_m^{l(m)}\right|n_m.$$

We estimate the norm $J(T) = \|T^{-1}\sum_{t=1}^{T} f(x_t) - \bar{W}\|$. Indeed,

$$TJ(T) = \left\|\sum_{t=1}^{T} f(x_t) - \bar{W}\right\| \le \sum_{m=1}^{q}\left\|\sum_{i=\tau_m''+1}^{\tau_m'}(f(x_i) - \bar{W})\right\|$$

$$+ \sum_{m=1}^{q}\left\|\sum_{i=\tau_{m-1}''+1}^{\tau_m'}(f(x_i) - \bar{W})\right\| + \left\|\sum_{i=\tau_q'+1}^{\min(T,\tau_{q+1}'')}(f(x_i) - \bar{W})\right\|$$

$$+ \left\|\sum_{i=\min(T,\tau_q''+1)}^{T}(f(x_i) - \bar{W})\right\| = S_1 + S_2 + S_3 + S_4 \tag{1}$$

where $q = \max\{n : \tau_n' \le T\}$. We now estimate every term on the right-hand side. We write the first of them as $S_1 = \sum_{m=1}^{q} S_{1,m}$, where

$$S_{1,m} = \left\|\sum_{i=\tau_m''+1}^{\tau_m'}(f(x_i) - \bar{W})\right\| \le \left\|\sum_{i=\tau_m''+1}^{\tau_m'}(f(x_i) - \bar{W}(\tilde{\sigma}_m))\right\|$$

$$+ \left\|\sum_{i=\tau_m''+1}^{\tau_m'}(W(\tilde{\sigma}_m) - W(\sigma_m^{(l(m))}))\right\| + \left\|\sum_{i=\tau_m''+1}^{\tau_m'}(W(\sigma_m^{l(m)}) - \bar{W})\right\|.$$

Here $\tilde{\sigma}_m$ denotes the stationary strategy having depth m corresponding to the maximun value of the r.v. Z. Hence $W(\tilde{\sigma}_m)$ is the average reward on the time interval $[\tau_m'' + 1, \tau_m']$. So, we have

$$S_{1,m} \le \left\|\sum_{i=\tau_m''+1}^{\tau_m'}(f(x_i) - \bar{W}(\tilde{\sigma}_m))\right\| + \left\|W(\tilde{\sigma}_m) - W(\sigma_m^{(l(m))})\right\|$$

$$+ N_m\left\|W(\sigma_m^{l(m)}) - \bar{W}\right\|.$$

By Lemma 5 from Sec. 3 for any past history x^0, u^0 the inequality

$$\mathbf{E}\left(\left[N_m^{-1}\sum_{i=\tau_m''+1}^{\tau_m'}(f(x_i)-\bar{W}(\tilde{\sigma}_m))\right]^2\Big|x^0,u^0\right)\le\Psi_m(N_m)$$

is satisfied, i.e.

$$\left\|\sum_{i=\tau_m''+1}^{\tau_m'}(f(x_i)-\bar{W}(\tilde{\sigma}_m))^2\right\|\le N_m\Psi_m^{1/2}(N_m).$$

Using Lemma 6, we obtain

$$S_{1,m}\le N_m\Psi_m^{1/2}(N_m)+2N_m[|D_m^{l(m)}|\Psi_m(n_m)]^{1/2}+N_m|W(\sigma_m^{l(m)})-\bar{W}|.$$

For the other terms in (1) the following estimates can be obtained without special difficulties

$$S_2\le 2n_m|D_m^{l(m)}|F,\quad(F=\max_x|f(x)|)$$

$$S_3=\left\|\sum_{i=\tau_q'+1}^{\min(T,\tau_q''+1)}(f(x_i)-\bar{W})\right\|\le 2n_{q+1}\big|D_{q+1}^{l(q+1)}\big|F,$$

$$S_4=\left\|\sum_{i=\tau_q''+1}^{T}(f(x_i)-\bar{W})\right\|\le 2(T-\tau_{q+1}')\left[\big|D_{q+1}^{l(q+1)}\big|\Psi_{q+1}(n_{q+1})\right]^{1/2}$$

$$+(T-\tau_{q+1}'')\Psi_{q+1}^{1/2}(T-\tau_{q+1}'')+(t-\tau_{q+1}'')\big|W\big(\sigma_{q+1}^{l(q+1)}\big)-\bar{W}\big|.$$

We notice that if $\tau_{q+1}''>T$ then $S_4=0$. Substituting the estimates obtained in (1), we have

$$J(T)\le T^{-1}\left\{\sum_{m=1}^{q}\left[N_m\Psi_m^{1/2}(N_m)+2N_m(|D_m^{l(m)}|\Psi_m(n_m))^{1/2}\right.\right.$$

$$+N_m|W(\sigma_m^{l(m)})-\bar{W}|\Big]+2F\sum_{m=1}^{q+1}n_m|D_m^{l(m)}|$$

$$+\gamma(T-\tau_{q+1}'')\left[\Psi_{q+1}^{1/2}(T-\tau_{q+1}'')+2\left[\big|D_{q+1}^{l(q+1)}\big|\Psi_{q+1}(n_{q+1})\right]^{1/2}\right.$$

$$\left.\left.+\Big|W\big(\sigma_{q+1}^{l(q+1)}\big)-\bar{W}\Big|\right]\right\}\tag{2}$$

where

$$\gamma=\begin{cases}1,&\text{if }\tau_{q+1}''\le T,\\0,&\text{if }\tau_{q+1}''>T.\end{cases}$$

It remains to make sure that $\lim_{t\to\infty}J(t)=0$.

Let us verify that the first and second sums in the braces (and divided by T) tend to zero as $q\to\infty$.

This follows immediately from the conditions of the theorem, the inequality $T \geq \sum_{m=1}^{q} N_m$, the Toeplitz Theorem (as $x_n \to 0$, see Sec. 3, Chap. 4) and the equality $\lim_{m \to \infty} W(\sigma_m^{l(m)}) = \bar{W}$, where $W(\sigma_m^{l(m)}) = \sup_{\sigma \in D_m^{l(m)}} W(\sigma)$. To verify the last equality we fix an integer r and consider the quantity $|W(\sigma_m^{l(m)}) - \bar{W}|$ for all $m \geq r$. Noting that

$$W(\sigma_m^{l(m)}) \geq W(\sigma_r^{l(m)}), \quad \bar{W} \geq W(\sigma_m^{l(m)})$$

and using Theorem 1, Sec. 3 we see that $\lim_{l(k) \to \infty} W(\sigma_r^{l(m)}) = W_r$. Hence

$$\varlimsup_{m \to \infty} |W(\sigma_m^{l(m)}) - \bar{W}| \leq |\bar{W} - W_r|$$

for any r. Taking into account the equality $\lim_{r \to \infty} W_r = \bar{W}$ we obtain the required assertion. This relationship (in the limit as $q \to \infty$) with respect to the first and second summands on the right-hand side of (3) can obviously be extended to the case when $T \to \infty$. Indeed, in this case $q \to \infty$.

We now show that the other summands on the right-hand side of (2) containing γ tend to 0 as $T \to \infty$. It is easy to verify that the equalities

$$\lim_{T \to \infty} \gamma \frac{T - \tau_{q+1}''}{T} \left| D_{q+1}^{l(q+1)} \right| \Psi_{q+1}(n_{q+1}) = \lim_{T \to \infty} \left| D_{q+1}^{l(q+1)} \right| \Psi_{q+1}(n_{q+1}) = 0, \quad (3)$$

$$\lim_{T \to \infty} \gamma \frac{T - \tau_{q+1}''}{T} \left| W\left(\sigma_{q+1}^{l(q+1)} \right) - \bar{W} \right| = \lim_{T \to \infty} \left| W\left(\sigma_{q+1}^{l(q+1)} \right) - \bar{W} \right| = 0 \quad (4)$$

are fulfilled. Because $\Psi_{q+1}(n) \leq \varkappa$ we obtain the estimate

$$\gamma \frac{T - \tau_{q+1}''}{T} \Psi_{q+1}^{1/2} (T - \tau_{q+1}'') \leq \begin{cases} \dfrac{n_{q+1}}{N_q}, & \text{if } T - \tau_{q+1}'' < n_{q+1}, \\[2mm] \Psi_{q+1}^{1/2}, & \text{if } T - \tau_{q+1}'' \geq n_{q+1}. \end{cases}$$

Since $n_{q+1}/N_q \to 0$ as $t \to 0$, these estimates lead to the equality

$$\lim_{T \to \infty} \gamma \frac{T - \tau_{q+1}''}{T} \Psi_{q+1}^{1/2} (T - \tau_{q+1}'') = 0. \quad (5)$$

It follows that $\lim_{t \to \infty} J(t) = 0$. $\qquad \square$

We now consider another aim of control for the class of CSP considered: *provide the fulfilment of the equality*

$$\lim_{T \to \infty} \mathbf{E}(f(x_T)|x^0, u^0) = \bar{W}$$

for every process from $\mathcal{S}(\varepsilon, \delta_1, \delta_2)$ and for an arbitrary fixed past history x^0, u^0.

We remark that the strategy $A(n, N)$ cannot provide the stated aim. Indeed, at the moment of time far from the moment we started to control, the process with the efficiency of the rules having the same depth is evaluated and nonoptimal rules are used. So, the required limit may not exist. We now describe another strategy denoted by $\tilde{A}(n, p)$.

The structure is similar to the previous one. The course of control is divided into successive stages. In the mth stage rules with the same depth from the set D_m are used. Each of them is applied to the process n_m times in succession. The best of them is used a random number ν_m times. Using the best rule can be stopped at any time with probability p_m independently of the past history. Thereafter, stage $m+1$ begins. This completes the description of the strategy $\tilde{A}(n,p)$.

Theorem 2. *With respect to the class of CSP the strategy $\tilde{A}(n,p)$ secures the attainment of the aim*

$$\lim_{t\to\infty} \mathbf{E} f(x_t|x^0, u^0) = \bar{W}$$

if the parameters specifying the strategy are related by the conditions

$$\sum_{m=1}^{\infty} p_1 p_2 \cdots p_m |D_m^{l(m)}| n_m < \infty, \quad \lim_{m\to\infty} |D_m^{l(m)}| \Psi_m(n_m) = 0,$$

where $l(m)$ is a subsequence of the positive integers.

The proof is similar to that of the previous theorem and hence is omitted.

The strategies considered above are of the searching type. They do not implement the identification which would consist of estimating the densities $p(\cdot|x^0)$. These strategies are neither stationary nor Markov. Their realization is based on the enumeration of the elements of all the sets D_m and on the determination of some probability distributions. The next rules are formed with the help of the observations during the course of the controlled process.

The complexity of the algorithms considered are connected with the generality of the classes of controlled models. In the previous (and next) chapters for special classes of models the strategies possess simpler structure.

CHAPTER 9

FINITE-CONVERGING PROCEDURES FOR CONTROL PROBLEMS WITH INEQUALITIES

In this chapter we consider recurrent procedures for solving countable systems of inequalities. These procedures are conventionally called *finite-converging*. Sufficient conditions for their existence have been obtained. On the base of these procedures adaptive stabilizing strategies are constructed for the classes of linear difference equations. The realization simplicity makes them useful in some practical problems.

9.1. Formulation of the Problem

Let us consider a class L_l of scalar linear difference equations of order l

$$x_t = \sum_{i=1}^{l}(a_i x_{t-i} + b_i u_{t-i}) + f(t), \quad z_t = c x_t + d$$

with some initial conditions. Here x_t, u_t, z_t are a "state", a control and an observation at time t, respectively, and $f(t)$ is an additive noise (or a disturbance). The set of coefficients of the equation $\theta = \{a_1, \ldots, a_l, b_1, \ldots, b_l\}$ defines an element from L_l. For arbitrary initial conditions the solution x_t must belong to an interval given by $|x_t| < r$ starting from some moment. Under the given value of the parameter θ and appropriate conditions concerning the equation the solution is rather simple. For future aims we note the functional form of the "optimal" control

$$u_t = \sum_{j=1}^{m} \alpha_j(\theta) z_{t-j+1},$$

i.e. it is linear with respect to its arguments (the observations z_t) and the coefficients α_j of this form depend on the parameter θ in a known manner.

We now describe the problem outlined above in general form. The control aim of the processes from the class $\mathcal{K}_\Theta$ consists of fulfilling the inequality $\varphi(t) > 0$ starting from some moment t_0, where $\varphi(t)$ is some functional defined on the paths of the considered process. If for the process $x_t(\theta) \in \mathcal{K}_\Theta$ the form of the "optimal" law $u_t = h(\cdot, \theta)$ is known, the described problem will be solved. We consider the adaptive variant when the parameter θ is unknown.

We seek the required "optimal" law in the form $h(z^t, \theta_t)$, where the function h depends on the estimate of an unknown parameter. By observing the evolution of the process and keeping the inequality $\varphi_t > 0$ the estimates θ_t are calculated. We can figuratively say that "θ_t is the estimate of the true value of the parameter θ"

but this does not mean the $\theta_t \to \theta$ convergence. We now try to write down the inequalities we are interested in so that it would be possible to compute the values of the estimates θ_t.

For the functional $\varphi_t = \varphi(x^t, u^t)$, we see that at time t

$$\varphi(x^t, u^{t-1}, u_t) = \varphi(x^t, u^{t-1}, h(z^{t-1}, \theta_{t-1})) = \varphi_t(\theta_{t-1}).$$

The solution of the inequality $\varphi_t(\theta_t) > 0$ with respect to θ_t produces the solutions of two subproblems, namely,

(1) It ensures the fulfilment of the goal inequality at time t;
(2) It gives the next control $u_{t+1} = h(z^t, \theta_t)$.

In the case of deterministic processes defined by difference equations of order l, $x_t = g(x_{t-1}, \ldots, x_{t-l}; u_{t-1}, \ldots, u_{t-l})$, there is one peculiarity for the functional of the form $\varphi(x^t)$ not depending on the control. In particular, when

$$\varphi(x^{t-1}, x_t) = \varphi\big(x^{t-1}, g\big(x_{t-l}^{t-1}; u_{t-l}^{t-1}\big)\big) = \varphi\big(x^{t-1}, g\big(x_{t-l}^{t-1}; u_{t-l}^{t-2}, u_{t-1}\big)\big)$$
$$= \varphi\big(x^{t-1}, g\big(x_{t-l}^{t-1}; u_{t-l}^{t-2}, h(z^{t-2}, \theta_{t-2})\big)\big) = \varphi_t(\theta_{t-2}),$$

either the value or the sign of the functional will be recorded after the parameter θ_{t-2} is fixed.

Thus, at each moment either the value or the sign of the functional is known. However, the future values of the functional depend on the controls chosen in the future which are defined by future values of the parameters $\theta_{t+1}, \theta_{t+2}, \ldots$. These parameters will be determined as soon as the next values of the functional are known. In other words, we need to solve a countable system of "hidden" inequalities for a finite number of steps, i.e. it is necessary to ensure starting from some moment t^* the fulfilment of inequalities $\psi_t > 0$ for all $t > t^*$. This signifies, in fact, the achievement of the given aim of control.

In what follows the task is to investigate the systems of inequalities $\psi(\theta_t, x_t) > 0$, $t \geq 1$.

The procedure of solving a similar system is called the *finite-converging procedure* (for short FCP) if there exists a finite t^* such that $\psi(\theta_t, x) > 0$ for all $t > t^*$ and every x from the admissible domain of the space X. The number of moments the opposite inequality holds ($\psi_t \leq 0$) is called the *number of corrections of the procedure*. Owing to the realization simplicity, recurrent finite-converging procedures are of special interest. The procedures of this type and the adaptive strategies based upon them are the subject-matter of this chapter.

For the sake of completeness we remark that under the additional assumption $\theta_{t^*+1} = \theta_{t^*+2} = \cdots = \theta_\infty$ with respect to the solution of the system of inequalities, the FCP are sometimes referred to as finite-converging *algorithms*.

9.2. Finite-converging Procedures of Solving A Countable System of Inequalities

Here and in the next section the existence conditions for the finite-converging procedures are studied and some examples are given irrespective of control problems.

Let us associate a sequence $\{x_t\}$ from a proper subset M of the set X with the system of inequalities $\varphi(\theta, x_t) > 0$, $t = 1, 2, \ldots$ where $\theta \in \Theta$ and Θ is a real Hilbert space. The following condition is always assumed.

There exists a number $\varepsilon_* > 0$ and an element $\theta_* \in \Theta$ such that for every $x \in M$ the inequality $\varphi(\theta_*, x) \geq \varepsilon_*$ is satisfied.

The values of ε_* and θ_* are, of course, unknown. Actually this condition means that the system of inequalities $\varphi(\theta, x_t) > 0, t = 1, 2, \ldots$ is solvable. The next theorem gives a sufficient condition for the existence of a recurrent procedure of the following form

$$\theta_{t+1} = g_t(\theta_t, x_t), \quad t \geq 1, \ \theta_1 = \tilde{\theta} \in \Theta. \tag{1}$$

Theorem 1. *Let $V(\theta)$ be a non-negative function, $\{\varepsilon_n\}$ be a sequence of positive numbers such that $\sum_{n=1}^{\infty} \varepsilon_n = \infty$ and procedure (1) possess the property*

$$V(\theta_t) - V(\theta_{t+1}) \geq \varepsilon_t, \quad \text{if } \varphi(\theta_t, x_t) \leq 0.$$

Then the following procedure

$$\theta_{t+1} = \begin{cases} \theta_t, & \text{if } \varphi(\theta_t, x) > 0, \\ g_t(\theta_t, x_t), & \text{if } \varphi(\theta_t, x) \leq 0, \end{cases} \tag{2}$$

is FCP. The number of corrections r for this procedure satisfies the inequality $\varepsilon_1 + \cdots + \varepsilon_r \leq V(\theta_1)$.

Proof. We cross out all repetitions from the sequence $\{\theta_t\} = \theta_1, \theta_2, \ldots, \theta_n, \ldots$ obtained with the help of (2). For the new sequence we keep the old notation $\{\theta_t\}$. Thus for all t the inequality $V(\theta_t) - V(\theta_{t+1}) \geq \varepsilon_t$ holds. Summing up these inequalities $\varepsilon_j \leq V(\theta_j) - V(\theta_{j+1})$ over j from 1 to t we obtain

$$\varepsilon_1 + \cdots + \varepsilon_t \leq \sum_{j=1}^{t} \left[V(\theta_j) - V(\theta_{j+1}) \right] = V(\theta_1) - V(\theta_{t+1}) \leq V(\theta_1).$$

Since $\sum_{n=1}^{\infty} \varepsilon_n = \infty$, the sequence $\{\theta_t\}$ must be finite. This proves our theorem. $\square$

The assumptions of Theorem 1 are fulfilled if the inequality $V(\theta) - V(g(\theta, x)) \geq \varepsilon$ holds at some ε as soon as $g(\theta, x) \leq 0$. We notice that the function V may be treated as a Lyapunov function for the move defined by equation (1). We now use the above theorem to verify finite convergence of the recurrent procedures for solving systems of inequalities. Later on we show that $\Theta = X$.

First, we choose the function $\varphi(\theta, x)$ as follows

$$\varphi(\theta, x) = (\theta, x).$$

Here $\|x\| \le a$ and $(\cdot, \cdot)$ stands for the scalar product. We define the numbers b_t, c_t, c', c'' so that

$$0 < c' \le c_t < c'', \quad 0 \le b_t \le 2.$$

The recurrent procedure is described by the relationship

$$\theta_{t+1} = \begin{cases} \theta_t, & \text{if } (\theta_t, x_t) \ge 0, \\ \theta_t + d_t x_t, & \text{if } (\theta_t, x_t) < 0 \end{cases} \tag{3}$$

where $d_t = c_t - b_t(\theta_t, x_t)\|x_t\|^{-2}$, i.e. it is assumed that $x_t \ne 0$.

Proposition 1. *Procedure* (3) *is a FCP of solving the system of inequalities* $(\theta_t, x_t) \ge 0$. *At* $\theta_1 = 0$ *the number of corrections satisfies the condition*

$$r \le \frac{c' a^2}{c'' \varepsilon_*^2} \|\theta_*\|^2.$$

Indeed, put $V(\theta) = \|\theta - h\theta_*\|^2$ with $h > h_0 = (2\varepsilon_*)^{-1} a^2 c''$. Hence

$$\begin{aligned} V(\theta_t) - V(\theta_{t+1}) &= \|\theta_t - h\theta_*\|^2 - \|\theta_{t+1} - h\theta_*\|^2 \\ &= -2d_t(\theta_t, x_t) - d_t^2(x_t, x_t) + 2hd_t(\theta_*, x_t) \\ &\ge \varepsilon(h) = 2\varepsilon_* c'(h - h_0). \end{aligned}$$

According to the above theorem procedure (3) is a FCP and from the inequality $r \le \min_{h > h_0} h^2 \varepsilon^{-1}(h)\|\theta_*\|^2$ the estimate of r follows.

We now consider non-homogeneous inequalities. Let

$$\varphi(\theta, x_t) = (\theta, x_t) + \gamma_t, \quad \|x\| \le a$$

and the sequences $\rho_t > 0$ and b_t satisfy the following conditions

$$\lim_{t \to \infty} \rho_t = 0, \quad \sum_{n=1}^{\infty} b_n = \infty, \quad b_t \in [0, 2],$$

moreover, the initial value θ_1 is chosen arbitrarily and $\varkappa(1) = 0$. The recurrent procedure is given by the relationship

$$\theta_{t+1} = \begin{cases} \theta_t, & \varkappa(t+1) = \varkappa(t), & \text{if } g_t > 0, \\ \theta_t + \lambda_t x_t, & \varkappa(t+1) = \varkappa(t) + 1, & \text{if } g_t \le 0 \end{cases} \tag{4}$$

with $\lambda_t = \rho_{\varkappa(t)} - b_t g_t \|x\|^{-2}$.

Proposition 2. *Procedure* (4) *is a FCP of solving the system of inequalities* $(\theta_t, x_t) + \gamma_t > 0$.

Indeed, for the function $V(\theta) = \|\theta - \theta_*\|^2$ and all sufficiently large t the following inequalities hold

$$V(\theta_t) - V(\theta_{t+1}) = \lambda_t \left[-\lambda_t(2 - b_t)\varphi_t + (\theta_t, x_t) + \gamma_t - \rho_t \|x_t\|^2 \right] \ge \varepsilon_* \rho_t.$$

Taking the above into account, the required assertion follows from Theorem 1.

Let the "strip"-type inequalities

$$\varphi_t = |\psi_t| = |(\theta_t, x_t) + \gamma_t| < \varepsilon, \quad \|x\| \le a$$

be given. We solve them by using the recurrent procedure

$$\theta_{t+1} = \begin{cases} \theta_t, & \text{if } \varphi_t < \varepsilon, \\ \theta_t - \psi_t x_t \|x_t\|^{-2}, & \text{if } \varphi_t \ge \varepsilon. \end{cases} \tag{5}$$

Proposition 3. *Procedure (5) is a FCP of solving the system of inequalities $\varphi_t < \varepsilon$ at an arbitrary initial value θ_1. The number of corrections may be estimated as follows*[a]

$$r \le a^2 \varepsilon^{-1} (\varepsilon - 2\varepsilon_*) \|\theta_1 - \theta_*\|^2.$$

Indeed, for the function $V(\theta) = \|\theta - \theta_*\|$ we have the following

$$V(\theta_t) - V(\theta_{t+1}) = \psi_t \frac{\psi_t - 2\nu_t}{\|x_t\|^2}, \quad \nu_t = (\theta_*, x_t) - \gamma_t.$$

The required assertion follows immediately. There exists a simple geometric interpretation demonstrating clearly the process of finding the solution of the systems of inequalities considered. Interested readers can find this interpretation without difficulty.

Consider one more procedure for solving the "strip"-type inequalities $\varphi_t = |\beta_t(\theta, x_t) + \gamma_t| < \varepsilon_t$ on the assumption that

$$\|x\| \le a, \quad 0 < |\beta_t| \le B, \quad 0 < \varepsilon \le \varepsilon_t.$$

We also strengthen the condition stated at the beginning of this section, namely,

there exist θ_* and $\rho \in (0, 1)$ such that the inequalities $|\beta_t(\theta_*, x_t) + \gamma_t| < \rho\varepsilon_t$ are satisfied for all t.

In addition, the following values are assumed to be known

$$x_t, \quad \mathbf{sign}\,\beta_t, \quad \psi = \beta_t(\theta_t, x_t) + \gamma_t, \quad \mathbf{sign}\,(\varepsilon_t - |\psi_t|), \quad \rho, \quad B.$$

The recurrent procedure has the form

$$\theta_{t+1} = \begin{cases} \theta_t, & \text{if } \psi_t \le \varepsilon, \\ \theta_t - (\mathbf{sign}\,\beta_t)(1 - \rho)\psi_t x_t \|x_t\|^{-2}, & \text{if } \psi_t > \varepsilon. \end{cases} \tag{6}$$

Proposition 4. *Under the given assumptions procedure (6) is a FCP of solving the system of "strip"-type inequalities. The number of corrections r can be estimated as follows*

$$r \le \frac{aB}{(1 - \rho)\varepsilon} \|\theta_1 - \theta_*\|^2.$$

We omit the standard proof.

---

[a] The quantities θ_* and ε_* have been characterized by the condition at the beginning of this section. Here, we assume additionally that $\varepsilon_* < \varepsilon/2$.

The next system of "strip"-type inequalities has a more complicated form

$$\varphi_t = |\psi_t| = |(\theta, x_t) + \gamma_t| < \mu + \nu\|x_t\|.$$

Suppose it is solvable and θ_* is one of its solutions. Moreover, let M be some closed convex set and $\theta_* \in M$ and P_M be a projection operator on it. We see that

$$\theta_{t+1} = \begin{cases} \theta_t, & \text{if } \varphi_t < \mu + \nu\|x_t\|, \\ P_M[\theta_t - \psi_t x_t \|x_t\|^{-2}], & \text{if } \varphi_t \geq \mu + \nu\|x_t\|. \end{cases} \tag{7}$$

Proposition 5. *For every $\nu > 0$ and arbitrary initial value θ_1 the recurrent procedure (7) is a FCP of solving the system of "strip"-inequalities of the second type. The number of corrections r may be estimated as follows*

$$r \leq \nu^{-2}\|\theta_1 - \theta_*\|^2.$$

The proof is standard.

We now consider the system of quadratic inequalities of the form

$$\varphi_t = (A_t(\theta - x_t), \theta - x_t) < \varepsilon_t, \quad t \geq 1.$$

Here A_t are some positive definite matrices. The condition of strengthened solvability of the given system is formulated below.

There exist θ_* and $\varepsilon \in (0,1)$ such that the inequality $\varphi(t) < \rho^2 \varepsilon_t$ is satisfied for all t.

A matrix $B > 0$ is chosen so that $B \geq A_t$ for all t and the recurrent procedure is defined by the following relationship

$$\theta_{t+1} = \begin{cases} \theta_t, & \text{if } \varphi_t(\theta_t) < \varepsilon_t, \\ \theta_t - b_t B^{-1} A_t(\theta_t - x_t), & \text{if } \varphi_t(\theta_t) \geq \varepsilon_t. \end{cases} \tag{8}$$

Proposition 6. *If $0 < \varepsilon \leq \varepsilon_t, 0 < b' \leq b_t \leq b'' < 2(1-\rho)$ then for an arbitrary initial value θ_1, procedure (8) is a FCP of solving the system of quadratic inequalities. The number of corrections r may be estimated as follows*

$$r \leq \delta^{-1}(B(\theta_1 - \theta_*), \theta_1 - \theta_*)$$

with $\delta = 2\varepsilon b'(1 - \rho - b''/2)$.

Indeed, denoting

$$y_t = \theta_t - x_t, \quad z_t = \theta_* - x_t$$

we put $V(\theta) = (B(\theta - \theta_*), \theta - \theta_*)$ and $\triangle V_t = V(\theta_t) - V(\theta_{t+1})$. Making use of the relationship $A_t B^{-1} A_t \leq A_t$, we obtain

$$\frac{\triangle V_t}{2b_t} = (A_t y_t, y_t) - (A_t y_t, z_t) - 2^{-1}b_t(B^{-1}A_t y_t, A_t y_t)$$

$$\geq (1 - b_t/2)\varphi_t(\theta_t) - [(A_t y_t, y_t)(A_t z_t, z_t)]^{1/2}$$

$$\geq (1 - b_t/2)\varepsilon_t - \rho\varepsilon_t = (1 - \rho - b_t/2)\varepsilon_t.$$

Taking into account the constraints on ε_t and b_t, we have

$$\triangle V_t \geq \delta = 2\varepsilon\delta'(1 - \rho - b''/\varepsilon).$$

The above implies the required assertion.

The last procedure completes the list of the best known and frequently used FCP; their application to control problems is discussed below. In conclusion we rewrite the usual form of the FCP

$$\theta_{t+1} = \begin{cases} \theta_t, & \text{if at time } t \text{ the goal inequality holds,} \\ \theta_t + h_t(\theta_t, x_t), & \text{otherwise,} \end{cases}$$

in the more compact form

$$\theta_{t+1} = \theta_t + \alpha_t h_t(\theta_t, x_t) \tag{9}$$

to be used later on. Here we have used the notation

$$\alpha_t = \begin{cases} 0, & \text{if at time } t \text{ the goal inequality holds,} \\ 1, & \text{otherwise.} \end{cases}$$

9.3. Sufficient Conditions for Existence of FCP

In the preceding section we considered examples of FCP for solving several concrete types of systems of inequalities. Here we are interested in finding more general conditions for existence of procedures of this type. This is important because the conditions established will be used later on for solving adaptive control problems.

Let the function $\varphi_t(\theta)$ be given on $\Theta = R^n$. We need to solve the system of inequalities $\varphi_t(\theta) < \varepsilon_t$, $t \geq 1$. Let also $\varepsilon_t \geq \varepsilon > 0$ and let there exist θ_* and $\rho \in (0, 1)$ such that the strengthened inequalities $\varphi_t(\theta_*) < \rho^2\varepsilon_t$ hold for all t. The functions φ_t satisfy the following restrictions:

(a) the functions $\varphi_t(\theta)$ are twice continuously differentiable;
(b) there exists a number $k > 0$ and a matrix $A > 0$ such that

$$(A^{-1}\nabla\varphi_t(\theta), \nabla\varphi_t(\theta)) \leq k\varphi_t(\theta)$$

whenever $\varphi_t(\theta) \geq 0$;
(c) $\nabla^2\varphi_t \geq 0$, i.e. the matrices of the second derivatives of the functions φ_t are non-negative definite.

Using notation (9) from the previous section and putting $b = (1 - \rho^2)k^{-1}$, we define the following recurrent procedure

$$\theta_{t+1} = \theta_t + \alpha_t bA\nabla\varphi_t(\theta_t). \tag{1}$$

Theorem 1. *We assume that conditions* (a)–(c) *and the above-mentioned assumptions about the system of inequalities* $\varphi_t(\theta) < \varepsilon_t$ *are satisfied. Then* (1) *is a FCP* *at an arbitrary initial value θ_1. The number of corrections r can be estimated as follows*

$$r \leq (A(\theta_1 - \theta_*), \theta_1 - \theta_*)(\varepsilon k b^2)^{-1}.$$

Proof. In the sequence $\theta_1, \theta_2, \ldots$ we cross out the elements θ_n for which the inequalities $\varphi_t(\theta_t) < \varepsilon_t$ hold. The remaining elements are indexed anew in succession. We consider further only this sequence. Putting $V(\theta) = (A(\theta - \theta_1), \theta - \theta_1)$ we see that

$$\triangle V_t = V(\theta_t) - V(\theta_{t+1}) = b(\nabla \varphi_t(\theta_t), 2\theta_t - 2\theta_* - bA^{-1}\nabla \varphi_t(\theta_t))$$
$$= b[2\varphi_t(\theta_t) - 2\varphi_t(\theta_*) + (\nabla^2 \varphi_t(\theta(\lambda_0)))(\theta_t - \theta_*), \theta_t - \theta_*) - b\|A^{-1/2}\nabla \varphi_t(\theta_t)\|^2].$$

We explain this equality. Let θ' and θ'' be chosen arbitrarily and $\theta(\lambda) = (1 - \lambda)\theta' + \lambda\theta''$. We put $\Phi(\lambda) = \varphi_t(\theta(\lambda))$. According to the Taylor formula we have

$$\Phi(1) - \Phi(0) = \varphi_t(\theta'') - \varphi_t(\theta') = (\nabla \varphi_t(\theta'), \theta'' - \theta')$$
$$+ \frac{1}{2}(\nabla^2 \varphi_t(\theta(\lambda_0)))(\theta'' - \theta'), \theta'' - \theta'), \quad \lambda_0 \in [0, 1].$$

In view of conditions (b) and (c), the inequalities $kb < 2$ and $\varphi_t(\theta) < \rho^2 \varepsilon_t$ we have

$$\triangle V_t \geq [(2 - kb)\varphi_t(\theta_t) - \rho^2 b\varepsilon_t]b \geq \varepsilon b(1 - \rho^2).$$

From the above and Theorem 1 from Sec. 2 the required assertion follows. $\square$

We now formulate conditions for existence of recurrent FCP having the general form $\theta_{t+1} = g(\theta_t, x_t)$.

We assume that the elements of a sequence $\{x_t\}$ belong to a bounded subset X_0 of a Hilbert space X. The quantity θ belongs to a Hilbert space Θ with the scalar product (θ', θ'') and the norm $\|\theta\| = (\theta, \theta)^{1/2}$.

Theorem 2. *Let a function $\varphi(\theta, x)$ satisfy the following conditions:*

(a) *it is defined on $\Theta_0 \times X_0$ where Θ_0 is a convex set;*
(b) *it is differentiable with respect to θ and its derivative in the Fréchet sense $\nabla_\theta \varphi(\theta, x)$ is bounded on $\Theta_0 \times X_0 \cap \{(\theta, x) : \varphi(\theta, x) \leq 0\}$;*
(c) *there exists an element $\theta_* \in \Theta$ and a number $\varepsilon_* > 0$ such that $\varphi(\theta_*, x) \geq \varepsilon_*$ for all $x \in X_0$;*
(d) *for all $(\theta, x) \in \Theta_0 \times X_0 \cap \{(\theta, x) : \varphi(\theta, x) \leq 0\}$ the inequality*

$$(\nabla_\theta \varphi(\theta, x), \theta_* - \theta) \geq \varphi(\theta_*, x) - \varphi(\theta, x)$$

holds.

*Let $\{x_t\}$ be a countable sequence from X_0. The recurrent procedure is defined by
the relationship*

$$\theta_{t+1} = \begin{cases} \theta_t, & \text{if } \varphi(\theta_t, x_t) > 0, \\ P[\theta_t + a(t)\nabla_\theta\varphi(\theta_t, x)], & \text{if } \varphi(\theta_t, x_t) \leq 0 \end{cases} \qquad (2)$$

*where P denotes a projection operator on Θ_0, θ_1 is chosen arbitrarily, $b(t) \in [0, 2]$
and*

$$a(t) = \frac{1}{n(t)} - b(t)\frac{\varphi(\theta_t, x_t)}{\|\nabla_\theta\varphi(\theta_t, x_t)\|^2}.$$

*Here $n(t)$ is the number of corrections of the parameter θ_t up to time t. It is deter-
mined as follows: $n(1) = 1$ and*

$$n(t+1) = \begin{cases} n(t), & \text{if } \varphi(\theta_t, x_t) > 0, \\ n(t) + 1, & \text{if } \varphi(\theta_t, x_t) \leq 0. \end{cases} \qquad (3)$$

*The procedure (2) is a FCP of solving the system of inequalities $\varphi(\theta_t, x_t) > 0$ on
an arbitrary countable set $\{x_t\}$.*

Proof. To prove this, we have to show that $\lim_{t\to\infty} n(t) < \infty$. We consider the
quantities α_t which have been defined at the end of Sec. 2. According to procedure
(2) and the assumptions, we obtain

$$\begin{aligned} \|\theta_{t+1} - \theta_*\|^2 &= \|P[\theta_t + \alpha_t a(t)\nabla_\theta\varphi(\theta_t, x_t)] - \theta_*\|^2 \\ &\leq \|\theta_t - \theta_* + \alpha_t a(t)\nabla_\theta\varphi(\theta_t, x_t)\|^2 \\ &= \|\theta_t - \theta_*\|^2 + 2\alpha_t a(t)(\nabla_\theta\varphi(\theta_t, x_t), \theta_t - \theta_*) + \alpha_t a(t)\|\nabla_\theta\varphi(\theta_t, x_t)\|^2. \end{aligned}$$

Hence

$$\|\theta_t - \theta_*\|^2 - \|\theta_{t+1} - \theta_*\|^2 \geq \alpha_t a^2(t)[2(\nabla_\theta\varphi(\theta_t, x_t), \theta_* - \theta_1) - a(t)\|\nabla_\theta\varphi(\theta_t, x_t)\|^2].$$

Using the representation of $a(t)$ it is easy to show that

$$\alpha_t a(t)[2\varphi(\theta_*, x_t) - 2\varphi(\theta_t, x_t) - a(t)\|\nabla_\theta\varphi(\theta_t, x_t)\|^2] \geq \frac{\varepsilon_*\alpha_t}{n(t)}$$

for all sufficiently large $n(t)$. These inequalities imply the fulfilment of the following
inequality

$$\|\theta_t - \theta_*\|^2 - \|\theta_{t+1} - \theta_*\|^2 \geq \frac{\varepsilon_*\alpha_t}{n(t)}.$$

After summation, we obtain

$$\|\theta_1 - \theta_*\|^2 - \|\theta_{t+1} - \theta_*\|^2 \geq \varepsilon_* \sum_{j=1}^{t} \frac{\alpha_j}{n(j)} \geq \varepsilon_* \sum_{j=1}^{n(t)} j^{-1}.$$

From this the desired condition $\lim_{t\to\infty} n(t) < \infty$ follows immediately. $\qquad \square$

In the deterministic case we have obtained rather general conditions for existence
of FCP. These procedures also exist in the stochastic case. However, we will not
consider them here.

9.4. Stabilization of Solutions of Linear Difference Equations: Part I

We consider a class of processes $\mathcal{L}_l$ represented by the solution of the ARMAX-type equation

$$x_t = A_1 x_{t-1} + \cdots + A_l x_{t-l} + B_1 u_{t-1} + \cdots + B_l u_{t-l} + \zeta(t). \tag{1}$$

Here x_t and u_t belong to R^m, A_i and B_i are $m \times m$-matrices[b] and $\zeta(t)$ is a disturbance (the "external force"). The control u_t is assumed to satisfy the condition $\|u_t\| \le Q$. The states (or the solutions) x_t are not only observed but also satisfy the condition $\|x_t\| \le R$. The numbers Q and R are given.

The aim of control is, starting from some t, to ensure the fulfilment of the inequality

$$\|x_t\| < r < R$$

or, in other words, $\overline{\lim} \lim_{t \to \infty} \|x_t\| < r$. It is easy to understand that r depends on the restrictions on the disturbance $\zeta(t)$, i.e. it is at least necessary that the condition $\|\zeta(t)\| < r$ be fulfilled. First, we assume that the coefficients of Eq. (1) are known. Assuming the matrix B_1 is non-degenerate and ignoring temporarily the condition $\|u\| \le Q$, the control at time t is chosen as the solution of the equation

$$A_1 x_t + \cdots + A_l x_{t-l+1} + B_1 u_t + \cdots + B_l u_{t-l+1} = 0, \tag{2}$$

i.e.

$$u_t = -B_1^{-1} A_1 x_t - \cdots - B_1^{-1} A_l x_{t-l+1} - B_1^{-1} B_2 u_{t-1} - \cdots - B_1^{-1} B_l u_{t-l+1}. \tag{3}$$

This relationship can be written in a compact form if we introduce the $n \times 2ln$-matrix $M = (A_1, \ldots, A_l; B_1, \ldots, B_l)$ and the $2ln$-vector $v_t = (x_t^T, \ldots, x_{t-l+1}^T, u_t^T, \ldots, u_{t-l+1}^T)$. Here, the symbol "$T$" stands, as usual, for transposition. Now, we can rewrite formula (3) in the form

$$B_1^{-1} M v_t = 0.$$

Since $x_{t+1} = \zeta_t$, the given aim is attained, namely, $\|x_{t+1}\| = \|\zeta_t\| < r$.

These arguments explain the structure of the optimal control and suggest a method of construction of adaptive control for equations from $\mathcal{L}_l$ in the case when the coefficients of Eq. (1) are unknown. There is a problem with the appropriate identification because of non-observability of the disturbance $\zeta(t)$. In this case we cannot use the equality (3) containing the unknown matrices A_i.

Let the "estimates" $A_i(t)$, $B_i(t)$ of the matrices A_i, B_i be obtained by some method (to be discussed later on). We form the $n \times 2ln$-matrix

$$M_t = (A_1(t), \ldots, A_l(t); B_1(t), \ldots, B_l(t)).$$

---

[b] For a matrix $M = (m_{ij})$ the number $\|M\| = (\sum_{i,j} m_{ij}^2)^{1/2}$ is the norm.

At time $t+1$ we can observe the vector $g_{t+1} = x_{t+1} - M_t v_t$. If

$$\|g_{t+1}\| = \|x_{t+1} - M_t v_t\| < \gamma$$

and the control u_t is the solution[c] of the equation

$$M_t v_t = 0,$$

then the goal inequality $\|x_{t+1}\| < r$ is fulfilled at time $t+1$.

We can now give the adaptive strategy ensuring the stability-type aim with respect to the appropriate subclass of L_l.

The controls u_t are computed by using the current estimates M_t depending on whether the matrix $B_1(t)$ is degenerate or not. If it is non-degenerate, then

$$u_t = \begin{cases} U_t(M_t), & \text{if } \|U_t(M_t)\| < Q, \\ Q\dfrac{U_t(M_t)}{\|U_t(M_t)\|}, & \text{if } \|U_t(M_t)\| \geq Q \end{cases} \tag{4}$$

where $U_t(M_t) = -B_1^{-1}(t) M_t v_t + u_t$. Otherwise, we chose a vector e_t with norm one which satisfies the condition $B_1(t) e_t = 0$ and let

$$u_t = Q e_t. \tag{5}$$

This formula can be considered as the limiting one for the previous formula provided **det** $B_1(t) \to 0$.

The phase restriction $\|x_t\| \leq R$ has to be fulfilled during the entire time of control. We show how this can be done. The time axis is divided into the nonintersecting intervals I_1, I_2, ... of length greater than l. At the first $l-1$ moments of the interval I_m, the initial values of x_t are chosen so that the restriction $\|x_t\| < r$ holds and the controls are equal to zero. Beginning from the lth moment of I_m the trajectory progresses according to equation (1). If either the restriction $\|x_t\| \leq R$ fails or the current interval I_m ends then the movement will be interrupted and the procedure described above should be repeated on the next interval I_{m+1}. This procedure is called "shaking" the controlled object. Of course, it can not always be used. For example, this is true if the violation of the restriction leads to an explosion. However, "shaking" is admissible and widespread. It is natural, for example, in teaching a robot to cycle. The bicycle is the non-stable object of control and a violation of the restriction means falling. If this has happened we have to put the bicycle in the vertical position again.

The matrix M_t which is an "estimate" of M is determined by the recurrent procedure

$$M_{t+1} = \begin{cases} M_t, & \text{if } \|g_{t+1}\| < r \\ M_t + g_{t+1} v_t^T \|v_t\|^{-2}, & \text{if } \|g_{t+1}\| \geq r, \end{cases} \tag{6}$$

---

[c]To compute u_t we have to assume that the matrix $B_1(t)$ is non-degenerate without paying attention to the restrictions on the control.

with the initial value M_1 chosen arbitrarily. Relations (4)–(6) together with the "shaking" procedure define some strategy for the class $\mathcal{L}_l$. It remains to select a subclass $\tilde{\mathcal{L}}$ of this strategy ensuring the achievement of the aim.

Let $\mathcal{L}_{l,\rho} \subset \mathcal{L}_l$ denote a class of linear equations of the form (1), whose disturbances satisfy the following condition

$$\|\zeta(t)\| \leq \rho r, \quad \rho \in (0, 1/2).$$

Lemma 1. *Procedure* (6) *of solving the system of inequalities*

$$\|x_{t+1} - M_t v_t\| < r$$

is a FCP with respect to the class $\mathcal{L}_{l,\rho}$ provided the controls are chosen according to (4)–(5). *The number of corrections r may be estimated as follows*

$$r \leq l \frac{R^2 + Q^2}{(1 - 2\rho)r^2} \|M - M_1\|^2.$$

Proof. We have to verify that $\|M_t - M\|^2$ decreases as $t \to \infty$. Let

$$\gamma_t = \|M_t - M\|^2 - \|M_{t+1} - M\|^\delta.$$

If M_t is computed according to the second formula in (6), then

$$\gamma_t \geq \delta = \frac{(1 - 2\rho)r^2}{l(R^2 + Q^2)}$$

i.e. $\|M_t - M\|^2$ decreases by no less than δ on each step. Hence, from M_1 to M there are no more than $\|M - M_1\|^2/\delta$ steps.

We prove that $\gamma_t \geq \delta$. We use the identity

$$\|N + h_1 h_2^T\|^2 = \|N\|^2 + \|h_1\|^2 \|h_2\|^2,$$

where N is an $n \times n$-matrix, h_1, h_2 are vectors having order n and $Nh_1 = 0$. We also use the notation

$$g_{t+1}(\Gamma) = x_{t+1} - \Gamma v_t$$

where the matrix Γ has the same form as M. Let us put

$$M' = M_{t+1} + [g_{t+1}(M_t) - g_{t+1}(M)] \frac{v_t^T}{\|v_t\|^2}.$$

From the easily verified equality $g_{t+1}(M') = g_{t+1}(M)$ the equality $(M' - M)v_t = 0$ follows. Introducing

$$N = M' - M, \quad h = \frac{v_t^T}{\|v_t\|^2}$$

(i.e. $Nh = 0$) and using the equality mentioned above, we obtain

$$\|M_{t+1} - M\|^2 = \|N + (g_{t+1}(M) - g_{t+1}(M_t))h^T\|^2$$
$$= \|N\|^2 + \|g_{t+1}(M) - g_{t+1}(M_t)\|^2 \|h\|^2$$

and additionally

$$\|M_{t+1} - M\|^2 = \|N + g_{t+1}(M))h^T\|^2 = \|N\|^2 + \|g_{t+1}(M)\|^2\|h\|^2.$$

These relationships lead to the equality

$$\gamma_t = [\|g_{t+1}(M) - g_{t+1}(M_t)\|^2 - \|g_{t+1}(M)\|^2]\|h\|^2$$
$$= \frac{1}{\|v_t\|^2}[\|g_{t+1}(M_t)\|^2 - 2g_{t+1}^T(M_t)g_{t+1}(M)].$$

Under the above assumptions this implies the following inequalities

$$\|v_t\|^2 = \sum_{j=1}^{l-1}(\|x_{t-j}\|^2 + \|u_{t-j}\|^2) \le (R^2 + Q^2)l,$$
$$\|g_{t+1}(M_t)\| \ge r, \quad \|g_{t+1}(M)\| \le \rho r.$$

Hence we can easily complete the proof:

$$\gamma_t = \|g_{t+1}(M_t)\|[\|g_{t+1}(M_t)\|^2 - 2\|g_{t+1}(M)\|]\|v_t\|^{-2} \ge r^2(1-2\rho)[(Q^2 + R^2)l]^{-1}. \quad \square$$

We now consider the linear equation

$$B_1 u_t + \cdots + B_l u_{t-l+1} = z_t \tag{7}$$

with the same coefficients as in (1). The characteristic polynomial of this equation has the form

$$\beta(\lambda) = \mathbf{det}\,[B_1\lambda^{l-1} + B_2\lambda^{l-2} + \cdots + B_l].$$

Definition 1. Equation (7) is called *minimum phase* if all roots of the equation $\beta(\lambda) = 0$ belong to the interior of the unit circle with the center at the origin.

This definition can be extended to a wider class of linear equations. However, in this case it will have a much more complicated form. In the next section we shall state the minimum phase notion for other classes of linear equations.

Lemma 2. *The solution of the minimum phase equation has the following property: there exists $\nu > 0$ such that the inequalities*

$$\|z_{t+m}\| < C, \quad m = 0, 1, 2, \ldots$$

and the initial conditions $u_{t-1} = u_{t-2} = \cdots = u_{t-l+1} = 0$ imply

$$\|u_{t+m}\| \le C\nu$$

for every t and $C > 0$.

We leave the proof of this assertion to the reader.

The quantity ν in Lemma 2 depends on the matrices B_i. The least among these quantities is denoted by ν.

We introduce some additional constants:

$$C_1 = \|M\|\sqrt{l} + 1 + \rho, \qquad C_2 = \|M - M_0\|\sqrt{l^2 + \nu^2 C_1^2(l-1)} + 1 + \rho,$$

$$C_3 = (\|M - M_0\| + \|M\|)\sqrt{l^2 + \nu^2 C_1^2(l-1)}, \quad C_4 = \varkappa^{-1}(C_2 + C_3),$$

$$q = r\max(\nu C_1, C_4), \quad \varkappa = \min_{z:\|z\|=1} \|B_1 z\|.$$

We can now define the class of linear equations for which the above strategy ensures the fulfilment of the control aim.

The symbol $\tilde{\mathcal{L}} = \mathcal{L}_l(\rho, r, \varkappa, q)$ denotes the set of all linear difference equations satisfying the conditions:

(1) the disturbances are bounded, i.e. $\|\zeta(t)\| \le \rho r$, $0 < \rho < 1/2$, $\forall t$;
(2) all roots of the equation $\beta(\lambda) = 0$ belong to the interior of the unit circle, i.e. equation (7) is minimum phase;
(3) $\varkappa > 0, q < Q$.

Theorem 1. *The strategy defined above leads to achievement of the control aim $\|x_t\| < r$ for all sufficiently large t with respect to the class $\tilde{\mathcal{L}}$. For the number of corrections the estimate from Lemma 1 remains valid.*

Proof. According to the assumptions all intervals I_m are finite. Let FCP (6) be finished at time $t^0 \in I_{k-1}$. It is necessary to show that $\|x_t\| < r$ for all $t > t_* \in I_k$. This is true for the first $l - 1$ moments of I_k. Taking into account the inequality $t_* > t^0$, the FCP is no longer used to recalculate the matrices M_t, but the controls continue to be computed according to (4), (5). Therefore we can assume that t_* is the first moment of their computation. Then it assumes either the strict inequality $\|u_{t_*}\| < Q$, or the equality $\|u_{t_*}\| = Q$. In the first case the goal condition holds. Indeed, since the control is defined by the first equality in (4), the matrix $B_1(t)$ is non-degenerate and $M_t v_t = 0$. From this and the inequality $\|g_{t+1}\| < r$, $\forall t > t_0$, the required inequality follows. Moreover, by Lemma 2, it is easy to show that $\|u_t\| \le q$.

It remains to make sure that the equality $\|u_{t_*}\| = Q$ is impossible. We prove this by contradiction, i.e. suppose the control has been computed according to:

(a) the second equality in (4) if the matrix $B_1(t)$ is non-degenerate;
(b) formula (5) provided $\det B_1(t) = 0$.

Putting $h_t(M - M_t) = (M - M_t)v_t - (B_1 - B_1(t))u_t$, we rewrite the inequality $\|g_{t+1}(M)\| < r$ in the form

$$\|h_t(M - M_t) + (B_1 - B_1(t))u_t + \zeta(t)\| < r.$$

For h_t we obtain the estimate

$$\|h_t\|^2 \leq \left[\sum_{j=1}^{l}(\|A_j - A_j(t)\|^2 + \|B_j - B_j(t)\|^2)\right]\left[\sum_{j=1}^{l}(\|x_{t-j}\|^2 + \|u_{t-j}\|^2)\right]$$

and, as mentioned above,

$$\|M - M_t\| \leq \|M - M_0\|, \quad \|x_{t-j}\| \leq r, \quad j = 1, \ldots, l;$$
$$\|u_{t-j}\| = 0 < r\nu c_1, \qquad\qquad j = 2, \ldots, l.$$

Hence

$$\|h_t\| \leq r\|M - M_0\|(l^2 + \nu^2 c_1^2(l-1))^{1/2}.$$

This leads to the following inequality

$$\|(B_1 - B_1(t))u_t\| \leq rC_2. \tag{8}$$

Let us estimate the norm of $B_1(t)u_t$ in both variants of the control choice considered.

In the case (a) we obtain

$$B_1(t)u_t = a(t)[B_1(t)u_t - M_t v_t], \quad a(t) = \frac{Q}{\|U_t(M_t)\|} \leq 1,$$

but in the case (b) the equality $B_1(t)u_t = 0$ is valid. By analogy with the previous arguments, we obtain the following estimate

$$\|h_t(M_t)\| \leq r[\|M - M_0\| + \|M\|](l^2 + \nu^2 c_1^2(l-1))^{1/2} = rC_3.$$

This means that the inequality $\|B_1(t)u_t\| \leq rC_3$ holds in both variants. From inequality (8) it follows that $\|u_t - B_1^{-1}B_1(t))u_t\| \leq rC_2/\varkappa$, and hence, the inequality $\|u_t\| \leq r(C_2 + C_3)/\varkappa = rC_4 < Q$. So, we arrive at a contradiction. The result obtained remains valid for the next moments of the interval I_k. For the subsequent intervals the arguments are the same. $\qquad\square$

We have obtained the estimate of the number of corrections but not of the time starting from which the goal inequality holds. In the general case this is impossible because the "strong" disturbances of the movement can occur at any time in the future. The real values (not upper estimates) of the number of corrections would be found by modelling the controlled process on a computer.

Besides, the FCP lead to estimates of the coefficients of Eq. (1) which differ significantly from their true values, i.e. there is no real identification.

9.5. Stabilization of Solutions of Linear Difference Equations: Part II

The adaptive strategy described in Sec. 4 has been based on "shaking" the controlled object, i.e. if the phase restriction fails, the movement is stopped and new initial conditions are chosen. We now discuss another method of constructing a strategy for which uniform boundedness of the states is not essential. For simplicity we consider the scalar deterministic processes given by the solution of the linear difference equation

$$x_t + a_1 x_{t-1} + \cdots + a_l x_{t-l} = b_1 u_{t-1} + \cdots + b_l u_{t-l} + \zeta(t) \tag{1}$$

where $\sum_{j=1}^{l} |b_j| \neq 0$, the disturbances are bounded, i.e. $|\zeta(t)| \leq C_\zeta$, with a known constant. The dissipativity aim is chosen as the control aim, i.e.

$$\overline{\lim_{t \to \infty}} \, (|x_t| + |u_t|) \leq C_{xu}$$

for arbitrary initial values and a constant C_{xu} which depends on C_ζ (otherwise the aim may be unrealizable: if C_ζ is large, then C_{xu} cannot be too small).

It is not difficult to ensure the fulfilment of this aim in the classical statement of the problem considered. We define the control as the solution of the following linear equation (it is called the linear regulator)

$$u_t + \alpha_1 u_{t-1} + \cdots + \alpha_m u_{t-m} = \beta_1 x_t + \beta_1 x_{t-1} + \cdots + \beta_m x_{t-m} \tag{2}$$

whose coefficients are defined by the equality

$$a(\lambda)\alpha(\lambda) - b(\lambda)\beta(\lambda) = g(\lambda) \tag{3}$$

where

$$a(\lambda) = 1 + a_1\lambda + \cdots + a_l\lambda^l, \qquad b(\lambda) = b_1\lambda + \cdots + a_l\lambda^l,$$
$$\alpha(\lambda) = 1 + \alpha_1\lambda + \cdots + \alpha_m\lambda^m, \quad \beta(\lambda) = \beta_0\lambda + \cdots + \beta_m\lambda^m$$

and $g(\lambda)$ is some *stable* polynomial given in advance, i.e. the moduli of all its roots are greater than one and $g(0) = 1$. The polynomials $a(\lambda)$ and $b(\lambda)$ are assumed to be irreducible. The polynomial relationship (3) generates a system of linear algebraic equations with respect to $(\alpha_1, \ldots, \alpha_m; \beta_0, \beta_1, \ldots, \beta_m)$ which are consistent provided the degree m is the smallest.

The polynomial $g(\lambda)$ is the characteristic polynomial of the system (1), (2). Its stability and the irreducibility of $a(\lambda)$, $b(\lambda)$ imply the fulfilment of the given aim with some constant C_{xu}. The constants C_ζ and C_{xu} are related by a relationship we explain below.

We write the equations (1), (2) in the form of a system of equations of the first order with respect to the vector of "phase variables" $z_t = (x_t, x_{t-1}, \ldots, x_{t-n}; u_t, \ldots, u_{t-n})$, $n = \max(l, m) - 1$. It is easy to show that

$$z_{t+1} = A z_t + b\zeta(t), \tag{4}$$

where the matrix A and the vector b can be expressed in terms of the initial coefficients. Nonzero eigenvalues λ_i of the matrix A satisfy the algebraic equation $g(\lambda_i^{-1}) = 0$, i.e. $|\lambda_i| < 1$. Hence, there exists $a > 0$ and $\rho \in (0,\ 1)$ such that $\|A^t\| \leq a\rho^t$ for all t.

Lemma 1. *If z_t is a solution of* (4), *then*

$$\overline{\lim_{t \to \infty}} \, \|z_t\| \leq \frac{a\|b\|}{1 - \rho} C_\zeta.$$

Proof. Iterating (4) we arrive at an explicit form of the solution

$$z_{t+1} = A^t z_1 + \sum_{k=1}^{t} A^{t-k} b \zeta(k+1).$$

From this the required assertion follows immediately. $\qquad\square$

The result obtained answers our question, namely,

$$C_{xu} \leq \frac{a\|\bar{b}\|}{1 - \rho} C_\zeta.$$

Definition 2. In agreement with the definition stated in Sec. 4, Eq. (1) is *minimum phase* if $b(\lambda) = \lambda^k b_+(\lambda)$, $k \in [1, \, l]$ and the polynomial $b_+(\lambda)$ is stable.

For minimum phase equations we can explicitly compute the coefficients of the optimal linear regulator (2) minimizing the objective function

$$W(\alpha, \beta) = \sup_{\zeta:|\zeta| \leq C_\zeta} \overline{\lim_{t \to \infty}} \, |x_t|.$$

The coefficients of the optimal linear regulator (2), which are, in fact, the coefficients of the polynomials $\alpha(\lambda)$, $\beta(\lambda)$ prove to be defined uniquely by the relationship (3) with $g(\lambda) = b_+(\lambda)$. Then

$$\alpha(\lambda) = \gamma(\lambda) b_+(\lambda), \quad \beta(\lambda) = \lambda^{-k}[\gamma(\lambda) a(\lambda) - 1] \tag{5}$$

where the polynomial $\gamma(\lambda)$ having degree r is determined by the equalities

$$\gamma(0) = 1, \quad \frac{d^p}{d\lambda^p}(\gamma(\lambda) a(\lambda)) \, |_{\lambda=0} = 0, \quad p = 1, \ldots, k-1.$$

From equalities (1), (2), (5) it follows that

$$x_t = \zeta(t) + \gamma_1 \zeta(t-1) + \cdots + \gamma_k \zeta(t-k),$$

where $\gamma_1, \ldots, \gamma_k$ are the coefficients of the polynomial $\gamma(\lambda)$. Hence, we conclude that

$$\min_{\alpha, \beta} W(\alpha, \beta) = C_\zeta \sum_{h=1}^{k} |\gamma_h|,$$

i.e. we have obtained another expression for the smallest admissible value of the constant C_{xu}.

For non-minimum phase equations the algorithm described implies unlimited growth of the controls u_t as $t \to \infty$.

We return to the original aim of control identifying the parameter space with collections $\theta = (\bar{a}, \, \bar{b}) \in \Theta$ of coefficients of equation (1). We select a subset $\tilde{\Theta}$

of Θ (i.e. a class $L(\tilde{\Theta})$ of equations having the form (1)) defined by the following conditions:

(1) $\tilde{\Theta}$ is a closed, convex set;

(2) for every $\theta \in \tilde{\Theta}$ the corresponding polynomials $a(\lambda, \theta)$ and $b(\lambda, \theta)$ have no common zeros inside the unit circle.

     We denote the two polynomials in λ (having the least degree) defined by the relationship (3) with $g(\lambda) = b_+(\lambda, \theta)$, the stable part of $b(\lambda, \theta)$, by $\alpha(\lambda, \theta)$ and $\beta(\lambda, \theta)$. With the help of the polynomials $a(\lambda, \theta)$, $b(\lambda, \theta)$, $\alpha(\lambda, \theta)$, and $\beta(\lambda, \theta)$ we can compute the matrix $A(\theta)$ and the vector $b(\theta)$ in Eq. (4).

(3) There exist positive numbers $a(\theta)$ and $\rho(\theta)$ such that

$$\|A^t(\theta)\| \le a(\theta)\rho^t(\theta), \qquad \sup_{\theta \in \tilde{\Theta}} \rho(\theta) < 1$$

for all t.

(4) The quantity

$$\varepsilon = \frac{1 - \sup_{\theta \in \tilde{\Theta}} \rho(\theta)}{\sup_{\theta \in \tilde{\Theta}} a(\theta)\|b(\theta)\|}$$

is known.

We now give a method of constructing an adaptive strategy which will be denoted by F. It alternates two procedures which consist of obtaining the estimates θ_t of the unknown parameters and calculating the appropriate controls by using the estimates obtained. We consider one of these procedures in detail.

     The estimates will be obtained as the solutions of some inequalities. To write them down, we use the notation

$$h_t = (x_{t-1}, \ldots, x_{t-l}; u_{t-k}, \ldots, u_{t-l}),$$
$$\theta_* = (a_1, \ldots, a_l; -b_k, \ldots, -b_l),$$
$$\theta_t = (a_1(t), \ldots, a_l(t); -b_k(t), \ldots, -b_l(t))$$

where $a_i(t)$, $b_i(t)$ serve for the estimates of a_i, b_i. Eq. (1) has the form

$$x_t + (\theta_t, h_t) = (\theta_t - \theta_*, h_t) + \zeta(t).$$

Taking into account the inequalities $\|h_t\| \le \|z_t\|$, where z_t is the solution of Eq. (4), we obtain the following system of the inequalities

$$|x_t + (\theta, h_t)| \le 2C_\zeta + \varepsilon\|z_t\|, \quad 0 < \varepsilon < \varepsilon_0.$$

We apply to this system the FCP of the form (7) from Sec. 2 (now P_M is the projection operator on the set $\tilde{\Theta}$). The solutions obtained are the estimates θ_t.

     The controls are calculated by the following algorithm. The polynomials $a_t(\lambda) = a(\lambda, \theta_t)$, $b_t(\lambda) = b(\lambda, \theta_t)$ are defined by using θ_t. Making use of the equation

$a_t(\lambda)\alpha_t(\lambda) - b_t(\lambda)\beta_t(\lambda) = g_t(\lambda)$ (here $g_t(\lambda) = b_+(\lambda, \theta_t)$) we can find the polynomials $\alpha_t(\lambda)$ and $\beta_t(\lambda)$ whose coefficients are used to construct the current linear regulator

$$\alpha_t(\triangle)u_t = \beta_t(\triangle)x_t \tag{6}$$

where $\triangle$ denotes the shift-back operator of the form $\triangle x_t = x_{t-1}$ and $\triangle^k = \triangle^{k-1}(\triangle)$, $k \geq 2$. This completes the description of the strategy F.

Theorem 1. *The strategy F ensures the attainment of the control aim*

$$\varlimsup_{t \to \infty} (|x_t| + |u_t|) \leq C_{xu} = \frac{2C_\zeta}{\varepsilon_0 - \varepsilon}$$

with respect to the class $L(\tilde{\Theta})$. The number of corrections of the coefficients in (6) is no greater than $\varepsilon^{-1}\|\theta_1 - \theta_*\|^2$.

Proof. Because the estimates obtained by means of FCP belong to $\tilde{\Theta}$, Eq. (3) defines the polynomials $\alpha_t(\lambda)$, $\beta_t(\lambda)$ uniquely. After a finite number of steps from the time t_0, the FCP produces the final value of the parameter θ' (it is not necessarily true). For $t \geq t_0$ the equation for z_t (compare with (4))

$$z_{t+1} = A'z_t + b'[\zeta(t) + (\theta' - \theta_*, h_t)]$$

holds, with

$$|\zeta(t) + (\theta' - \theta_*, h_t)| \leq 2C_\zeta + \varepsilon\|h_t\| \leq 2C - \zeta + \varepsilon\|z_t\|.$$

According to the condition $\|A'^t\| \leq a'\rho'^t$ and rewriting the previous equation as follows

$$z_{t+1} = A'^{t-t_0+1}z_{t_0} + \sum_{n=t_0}^{t} A'^{t-n}b'[\zeta(n+1) + (\theta' - \theta_*, h_n)]$$

we arrive at the estimates

$$\|z_{t+1}\| \leq a'\rho'^{t-t_0+1}\|z_{t_0}\| + \|b'\| \sum_{n=t_0}^{t} a'\rho'^{t-n}(2C_\zeta + \varepsilon\|z_n\|).$$

The following relationship define a numerical sequence d_n, $n \geq 0$. Here $\rho' + \varepsilon a'\|b'\| < 1$.

$$d_{t+1} = (\rho' + \varepsilon a'\|b'\|)d_t + 2C_\zeta a'\|b'\|$$

$$= (\rho' + \varepsilon a'\|b'\|)^t d_0 + 2C_\zeta a'\|b'\| \sum_{k=0}^{t} (\rho' + \varepsilon a'\|b'\|)^k$$

$$= (\rho' + \varepsilon a'\|b'\|)^t d_0 + 2C_\zeta a'\|b'\| \frac{1 - (\rho' + \varepsilon a'\|b'\|)^{t+1}}{1 - \rho' - \varepsilon a'\|b'\|}.$$

Comparing the last equality with the previous inequality, we conclude that $\|z_t\| \leq d_t$ for all $t > t_0$. Since $\rho' + \varepsilon a'\|b'\| < 1$, we see that

$$\varlimsup_{t \to \infty} \|d_t\| \leq \frac{2C_\zeta a'\|b'\|}{1 - \rho' - \varepsilon a'\|b'\|}.$$

Therefore, the desired estimate

$$\varlimsup_{t\to\infty} \|z_t\| \leq \frac{2C_\zeta \sup_\theta a(\theta)\|b(\theta)\|}{1 - \sup_\theta \rho(\theta) - \varepsilon \sup_\theta a(\theta)\|b(\theta)\|} = \frac{2C_\zeta}{\varepsilon_0 - \varepsilon}$$

takes place, i.e. the given aim of control is attained. From the appropriate estimate for FCP (7) in Sec. 2, the estimate of the number of corrections of the coefficients of (6) follows. $\qquad\square$

CHAPTER 10

CONTROL OF LINEAR DIFFERENCE EQUATIONS

The present chapter is concerned with adaptive control for difference equations. We study the problems of tracking, of reference models, modal control, the linear-quadratic problem (LQP) and, finally, strong consistency of parameter estimates. The identification strategies based on the least square method and stochastic gradient method are used. For the first approach the convergence rate of the estimates to the true values of the unknown parameters has been obtained.

10.1. Auxiliary Results

Here we give some necessary results on solutions of linear difference equations with constant coefficients and with controls $u \in \mathrm{R}^l$ in their description. The state (physical) space is R^n with $n \geq l$. The equation in the state space has the form (here x_t, u_t and ζ_t are column vectors)

$$x_{t+1} = Ax_t + Bu_t + \zeta_t, \quad z_t = Cx_t \tag{1}$$

where $\zeta_t \in \mathrm{R}^n$ is a disturbance (a noise or an external action), the quantity $z_t \in \mathrm{R}^m, m \leq n$, is observed, A, B, C are numerical matrices of the corresponding dimensions. The control u_t and the observation z_t are called *input* and *output* respectively. If the input is scalar then B is a vector and such an equation is called *one-input*. From Eq. (1) we can pass to the equation of order n represented by the "input-output" relationship

$$y_t + A_1 y_{t-1} + \cdots + A_n y_{t-n} = B_1 u_{t-1} + \cdots + b_m u_{t-m} + \zeta_t$$
$$+ C_1 \zeta_{t-1} + \cdots + C_k \zeta_{t-k} \tag{2}$$

where the coefficients can be expressed in terms of elements of the matrices A, B, and C. In turn, Eq. (2) can be reduced to the form (1).

Equation (2) can be rewritten in the following compact form. We introduce the shift-back operator $\triangle : \triangle f_t = f_{t-1}$, $\triangle^n = \triangle(\triangle^{n-1})$. Making use of operator polynomials (here I means the identity operator)

$$a(\triangle) = I + A_1 \triangle + \cdots + A_p \triangle^p,$$
$$b(\triangle) = B_1 \triangle + \cdots + B_q \triangle^q,$$
$$c(\triangle) = I + C_1 \triangle + \cdots + C_r \triangle^r$$

287

we can rewrite Eq. (2) in the form

$$a(\triangle)y_t = b(\triangle)u_{t-1} + c(\triangle)\zeta_t. \tag{3}$$

Equation (2) (or (3)) is often called the *Autoregression Moving Average one* (we shall use ARMAX for short). The behavior of the solutions of Eqs. (1), (2) is defined by the properties of the corresponding homogeneous equations

$$x_{t+1} = Ax_t + Bu_t, \quad z_t = Cx_t, \tag{4}$$

$$a(\triangle)y_t = b(\triangle)u_{t-1}. \tag{5}$$

For Eq. (1), we introduce the following matrix function of the complex argument λ

$$\Pi(\lambda) = C(\lambda I - A)^{-1}B$$

which is called the *transfer matrix*. It is a ratio between the discrete Laplace transformation of the input u and that of the output z. For Eq. (5) such a function has the form

$$\Pi(\lambda) = a(\lambda)^{-1}b(\lambda).$$

All elements of the transfer matrix are rational functions of λ with the denominator $\det(\lambda I - A)$, the characteristic polynomial of the matrix A. In the scalar case the transfer function of (5) has the following form

$$\Pi(\lambda) = \frac{b(\lambda)}{a(\lambda)} = \frac{b_1\lambda + \cdots + b_q\lambda^q}{1 + a_1\lambda + \cdots + a_p\lambda^p}.$$

Definition 1. System (4) is called *T-controllable* if for all points x', $x'' \in \mathrm{R}^n$ there exists a control $\bar{u} = (u_0, u_1, \ldots, u_{T-1})$ such that under the control $\bar{u}$ starting from $x_{t_0} = x'$ we obtain $x_{t_0+T} = x''$.

There are several equivalent formulations of this notion. The following algebraic form is commonly used.

Definition 1'. System (4) is called *T-controllable* if the rank of the $n \times nm$-matrix $(B, AB, \ldots, A^{n-1}B)$ is equal to n.

There is one more equivalent definition

Definition 1''. System (4) is *T*-controllable if

$$\sum_{i=0}^{T-1} A^i BB^T (A^T)^i > 0,$$

i.e. the matrix on the left-hand side of the inequality is positive definite.

If a system is *T*-controllable then it is *s*-controllable for $s \geq \min(n, T)$, where n is the dimension of the state vector.

If a system of order n is *T*-controllable at some $T \geq n$ then it is *n*-controllable. If this system is, in turn, *n*-controllable then it is *T*-controllable for all $T \geq n$. For this reason an *n*-controllable system will simply be called *a controllable system*.

Since controllability is defined in terms of algebraic notions one often speaks of a controllable pair of matrices (A, B).

To formulate the next property of controllable systems we consider a linear transformation of variables $x = Px'$ with a non-degenerate matrix P. System (4) takes the form

$$x'_{t+1} = A'x'_t + B'u_t, \quad y_t = C'x'_t$$

where

$$A' = P^{-1}AP, \quad B' = P^{-1}B, \quad C' = P^TC.$$

The controllability condition equivalent to the above may be formulated as follows.

There does not exist a linear non-singular transformation of variables x_i, $i = 1, \ldots, n$, such that the matrices A', B' have the form

$$A' = \begin{pmatrix} A_{11} & A_{12} \\ O & A_{22} \end{pmatrix} \begin{matrix} \}k \\ \}n-k \end{matrix}, \qquad B' = \begin{pmatrix} A_1 \\ O \end{pmatrix} \begin{matrix} \}k \\ \}n-k \end{matrix},$$
$$\underbrace{\phantom{A_{11}}}_{k} \quad \underbrace{\phantom{A_{12}}}_{n-k}$$

i.e. system (4) cannot be decomposed into two subsystems with respect to the variables $x = (x', x'')$ where

$$\begin{cases} x'_{t+1} = A_{11}x'_t + A_{12}x''_t + B_1u, \\ x''_{t+1} = A_{22}x''_t. \end{cases} \tag{6}$$

We remark that the pair (A_{11}, B_1) is controllable and, perhaps, $k = n$. From (6) it follows that the variables included in x''_t are uncontrollable, i.e. there exists a basis of the state space in which the control action B_1u has no more than k nonzero components. This enables us to say that the number of controllable equations is "greater" than the number of the uncontrollable ones. More precisely, any uncontrollable system (4) can be made controllable by changing the matrices A and B arbitrarily little. The inverse assertion is false. Interpreting the composed matrix $[A, B]$ of dimension $n \times (n + m)$ as a point of $\mathrm{R}^{n(n+m)}$ we conclude that an open, everywhere dense set of $\mathrm{R}^{n(n+m)}$ corresponds to the controllable pairs of matrices (and hence to the controllable systems as well). This topological fact is important in adaptive control.

We now focus our attention on stability of solutions of system (4) considering this property as the control aim. First, we give an asymptotic stability condition for the equation $x_{t+1} = Ax_t$:

it is necessary and sufficient that all eigenvalues of the matrix A lie inside the unit circle of the complex plain.

Such a matrix is called *stable* (or a Hurwitz' matrix). In other words, it is required that all roots λ_i of the equation $\det(\lambda I - A) = 0$ should satisfy the condition $|\lambda_i| < 1$. An equivalent stability condition is the following inequality

$$\varlimsup_{t \to \infty} \|x_t\| \le c_1 \sup_{t>0} \|u_t\| + c_2, \quad c_1, \ c_2 > 0$$

which is useful in the weak form (as $T \to \infty$)

$$\varlimsup_{T \to \infty} T^{-1} \sum_{t=1}^{T} \|x_t\| \le c_1 \varlimsup_{T \to \infty} T^{-1} \sum_{t=1}^{T} \|u_t\| + c_2, \quad c_1, \ c_2 > 0.$$

There is an often used special form of stability called — "stability with respect to the control" — which is defined by the following inequality

$$\varlimsup_{t \to \infty} \|u_t\| \le c_1 \sup_{t>0} \|x_t\| + c_2, \quad c_1, \ c_2 > 0,$$

which can also be written in the weak form. For $n = l$ this simply means that the function $\det B(\lambda)$ has no roots in the circle $|\lambda| < 1$.

The equations having this property are called *minimum phase*. These equations will be the subject of our special attention.

In the case of the homogeneous equation $x_{t+1} = Ax_t + Bu_t$ a *linear feedback* $u_t = Kx_t$ is both the simplest law of control and the one most often used. Now, the equation takes the form

$$x_{t+1} = (A + BK)x_t = \tilde{A}x_t.$$

Its solution is asymptotically stable if it is possible to find a "gain matrix" K such that $\tilde{A} = A + BK$ is a Hurwitz' matrix. This leads to the following definition.

Definition 2. System (4) is called *stabilizable* if there exists a matrix K such that $A + BK$ is a Hurwitz matrix.

The controllable equations are stabilizable and, moreover, a stronger assertion holds.

Proposition 1. *Let Eq. (4) be controllable, $\alpha(\lambda) = \lambda^n + \alpha_1 \lambda^{n-1} + \cdots + \alpha_n$ be an arbitrary polynomial with real coefficients and with roots $\lambda_1, \ldots, \lambda_n$. There exists a gain matrix K such that $\det(\lambda E - A - BK) = \alpha(\lambda)$.*

This means that by choosing the matrix K properly it is possible to make the eigenvalues of the closed loop system $x_{t+1} = Ax_t + Bu, \ u_t = Kx_t$ coincide with the numbers $\lambda_1, \ldots, \lambda_n$ given beforehand.

Definition 3. The problem that consists of finding a matrix K such that either all or some eigenvalues of the closed loop system coincide with the values given beforehand is called a *modal control problem*.

The stabilizability of uncontrollable equations takes place if the matrix A_{22} of the subsystem with respect to the variable x'' in (6) is a Hurwitz matrix.

We now consider one more notion of linear system theory.

Definition 4. System (4) is called *r-observable* if its initial state x_0 can be calculated by using the known values of observations and controls $z_1, \ldots, z_r; u_0, \ldots, u_{r-1}$.

For system (4) to be observable it is necessary and sufficient that either of the following two conditions holds:

$$\mathbf{rank}\,[C^T\ A^T C^T \cdots (A^T)^{r-1} C^T] = n, \quad \sum_{i=0}^{r-1} (A^T)^i C^T A^i > 0.$$

If an equation is r-observable, then it is s-observable for $s \geq \min(n, r)$.

System (4) is called just *observable* if it is n-observable. This property is formulated in terms of matrices. For this reason we speak of observability of a pair of matrices (C, A). The observability of the pair (C, A) is equivalent to the controllability of the pair (A^T, C^T).

The property of Eq. (4) to be controllable and observable simultaneously can be expressed as follows. "The triplet of matrices (A, B, C) is controllable and observable". We now formulate this condition using the transfer matrix (or function) of Eqs. (4) or (5). In the simplest case $n = l = 1$, a transfer function $\Pi(\lambda)$ is called *non-degenerate* if its representation having the form of a rational function has the following properties: degree of the denominator is equal to n $(= \dim X)$, the denominator and numerator being irreducible. In the general case, the transfer matrix is represented by the ratio of two polynomials: the matrix one (having degree $n - 1$) and the scalar one $\mathbf{det}\,(\lambda E - A)$. Such a matrix is called non-degenerate if it cannot be written in the form of a ratio of two irreducible polynomials, the degree of the denominator being less than n.

Proposition 2. *For Eq. (4) to be observable and controllable (in the state space) it is necessary and sufficient that its transfer function be non-degenerate.*

For Eq. (5) (in the form of the "input-output" relationship) to be observable and controllable it is necessary and sufficient that the matrix polynomials $a(\lambda)$ and $b(\lambda)$ have no common left factors.

Definition 5. System (4) is called *detectable* if there exists an $n \times l$-matrix H such that $A + HC$ is a Hurwitz matrix.

The relation between this notion and observability is the same as between stabilizability and controllability.

The stabilizability and detectability of Eq. (4) may also be expressed in terms of "stabilizability and detectability of the triplet of matrices (A, B, C)". So, some main notions concerning homogeneous linear systems

$$x_{t+1} = Ax_t + Bu, \quad z_t = Cx_t$$

and

$$a(\triangle)y_t = b(\triangle)u_{t-1}$$

have been considered. We return to the original equations containing an additive noise. Throughout this chapter the noise is assumed to be stochastic and to satisfy some restrictions.

Let $(\Omega, \mathcal{F}, \mathbf{P})$ be a probability space with a fixed increasing flow of σ-algebras $\mathcal{F}_t$, i.e. $\mathcal{F}_0 \subseteq \mathcal{F}_1 \subseteq \cdots \subseteq \mathcal{F}_r \subseteq \mathcal{F}$. Let a sequence of $\mathcal{F}_{i-1}$-measurable r.v. ξ_i on this space be given. We use the following notation $\xi = \{\xi_i, \mathcal{F}_i, i \geq 1\}$ for this sequence. Under the conditions

$$\mathbf{E}|\xi_i| < \infty, \quad \mathbf{E}(\xi_{i+1}|\mathcal{F}_i) = 0,$$

the sequence ξ_i is a martingale-difference. A sequence of independent r.v. with zero means and finite absolute moments is a particular case of such a sequence. From now on, we always assume that $\mathbf{E}(\xi_{i+1}^2|\mathcal{F}_i) < \infty$. In a number of cases we shall be forced to impose stronger restrictions, i.e. $\mathbf{E}(\xi_{i+1}^\alpha|\mathcal{F}_i) < \infty$ where either $\alpha > 2$ or even $\alpha \geq 4$. In the case of the vector martingale-difference these conditions have the form $\mathbf{E}(\|\xi_{i+1}\|^\alpha|\mathcal{F}_i) < \infty$. We consider either a martingale-difference or its moving average

$$\zeta_t = C(\nabla)\xi_t = \xi_t + c_1\xi_{t-1} + \cdots + c_r\xi_{t-r},$$

as the noise ζ_i. On the scalar or matrix polynomial $C(\lambda)$ a restriction will be imposed. This restriction is called *strictly positive reality,* defined only for a rational quadratic matrix $C(\lambda)$. It is the following condition (for some $\varkappa > 0$)

$$C(e^{i\omega}) + C^T(e^{-i\omega}) - \varkappa I = 2\text{Re } C(e^{i\omega}) - \varkappa I > 0,$$

i.e. the left-hand side is a positive definite Hurwitz matrix. This property is equivalent to the following. Let $C(\lambda)$ be a transfer matrix (function) that relates the sequences z_t, x_t in the following way

$$z_t = [C(\lambda) - \varkappa' I]x_t, \quad \varkappa' > 0.$$

The matrix $C(\lambda)$ is strictly positive real (SPR for short) if and only if the inequality

$$\sum_{n=1}^{t} z_n^T x_n + \varkappa \geq 0, \quad \forall t$$

is satisfied. We note again that the matrix $C(\lambda)$ is SRP if all zeros of $\det C(\lambda)$ are outside the closed unit circle of the complex plain.

For Eqs. (1) or (2) the limiting average reward (per step) is chosen as the objective function having the form of a quadratic functional. In the case of the weak aim

$$W(u, x_0) = \lim_{T \to \infty} T^{-1} \sum_{t=1}^{T} \mathbf{E}_{u,x_0}\left[x_t^T P x_t + u_t^T Q u_t\right], \quad P \geq 0, \ Q > 0,$$

and in the case of the strong aim

$$V(u, x_0) = \lim_{T \to \infty} T^{-1} \sum_{t=1}^{T} \left[x_t^T P x_t + u_t^T Q u_t\right].$$

The argument u on the left-hand sides serves as a symbol of the strategy σ. We are interested in answering the following question: for which classes of equations are the proposed objective functions suitable, i.e. these functions are finite (the second one a.s.) at every x_0 and with disturbances of rather general form (for example, at a martingale-difference with a finite second moment)? The linear feedbacks $u_t = Kx_t$ with a constant amplification matrix K of dimension $m \times n$ are chosen as the control strategy.

Theorem 1. *The functions $W(u, x_0)$ and $V(u, x_0)$ are finite (the second a.s.) if and only if the equation $x_{t+1} = Ax_t + Bu_t$ is stabilizable.*

In adaptive control of linear equations the parameters (coefficients) of equation are unknown in advance. Many methods of constructing adaptive strategies are based on using estimates of these parameters. The question of choice of convenient and effective method of estimating, i.e. identification arises. A brief survey of identification methods is given below.

Consider the vector equation of ARMAX-type

$$\bar{x}_t = A_1\bar{x}_{t-1} + \cdots + A_p\bar{x}_{t-p} + B_1\bar{u}_{t-1} + \cdots + B_q\bar{u}_{t-q} + \bar{\zeta}_t \tag{7}$$

where $\bar{\zeta}_t$ ($\in \mathrm{R}^n$) is a martingale-difference defined on a probability space $(\Omega, \mathcal{F}, \mathbf{P})$ with an increasing flow of σ-algebras $\mathcal{F}_t = \sigma(\zeta_i, \ i \leq t)$ and let $\mathbf{E}(\bar{\zeta}_t\bar{\zeta}_t^T|\mathcal{F}_{t-1}) < \infty$. The control $u_t \in \mathrm{R}^l$ is a non-anticipated function depending on the past history, in other words, it is $\mathcal{F}_{t-1}$-measurable. For the parameters of the equation and finite history up to time t we introduce the following notation

$$\theta = [A_1, \ldots, A_p; B_1, \ldots, B_q]^T, \quad \Phi_t = [x_t^T, \ldots, x_{t-p+1}^T; u_t^T, \ldots, u_{t-q+1}^T]^T,$$
$$X_t = [x_1^T, \ldots, x_t^T]^T, \quad \Psi_t = [\Phi_0^T, \ldots, \Phi_t^T], \quad H_{t-1} = [\zeta_1^T, \ldots, \zeta_t^T],$$

Φ_{-1} being chosen arbitrarily. The collection Φ_t is called the *stochastic regressor*. The original equation can be written in one of two forms:

$$x_t = \theta^T\Phi_{t-1} + \zeta_t, \quad X_t = \Psi_{t-1}\theta^T + H_{t-1}.$$

An optimality criterion of the estimate θ_t of the matrix θ can be defined as follows

$$\|X_t - \Psi_{t-1}\theta_t\|^2 = \min_{\theta} \|X_t - \Psi_{t-1}\theta\|^2,$$

so the desired estimate θ_t minimizes the "mean square error", i.e. it is the estimate in the *least square* sense. We denote this estimate by LSE. Naturally, the required estimate θ_t is a function of the observable history of the process. Always assuming that

$$\Psi_t^T\Psi_t = \sum_{i=1}^{t} \Phi_i\Phi_i^T > 0 \tag{8}$$

we obtain the LSE θ_t in explicit form

$$\theta_t = (\Psi_{t-1}^T \Psi_{t-1})^{-1} \Psi_{t-1}^T X_t = \left(\sum_{j=1}^{t-1} \Phi_j \Phi_j^T\right)^{-1} \left(\sum_{j=0}^{t-1} \Phi_j x_{j+1}^T\right). \tag{9}$$

The basic condition (8) means that the "information matrix" $\Psi_t^T \Psi_t$ must be non-degenerate. In constructing the consistent LSE for linear equations containing stochastic disturbances the main difficulty is the fulfilment of this condition.

The proposed scheme of calculating the LSE has one disadvantage that prevents from using it in applications. Indeed, at every step we need to recompute all the elements of the two matrices and to invert one of them. For this reason it is lucky that we can write the relationship (9) in the easily verified recurrent form:

$$\begin{aligned}
\theta_{t+1} &= \theta_t + a_t P_t \Phi_t (x_{t+1}^T - \Phi_t^T \theta_t) = \theta_t + P_t \Phi_{t+1}(x_{t+1}^T - \Phi_t^T \theta_t), \\
P_{t+1} &= P_t - a_t P_t \Phi_t \Phi_t^T P_t, \quad P_0 > 0
\end{aligned} \tag{10}$$

where

$$P_t = \left(\sum_{i=0}^{t-1} \Phi_i \Phi_i^T\right)^{-1}, \quad a_t = (1 + \Phi_t^T P_t \Phi_t)^{-1}.$$

An arbitrary deterministic matrix θ_0 of appropriate dimension and $P_0 = (np + lq)I$ are chosen as the initial values. Then the matrix P_t takes the following form

$$P_t = \left(\sum_{i=1}^{t-1} \Phi_i \Phi_i^T + (np + lq)^{-1} I\right)^{-1}.$$

We introduce the quantity

$$r_t = \mathbf{tr}\, P_{t-1}^{-1} = 1 + \sum_{i=0}^{t-1} \|\Phi_i\|^2.$$

It is a nondecreasing function as $t \to \infty$. For the LSE to be consistent it is important that the quantity r_t should increase unboundedly. Otherwise, the estimates θ_t tend to some value depending on the initial value θ_0. We now focus attention on obtaining conditions for strong consistency of the LSE.

To understand the obstacles connected with this we consider the scalar version of Eq. (7) where the noise ζ_t is a martingale-difference. The LSE (9) can be written as follows

$$\theta_t = \theta + P_{t-1} \sum_{i=0}^{t-1} \Phi_i \zeta_{i+1}.$$

Hence its limiting properties (as $t \to \infty$) are defined by the second term on the right-hand side of this equality. We consider this in detail.

Let the symbols $\lambda_{\max}(A)$ and $\lambda_{\min}(A)$ denote the maximal and minimal eigenvalues of the matrix A. Consider the measurable set

$$\bar{\Omega} = \left\{\omega : \lim_{t\to\infty} \lambda_{\min}(P_t^{-1}) = \infty;\ \frac{P_t^{-1}}{\mathbf{tr}\, P_t^{-1}} \geq \delta I\right\} \subset \Omega$$

were the matrix inequality takes place for some $\delta > 0$ and for all sufficiently large t.

Theorem 2. *If* $\mathbf{E}(\xi_t^2|\mathcal{F}_{t-1}) \leq d < \infty$ *then for the scalar Eq. (7), the LSE is strongly consistent on the set* $\bar{\Omega}$.

Proof. We write the LSE θ_t in the form

$$\theta_t = \theta + \left(\frac{P_{t-1}^{-1}}{r_{t-1}}\right)^{-1} \frac{\sum_{i=0}^{t-1} \Phi_i \zeta_{i+1}}{r_{t-1}}. \tag{11}$$

By the assumptions, the first factor on the right-hand side is separated from zero. We show that the second factor tends to zero. We consider the random vector $\sum_{i=0}^{t-1} \Phi_i \zeta_{i+1}$. Let Φ_l^i, $i = 1, \ldots, p+q$, be the components of the vector Φ_l. We put

$$r_t^i = 1 + \sum_{l=0}^{t} (\Phi_l^i)^2, \quad r_0^i = r_{-1}^i = 0,$$

$$\eta_t^i = \sum_{l=0}^{t-1} \frac{\Phi_l^i \zeta_{l+1}}{r_l^i}, \quad t \geq 1, \quad i = 1, \ldots, p+q.$$

It is clear that the r.v. r_t^i, η_t^i are $\mathcal{F}_t$-measurable. We have

$$\mathbf{E}(\eta_{t+1}^i|\mathcal{F}_t) = \mathbf{E}\left(\eta_t^i + \frac{\Phi_t^i \zeta_{t+1}}{r_t^i}\bigg|\mathcal{F}_t\right) = \eta_t^i$$

and, hence, $(\eta_t^i, \mathcal{F}_t)$ are martingales. Moreover, it is easy to prove that they are square integrable, i.e. $\mathbf{E}(\eta_t^i)^2 \leq d < \infty$. According to a well-known theorem on convergence of martingales, the following finite limit exists

$$\lim_{t\to\infty} \sum_{l=0}^{t} \frac{\Phi_l^i \zeta_{l+1}}{r_l^i} < \infty \quad \text{a.s.} \tag{12}$$

The equality $\lim_{t\to\infty} r_t^i = \infty$ is satisfied on the set $\bar{\Omega}$. Applying Kronecker Lemma to (12) we obtain

$$\lim_{t\to\infty} (r_t^i)^{-1} \sum_{l=1}^{t} \Phi_l^i \zeta_{l+1} = 0, \quad 1 \leq i \leq p+q.$$

Noting that $r_t^i = 1 + \sum_{l=0}^{t}(\Phi_l^i)^2 \leq 1 + \mathbf{tr}\left(\sum_{l=0}^{t} \Phi_l \Phi_l^T\right)$, we finally see that

$$\lim_{t\to\infty} r_t^{-1} \sum_{l=0}^{t} \Phi_l^i \zeta_{l+1} = 0, \quad 1 \leq i \leq p+q.$$

According to (11), this means strong consistency of the LSE. $\qquad\square$

This theorem giving one of many sufficient conditions for strong consistency of LSE suggests a way of attaining it. For the LSE to be strongly consistent it is necessary that the measure of the set $\bar{\Omega}$ is equal to one. There are two ways to achieve this. The first consists in narrowing the class of the equations considered. The second is to randomize the control (input) signals. Examples are given below.

We now state one more result on convergence of the LSE for Eq. (7) with $\zeta_t = Fw_t$, where F is a constant matrix, w_t is a martingale-difference with respect to the non-decreasing flow of σ-algebras $\mathcal{F}_t(w_s, s \leq t)$ such that

$$\mathbf{E}(w_t|\mathcal{F}_{t-1}) = 0, \quad E(w_t w_t^T|\mathcal{F}_{t-1}) = I.$$

Let $\lambda_{\min}$ $(\lambda_{\max})$ be the minimal (maximal) eigenvalue of the matrix P_t and $r_t = \mathbf{tr}\, P_t^{-1}$.

Theorem 3. *Let* $\lim_{t \to \infty} r_t = \infty$ *a.s.*, $\lambda_{\max} \leq r_t^{a-1}\gamma, a \in [0, 1/2)$, *where* γ *is a finite r.v. If the initial matrix* $P_0 = (np + lq)I$, *then the LSE* θ_t *is strongly consistent and*

$$\|\theta_t - \theta\| = O\big(r_t^{b-1/2}\big), \quad b \in (a, 1/2].$$

We turn to ARMAX type equations containing the correlated ("coloured") disturbance

$$A(\triangle)x_t = B(\triangle)u_{t-1} + C(\triangle)\xi_t$$

where

$$\begin{aligned}
A(\lambda) &= I - A_1\lambda - \cdots - A_p\lambda^p, \\
B(\lambda) &= B_1 + B_2\lambda + \cdots + B_q\lambda^{q-1}, \\
C(\lambda) &= I + C_1\lambda + \cdots + C_r\lambda^r.
\end{aligned}$$

The disturbance ξ_t has the known form $\xi_t = Fw_t$, but it is unknown beforehand and unobserved. We need to estimate the matrix coefficients of these three polynomials, i.e. the components of the following vector

$$\theta = [A_1 \cdots A_p \, B_1 \cdots B_q \, C_1 \cdots C_p].$$

Its estimate θ_t at time t is based on the observable history of the process.

$$\Phi_t = \begin{cases} \left[x_t^T \cdots x_{t-p+1}^T u_t^T \cdots u_{t-q+1}^T\right]^T & \text{for } r = 0, \\ \left[x_t^T \cdots x_{t-p+1}^T u_t^T \cdots u_{t-q+1}^T \, x_t^T - \Phi_{t-1}^T\theta_t \cdots x_{t-r+1}^T - \Phi_{t-r}^T\theta_{t-r+1}\right]^T \\ \hspace{8cm} \text{for } r \geq 1. \end{cases} \tag{13}$$

In the case $r \geq 1$, the last r components of Φ_t imitate the values of the unobservable disturbance on the time interval $[t, t - r + 1]$. The LSE of the parameters for ARMAX are computed by using the recurrent formulas (10). The initial value $P_0 = [np + lq + nr]I$, the sense and dimension of the vector Φ_t are only altered. The convergence of this procedure will be studied in the next sections devoted to concrete problems of adaptive control of linear difference equations. This procedure is referred to as the *extended least square* (ELS) method. Besides the least square method, other methods are used as well. Here we consider only the *stochastic gradient method* (SGM) given by the following recurrent procedure

$$\theta_{t+1} = \theta_t + \frac{\varphi_t}{r_t}\left[x_{t+1}^T - \varphi_t^T\theta_t\right]$$

where the stochastic regressor

$$\varphi_t = [x_t^T \cdots x_{t-p+1}^T u_t^T \cdots u_{t-q+1}^T \; x_t^T - \varphi_{t-1}^T \theta_{t-1} \cdots x_{t-r+1}^T - \varphi_{t-r}^T \theta_{t-r}]^T$$

denotes the history of the process and

$$r_t = 1 + \sum_{i=1}^{t} \|\varphi_i\|^2, \quad r_0 = 1.$$

The advantage of this method is simplicity. In fact, it is not required to calculate the matrices P_t. As shown below, the estimates obtained by using SGM are strongly consistent under the appropriate conditions. However, more profound results connected with the use of this method (in contrast with the ELS method) are unknown yet.

10.2. Control of Homogeneous Equations $x_{t+1} = Ax_t + Bu_t$

Linear equations containing controls are an important type of objects in the classical theory of regulating. The simplest objects are represented by the homogeneous equations

$$x_{t+1} = Ax_t + Bu_t \tag{1}$$

with an initial value x_0. The solution x_t is assumed to be observed. In studying these equations external disturbances and noise influencing the behavior of the controlled object are not taken into account. Moreover, we are not trying to provide insensibility (robustness) of the methods proposed here with respect to these factors.

To solve the adaptive control problems we are interested in, the identification approach will be used. First, we assume that u_t is scalar, i.e. we deal with a one-input equation having the form

$$x_{t+1} = Ax_t + bu_t, \tag{2}$$

where $b = (b_1, \ldots, b_n)^T$ is a vector ($m = 1$). By using the current observations x_t, we need to compute $n^2 + n$ numbers a_{ij} and b_l which are the elements of the matrix A and the vector b, respectively. This can be done under appropriate assumptions.

Let $u_t = 0$. By (2), one finds the following vectors

$$x_1 = Ax_0, \quad x_2 = Ax_1 = A^2 x_0, \ldots, x_{n_1-1} = Ax_{n_1-2} = A^{n_1-1} x_0, \tag{3}$$

where n_1 is the maximal number of linearly independent vectors $x_0, x_1, \ldots, x_{n_1-1}$. In the case when these iterations generate the whole space R^n (i.e., $n_1 = n$), we can find the elements of the matrix A by solving the system of linear (with respect to a_{ij}) algebraic equations $x_i = Ax_{i-1}$, $i = 1, 2, \ldots, n$. The control u_t is assumed to be equal to zero on the interval $[0, n-1]$ but at $t = n$ we put $u_n = 1$. Then we can find the vector b by (2). This procedure is repeated $n + 1$ times. This number cannot be reduced.

Now let $n_1 < n$. The matrix A can be defined on a subspace R^{n_1} generated by the base $x_0, Ax_0, \ldots, A^{n_1-1} x_0$. It remains to explain how and under what conditions it can be defined on the complementary subspace R^{n-n_1}. Obviously, the matrix A may be completely calculated, i.e. identified, if and only if the vectors

$$b, \ Ab, \ldots, A^{n-n_1-1}b \tag{4}$$

form a base of R^{n-n_1}. To define A on the complementary subspace we should put $u_t = 1$ starting from $t = n_1$. Then explicit expressions of x_t will contain for $t \geq n_1$ the terms b, Ab etc. This gives the missing groups of linear algebraic equations for defining the elements of the matrix A. This procedure is repeated $n + 2$ times, this value being minimal. The following assertion is evident.

Theorem 1. *Let $n_1 \geq 1$ be the dimension of the space generated by the vectors $\{A^l x_0\}$. A necessary and sufficient condition of identifying the parameters of the one-input equation (2) is the linear independence of the vectors:*

(a) $x_0, \ Ax_0, \ldots, A^{n_1-1}x_0, \ b, \ Ab, \ldots, A^{n-n_1-1}b$ *if $n_1 < n$. The identification procedure is repeated $n + 2$ times.*

(b) $x_0, \ Ax_0, \ldots, A^{n-1}x_0$ *if $n_1 = n$. The identification procedure is repeated $n + 1$ times.*

The number of repetitions cannot be reduced in either case.

The initial state x_0 can be arbitrary. Therefore it may belong to any invariant subspace R^{n_i} of the operator A. Thus the vectors from the collection (4) must be linearly independent on the complement to any of these subspaces. If $n_1 = 0$ this condition is linear independence of the n vectors $b, \ Ab, \ldots, A^{n-1}b$. As concerns Eq. (2), this means the presence of the controllability property. The main result follows.

Theorem 2. *Equation (2) is identifiable at any initial value if and only if it is controllable.*

We now consider the multi-input equation (1), where $B = [b_1 \ b_2 \ \cdots \ b_m]$ is some $n \times m$-matrix having columns b_i. The initial stage in estimating the parameters is as in the case of the one-input equation. A difference is that a control may be chosen in various manners. Having finished forming a subspace generated by the vectors $\{A^l x_0\}$ and made sure that the vector $A^{n_1}x_0$ belongs to the linear span of the preceding vectors we designate the control not equal to zero, namely, $u' = (1, 0, \ldots, 0)^T$. This control is used only once as the input of the linear system $x_{t+1} = Ax_t + Bu'$ that operates thereafter with controls equal to zero. The solution of such a system has the form $x_{t+1} = A^t x_0 + A^{t-t'} b_1$ where b_1 is the first column of the matrix B. At this time a new group of linear algebraic equations with respect to the elements of the matrices A and B which are remaining unknown is formed. At some instant t'' a value $x_{t''}$, that depends linearly on the preceding ones, appears. Then we designate a new control $u'' = (0, 1, 0, \ldots, 0)^T$ and the procedure described above has to be repeated again. We act m times in this manner. The latest control (in the mth stage) has the form $u^{(m)} = (0, \ldots, 0, 1)^T$. If we obtain $n + m$ groups of compatible and independent linear algebraic equations, we can define all elements of the matrices A and B. The procedure described is repeated the number of times that lies inside the boundaries, which are not possible to improve, from $n + m$ to $n + 2m - 1$.

We shall soon specify the conditions under which identification will take place, but first we explain the order of calculation by considering an initial stage. On forming the group of linear equations $x_1 = Ax_0, \ldots, x_{n_1-2} = Ax_{n_1-3}, x_{n_1-1} = Ax_{n_1-2}$ we make sure that at time $t = n_1$ the vector x_{n_1} belongs to the linear span of these vectors. This takes the first period of time. In the next one, we designate some nonzero control u'. As a result of this, the observable vector $x_{n_1+1} = Ax_{n_1} + b_1$ is obtained. Since the term Ax_{n_1} is known, the vector $b_1 = x_{n_1+1} - Ax_{n_1}$ is calculated exactly. In a similar way we spend one period of time in the following stages to verify that the set of vectors $x_0, Ax_0, \ldots, A^{n_1-1}x_0; b_1, Ab_1, \ldots, A^{n_2-1}b_1; b_2, Ab_2, \ldots, A^{n_3-1}b_2;$ and so on obtained on the preceding stages does not enable us to construct a basis of the whole space R^n and to calculate the corresponding column of the matrix B. The number of periods of time which are necessary to identify the parameters' equation will be least if the vector x_0 generates the whole space R^n, i.e. the vectors $x_0, Ax_0, \ldots, A^{n-1}x_0$ are linearly independent. Their construction takes n periods of time. Further, for the elements of the matrix B to be calculated it is necessary to spend m periods of time more. If the vectors mentioned above do not form a basis of R^n, then when supplementing them with the vectors b_i, $Ab_i, \ldots$, we shall use two additional periods every time (see the explanation on the initial stage). In the last stage when the matrix A is already known, only one period of time is needed to compute the last column of the matrix B, i.e. the vector b_m.

It remains to specify the conditions under which the described procedure enables us to identify the matrices A and B in Eq. (1). These conditions have been stated in Theorem 2 which also remains valid in the case of a multi-input Eq. (1).

It is important to emphasize that if the algorithm described above cannot be used, then Eq. (1) is uncontrollable, i.e. *no* algorithm allows calculating exactly all elements of the matrices A and B (for any initial values). The alternative, workable algorithms are asymptotical, i.e. they produce the desired values only in the limit as $t \to \infty$. The procedure proposed above is finite. Therefore, it is much simpler and more economical.

We now consider the case when the states x_t are not observed. Instead of the state vector x_t, the l-vector $z_t = Cx_t$, where C is some known $n \times l$-matrix, is observed. Hence we deal with a partially observable model. The question is to find the conditions under which this model could be identified. The following theorem answers this question.

Theorem 3. *The partially observable equation* (1) *is identifiable (at arbitrary initial values) if and only if the following two conditions hold*

1. *Equation* (1) *is controllable.*
2. *The matrix C is quadratic and non-singular.*

Sufficiency of these conditions is obvious but the proof of necessity is left to the reader.

Let us briefly consider the situation connected with the non-homogeneous equation

$$x_{t+1} = Ax_t + Bu_t + f_{t+1}$$

where f_t denotes an external disturbance which is assumed either to be completely known or to be observed in the course of control.

This procedure can be extended in a natural way to the equation

$$x_t + a_1 x_{t-1} + \cdots + a_p x_{t-p} = b_1 u_{t-1} + \cdots + b_q u_{t-q}, \quad p > q \geq 1.$$

For simplicity we restrict attention to the case of a scalar equation with initial values $x_0, x_1, \ldots, x_{p-1}$. To solve the identification problem, we must preliminarily reduce this equation to the one in the state space. We can use the procedure described above provided the new equation turns out to be controllable. This takes place if the polynomials $a(\lambda) = 1 + a_1\lambda + \cdots + a_p\lambda^p$ and $b(\lambda) = b_1 + b_2\lambda + \cdots + b_q\lambda^{q-1}$ are irreducible.

We now study the main topic of our consideration, adaptive control of homogeneous linear equations.

The first problem consists of moving from the point x' to the point x'' of the states space in a given time T. This problem cannot evidently be solved by using the asymptotic methods mentioned above. But it can be done by using the finite procedure described if time T is not too small. The value T must be greater than the identification time (i.e. $T \geq n + 2m - 1$). The guaranteed time to reach the point x'' is $T = 2n + 2m - 1$.

The next problem consists of seeking a feedback $u_t = Lx_t$ such that the matrix $A + BL$ has a given collection of eigenvalues. This is the so-called pole-zero assignment problem. On calculating the elements of the matrices A and B at most in $n + 2m - 1$ time periods it is solved with the help of a well-known procedure from linear systems theory.

The third problem is known either as the *control with model reference* or as the *tracking problem*. Let z_t be a given path belonging to the space R^n. It can be either the solution of the equation $z_{t+1} = A'z_t + B'\bar{u}_t$ where the coefficients A', B' and control $\bar{u}$ are known, the matrix A' is stable or arbitrary, but the sequence of the vectors $z_0, z_1, \ldots, z_t, \ldots$ is bounded. In both cases we need to choose the control so that the solution of the equation considered and the given path of the "model" should approach each other, i.e. the error $e_t = x_t - z_t$ should tend to zero as $t \to \infty$.

Let us assume that the original equation has the form

$$x_{t+1} = (A + A_1(t))x_t + (B + B_1(t))u_t.$$

Here A_1, B_1 denote the parametric control (acting on the coefficients of the equation) and A, B are some constant unknown matrices forming a controllable pair.

To construct the adaptive control with respect to the class of such equations it is necessary to identify the matrices A and B on the assumption that $A_1 = B_1 = 0$. After calculating the matrices A, B (at most for $n + 2m - 1$ periods of time) we can solve the problem of transition from the state $x_{t'}$ into the state $z_{t'+n}$ of the reference model. At time $t'' = t' + n$ when $x_{t'+n} = z_{t'+n}$ we shall put $A_1 = A' - A$, $B_1 = B' - B$ and $u_t = \bar{u}_t$. Thus $x_t = z_t$ for $t > t'$ and the desired aim will be reached at $u_t = \bar{u}_t$.

The last problem is optimization. It consists of seeking a control u^0 such that the following objective function

$$W(u, x_0) = \overline{\lim_{T \to \infty}}\, T^{-1} \sum_{t=0}^{T} (x_t P x_t + u_n t u_t), \quad P \geq 0,\ Q > 0$$

has its minimal value for Eq. (1). Within the framework of the classical theory the solution of this problem is well-known. Therefore, we should estimate the matrices A, B for a finite time, and use the classical optimization methods for Eq. (1).

In conclusion we mention some nonlinear equations to which the identification procedure described above can be extended. In various technological and medical problems bilinear equations are used. In the simplest cases they have the form

$$x_{t+1} = A x_t + b u_t + \sum_{j=1}^{L} u_{t-j} C_j x_j + f_{t+1}$$

where x_t is a vector of order n, u_t is a scalar control, f_t is an external influence, b is a column-vector, $A, C_1, \ldots, C_l$ are some $n \times n$-matrices. Sometimes, we have to apply the adaptive approach to control classes of such equations. As a result of this different assumptions about the sets of unknown parameters are made. We briefly consider the situation when A and b are unknown, x_t is observed exactly and $f_t \equiv 0$ (though we could assume that f_t is some function known beforehand.)

First, we put $u_t = 0$ and form groups of linear equations $x_{j+1} = A x_j$, $j = 0, 1, \ldots, n' - 1$, up to the moment $t = n'$, when the next state $x_{n'}$ belongs to the span of the preceding vectors x_j, $j \leq n' - 1$. At time $t = n'$ we designate the control $u_{n'} = 1$. This makes possible finding the vector b. Further we continue to track the solution. It enables us to find the missing equations to compute the matrix A completely. This can be done for arbitrary initial values if the pair (A, b) is controllable. We note that the terms $C_j x_t$ are known. Hence there are no obstacles to the identification process.

In a similar way we could consider the more complicated (nonlinear) equations

$$x_{t+1} = a x_t + b u_t + h(x_t, u_t; \theta)$$

containing the parameters θ in such a manner that the algebraic (or transcendental) equations

$$h(x, u; \theta) = h_0$$

are solvable with respect to θ. But this is beyond the scope of the general theory. Here we only discuss some special cases (bilinearity).

10.3. Optimal Tracking Problem for ARMAX

We discuss the following problem. Let a movement x_t be defined by the equation $x_t = h(x_{t-n}^{t-1}, u_{t-n}^{t-1})$ in the space R^k where u_t is a control. Besides, there is a deterministic trajectory (a reference) $\{x_t^*\}$ defined either by the appropriate equation of a model reference or be arbitrary. We assume that the model reference satisfies the condition

$$|x_t^*| \le g \quad \forall t.$$

The tracking problem consists of finding a control such that the movement x_t approaches (or coincides accurately) with the reference x_t^*, i.e.

$$\lim_{t \to \infty} \|x_t - x_t^*\| = 0.$$

In the general case the solution of this problem is difficult. For this reason we restrict attention to the linear equation

$$x_t = a_1 x_{t-1} + \cdots + a_p x_{t-p} + b_1 u_{t-1} + \cdots + b_q u_{t-q}, \quad b_1 \ne 0.$$

In the scalar case $(k = 1)$ this problem has a simple solution, namely, the control u_{t-1} can be computed from the following relationship

$$a_1 x_{t-1} + \cdots + a_p x_{t-p} + b_1 u_{t-1} + \cdots + b_q u_{t-q} = x_t^*. \tag{1}$$

Here we assume that the values $x_{t-1}, \ldots, x_{t-p}, u_{t-2}, \ldots, u_{t-q}$ are known. This procedure implies the coincidence of the given movement with the reference path. In the multi-dimensional case, Eq. (1) should satisfy some restrictions to guarantee its solvability in u_{t-1}.

Some difficulties arise when the equation describing the initial movement contains unpredictable or unobservable random factors. For instance, this is the case for the ARMAX equation

$$x_t = a_1 x_{t-1} + \cdots + a_p x_{t-p} + b_1 u_{t-1} + \cdots + b_q u_{t-q} + \zeta_t$$

where the random variables ζ_t are independent, identically distributed with finite variance σ^2. The choice of control at time $t-1$ in accordance with relation (1) leads to the state value $x_t = x_t^* + \zeta_t$ at time t. In view of the equality $\mathbf{E}(x_t - x_t^*)^2 = \mathbf{E}\zeta_t^2 = \sigma^2 > 0$, the paths x_t and x_t^* differ for all t within the given variance. Hence, they cannot be very close. So, the aim of control in the tracking problem has to be stated in another form. The following is rather natural.

The equality

$$\lim_{T \to \infty} T^{-1} \sum_{t=0}^{T} (x_t - x_t^*)^2 = \sigma^2.$$

holds with probability one.

In all cases mentioned above this aim has been attained (in the deterministic case $\sigma^2 = 0$).

We now turn to the adaptive version of the tracking problem. We consider a *scalar* linear ARMAX equation. In the operator form it can be written as follows

$$A(\triangle)x_t = \triangle^d B(\triangle)u_t + C(\triangle)\zeta_t, \qquad (2)$$

where $d \geq 1$ is the delay by the control. The polynomials A, B, C have the following form

$$A(\lambda) = 1 - a_1\lambda - \cdots - a_p\lambda^p, \quad B(\lambda) = b_0 + b_1\lambda + \cdots + b_q\lambda^q,$$
$$C(\lambda) = 1 + c_1\lambda + \cdots + c_r\lambda^r,$$

their coefficients being known.

The noise ζ_t is defined on a probability space $(\mathbf{\Omega}, \mathcal{F}, \mathbf{P})$ with a fixed, non-decreasing flow of σ-algebras $\{\mathcal{F}_t\}$, ζ_t being $\mathcal{F}_t$-measurable. Let us suppose that ζ_t forms a martingale-difference and $\mathbf{E}(\zeta_t^2|\mathcal{F}_{t-1}) = \sigma^2$ a.s.

The aim of control may take two forms:

1. (**weak**) *For any ARMAX equation from a class $\mathcal{K}$ the following relationships hold*

$$\varlimsup_{T\to\infty} T^{-1}\sum_{t=1}^{t} x_t^2 < \infty, \quad \varlimsup_{T\to\infty} T^{-1}\sum_{t=1}^{t} u_t^2 < \infty,$$

$$\lim_{T\to\infty} T^{-1}\sum_{t=1}^{T} \mathbf{E}\{(x_t - x_t^*)^2|\mathcal{F}_{t-d}\} = \sigma^2.$$

2. (**strong**) *It coincides with the aim stated above except for the last equality which should be replaced by*

$$\lim_{T\to\infty} T^{-1}\sum_{t=1}^{T}(x_t - x_t^*)^2 = \sigma^2 \quad \text{a.s.}$$

Let us find a strategy which ensures the attainment of these aims with respect to some class. First, we introduce some notation. The column vector θ denotes the collection of unknown coefficients of Eq. (2)

$$\theta = (-\alpha_1,\ldots,-\alpha_p; \ b_0,b_1,\ldots,b_q; \ c_1,\ldots,c_r).$$

and

$$\theta_t = (-\alpha_1(t),\ldots,-\alpha_p(t); \ \beta_0(t),\beta_1(t),\ldots,\beta_q(t); \ \gamma_1(t),\ldots,\gamma_r(t)).$$

is an estimate of this vector. The column vector Φ_t of order $p + q + r$, called the stochastic regressor, denotes the observations at time t. At time $t - 1$ it has the form

$$\Phi_{t-1} = [x_{t-1},\ldots,x_{t-p}; \ u_{t-1},\ldots,u_{t-q}; \ -\bar{x}_{t-1},\ldots,-\bar{x}_{t-r}],$$

where the notation $\bar{x}_t = \Phi_{t-1}^T \theta_t$ is used. The control u_{t-1} at time $t-1$ is calculated by the equation

$$\Phi_{t-1}^T \theta_{t-1} = x_t^*, \tag{3}$$

which is similar to Eq. (1) but does not coincide with it in the non-adaptive version of the original deterministic or stochastic equation. It remains to give the transformation rules of the vectors Φ_t, θ_t. For the vector Φ_t it consists of shifting current information in time. The estimates θ_t can be found by the *modified least square method*, i.e.

$$\theta_{t-1} = \theta_t + \frac{P_{t-1}\Phi_t}{1 + \Phi_t^T P_{t-1}\Phi_t}[x_{t+1} - x_t^*], \tag{4}$$

and the matrices P_t are transformed in the following way[a]

$$P_t = c_{t-1}\left[P_{t-1} - \frac{P_{t-1}\Phi_t\Phi_t^T P_{t-1}}{1 + \Phi_t^T P_{t-1}\Phi_t}\right], \quad P_1 > 0. \tag{5}$$

The numbers c_t are defined by the equality

$$c_t = \frac{L}{\max\{L, \tilde{r}(t)\lambda_{\max}[P_t']\}}, \quad L > 0 \tag{6}$$

where the number L is given in advance, $\tilde{r}(t) = \prod_{l=1}^{T}(1 + \Phi_l^T P_{l-1}\Phi_l)$ and $\lambda_{\max}[P_t']$ means the maximum eigenvalue of the matrix

$$P_t' = P_{t-1} - \frac{P_{t-1}\Phi_t\Phi_t^T P_{t-1}}{1 + \Phi_t^T P_{t-1}\Phi_t} = c_{t-1}^{-1}P_t.$$

Obviously, $0 < c_t \le 1$ and it is easy to show that

$$\tilde{r}_t\lambda_{\max}[P_t] \le L, \quad \tilde{r}_t \le Lr_t. \tag{7}$$

The equalities (3)–(6) define a strategy denoted by GS. It depends on the parameters p, q, r of the original Eq. (2).

We now describe a class of linear equations for which the strategy GS is optimal with respect to both aims stated above. We denote by $\mathcal{K}(d, p, q, r)$ the class of scalar ARMAX equations of the form (2) with known parameters d, p, q and r. Let the following conditions hold.

1. The equations from this class are minimum phase, i.e. moduli of all roots of the polynomial $B(\lambda)$ are greater than one.
2. Either $r = 0$ or $r \ge 1$ and the function $C^{-1}(\lambda) - 1/2$ is strictly positive real (SPR) but moduli of all roots of the polynomial $C(\lambda)$ are greater than one.

The final result is as follows.

---

[a]This complication is designed to reduce the destructive effect of the unbounded ratio of the maximal $(\lambda_{\max}(t))$ and minimal $(\lambda_{\min}(t))$ eigenvalues of the matrix P_t respectively, i.e.

$$\varlimsup_{t\to\infty} \frac{\lambda_{\max}(t)}{\lambda_{\min}(t)} = \infty.$$

If the above limit were finite, the LSM would have the usual form.

Theorem 1. *The strategy GS ensures the attainment of the weak aim in the optimal tracking problem for bounded deterministic paths with respect to the class* $\mathcal{K}(d, p, q, r)$. *In addition, if* $\mathbf{E}(\zeta_t^4 | \mathcal{F}_{t-1}) < \infty$ *a.s., then the strong aim is attainable as well.*

Proof. The following r.v.

$$\varkappa = \sum_{t=1}^{\infty} \frac{[e(t) - \zeta_t]^2}{r(t-1)}, \quad r(t) = r(t-1) + \Phi_t^T \Phi_t, \quad r(-1) = 0 \tag{8}$$

plays the main role in the proof. We assume the following:

(a) There is a sequence of $\mathcal{F}_{t-1}$-measurable r.v. ζ_t satisfying

$$\mathbf{E}(\zeta_t | \mathcal{F}_{t-1}) = 0, \quad \mathbf{E}(\zeta_t^2 | \mathcal{F}_{t-1}) = \sigma^2, \quad \overline{\lim_{T \to \infty}} \, T^{-1} \sum_{t=1}^{T} \zeta_t^2 < \infty, \quad \text{a.s.}$$

(b) $e(t) = x_t - x_t^*$ where x_t^* and x_t are $\mathcal{F}_{t-1}$-measurable.
(c) $\{r(t)\}$ is a non-negative, non-decreasing sequence, $r(t)$ being $\mathcal{F}_t$-measurable.

From these assumptions we see that we may digress from the discussed problem and consider a more general situation. As concerns the r.v. $\varkappa$ we are interested in finding the conditions under which it is proper, i.e. $\mathbf{P}\{\varkappa < \infty\} = 1$. If this is the case for the tracking problem then the assertion of Theorem 1 will follow immediately. $\square$

Lemma 1. *Let the following conditions hold:*

(i) *r.v.* $\varkappa$ *is proper and assumptions* (a), (b), (c) *take place;*
(ii) *for some constants* C', C'', T_0 *we have*

$$T^{-1} r(T-1) \leq C' + C'' T^{-1} \sum_{t=1}^{T} [e(t) - \zeta_t]^2, \quad T \geq T_0, \quad \text{a.s.}$$

Then the equalities

(1) $\lim_{T \to \infty} T^{-1} \sum_{t=1}^{T} [e(t) - \zeta_t]^2 = 0$;

(2) $\overline{\lim}_{T \to \infty} T^{-1} r(T-1) < \infty$;

(3) $\lim_{T \to \infty} T^{-1} \sum_{t=1}^{T} \mathbf{E}\{(x_t - x_t^*)^2 | \mathcal{F}_{t-1}\} = \sigma^2$

are satisfied a.s.

If we add the condition $\mathbf{E}(\zeta_t^4 | \mathcal{F}_{t-1}) < \infty$ *a.s. to the restrictions imposed on the noise then we can add the following equality to* (1)–(3) *above*

(4) $\lim_{T \to \infty} T^{-1} \sum_{t=1}^{T} (x_t - x_t^*)^2 = \sigma^2$ *a.s.*

From this lemma the assertions of Theorem 1 follow immediately. It remains to verify that assumptions (i) and (ii) are satisfied. This is the essence of the following lemmas.

Lemma 2. *Under the assumptions of Theorem 1, the r.v $\varkappa$ is proper.*

Lemma 3. *There exist positive constants C_1–C_8 such that*

(1) $\lim_{T\to\infty} T^{-1} \sum_{t=1}^{T} u_t^2 \leq C_1 \lim_{T\to\infty} T^{-1} \sum_{t=1}^{T} x_t^2 + C_2$.

(2) $\lim_{T\to\infty} T^{-1} \sum_{t=1}^{T} x_t^2 \leq C_3 \lim_{T\to\infty} T^{-1} \sum_{t=1}^{T} [e(t) - \zeta_t]^2 + C_4$.

(3) $\lim_{T\to\infty} T^{-1} \sum_{t=1}^{T} \bar{x}_t^2 \leq C_5 \lim_{T\to\infty} T^{-1} \sum_{t=1}^{T} [e(t) - \zeta_t]^2 + C_6$.

(4) $T^{-1} r(T-1) \leq C_7 T^{-1} \sum_{t=1}^{T} [e(t) - \zeta_t]^2 + C_8$.

Proof of Lemma 3. Property (1) is, as it has been noted in Sec. 1, equivalent to the minimum phase property provided

$$P\left\{ \varlimsup_{T\to\infty} T^{-1} \sum_{t=1}^{T} \zeta_t^2 < \infty \right\} = 1.$$

From the equality $x_t = e(t) + x_t^* = (e(t) - \zeta_t) + \zeta_t + x_t^*$ it follows that

$$x_t^2 \leq 3(e(t) - \zeta_t)^2 + 3\zeta_t^2 + 3(x_t^*)^2.$$

It remains to take into account the relation $|x_t^*| \leq g$ and properties of the noise ζ_t stated above. This proves property (2). To verify property (3), let us introduce the quantity $f_t = x_t - \bar{x}_t = x_t - \Phi_{t-1}^T \theta_t$. It is easy to show that

$$f_t = \frac{e(t)}{1 + \Phi_{t-1}^T P_{t-2} \Phi_{t-1}}.$$

We have

$$\bar{x}_t = x_t - f_t = e(t) + x_t^* - \frac{e(t)}{1 + \Phi_{t-1}^T P_{t-2} \Phi_{t-1}}$$

$$= \frac{\Phi_{t-1}^T P_{t-2} \Phi_{t-1}}{1 + \Phi_{t-1}^T P_{t-2} \Phi_{t-1}} (e(t) - \zeta_t) + \frac{\Phi_{t-1}^T P_{t-2} \Phi_{t-1}}{1 + \Phi_{t-1} P_{t-2} \Phi_{t-1}} \zeta_t + x_t^*$$

$$\leq (e(t) - \zeta_t) + \zeta_t + x_t^*.$$

This leads to the inequality

$$\bar{x}_t^2 \leq 3(e(t) - \zeta_t)^2 + 3\zeta_t^2 + 3(x_t^*)^2.$$

From this property (3) follows immediately. Property (4) follows from the preceding ones and the definition of the quantities Φ_t, $r(t)$ (see (8)).

So, we have proved condition (ii) of Lemma 1. We now turn to the proof of Lemmas 1 and 2.

Proof of Lemma 1. We begin by proving (1). If the sequence $\{r(t)\}$ is bounded $(r(t) \leq c < \infty)$ then condition (i) will imply $\lim_{T\to\infty} T^{-1} \sum_{t=1}^{T} [e(t) - \zeta_t]^2 < \infty$ a.s.

If $r(t)$ increases unboundedly, then by the Kronecker theorem (see Sec. 3, Chap. 4) we have

$$\lim_{T \to \infty} \frac{T}{r(T)} \frac{1}{T} \sum_{t=1}^{T} [e(t) - \zeta_t]^2 = 0 \quad \text{a.s.}$$

Substituting the estimate of $T^{-1}r(T-1)$ from condition (ii), we obtain

$$\lim_{T \to \infty} \frac{T^{-1} \sum_{t=1}^{T} [e(t) - \zeta_t]^2}{C_1 + C_2 T^{-1} \sum_{t=1}^{T} [e(t) - \zeta_t]^2} = 0 \quad \text{a.s.}$$

From this assertion (1) follows. From it and condition (ii), assertion (2) follows. We now consider assertion (3). We have

$$\mathbf{E}\{[x_t - x_t^*]^2 | \mathcal{F}_{t-1}\} = \mathbf{E}\{[(x_t - x_t^* - \zeta_t) + \zeta_t]^2 | \mathcal{F}_{t-1}\}$$
$$= \mathbf{E}\{[x_t - x_t^* - \zeta_t]^2 + 2[x_t - x_t^* - \zeta_t]\zeta_t + \zeta_t^2 | \mathcal{F}_{t-1}\}$$
$$= [e(t) - \zeta_t]^2 + \mathbf{E}(\zeta_t^2 | \mathcal{F}_{t-1}),$$

where the last equality follows from $\mathcal{F}_{t-1}$-measurability of the r.v. $x_t - \zeta_t$ and x_t^*. This proves assertion (3). It remains to prove assertion (4). We need two results from the theory of random processes.
(α) Let $(X_t, \mathcal{F}_t)$ be a martingale-difference and

$$\sum_{t=1}^{\infty} t^{-2} \mathbf{E}(X_t^2 | \mathcal{F}_{t-1}) < \infty \quad \text{a.s.}$$

Then $\lim_{T \to \infty} \sum_{t=1}^{T} X_t = 0$ a.s.
(β) Let $(X_t, \mathcal{F}_t)$ be a martingale-difference. If

$$\mathbf{E}(X_t^2 | \mathcal{F}_{t-1}) = \sigma^2, \quad \mathbf{E}(X_t^4 | \mathcal{F}_{t-1}) \le q < \infty, \quad \text{a.s.}$$

then

$$\lim_{T \to \infty} T^{-1} \sum_{t=1}^{T} X_t^2 = \sigma^2 \quad \text{a.s.}$$

The second result follows from the first in an obvious way. Making use of (β), we have

$$\lim_{T \to \infty} T^{-1} \sum_{t=1}^{T} [x_t - x_t^*]^2 = \lim_{T \to \infty} T^{-1} \sum_{t=1}^{T} [x_t - x_t^* - \zeta_t + \zeta_t]^2$$
$$= \lim_{T \to \infty} T^{-1} \sum_{t=1}^{T} \{[x_t - x_t^* - \zeta_t]^2 + 2[x_t - x_t^* - \zeta_t]\zeta_t + \zeta_t^2\}$$
$$= \sigma^2 + 2 \lim_{T \to \infty} T^{-1} \sum_{t=1}^{T} [x_t - x_t^* - \zeta_t]\zeta_t.$$

We consider the second summand on the right-hand side of the last equality. The relationships

$$\mathbf{E}(\zeta_t^2|\mathcal{F}_{t-1}) = \sigma^2, \quad \lim_{t\to\infty} T^{-1}r(T-1) < \infty$$

and finiteness (a.s.) of the r.v. $\varkappa$ imply

$$\sum_{t=1}^{\infty} t^{-2}\mathbf{E}\{[x_t - x_t^* - \zeta_t]^2\zeta_t^2|\mathcal{F}_{t-1}\} < \infty.$$

This and (α) lead to the equality

$$\lim_{T\to\infty} T^{-1}\sum_{t=1}^{T}[x_t - x_t^* - \zeta_t]\zeta_t = 0 \quad \text{a.s.}$$

which completes our proof. $\qquad\qquad\qquad\qquad\qquad\qquad\qquad\qquad\square$

Proof of Lemma 2. Let us first introduce some notation. Let $\bar{\theta}_t = \theta_t - \theta$ be the deviation of the estimate from the true value of the parameter θ, $f_t = x_t - \bar{x}_t = x_t - \Phi_{t-1}\theta_t$ or, which is the same,

$$f_t = \frac{e(t)}{1 + \Phi_{t-1}^T P_{t-2}\Phi_{t-1}}, \quad e(t) = x_t - x_t^*.$$

Finally, put $z_t = f_t - \zeta_t$, $g_t = z_t - 1/2h_t$, $h_t = -\Phi_{t-1}\bar{\theta}_t$. It is easy to verify that

$$C(\triangle)z_t = h_t, \quad (C^{-1}(\triangle) - 1/2)h_t = g_t, \qquad (9)$$

and

$$\mathbf{E}(h_t\zeta_t|\mathcal{F}_{t-1}) = -\Phi_{t-1}^T P_{t-1}\Phi_{t-1}\sigma.$$

We write the original recurrent relation for calculating LSE in the form

$$\bar{\theta}_t - P_{t-2}\Phi_{t-1}f_t = \bar{\theta}_{t-1}.$$

Multiplying both sides of the last equality by $\bar{\theta}_t^T P_{t-2}^{-1}$, we obtain

$$\bar{\theta}_t P_{t-2}^{-1}\bar{\theta}_t - 2\Phi_{t-1}^T\bar{\theta}_t f_t + \Phi_{t-1}^T P_{t-2}\Phi_{t-1}f_t^2 = \bar{\theta}_{t-1}^T P_{t-2}^{-1}\bar{\theta}_{t-1}.$$

For the auxiliary function $V(t) = \bar{\theta}_t P_{t-1}^{-1}\bar{\theta}_t$ we have the following recurrent relation

$$\begin{aligned}
V(t) &= V(t-1) + h_t^2 - 2h_t f_t - \Phi_{t-1}^T P_{t-2}\Phi_{t-1}f_t^2 \\
&= V(t-1) + h_t^2 - 2h_t z_t - 2h_t\zeta_t - \Phi_{t-1}^T P_{t-2}\Phi_{t-1}f_t^2.
\end{aligned}$$

The conditional mathematical expectation of this function can be written as follows

$$\begin{aligned}
\mathbf{E}\{V(t)|\mathcal{F}_{t-1}\} = V(t-1) &+ \mathbf{E}\{h_t^2 - 2h_t z_t|\mathcal{F}_{t-1}\} \\
&+ 2\Phi_{t-1}^T P_{t-1}\Phi_{t-1}\sigma^2 - \mathbf{E}\{\Phi_{t-1}^T P_{t-2}\Phi_{t-1}f_t^2|\mathcal{F}_{t-1}\}. \quad \text{a.s.}
\end{aligned}$$

Taking into account the relation between h_t and g_t, we can write this equality in another form

$$\begin{aligned}
\mathbf{E}\{V(t)|\mathcal{F}_{t-1}\} = V(t-1) &- 2\mathbf{E}(h_t g_t|\mathcal{F}_{t-1}) \\
&- E\{\Phi_{t-1}^T P_{t-2}\Phi_{t-1}f_t^2|\mathcal{F}_{t-1}\} + 2\Phi_{t-1}^T P_{t-1}\Phi_{t-1}\sigma^2.
\end{aligned}$$

We introduce the following r.v. ($t \geq 1$)

$$S(t) = 2 \sum_{t=1}^{T} \left[g_t h_t - \frac{\lambda_1}{2} g_t^2 - \frac{\lambda_2}{2} h_t^2 \right] + K, \quad 0 < \lambda_1, \lambda_2, \ K < \infty.$$

According to (9), the quantities g_t and h_t are related by the strictly positive real transfer function $C^{-1}(\triangle) - 1/2$, hence the constants λ_1, λ_2 and K can be chosen so that $S(t) \geq 0$. We consider the following random sequence

$$\eta_t = \frac{V(t)}{\tilde{r}(t-1)} + \frac{S(t)}{\tilde{r}(t-2)} + \lambda_1 \sum_{t=1}^{T} \frac{g_t^2}{\tilde{r}(t-2)} + \lambda_2 \sum_{t=1}^{T} \frac{h_t^2}{\tilde{r}(t-2)}$$
$$+ \sum_{t=1}^{T} \frac{\Phi_{t-1}^T P_{t-2} \Phi_{t-1}}{\tilde{r}(t-2)} f_t^2 + \sum_{t=1}^{T} \frac{\Phi_{t-1}^T \Phi_{t-1}}{\tilde{r}(t-2)} \frac{V(t)}{\tilde{r}(t-1)}.$$

It is not difficult to show that

$$\mathbf{E}(\eta_t | \mathcal{F}_{t-1}) \leq \eta_{t-1} + \frac{\Phi_{t-1}^T P_{t-1} \Phi_{t-1}}{\tilde{r}(t-1)} \sigma^2, \quad \text{a.s.}$$

and the series formed by the second terms on the right-hand side of this inequality is convergent.[b] So, we can use the Martingale lemma which guarantees that $\eta_t \to \eta < \infty$ a.s. as $t \to \infty$. From this follows the convergence of the following series

$$\sum_{t=1}^{\infty} \frac{g_t^2}{\tilde{r}(t-2)} < \infty, \quad \sum_{t=1}^{\infty} \frac{h_t^2}{\tilde{r}(t-2)} < \infty, \quad \sum_{t=1}^{\infty} \frac{\Phi_t^T P_{t-1} \Phi_t}{\tilde{r}(t-2)} f_t^2 < \infty, \quad \text{a.s.} \qquad (10)$$

Hence

$$\sum_{t=1}^{\infty} \frac{z_t^2}{\tilde{r}(t-2)} < \infty \quad \text{a.s.}$$

First, we write the recurrent relation for $\bar{\theta}_t = \theta_t - \theta$ as follows

$$\bar{\theta}_t = \bar{\theta}_{t-1} + P_{t-1} \Phi_{t-2} f_t.$$

Therefore

$$x_t - \Phi_{t-1}^T \bar{\theta}_t - \zeta_t + \Phi_{t-1}^T P_{t-2} \Phi_{t-1} f_t = x_t - \Phi_{t-1}^T \bar{\theta}_{t-1} - \zeta_t$$

or, which is the same,

$$e(t) - \zeta_t = z_t + \Phi_{t-1} P_{t-2} \Phi_{t-1} f_t.$$

By Schwartz inequality

$$[e(t) - \zeta_t]^2 \leq 2 z_t^2 + 2 [\Phi_{t-1}^T P_{t-2} \Phi_{t-1}]^2 f_t^2$$

[b]Indeed, for any term we have

$$\frac{\Phi_t^T P_t \Phi_t}{\tilde{r}(t-2)} = \frac{\Phi_t^T P_{t-1} \Phi_t}{\tilde{r}(t-2)[1 + \Phi_t^T \mathbf{P}_{t-1} \Phi_t]} \leq \frac{\Phi_t^T P_{t-2}^2 \Phi_t}{1 + \Phi_t^T P_{t-1} \Phi_t}.$$

or

$$\frac{[e(t) - \zeta_t]^2}{r(t-1)} \leq 2\frac{z_t^2}{r(t-1)} + 2\frac{\tilde{r}(t-1)\Phi_{t-1}^T P_{t-2}\Phi_{t-1}}{r(t-2)}\left[\frac{\Phi_{t-1}^T P_{t-2}\Phi_{t-1}}{r(t-2)}f_t^2\right]$$

$$\leq 2\frac{z_t^2}{r(t-1)} + 2\frac{\Phi_{t-1}^T P_{t-2}\Phi_{t-1}}{r(t-2)}f_t^2.$$

In view of (10) and the second inequality in (7) we obtain

$$\sum_{t=1}^{\infty}\frac{[e(t) - \zeta_t]^2}{r(t-1)} < \infty \quad \text{a.s.}$$

Thus, the r.v. $\varkappa$ is proper. This completes the proof of Lemma 2 and Theorem 1 as well.

We make some remarks about this result. First, let us say a few words about the condition imposed on the polynomial $C(\lambda)$ according to which the function $C^{-1}(\lambda) - 1/2$ must be strictly positive real. This has been used only in proving Lemma 2, when it was necessary to show that the sequence η_t is a semi-martingale, i.e. this condition is purely technical and we are forced to introduce it in all optimal control problems connected with linear equations. In the simplest case, i.e. when the noise in an ARMAX equation is a martingale-difference and $C(\lambda) \equiv 1$, this condition is fulfilled automatically. Later we shall see that the form of this condition is caused by the indentification method proposed. When using the stochastic gradient method this constraint is imposed on the function $C(\lambda) - 1/2$ but not on $C^{-1}(\lambda) - 1/2$. Hence, the class of models considered depends on the identification method used.

In conclusion we discuss the convergence of the estimates θ_t. This has not been claimed in the theorems above because the necessary assumptions under which the LSE are strongly consistent have not been formulated as well. This general remark may be illustrated by concrete examples when these estimates are not consistent under the assumptions stated above. Further, we consider this problem in detail.

10.4. Optimal Tracking and Consistency of Estimates for ARMAX

The results of the previous sections give a solution of the adaptive tracking problem in the framework of the assumptions stated there. But the results are incomplete for the following reasons:

(1) we have used only one identification method which, generally speaking, is not consistent;
(2) we have no estimates of the convergence rate for the considered functionals;
(3) the question of achieving optimal tracking and strong consistency of estimates of parameters simultaneously has been left without answer.

This justifies that the control aim be changed. In this section it takes the following form:

for a class $\mathcal{K}$ of ARMAX equations find an optimal adaptive strategy σ^0 such that for every equation from $\mathcal{K}$ the following conditions hold:

(α)

$$\lim_{T\to\infty} T^{-1} \sum_{t=1}^{T} (\|x_t\|^2 + \|u_t\|^2) < \infty \quad \text{a.s.}$$

$$V(\sigma^0) \to \min,$$

for any bounded path x_t. Here

$$V(\sigma) = \varlimsup_{t\to\infty} T^{-1} \sum_{t=1}^{T} (x_t - x_n^*)^T (x_n - x^*);$$

(β)

$$P\left\{ \lim_{t\to\infty} \theta_t = \theta \right\} = 1.$$

We denote this aim by OTJ and first consider a sufficient condition for (β) to hold, i.e. consider strong consistency of the estimates. We will do this with respect to two identification algorithms: the modified LSM (we keep the former abbreviation LSM) and stochastic gradient method (SGM).

We now consider the following question. We have a control u_t^0 which is assumed to be measurable with respect to the σ-algebra $\mathcal{F}_t$ generated by the history of the considered process up to time t. The real control u_t is formed with the help of two components, namely, by the control u_t^0 and some stochastic component v_t, i.e.

$$u_t = u_t^0 + v_t.$$

Sometimes the method that consists of randomizating the control is called the "*trial signal method*" or the "*damping excitations method*". More exactly it consists of the following: let W_t be some sequence of independent (of each other and of the noise ζ_t) random vectors with continuous distributions satisfying the conditions

$$\mathbf{E}W_t = 0, \quad \mathbf{E}W_t W_t^T = \mu I, \quad \mu > 0, \quad \|W_t\| \leq \bar{W} < \infty, \quad \forall t.$$

The trial signal is formed by using these random vectors as follows

$$v_t = \varphi(t) W_t$$

where $\lim_{t\to\infty} \varphi(t) = 0$. Thus the control at time t has the form

$$u_t = u_t^0 + \varphi(t) W_t. \tag{1}$$

For large t the vectors u_t and u_t^0 differ slightly from each other and coincide in the limit. The choice of the damping function $\varphi(t)$ depends on the identification method being used.

The use of a trial signal leads, as we will see latter, to strong consistency of the estimates for linear difference equations of ARMAX-type

$$A(\triangle)x_t = \triangle^d B(\triangle)u_t + C(\triangle)\zeta_t$$

where the matrix polynomials $A(\lambda)$, $B(\lambda)$ and $C(\lambda)$ of orders p, q and r are unknown.

First, we consider consistency of the LSE by choosing the damping function as follows

$$\varphi(t) = t^{-h/2}, \quad h \in [0, (2(\chi + 1))^{-1}], \quad \chi = \max(p, q, v) + np - 1.$$

It remains to define the class of equations using conditions imposed on their coefficients and the noise ζ_t.

Definition 1. The class $\mathcal{K}_{LSM}(n, l; p, q, r)$ of ARMAX equations with dim $X = n$ and dim $U = l$ is defined by the following conditions:

A. The matrices $A(\lambda)$, $B(\lambda)$, $C(\lambda)$ have no common left factors and the matrix A_p has full rank n.
B. The transfer matrix $C^{-1}(\lambda) - 1/2I$ is strictly positive real (SPR).
C. $\sup_t \mathbf{E}(\|\zeta_t\|^b | \mathcal{F}_{t-1}) = \sigma^2 < \infty$ a.s. $b \geq 2$.
D. There exists a matrix $R > 0$ such that

$$\lim_{T \to \infty} T^{-1} \sum_{t=1}^{T} \zeta_t \zeta_t^T = R, \quad \text{a.s.}$$

Condition **A** means, in particular, that the equation $y_{t+1} = Ay_t + Bu_t + C\zeta_t$ is controllable. Condition **B** will be fulfilled automatically if the random disturbances are not correlated, and then $C(\lambda) = I$. The main result is formulated below.

Theorem 1. *If the control* (1) *with* $\varphi(t) = t^{-h/2}$ *is applied to the equation from* $\mathcal{K}_{LSM}(h, l; p, q, r)$ *and*

$$T^{-1} \sum_{t=1}^{T} \|u_t^0\|^2 = O(T^\delta) \quad \text{a.s. } \delta \in \left(0, \frac{1 - 2h(x + 1)}{2\chi + 3}\right]$$

as $T \to \infty$ *then the LSE will be strongly consistent and, moreover,*

$$\|\theta_t - \theta\|^2 = \begin{cases} O(t^{-a} \ln t), & \text{if } b > 2, \\ O(t^{-a}(\ln \ln t)^g) \ln t \quad \forall g > 1, & \text{if } b = 2 \end{cases}$$

for all $a \in ((1 + \delta)/2, 1 - (1 + \chi)(h + \delta))$.

Section 6 is devoted to the proof of this theorem.

From this theorem it follows that the convergence rate of the estimates has the order less than $t^{-1/2+\varepsilon}$ without logarithm in the numerator, i.e. it is somewhat worse than the usual rate $t^{-1/2}$.

We now turn to another method of estimating the parameters, namely to the stochastic gradient method. We define the damping function as follows

$$\varphi(t) = \ln^{-h/2} t, \quad \forall t \geq 2, \quad h \in \left(0, \; (4\chi_1(n+2))^{-1}\right), \quad \chi_1 = \max(p, q, r+1).$$

The control with the trial signal is defined in accordance with (1) but condition **B** must be replaced by the following:

B′. The transfer function $C(\lambda) - 1/2I$ is strictly positive real.

The main result concerning the SGM applies to the class of equations $\mathcal{K}_{SG}(n, l; p, q, r)$ which is defined similarly to $\mathcal{K}_{LSM}(n, l; p, q, r)$ but with condition **B′** instead of condition **B**.

Theorem 2. *If the control* (1) *with* $\varphi(t) = \ln^{-h/2} t$ *is applied to an equation from* $\mathcal{K}_{SG}(n, l; p, q, r)$ *and*

$$T^{-1} \sum_{t=1}^{T} \|u_t^0\| = O(\ln^\delta T) \quad \text{a.s. } \delta \in \left(0, \frac{1/4 - (n+2)\chi_1 h}{2 + (n+1)\chi_1}\right)$$

as $T \to \infty$, *then the SGE are strongly consistent.*

We give a sketch of proof in Sec. 6. The convergence rate of the SGE is not given in this theorem. This is the price for the simplicity of this algorithm, i.e. difficulties of its analysis prevent from obtaining the convergence rate of the appropriate functionals. Let us return to the adaptive problem of optimal tracking when the OTJ is the control aim and the LSM is in the extended form.

Let CG_{LSM} denote an adaptive strategy which we are going to define. The unknown parameter θ is estimated by the LSM and A_{it}, B_{jt}, C_{kt} denote the current (at time t) estimates of the parameters A_i, B_j, C_k. Assuming that the dimensions of the spaces X and U are the same ($n = l$) we define the control u_t' as follows

$$u_t' = \begin{cases} B_{1t}^{-1}(B_{1t}u_t - \theta_t^T \varphi_t + x_t^*), & \text{if } \det B_{1t} \neq 0, \\ 0, & \text{if } \det B_{1t} = 0. \end{cases}$$

It is easy to note that the right-hand side of the first equality does not contain u_t. We define the sequence of non-Markovian moments t_k, τ_k by the relations:

$$\tau_1 = 1;$$

$$t_k = \sup\left\{ t > \tau_k : \sum_{i=\tau_k}^{j-1} \|u_i'\|^2 \geq (j-1)^{1-\delta} + \|u_{\tau_k}'\|^2, \forall j \in (\tau_k, t] \right\};$$

$$\tau_{k+1} = \inf\left\{ t > t_k : \sum_{i=k+1}^{t_k-1} \|u_i'\|^2 \geq 2^{-k} t^{1+\delta/2}, \|u_t'\|^2 \leq t^{1+\delta} \right\}$$

where the number $\delta > 0$ is chosen as in Theorem 1. We take the main control in the form

$$u_t^0 = \begin{cases} u_t', & \text{if } t \in [\tau_k, t_k) \quad \text{for some } k, \\ 0, & \text{if } t \in [\tau_k, t_{k+1}) \quad \text{for some } k. \end{cases} \tag{2}$$

We note that this definition involves a partition of the time axis into alternate stages similar to those in Chaps. 7, 8, 9. Of course, this similarity is only visual. The required optimal adaptive control u_t is constructed with the help of u_t^0 and the trial signal $v_t = \varphi(t)W_t$ in the form (1) by the method described above. This completes the description of the identification strategy CG_{LSM}. It remains to define the ARMAX type equations. They form the class $\mathcal{K}_{LSM}^+(n, l, p, q, r)$ which differs from $\mathcal{K}_{LSM}$ by the following two additional conditions:

D$^+$. There exist a number $\rho > 0$ and a matrix $R > 0$ such that

$$\left\| T^{-1} \sum_{t=1}^{T} \zeta_t \zeta_t^T - R \right\| = o(T^{-\rho}).$$

E. The functions $\det A(\lambda)$ and $\det B(\lambda)$ have no zeros in $\{\lambda : |\lambda| \leq 1\}$.

The last condition means that the equations considered are stable and minimum phase.

Theorem 3. *The strategy CG_{LSM} is adaptive with respect to the class $\mathcal{K}_{LSM}^+(n, l; p, q, r)$ under the given aim OTJ and*

$$\left\| T^{-1} \sum_{t=1}^{T} (x_t - x_t^*)(x_t - x_t^*)^T - R \right\| = o(T^{-\varrho}) + o(T^{-h}),$$

$$\|\theta_t - \theta\|^2 = O(t^{-a}(\ln \ln t)^g \ln t), \quad \forall g > 1, \ a = 1 - (\chi + 1)h, \quad \text{a.s.}$$

Proof. First, we prove the assertions relating to the estimates of the parameters. Condition **E** implies that the matrix B_1 is non-degenerate. Hence, beginning from some t its estimate B_{1t} is also non-degenerate. Thus for all large t the control u_t' can be calculated according to the first formula entering its definition. For all equations from $\mathcal{K}_{LSM}^+(n, l; p, q, r)$ the assumptions of Theorem 1 are fulfilled except, maybe, the following one: $T^{-1} \sum_{t=1}^{T} \|u_t^0\|^2 = O(T^\delta)$ a.s. To verify this, we note that three cases only are possible:

(α) $\sigma_k < \infty$, $\tau_{k+1} = \infty$ for some k;
(β) $\tau_k < \infty$, $\sigma_k = \infty$ for some k;
(γ) $\sigma_k < \infty$, $\tau_k < \infty$ for all k.

In the first case, $u_t^0 \equiv 0$ if $t \geq \sigma_k$. Then the required assertion is obvious. In the second case, this assertion follows from the definitions of u_t^0 and σ_k. In the case γ, the desired assertion follows from the arguments below.

$$\sup_{\tau_k \le t < \tau_{k+1}} t^{-(1+\delta)} \sum_{i=1}^{t} \|u_i^0\|^2$$

$$= \sup_{\tau_k \le t \le \sigma_k} t^{-(1+\delta)} \sum_{i=1}^{t} \|u_i^0\|^2$$

$$= \sup_{\tau_k \le t < \sigma_k} t^{-(1+\delta)} \left[\sum_{\tau_1}^{\sigma_1 - 1} + \sum_{\tau_2}^{\sigma_2 - 1} + \cdots + \sum_{\tau_{k-1}}^{\sigma_{k-1} - 1} + \sum_{\tau_k}^{t} \right] \|u_i'\|^2$$

$$\le \tau_2^{-(1+\delta)} \sum_{\tau_1}^{\sigma_1 - 1} \|u_i'\|^2 + \cdots + \tau_k^{-(1+\delta)} \sum_{\tau_k}^{\sigma_{k-1} - 1} \|u_i'\|^2 + \sup_{\tau_k \le t < \sigma_k} t^{-(1+\delta)} \sum_{i=\tau_k}^{t} \|u_i'\|^2$$

$$\le \sum_{i=1}^{k-1} 2^{-i} + \sup_{\tau_k \le t \le \sigma_k} t^{-(1+\delta)}(t^{1+\delta} + \|u_{\tau_k}'\|^2) \le 2 + \frac{\|u_{\tau_k}'\|^2}{\tau_k^{1+\delta}} \le 3.$$

So, all assumptions of Theorem 1 are fulfilled and, therefore, the estimates of the parameters of the equations are strongly consistent. Since $\delta = 0$ the convergence rate of these estimates has the claimed form.

It is important to know which of the cases (α), (β), (γ) are possible (maybe two of them or even all three.) We show that only the case (β) is possible.

We consider first the case (α): the control $u_t = v_t$ at all $t \ge \sigma_k$. From the stability of $A(\lambda)$ and the form of u_t it follows that

$$\sum_{i=0}^{t} (\|x_i\|^2 + \|u_i\|^2) = O(t).$$

We put $\varphi_t = \varphi_t^0 + \tilde{\varphi}_t$, where $\varphi_t^0 = [x_t^T \cdots x_{t-p+1}^T u_t^T \cdots u_{t-q+1}^T \xi_t^T \cdots \xi_{t-r}^T]$. It is easy to verify that

$$\sum_{i=0}^{t} \|\tilde{\varphi}_i\|^2 = O((\ln r_t)(\ln \ln r_t)^2), \quad \forall g > 1; \quad r_t = 1 + \sum_{i=0}^{t} \|\varphi_i\|^2.$$

It follows that

$$r_t = 1 + \sum_{i=0}^{t} \|\varphi_i + \tilde{\varphi}_i\|^2 \le 1 + 2 \sum_{i=0}^{t} \|\varphi_i^0\|^2 + \sum_{i=0}^{t} \|\tilde{\varphi}_i\|^2$$

$$= O(t) + O((\ln r_t)(\ln \ln r_t)^g)$$

and, hence, $r_t = O(t)$. Therefore

$$\|\varphi_t\|^2 = O(t). \tag{3}$$

We rewrite Eq. (2) in the following form

$$B_1 u_t' = (B_1 - B_{1t})u_t' - (\theta_t^T \varphi_t - B_{1t})u_t + x_{t+1}^*.$$

Hence, by using (3) and the convergence $B_{1t} \to B_1$, we arrive at the estimate $\|u'_t\|^2 = O(\|u'_t\|^2) + O(t)$ leading to $\|u'_t\|^2 = O(t^{1/2})$ for all sufficiently large t. But the latter contradicts the assumption $\tau_{k+1} = \infty$, i.e. the case (α) is impossible.

We now consider the case (γ). Making use of condition **E** and the arguments used in the case (α) we can prove the following relations

$$\|u_t^1\|^2 = O\big(t_k^{1+\delta/2}\big), \qquad \sum_{i=\tau_k+1}^{t_k} \|u_i^1\|^2 \le O\big(t_k^{1+\delta/2}\big), \quad \text{a.s.}$$

which contradict the definition of the quantity t_k. Thus (γ) is impossible as well.

It follows that there exists a.s. some integer k_0 such that $u_t^0 \equiv u_t'$ for all $t \ge \tau_{k_0}$. From condition **E** it follows that the constructed control is admissible, i.e.

$$\overline{\lim_{T \to \infty}} \; T^{-1} \sum_{t=1}^{T} (\|x_t\|^2 + \|u_t\|^2) < \infty.$$

It remains to prove that it is optimal. This can be done by directly estimating the norm

$$\left\| T^{-1} \sum_{t=1}^{t} (x_t - x_t^*)(x_t - x_t^*)^T - R \right\|$$

that leads to the result stated in the theorem. $\qquad\qquad\qquad\qquad\square$

We now make some remarks on Theorem 3:

(1) The decreasing order of the tracking error is defined by the quantity $o(t^{-\rho}) + O(t^{-h})$. In the most interesting cases we have $\rho = 1/2$ and $h < (2(\chi + 1))^{-1}$, $\chi = \max(p, q, r + np - 1)$. In the simplest case $p = q = r = n = 1$ we have $\rho = 1/2$, $h < 1/4$.

(2) The stability assumption in condition **E** is not natural but forced by the proof. This assumption may be dropped by using a modified strategy (for example, using at certain times the SGM as the identification algorithm).

(3) We have restricted attention to the case $d = 1$. In the general case $d \ge 2$ the tracking error is $O(t^{-h/2})$.

In view of the length of the proof we are forced to omit these important and interesting details.

We now choose another identification algorithm for solving the optimal adaptive tracking problem, namely, instead of the LSM we shall use SGM. The control aim remains the OTJ. It changes the optimal strategy and the classes of the considered equations.

To estimate the parameter θ we can use the SGM in the form pointed to at the end of Sec. 1. As usual, θ_t denotes an estimate of this parameter at time t and φ_t denotes the history up to this moment. This information is used to calculate the admissible control u_t^0 at time t by the following equation

$$\theta_t^T \varphi_t = x_{t+1}^*. \tag{4}$$

It remains to define the class of ARMAX-type equations to be controlled. Let $\dim x_t = \dim \zeta_t = n$, $\dim u_t = l$ and the powers of the matrix polynomials $A(\lambda)$, $B(\lambda)$, $C(\lambda)$ be equal to p, $q-1$, r respectively. Let $\tilde{K}(n,l;p,q,r)$ denote the set of all equations with the characteristics stated above which satisfy the additional conditions:

$\mathbf{K_1}$: the sequence ζ_t is a martingale-difference such that a.s.

$$\sup_t \mathbf{E}(\|\zeta_t\|^2|\mathcal{F}_{t-1}) < \infty, \qquad \lim_{T\to\infty} T^{-1}\sum_{t=1}^{T} \zeta_t\zeta_t^T = R > 0;$$

$\mathbf{K_2}$: the transfer matrix $C(\lambda) - 1/2I$ is strictly positive real;

$\mathbf{K_3}$: $\det\left(B(\lambda)\begin{bmatrix} I_m \\ O \end{bmatrix}\right) \neq 0$ for $|\lambda| \leq 1$;

$\mathbf{K_4}$: the matrix polynomials $A(\lambda)$, $B(\lambda)$, $CX(\lambda)$ have no common factors and the matrix A_p has full rank.

Theorem 4. *For the equations from $\tilde{K}(n,l,p,q,r)$ the strategy $\widetilde{GS}$ is adaptive*, i.e. *it leads to the aim OTJ.*

Proof. We prove first that Eq. (4) is solvable with respect to u_t^0. To achieve this, we rewrite this equation in the form

$$B_{1t}u_t^0 + (\theta_t^T \varphi_t - B_{1t}u_t) = x_{t+1}^*$$

where B_{1t} stands for the estimate of the matrix B_1 at time t and

$$\theta_t = [-A_{1t} \cdots - A_{pt}\ B_{1t} \cdots B_{qt}\ C_{1t} \cdots C_{rt}].$$

We put

$$\theta_t^* = [-A_{1t} \cdots - A_{pt}\ OB_{2t} \cdots B_{qt}\ C_{1t} \cdots C_{rt}]$$

and Eq. (4) takes the form

$$B_{1t}u_t^0 = x_{t+1}^* - \theta_t^{*T}\varphi_t. \tag{5}$$

First, we investigate solvability of this equation when $n = l$. It has a unique solution if the initial values are chosen so that the matrix B_{1t} is non-degenerate. Indeed, if this is true, then

$$u_t^0 = B_{1t}^{-1}(x_{t+1}^* - \theta_t^{*T}\varphi_t).$$

The control u_t^0 is $\mathcal{F}_t$-measurable (here $\mathcal{F}_t = \sigma\{\zeta_j, v_j, j \leq t\}$). Using induction we prove that in the case considered the matrix B_{1t} can be made non-degenerate with probability one for all t.

Denoting $y_t = x_t - \zeta_t - \theta_{t-1}^T\varphi_{t-1}$ we have (taking into account the SGM relations)

$$B_{1\ t+1} = B_{1t} + r_t^{-1}(y_{t+1} + \zeta_{t+1})u_t^T. \tag{6}$$

It follows that B_{11} is non-degenerate if B_{10} is non-degenerate for $u_0 = 0$. Let $\det B_{1t} \neq 0$ a.s. We prove that $\det B_{1t+1} \neq 0$ a.s. To achieve this, we put

$M = \{\omega : \det B_{1t} = 0\}$, $N = \{\omega : \det B_{1t+1} = 0\}$ and show that $\mathbf{P}\{M\} = 0$ implies $\mathbf{P}\{NM^c\} = 0$. Let the contrary hold, i.e. $\mathbf{P}\{M\} = 0$ and $\mathbf{P}\{NM^c\} > 0$. Then from (6) we have $\det\left[B_{1t} + r_t^{-1}(y_{t+1} + \zeta_{t+1})u_t^T\right] = 0$ for $\omega \in NM^c$. According to our assumption $\det B_{1t} \neq 0$ for the same ω. We have

$$\det(I + r_t^{-1}B_{1t}^{-1}(y_{t+1} + \zeta_{t+1})u_t^T) = 0 = \det(I + r_t^{-1}u_t^T B_{1t}^{-1}(y_{t+1} + \zeta_{t+1}))$$

which means

$$u_t^T B_{1t}^{-1}(y_{t+1} + \zeta_{t+1}) = -r_t$$

and, hence, $u_t^T B_{1t}^{-1} \neq 0$ for all $\omega \in NM^c$.

Let $\alpha_i(\omega)$ and $\zeta_{t+1}^{(i)}$, $1 \leq i \leq m$, denote the components of the vectors $u_t^T B_{1t}^{-1}$ and ζ_{t+1}, respectively. From the above it follows that

$$\sum_{i=1}^{m} \alpha_i(\omega)\zeta_{t+1}^{(i)} + r_t + u_t^T B_{1t}^{-1}y_{t+1} = 0 \quad \forall \omega \in NM^c. \tag{7}$$

Since $u_t^T B_{1t}^T \neq 0$ and $\mathbf{P}\{NM^c\} > 0$, there are some $\alpha_i(\omega)$ and a set $N' \subset NM^c$ such that

$$\alpha_i(\omega) \neq 0, \quad \forall \omega \in N', \quad \mathbf{P}\{N'\} > 0. \tag{8}$$

Taking, for the sake of simplicity, $i = 1$ we define the following r.v.

$$z(\omega) = \begin{cases} \alpha_1^{-1}(\omega)\left[\displaystyle\sum_{j=2}^{m} \alpha_j(\omega) + r_t + u_t^T B_{1t}^{-1}y_{t+1}\right], & \omega \in N', \\ 0, & \omega \in N'^c \end{cases}$$

which does not depend on $\zeta_{t+1}^{(1)}$. Making use of the inequality

$$\sup_x \mathbf{P}\{\eta_1 + \eta_2 = x\} \leq \min\left(\sup_x \mathbf{P}\{\eta_1 = x\}, \sup_x \mathbf{P}\{\eta_2 = x\}\right),$$

which is true for all independent r.v. η_1 and η_2, we have $\mathbf{P}\{\zeta_{t+1}^{(1)} + z(\omega) = 0\} = 0$. From (7) and (8) it follows that $\mathbf{P}\{\zeta_{t+1}^{(1)} + z(\omega) = 0\} \geq \mathbf{P}\{N'\}$, contrary to the assumption. Hence the matrix B_{1t+1} is non-degenerate.

Now let $n < l$. Denote

$$B_{1t} = [\underbrace{B_{1t}^{(1)}}_{n}, \underbrace{B_{1t}^{(2)}}_{l-n}]\}n, \quad u_t^T = [\underbrace{u_t^{(1)^T}}_{n}, \underbrace{u_t^{(2)^T}}_{l-n}].$$

The first components of these vectors are related by the SGM as follows

$$B_{1\,t+1}^{(1)} = B_{1t}^{(1)} + r_t^{-1}(y_{t+1} + \zeta_{t+1})u_t^{(1)^T}.$$

This equation coincides with (6). Therefore choosing the non-degenerate matrix $B_{10}^{(1)}$ and $u_0' = 0$, we see that all matrices $B_{1t}^{(1)}$ are non-degenerate as well. So, Eq. (5) is equivalent to

$$\left[I, (B_{1t}^{(1)})^{-1}B_{1t}^{(2)}\right]u_t^0 = \left(B_{1t}^{(1)}\right)^{-1}(x_{t+1}^* - \theta_t^{*^T}\varphi_t)$$

or, in a more detailed form,

$$u_t^{(1)} + \left(B_{1t}^{(1)}\right)^{-1} B_{1t}^{(2)} u_t^{(2)} = \left(B_{1t}^{(1)}\right)^{-1}(x_{t+1}^* - \theta_t^{*T} \varphi_t).$$

Its solution is

$$u_t^0 = \begin{cases} \left(B_{1t}^{(1)}\right)^{-1}\left(x_{t+1}^* - \theta_t^{*T}\varphi_t - B_{1t}^{(2)} u_t^{(2)}\right), \\ u_t^{(2)} \end{cases}$$

where $u_t^{(2)}$ is an $l - n$-dimensional, $\mathcal{F}_t$-measurable random vector. Consequently, in the case $n < l$ the admissible control u_t^0 exists but is not unique.

It remains to verify that the strategy $\widetilde{GS}$ obtained leads to the aim OTJ. We now check that the strategy $\widetilde{GS}$ is admissible (the first condition in the formulation of the aim). From the minimum phase property of the equation it follows that

$$T^{-1} \sum_{t=1}^{T} \|u_t^0\|^2 \leq C_1 T^{-1} \sum_{t=1}^{T} \|x_t\|^2 + C_2$$

for some (in general random) constants C_1, C_2. It follows that

$$T^{-1} \sum_{t=1}^{T} \|u_t\|^2 \leq 2C_1 T^{-1} \sum_{t=1}^{T} \|x_t\|^2 + 2C_2 + 2T^{-1} \sum_{t=1}^{T} \|v_t\|^2,$$

with the last term tending to zero as $T \to \infty$. It remains to show that $\overline{\lim}_{T\to\infty} T^{-1} \sum_{t=1}^{T} \|x_t\|^2 = C < \infty$. To do this we use the equation

$$C(\nabla)y_t = \tilde{\theta}_{t-1}\varphi_{t-1}$$

where $y_t = x_t - \zeta_t - \theta_{t-1}^T \varphi_{t-1}$, $\tilde{\theta}_t = \theta - \theta_t$. According to the following relations ((see 12) and (13) from Sec. 6)[c]

$$\lim_{t\to\infty} r_t^{-1} \sum_{i=1}^{t} \|y_i\|^2 = 0 \quad \lim_{t\to\infty} r_t^{-1} \sum_{i=1}^{t} \|\tilde{\theta}_i\varphi_i\|^2 = 0,$$

condition **B′** implies stability of the matrix polynomial $C(\lambda)$. This means that

$$T^{-1} \sum_{t=1}^{T} \|x_t - \theta_{t-1}^T\varphi_{t-1}\|^2 \leq C'T^{-1} \sum_{t=1}^{T} \|\tilde{\theta}_{t-1}\varphi_{t-1}\|^2 + C''$$

for some positive constants C', C''. To use the relations mentioned above we should replace the factor t^{-1} by r_t^{-1}. This will be possible if we prove that $\overline{\lim}\, r_t/t < \infty$, which follows from (5) and the inequality

$$\frac{r_t}{t} \leq \frac{C}{t} \sum_{i=1}^{t} \|\tilde{\theta}_i\varphi_{i-1}\|^2 = C\frac{r_t}{t}\frac{1}{r_t} \sum_{i=1}^{t} \|\tilde{\theta}_i\varphi_{i-1}\|^2.$$

[c]This is not a vicious circle since the proof of Theorem 2 does not depend on the facts and proofs of the theorems which follow and, in particular, on the result proved here.

Noting that $\theta_{t-1}^T \varphi_{t-1} = x^*$, we find

$$\overline{\lim_{T\to\infty}} \, T^{-1} \sum_{t=1}^{T} \|x_t - x_t^*\|^2 \leq C''.$$

In view of the boundedness of the sequence $\{x_t^*\}$, we conclude that the outputs of the equation are bounded in the mean square sense. i.e.

$$\overline{\lim_{T\to\infty}} \, T^{-1} \sum_{t=1}^{T} \|x_t\|^2 < \infty.$$

Hence

$$\lim_{T\to\infty} T^{-1} \sum_{t=1}^{T} \|u_t\|^2 < \infty \qquad \overline{\lim_{T\to\infty}} \, T^{-1} \sum_{t=1}^{T} \|u_t^0\|^2 < \infty.$$

Thus, the strategy $\widetilde{GS}$ is admissible. This and our assumptions imply the assumptions of Theorem 2 and, hence, the SGE of parameters are strongly consistent.

It remains to prove that the strategy $\widetilde{GS}$ leads to the minimum of the mean square error of tracking. This fact can be verified directly by using the following equality

$$x_t - x_t^* = y_t - \zeta_t + B_{1t-1} v_{t-1}$$

and the asymptotic properties of the sequences y_t, ζ_t and v_t. $\square$

So, both identification algorithms (LSM and SGM) enable us to solve the optimal adaptive tracking problem but with different amount of information about the properties of the solutions. The results above can be extended to the case of random paths $\{x_t^*\}$ without difficulty. The boundedness condition is based on the fact that $\mathbf{P}\{\|x_t^*\| < R, \forall t\} = 1$. The variables x_t^* and ζ_t need to be independent but the sequence x_t^* must satisfy some non-singularity condition.

10.5. Adaptive Modal Control

The problem of modal control (or pole-zero assignment) plays a significant role in classical control theory. For linear equations in the state space

$$y_{t+1} = A y_t + B u_t$$

it can be described as follows (see Sec. 1). We must find a linear feedback $u_t = L y_t$ such that the matrix $A + BL$ has a given collection of eigenvalues $\lambda_1, \ldots, \lambda_n$. This problem can be solved without difficulty if the proposed equation is controllable. Then its solution has the desired transition regime and asymptotic behavior as $t \to \infty$. We are concerned with the adaptive version of this problem. In the deterministic case in Sec. 2, we have shown using the identification technique that it can be solved in a finite number of repetitions. For the equation in the "input-output" form (discussed below) this problem can be solved by reducing to an equation in the state space.

For the deterministic linear equation

$$A(\nabla x_t) = B(\nabla)u_t$$

with matrix polynomials $A(\nabla)$, $B(\nabla)$ of powers p and $q - 1$ respectively, some "reference" equation

$$A^0(\nabla x_t) = B^0(\nabla)\bar{u}_t$$

having the same type is given, where the matrix $A^0(\lambda)$ is stable and the "external" control $\bar{u}_t$ is ensures an aim defined for this equation. We need to find a control u_t which "transforms" the first equation into the second. This problem has an obvious solution if we can find the solution u_{t-1} of the following equation

$$B(\nabla)u_{t-1} = (A(\nabla) - A^0(\nabla))x_t + B^0(\nabla)\bar{u}_{t-1}.$$

We do not want to consider its solvability here. Instead, let us turn to the adaptive stochastic version of the modal control problem. Consider the ARMAX equation

$$A(\nabla)x_t = B(\nabla)u_{t-1} + C(\nabla)\zeta_t$$

with *a priori* unknown polynomials $A(\lambda)$, $B(\lambda)$ and $C(\lambda)$ (having degree equal to r) and some unobservable noise ζ_t. Under these conditions there is no hope to find an explicit solution of the problem in question. Some errors ε_t are inevitable. i.e.

$$A^0(\nabla)x_t = B^0(\nabla)\bar{u}_{t-1} + \varepsilon_t \tag{1}$$

where

$$\varepsilon_t = (A^0(\nabla) - A(\nabla))x_t + B(\nabla)u_{t-1} - B^0(\nabla)\bar{u}_{t-1} + C(\nabla)\zeta_t.$$

We make the following two assumptions:

(1) the additional external control $\bar{u}_{t-1}$ is $\mathcal{F}_t$-measurable;
(2) the noise ζ_t is a martingale-difference and a.s.

$$\sup_t \mathbf{E}(\|\zeta_t\|^b|\mathcal{F}_{t-1}) < \infty, \quad b > 2, \quad \lim_{T\to\infty} T^{-1} \sum_{t=1}^{T} \zeta_t\zeta_t^T = R > 0.$$

Let us estimate the average error from below

$$\overline{\lim_{T\to\infty}} \, T^{-1} \sum_{t=1}^{T} \mathbf{E}\{\|\varepsilon_t\|^2|\mathcal{F}_{t-1}\} \leq \overline{\lim_{T\to\infty}} \, T^{-1} \sum_{t=1}^{T} \mathbf{E}\{\|(A^0(\nabla) - A(\nabla))x_t + B(\nabla)u_{t-1}$$
$$- B^0(\nabla)\bar{u}_{t-1} + (C(\nabla)_I)\|^2 + \|\zeta_t\|^2|\mathcal{F}_{t-1}\}$$
$$\geq \mathbf{tr}\,R.$$

The properties of the errors are described more fully by the matrix inequality

$$\overline{\lim_{T\to\infty}} \, T^{-1} \sum_{t=1}^{T} \varepsilon_t\varepsilon_t^T \geq R.$$

We consider the following control aim.

(1) *to minimize the mean error* $\overline{\lim}_{T\to\infty} T^{-1} \sum_{t=1}^{T} \|\varepsilon_t\|^2$;
(2) *to ensure the consistency of the estimates of parameters of the equation.*

We denote this aim by PZA.

To solve this problem we use two identification strategies by combining the algorithms LSM and SGM. This leads to different techniques of choosing the controls.

The first strategy uses both LSM and SGM. The first estimates the coefficients of the polynomials $A(\lambda)$, $B(\lambda)$, and

$$C(\lambda) = I + C_1\lambda + \cdots + C_r\lambda^r, \quad r \geq 0$$

forming the vector of unknown parameters

$$\theta = [-A_1 \cdots - A_p \; B_1 \cdots B_q \; C_1 \cdots C_r]^T.$$

Its estimates (we denote them by θ_t at time t) can be calculated according to the formulas

$$\theta_{t+1} = \theta_t + a_t P_t \varphi_t (x_{t+1}^T - \varphi_t^T \theta_t), \quad a_t = (1 + \varphi_t^T P_t \varphi_t)^{-1},$$

$$P_{t+1} = P_t - a_t P_t \varphi_t \varphi_t^T P_t, \quad P_0 = dI, \quad d = n(p+r) + lr,$$

$$\varphi_t = [x_t^T \cdots x_{t-q+1}^T \; u_t^T \cdots u_{t-p+1}^T \; x_t^T - \varphi_{t-1}^T \theta_t \cdots x_{t-r+1}^T - \varphi_{t-r}^T \theta_{t-r+1}]^T.$$

An initial value θ_0 may be chosen arbitrarily

$$\theta_0 = [-A_{10} \cdots - A_{p0} \; B_{10} \cdots B_{q0} \; C_{10} \cdots C_{r0}]^T.$$

We now consider the following three polynomials whose coefficients we wish to estimate. The first of them is the matrix polynomial $D(\lambda)$ which is uniquely defined by the relation

$$(\det C(\lambda))I = (\mathbf{Adj}\, C(\lambda))A(\lambda) + \lambda D(\lambda)$$

and has the form

$$D(\lambda) = D_0 + D_1\lambda + \cdots + D_{p_1}\lambda^{p_1}, \quad p_1 = \max(nr - 1, (n-1)r + p - 1).$$

The second one has the form

$$E(\lambda) = (\mathbf{Adj}\, C(\lambda))B(\lambda) = E_0 + E_1\lambda + \cdots + E_{p_2}\lambda^{p_2}, \quad p_2 = (n-1)r + q$$

with the non-degenerate matrix $E_0 = B_1$. The third is the scalar polynomial

$$F(\lambda) = \det C(\lambda) = 1 + f_1\lambda + \cdots + f_{nr}\lambda^{nr}.$$

The collection of coefficients of these polynomials forms an unknown vector of parameters

$$\bar{\theta} = [D_0 \cdots D_{p_1} \; E_0 \cdots E_{p_2} \; f_1 \cdots f_{nr}]$$

which we estimate by using SGM. Let us write down this procedure in the form

$$\bar{\theta}_{t+1} = \bar{\theta}_t + \frac{a}{\bar{r}_t}\bar{\varphi}_t[x_{t+1}^T - \bar{\varphi}_t^T \bar{\theta}_t]$$

where we use the following notation

$$\bar{\varphi}_t = [x_t^T \cdots x_{t-p}^T \; u_t^T \cdots u_{t-p_2}^T - \bar{\varphi}_{t-1}^T \bar{\theta}_{t-1} \cdots - \bar{\varphi}_{t-r}^T \bar{\theta}_{t-r}^T]$$
$$\bar{r}_t = \bar{r}_{t-1} + \|\bar{\varphi}_t\|^2, \quad \bar{r}_{-1} = 1.$$

An initial value $\bar{\theta}_0$ may be chosen arbitrarily but the matrix E_0 must be non-degenerate. We pay special attention to the fact that the sequences of estimates $\{B_{0t}\}$ and $\{E_{0t}\}$ are different since they are calculated by different methods. But their limits will coincide and be equal to B_0 if the procedures LSM and SGM are consistent.

We now come to the construction of the controls. At time t the LSE θ_t and SGE $\bar{\theta}_t$ are calculated simultaneously. By using these estimates, two controls $u_t^{(1)}$ and $u_t^{(2)}$ are constructed. For $u_t^{(1)}$ we use the equation which is based on the LSE

$$B_{1t} u_t^{(1)} = (\det C_t(\nabla))(A^0(\nabla))^{-1}[B^0(\nabla)\bar{u}_t - C(\nabla)(x_t - \theta_t^T \varphi_t^T \varphi_{t-1})]$$
$$+ D_t(\nabla)x_t + (\mathbf{Adj}\, C_t(\nabla))B_t(\nabla)u_t + B_{1t}u_t \tag{2}$$

where the polynomial $D_t(\lambda)$ is found by the equality

$$(\det C_t(\lambda)I = (\mathbf{Adj}\, C_t(t))A_t(\lambda) + \lambda D_t(\lambda)$$

but not by its estimate obtained with the help of the SGM. The SGE are used to calculate the control $u_t^{(2)}$

$$E_{0t} u_t^{(2)} = -\bar{\theta}_t \bar{\varphi}_t - E_{0t} u_t. \tag{3}$$

We have already met with this equation which determines the optimal control in the tracking problem. Under our assumptions the trajectory found consists of one point $x_t^* \equiv 0$ only. Therefore we do not dwell on the solvability problem for Eqs. (2) and (3). We need only assume that the required modal control $\bar{u}_t$ is measurable.

We now state the alternation rules of the controls $u_t^{(1)}$ and $u_t^{(2)}$. To achieve this we define the stopping times $0 < \tau_1 < t_1 < \tau_2 < \cdots$ by the following inequalities:

$$t_k = \sup\left\{ t > \tau_k : \sum_{i=\tau_k}^{j-1} \|x_i\|^2 \le (j-1)^{1-\delta/2} + \|x_{\tau_k}\|^2, \forall j \in (\tau_k, t] \right\}$$

$$\tau_{k+1} = \inf\left\{ t > t_k : \sum_{i=t_k}^{t} \|x_i\|^2 \le 2^{-k} t \ln t, \; \sum_{i=\tau_k}^{t_k-1} \|x_i\|^2 \ge 2^{-k} t \ln t, \right.$$
$$\left. \sum_{i=t_k}^{t} \|u_i\|^2 \le 2^{-k} t \ln t, \; \sum_{i=\tau_k}^{t_k-1} \|u_i\|^2 \le 2^{-k} t \ln t \right\}$$

where

$$\delta \in \left[0, (2\chi+3)^{-1}(1 - 2h(\chi+1))\right), \quad \chi = np + \max(p, q, r), \quad h \in (0, [2(\chi+1)]^{-1}).$$

We also define the set

$$N = \left\{ i : \|u_i^{(1)}\|^2 \le i^{1+\delta} \right\}.$$

The alternation rule of the controls $u_t^{(1)}$ and $u_t^{(2)}$ is determined by means of the basic control u_t^a as follows

$$u_t^a = \begin{cases} u_t^{(1)}, & \text{if } t \in [\tau_k, t_k) \cap N \quad \text{for some } k, \\ 0, & \text{if } t \in [\tau_k, t_k) \cap N^c \text{ for some } k, \\ u_t^{(2)}, & \text{if } t \in [t_k, \tau_{k+1}) \qquad \text{for some } k. \end{cases}$$

The more complicated structure of the control is connected with a necessity to avoid the excessive increase of the output x_t and input u_t. For this purpose, we make the output approach zero in one cases and choose input being equal to zero in the other. The theorem below explains the efficiency of such actions. To complete the construction of the desired strategy it remains to add the trial signal, i.e.

$$u_t = u_t^a + \varphi(t)v_t.$$

This strategy is denoted by CZ.

As concerns the excitation source and function $\varphi(t)$, we assume that the first is formed by a sequence of independent (of each other and ζ_t) random vectors with independent components having continuous distributions such that

$$\mathbf{E}v_t = 0, \quad \mathbf{E}v_t v_t^T = g_1 T^{-1} I, \quad \|v_t\| = g_2 t^{-h/2}$$

and $\varphi(t) = t^{-h/2}$.

We define a class of ARMAX-type equations as follows.

Let $\mathcal{K}_1(n, l; p, q, r)$ denote the class of all equations such that $\dim X_t = n \leq l = \dim u_t$ and the numbers p, q, r represent the degrees of their matrix polynomials which satisfy the following conditions:

A. The matrices $A(\lambda)$, $B(\lambda)$, $C(\lambda)$ have no common left factors and rank of A_p is equal to n;

B. The matrix $C^{-1}(\lambda) - 1/2I$ is strictly positive real (SPR);

C. $\det C(\lambda) - a/2 > 0$;

D. All zeros of the function

$$\det\left(B(\lambda)\begin{bmatrix} I_t \\ 0 \end{bmatrix}\right)$$

lie outside the closed unit circle.

As usual, the noise ζ_t is supposed to be a martingale-difference and

E. $(\|\zeta_t\|^2|\mathcal{F}_{t-1}) \leq b < \infty, \quad \left\|T^{-1}\sum_{t=1}^{T}\zeta_t\zeta_t^T - R\right\| \xrightarrow[T\to\infty]{} 0, \quad R > 0 \quad \text{a.s.}$

Before stating the main result about the strategy CZ we give an auxiliary assertion of principal significance.

Lemma 1. *The strategy CZ applied to the equations from the class* $\mathcal{K}_1(n, l; p, q, r)$ *has the following property: there exists an integer* k_0 *finite (a.s.) such that* $\tau_{k_0} < \infty$, $t_{k_0} = 0$ *and the set* $N^c = \{i : \|u_t^{(1)}\|^2 > i^{1+\delta}\}$ *is finite.*

The proof of this lemma is similar to that of Theorem 3, Sec. 3.

In view of this lemma, beginning from a Markov moment the control is realized in accordance with the first condition entering into the definition of u_t^a. So, beginning from some finite moment τ_{k_0}, both the control $u_t^{(2)}$ and the SGM are not used. This gives ground to say that the strategy CZ is mainly based on the LSM.

Theorem 1. *For the equations from the class $\mathcal{K}_1(n, l; p, q, r)$ the strategy CZ leads (with probability one) to the following relationships:*

$$\overline{\lim_{T \to \infty}} \, T^{-1} \sum_{t=1}^{T} (\|x_t\|^2 + \|u_t\|^2) < \infty$$

$$\left\| T^{-1} \sum_{t=1}^{T} \varepsilon_t \varepsilon_t^T - T^{-1} \sum_{t=1}^{T} \zeta_t \zeta_t^T \right\| = O(T^{-h}),$$

$$\|\theta - \theta_t\| = O \left(\frac{(\ln \ln t)^g \ln t}{t^{1-(1-c)h}} \right)^{1/2}, \quad \forall g > 1$$

where $\varepsilon_t = (A^0(\nabla) - A(\nabla))x_t + \nabla B(\nabla)u_t - \nabla B^0(\nabla)u_t^0 + C(\nabla)\zeta_t$ means the error in solving the control modal problem.

This theorem can be proved by the same methods as similar results given in Sec. 3. So, we leave its proof to the reader. This theorem gives the convergence rate of the estimates θ_t and the mean square error ε_t.

The SGE are not, generally speaking, consistent with respect to the class $\mathcal{K}_1(n, l; p, q, r)$ and play only an auxiliary role in the initial stage of control.

It remains to consider another approach based on SGM to the modal control problem (or pole-zero assignment). It leads to other strategies and another class of equations. The definition of this strategy, which will be denoted by GC^*, is rather simple. The parameter $\theta = [-A_1 \cdots - A_p \, B_1 \cdots B_q \, C_1 \cdots C_r]^T$ is estimated by SGM and the estimates θ_t are successively used to compute the control u by the following relation

$$B_t(\nabla)u_t^0 = (A_t(\nabla) - A^0(\nabla))x_{t+1} + B^0(\nabla)\bar{u}_t - (C_t(\nabla) - I)(x_{t+1} - \theta_t^T \varphi_t). \quad (4)$$

It remains to add to $u_t^{(1)}$ some trial signal v_t with the properties:

$$\mathbf{E}v_t = 0, \quad \mathbf{E}v_t v_t^T = I, \quad \sup_t \mathbf{E}\|v_t\|^3 < \infty$$

Let $\varphi(t) = \ln^{-h/2} t, \; t \geq 2, \; h \in (0, [4c(n+2)]^{-1}), \; c = \max(p, q, r+1)$. Then the required adaptive control has the form already known:

$$u_t = u_t^0 + \varphi(t)v_t.$$

As concerns solvability of (4) new difficulties do not arise. So, we should define a class of ARMAX-type equations $\mathcal{K}_2(n, l; p, q, r)$. Namely,

A. The matrices $A(\lambda), \; B(\lambda), \; C(\lambda)$ have no common left factors and the rank of A_p is equal to n;

B. The matrix $C(\lambda) - 1/2I$ is strictly positive real (SPR);

D. All zeros of the function

$$\mathbf{det}\left(B(\lambda)\begin{bmatrix} I_t \\ 0 \end{bmatrix}\right)$$

lie outside the closed unit circle.

The aim of modal control is the same, i.e. PZA.

The external control $\bar{u}_t$ is assumed to be $\mathcal{F}_t$-measurable and to satisfy the condition $\overline{\lim}_{T\to\infty} T^{-1}\sum_{t=1}^{T}\|\bar{u}_t\|^2 < \infty$.

Theorem 2. *For the equations from the class $\mathcal{K}_2(n,l;p,q,r)$, the strategy GC^* leads to the following relationships*:

$$\overline{\lim_{T\to\infty}}\; T^{-1}\sum_{t=1}^{T}(\|x_t\|^2 + \|\bar{u}_t\|^2) < \infty,$$

$$\lim_{T\to\infty}\; T^{-1}\sum_{t=1}^{T}\|\varepsilon_t\|^2 = \mathbf{tr}\,R,$$

$$\lim_{T\to\infty}\; \theta_t = \theta.$$

The inequality and two equalities in the formulation of this theorem mean admissibility of the control and attainability of the control aim, respectively.

Proof. We first prove admissibility of the control. For this purpose, we transform the expression for ε_t. Substituting in it the explicit form of the controls $u_t = u_t^0 + v_t$ and u_t^0 (see (4)) we obtain

$$\begin{aligned}
\varepsilon &= (A^0(\nabla) - A(\nabla))x_{t+1} + (B(\nabla) - B_t(\nabla))u_t + B_t u_t^0 + B_t v_t \\
&\quad - B^0(\nabla)\bar{u}_t + C(\nabla)\zeta_{t+1} \\
&= (A^0 - A)x_{t+1} + (B - B_t)u_t + (A_t - A^0)x_{t+1} + B^0\bar{u}_t \\
&\quad - (C(\nabla) - I)(x_{t+1} - \bar{\theta}_t\varphi_t) + B_t v_t - B^0\bar{u}_t + C(\nabla)\zeta_{t+1} \\
&= \tilde{\theta}_t\varphi_t + B_t v_t - (C(\nabla) - I)y_{t+1} + \zeta_t,
\end{aligned} \tag{5}$$

where $y_t = x_t - \zeta_t - \theta_{t-1}^T\varphi_{t-1}$. After transformations we obtain

$$A^0(\nabla)x_{t+1} = B^0(\nabla)\bar{u}_t + \tilde{\theta}_t\varphi_t + B_t v_t - (C(\nabla) - I)y_{t+1} + \zeta_t. \tag{6}$$

At the end of Sec. 6, the following facts were stated

$$\sum_{t=1}^{\infty}\frac{\|y_{t+1}\|^2}{r_t} < \infty, \quad \sum_{t=1}^{\infty}\frac{\|\tilde{\theta}_t\varphi_t\|^2}{r_t} < \infty, \quad \tilde{\theta}_t = \theta - \theta_t.$$

For the sequence $r_t = 1 + \sum_{i=1}^{t}\|\varphi_i\|^2$ we have $r_t \to \infty$ (since $r_t \geq \sum_{t=1}^{T}\|u_t\|^2 \geq \alpha t(\ln^h t)^{-1}$). So, due to the Kronecker lemma

$$\sum_{t=1}^{T}\|y_t\|^2 = o(r_t), \quad \sum_{t=1}^{T}\|\tilde{\theta}_t\varphi_t\|^2 = o(r_t).$$

In view of stability of $A^0(\nabla)$, the boundedness of the SGE and preceding inequalities, we have

$$\sum_{t=1}^{T} \|x_t\|^2 = o(r_t) + O(t) \tag{7}$$

as $T \to \infty$.

From the minimum phase property of our equation it follows that (as $T \to \infty$)

$$\sum_{t=1}^{T} \|u_t\|^2 = o(r_t) + O(t). \tag{8}$$

The following estimate is obvious

$$\sum_{n=1}^{t} \|x_n - \theta_{n-1}^T \varphi_n\|^2 \leq 2 \sum_{n=1}^{t} \|y_n\|^2 + 2 \sum_{n=1}^{t} \|\zeta_t\|^2 = o(r_t) + O(t).$$

From the definition of r_t and last three estimates it follows that

$$r_t = o(r_t) + O(t)$$

which means $r_t = O(t)$. Thus (7) and (8) imply

$$\sum_{t=1}^{T} (\|x_t\|^2 + \|u_t\|^2) = O(T)$$

and, hence, the considered control is admissible. This and the obvious equality

$$\lim_{T \to \infty} T^{-1} \ln^h T \sum_{t=1}^{T} v_t v_t^T = I \quad \text{a.s.}$$

imply

$$\sum_{t=1}^{T} \|u_t^0\|^2 \leq 2 \sum_{t=1}^{T} (\|u_t\|^2 + \|v_t\|^2) = O(T).$$

So, **A, B, D** hold and the SGE are strongly consistent. It remains to show that the mean square error is minimal. Taking into account the expression (5) for the current error ε_t, the limiting relations for v_t and ζ_t and the fact that $r_t = O(t)$, we have

$$\lim_{T \to \infty} T^{-1} \sum_{t=1}^{T} \|\varepsilon_t\|^2 = \mathbf{tr}\, R$$

which completes the proof. $\qquad\square$

Two adaptive optimal control problems have been investigated in detail by using two identification approaches: tracking after some bounded path $\{x_t^*\}$ (Sec. 4) and modal control. Moreover, the consistency of the estimates of parameters of the equations has been obtained. In many cases the orders of convergence have been found as well. The following question is quite natural: is it possible to unite both these problems into one? In other words: whether we can find a control u_t acting on an ARMAX-type equation and an external control $\bar{u}_t$ in such a way that the equation takes the desired form but the mean square error and the mean square tracking error are minimal? We are also interested in the strong consistency of the estimates of parameters of these equations. The answer proves to be positive. We now consider this problem.

Let us define the control aim denoted by OTMJ:

find the controls u_t and $\bar{u}_t$ such that:

(1) *for all matrix polynomials $A^0(\lambda)$ and $B^0(\lambda)$ and $\mathcal{F}$-measurable control $\bar{u}_t$ the original equation can be reduced to the form (1) and*

$$\lim_{T\to\infty} T^{-1}\sum_{t=1}^{T}\|\varepsilon_t\|^2 = \mathbf{tr}\,R \quad \text{a.s.};$$

(2) *for every bounded sequence $\{x_t^*\}$ the following equality*

$$\lim_{T\to\infty} T^{-1}\sum_{t=1}^{T}\|x_t - x_t^*\|^2 = \mathbf{tr}\,R \quad \text{a.s.}$$

holds;

(3) *the estimates θ_t of the parameters of the ARMAX equation are strongly consistent.*

Let us describe the appropriate strategy GC^{**}. The parameters are estimated by the SGM, the controls u_t, u_t^0 are defined by the relation $u_t = u_t^0 + \varphi v_t$, where u_t^0 is a solution of (4) and $\bar{u}_t$ satisfies the equality

$$B^0(\nabla)\bar{u}_t = (A^0(\nabla) - I)x_{t+1} + x_{t+1}^*. \tag{9}$$

Theorem 3. *The strategy GC^{**} ensures the attainment of the aim OTMJ with respect to the class $\mathcal{K}_2(n, l; p, q, r)$.*

Proof. We verify that the given strategy ensures the fulfilment of (2). For this purpose, we return to (6). Substituting in it the expression (9) for $\bar{u}_t$, we have

$$x_{t+1} - x_{t+1}^* = \tilde{\theta}_t\varphi_t + B_t v_t - (C_t(\nabla) - I)y_t + \zeta_t = \varepsilon_t.$$

It follows immediately that assertion (2) holds. We note that this equality implies the mean square boundedness of x_t. Hence, from Theorem 2, (1) and (3) follow. $\qquad\square$

10.6. On Strong Consistency of LSE and SGE of Parameters

In this section we prove Theorems 1 and 2 from Sec. 4. First, we need some auxiliary assertions. Let $(\Omega, \mathcal{F}, \mathbf{P})$ be a probability space in which a non-decreasing flow of σ-algebras $\mathcal{F}_t$, $t \geq 0$, is given.

Lemma 1. *Let $(W_t, \mathcal{F}_t, t \geq 0)$ denote a matrix martingale-difference such that*

$$\sup_{t \geq 0} \mathbf{E}(\|W_{t+1}\|^\alpha | \mathcal{F}_t) = \gamma < \infty, \quad 1 \leq \alpha \leq 2$$

and M_t denote the random $\mathcal{F}_t$-measurable matrices having the same dimensions as W_t and such that $\|M_t\| < \infty$ a.s. Then for all large t we have

$$\left\| \sum_{l=0}^{t} M_l W_{l+1} \right\| = O\big(S_t(\alpha) \ln^{1/\alpha + \varepsilon}(S_t(\alpha) + e)\big), \quad \forall \varepsilon > 0, \quad \text{a.s.}$$

where $S_t(\alpha) = (\sum_{l=1}^{t} \|M_l\|^\alpha)^{1/\alpha}$, $t \geq 1$.

Proof. We consider the set $V = \{\omega : S_\infty < \infty\}$. By the Martingale Theorem on series (or Local Convergence Theorem)[d] the series $\sum_{l=0}^{t} M_l W_{l+1}$ converges on this set. On its complement V^c we have (we suppose that $M_1 = 0$ below)

$$\sum_{l=1}^{\infty} \mathbf{E}\left(\left\| \frac{M_l W_{l+1}}{S_l^\alpha \ln^{1/\alpha + \varepsilon}(S_l(\alpha) + e)} \right\|^\alpha \Big/ \mathcal{F}_l \right) \leq \gamma \sum_{l=1}^{\infty} \frac{\|M_l\|^\alpha}{S_l^\alpha \ln^{1/\alpha + \varepsilon}(S_l^\alpha + e)}$$

$$\leq \gamma \sum_{l=1}^{\infty} \frac{\int_{S_{l-1}^\alpha}^{S_l^\alpha} dx}{S_l^\alpha \ln^{1+\alpha\varepsilon}(x + e)}$$

$$\leq \gamma \sum_{l=1}^{\infty} \int_{S_{l-1}^\alpha}^{S_l^\alpha} \frac{dx}{x \ln^{1+\alpha\varepsilon}(x + e)}$$

$$= \gamma \int_{S_{l-1}^\alpha}^{\infty} \frac{dx}{x \ln^{1+\alpha\varepsilon}(x + e)} < \infty.$$

Using the Martingale theorem on series again, we conclude that

$$\sum_{l=1}^{\infty} \frac{M_l W_{l+1}}{S_l \ln^{1/\alpha + \varepsilon}(S_l + e)} < \infty$$

on the set V^c. An application of the Kronecker lemma completes the proof. $\qquad\square$

Lemma 2. *Let $(W_t, \mathcal{F}_t)$ be an n-dimensional martingale-difference with bounded α-moments, i.e. $\sup_{t \geq 0} \mathbf{E}(\|W_{t+1}\|^\alpha | \mathcal{F}_t) = \gamma < \infty$. Let T_t be random $\mathcal{F}_t$-measurable matrices of the same dimensions as W_t. Then for all t large enough we have*

$$\sum_{l=1}^{t} W_{l+1}^T T_l^T R_{l+1} T_l W_{l+1}$$

$$= \begin{cases} O(\ln(\boldsymbol{tr}\, R_T^{-1})), & \text{if } \alpha > 0, \\ O(\ln(\boldsymbol{tr}\, R_T^{-1}))[\ln \ln(\boldsymbol{tr}\, R_T^{-1})]^{1+\delta}, \ \forall \delta > 2, & \text{if } \alpha = 2, \quad \text{a.s.} \end{cases}$$

where $R_t = (\sum_{l=1}^{t-1} T_l T_l^T + dI)^{-1}$, $d > 0$.

---

[d]**Theorem.** (Local Convergence Theorem) *Let $\{\eta_i, \mathcal{F}_i\}$ be a martingale-difference and $\zeta_t = \sum_{i=1}^{n} \eta_i$. Then ζ_t converges as $n \to \infty$ on $A = \big\{\omega : \sum_{i=1}^{\infty} \mathbf{E}(|\eta_i|^p | \mathcal{F}_{i-1}) < \infty\big\}$, $1 \leq p \leq 2$.*

We omit the proof which is based on the Martingale lemma.

The following lemmas deal with properties of the LSE for the ARMAX equations. We show that strengthening the assumptions allows finding the order of decreasing of the error of the estimates. Let $\lambda_{\max}$ and $\lambda_{\min}$ denote the maximum and minimum eigenvalues of the matrix $P_{t+1}^{-1} = \sum_{l=1}^{t} \varphi_l \varphi_l^T + d^{-1} I$ where $d = np + lq + nr$. In what follows, these quantities play a decisive role.

Lemma 3. *If for an ARMAX equation the following conditions hold:*

(a) $\sup_t \mathbf{E}(\|W_{t+1}\|^{\alpha} | \mathcal{F}_t) < \infty, \quad \alpha \geq 2;$
(b) *the matrix $C^{-1} - 2^{-1}I$ is strictly positive real.*

Then for the LSM error we have the following asymptotic representations (as $T \to \infty$)

$$\|\theta_{t+1} - \theta\|^2 = \begin{cases} O(\lambda_{\min}^{-t} \ln \lambda_{\max}^t), & \text{if } \alpha > 2, \\ O(\lambda_{\min}^{-t} (\ln \ln \lambda_{\max}^t)^g) \ln \lambda_{\max}^t), \ \forall g > 1, & \text{if } \alpha = 2, \ \text{a.s.} \end{cases}$$

Proof. We put $\bar{\theta}_t = \theta - \theta_t$. From the explicit expression for P_{t+1}^{-1} (recall that $P_{t+1} = P_t - a_t P_t \Phi_t \Phi_t^T P_t$) we have

$$\|\bar{\theta}_{t+1}\|^2 < \lambda_{\min}^{-t} \mathbf{tr}\, \bar{\theta}_{t+1}^T P_{t+1}^{-1} \bar{\theta}_{t+1}.$$

According to this inequality it remains to verify that

$$\mathbf{tr}\, \bar{\theta}_{t+1}^T P_{t+1}^{-1} \bar{\theta}_{t+1} = \begin{cases} O(\ln \lambda_{\max}^t) & \text{if } \alpha > 2, \\ O((\ln \ln \lambda_{\max}^t)^g) \ln \lambda_{\max}^t, \ \forall g > 1 & \text{if } \alpha = 2, \ \text{a.s.} \end{cases}$$

To do this we note that (see Sec. 1)

$$x_{t+1}^T - \Phi_t^T \theta_{t+1} = a_t(x_{t+1}^T - \Phi_t^T \theta_t), \quad a_t = (1 + \Phi_t^T P_t \Phi_t)^{-1} \tag{1}$$

and

$$C(\nabla)\eta_{t+1} = \bar{\theta}_{t+1}\Phi_t \tag{2}$$

where $\eta_{t+1} = x_{t+1} - \xi_t - \theta_{t+1}^T \theta_t$.

Condition (b) implies the existence of the numbers $\varkappa_0 > 0$ and $\varkappa_1 \geq 0$ such that

$$S_t = \sum_{i=0}^{t} \Phi_i^T \bar{\theta}_{i+1} \left(\eta_{i+1} - \frac{1 + \varkappa_0}{2} \bar{\theta}_{i+1}^T \Phi_i \right) + \varkappa_1 \geq 0, \quad t \geq 0.$$

With the help of (1) and (2) we obtain

$$\begin{aligned}
\mathbf{tr}\, \bar{\theta}_{k+1}^T P_{k+1}^{-1} \bar{\theta}_{k+1} &= \mathbf{tr}\, \bar{\theta}_{k+1}^T \Phi_k \Phi_k^T \bar{\theta}_{k+1} + \mathbf{tr}\, \bar{\theta}_{k+1}^T P_{k+1}^{-1} \bar{\theta}_{k+1} \\
&= \|\Phi_k^T \bar{\theta}_{k+1}\|^2 - 2(\eta_{k+1}^T + \xi_{k+1}^T)(\bar{\theta}_{k+1} + P_k \Phi_k(\eta_{k+1}^T + \xi_{k+1}^T))^T \Phi_k \\
&\quad + \Phi_k^T P_k \phi_k \|\eta_{k+1} + \xi_{k+1}\|^2 + \mathbf{tr}\, \bar{\theta}_k P_k^{-1} \bar{\theta}_k \\
&\leq \mathbf{tr}\, \bar{\theta}_k^T P_k^{-1} \bar{\theta}_k - 2\Phi_k^T \bar{\theta}_{k+1}(\eta_{k+1} - 2^{-1}(1 + \varkappa_0)\bar{\theta}_{k+1}^T \Phi_k) \\
&\quad - \varkappa_0 \|\bar{\theta}_{k+1}^T \Phi_k\|^2 - 2\xi_{k+1}^T \bar{\theta}_{k+1}^T \Phi_k.
\end{aligned}$$

Summing over k and taking into consideration the inequality $S_t \geq 0$, we arrive at the following inequality

$$\mathbf{tr}\,\bar{\theta}_{t+1}^T P_{t+1}^{-1} \bar{\theta}_{t+1} \leq O(1) - \varkappa_0 \sum_{l=0}^{t} \|\bar{\theta}_{l+1}^T \Phi_l\|^2 - 2 \sum_{l=0}^{t} \xi_{l+1}^T \bar{\theta}_{l+1}^T \Phi_l. \tag{3}$$

We define an $\mathcal{F}_t$-measurable random vector $\bar{\eta}_t = x_{n+1} - \theta_t^T \Phi_t - \zeta_t$ and use Lemma 2 to estimate the last sum in (3).

$$\left| \sum_{l=0}^{t} \xi_{l+1}^T \bar{\theta}_{l+1}^T \Phi_l \right| = \left| \sum_{l=0}^{t} \xi_{l+1}^t (\bar{\theta}_l - a_l(\xi_{l+1} + \bar{\eta}_l)\Phi_l P_l))\Phi_l \right|$$

$$\leq \sum_{l=0}^{t} a_l \Phi_l^T P_l \Phi_l \|\xi_l\|^2 + O\left(\left(\sum_{l=0}^{t} \|\bar{\theta}_{l+1}^T \Phi_l\|^2 \right)^{\nu} \right)$$

$$+ O\left(\left(\sum_{l=0}^{t} (a_l \Phi_l^T P_l \Phi_l)^2 \|\xi_l\|^2 \right)^{\mu} \right)$$

$$= O\left(\left(\sum_{l=0}^{t} \|\bar{\theta}_{l+1}^T \Phi_l\|^2 \right)^{\nu} \right) + O\left(\sum_{l=0}^{t} a_l \Phi_l^T P_l \Phi_l \|\xi_l\|^2 \right), \quad 1/2 < \nu < 1.$$

This and (3) imply

$$\mathbf{tr}\,\bar{\theta}_{t+1}^T P_{t+1}^{-1} \bar{\theta}_{t+1} \leq O(1) - O\left(\sum_{l=0}^{t} a_l \Phi_l^T P_l \Phi_l \|\xi_l\|^2 \right)$$

$$= O(1) + O\left(\sum_{l=0}^{t} \xi_{l+1}^T \Phi_l^T P_{l+1} \Phi_l \xi_{l+1} \right). \tag{4}$$

Now Lemma 2 and the inequality $d^{-1}\,\mathbf{tr}\,P_{t+1}^{-1} \leq \lambda_{\max}^t \leq \mathbf{tr}\,P_{t+1}^{-1}$ lead to the required result. $\qquad\square$

Let us write the estimate of the decreasing order of $\theta - \theta_t$ in another form. Up to now to estimate the random disturbances ξ_t entering into Φ_t, i.e. finite history of the control process or the stochastic regressor we have used the quantities $x_t - \theta_t^T x_{t-1}$. Now, instead of these we use the disturbance ξ_t itself. The stochastic regressor takes the form

$$\Phi_t = [x_t^T \cdots x_{t-p+1}^T u_t^T \cdots u_{t-q+1}^T \xi_t^T \cdots \xi_{t-r+1}^T], \quad \Phi_j = 0,\ j \leq 0.$$

Let $\lambda_{\max}^{0t}$ and $\lambda_{\min}^{0t}$ denote the maximal and minimal eigenvalue of the matrix $\sum_{l=0}^{t-1} \Phi_l^0 \Phi_l^{0T} + d^{-1}I$ which is analogous to P_T^{-1}. The next lemma demonstrates the relation between these eigenvalues and the estimates of the decreasing rate of the errors $\theta - \theta_t$.

Lemma 4. *Under the assumptions of Lemma 3 we have*

$$\|\theta_{t+1} - \theta\|^2$$

$$= \begin{cases} O\left(\dfrac{\ln \lambda_{\max}^{0t}}{\lambda_{\min}^{0t}}\right), & \text{if } \alpha > 2, \ \ln \lambda_{\max}^{0t} = O(\lambda_{\min}^{0t}), \\[3ex] O\left(\dfrac{\ln \lambda_{\max}^{0t}(\ln \ln \lambda_{\max}^{0t})^g}{\lambda_{\min}^{0t}}\right), \ \forall g > 1, & \text{if } \alpha = 2, \ \ln \lambda_{\max}^{0t}(\ln \ln \lambda_{\max}^{0t})^c \lambda_{\min}^{0t} \\[3ex] & \qquad = O(\lambda_{\min}^{0t}), \ c > 1, \ \text{a.s.} \end{cases}$$

Proof. From linear system theory it is known that the strong positive reality of the matrix $C^{-1}(\lambda) - 2^{-1}I$ implies stability of the matrix $C^{-1}(\lambda)$. So, from (2) to (4) it follows that

$$\sum_{l=0}^{t} \|\eta_{l+1}\|^2 = O\left(\sum_{l=0}^{t} \|\bar{\theta}_{l+1}^T \Phi_l\|^2\right) = O(1) + O\left(\sum_{l=0}^{t} a_l \Phi_l^T P_l \Phi_l \|\xi_{l+1}\|^2\right)$$

$$= \begin{cases} O(\ln \lambda_{\max}^{0t}) & \text{a.s., if } \alpha > 2, \\ O(\ln \lambda_{\max}^{0t}(\ln \ln \lambda_{\max}^{0t})^g), \ \forall g > 1 & \text{a.s., if } \alpha = 2. \end{cases}$$

For $\alpha > 2$ we have $\lambda_{\min}^{t} = O(\lambda_{\max}^{0t})$ and $\lambda_{\min}^{0t} \leq 2\lambda_{\min}^{t} + O(\lambda_{\min}^{0t})$ which lead to the first assertion of the lemma. The second one is proved similarly. $\square$

The stated results reduce the estimation of the convergence rate of the LSE (and, in general, their convergence) to that of the maximal and minimal eigenvalues of the matrix P_T^{-1} which must be non-degenerate for all sufficiently large t. With this aim in view we use the randomized control described in Sec. 4. We can now prove the main result of this chapter.

Proof of Theorem 1 from Sec. 4. First, we note that the parameter a varies in some non-empty interval. Indeed, it is easy to verify that $(c+1)(1+\delta)+\delta/2 < 1/2$. Since $1 < (1-h)^{-1} < 2(1+c)(2c+1)^{-1} < 2$ we can choose some number $f \in ((1-h)^{-1}, 2)$ such that

$$\sum_{j=1}^{\infty} \mathbf{E}[j^{-(1-h)f)}\|v_j v_j^T - j^{-1}\mu I\|^f |\mathcal{F}_{j-1}] < \infty.$$

Due to properties of Martingales we have

$$\lim_{T \to \infty} T^{h-1} \sum_{t=1}^{t} v_t v_t^T = \frac{\mu}{1-h} I, \quad \text{a.s.} \tag{5}$$

This equality will be used later on.

From the assumptions it follows that $r_t^0 = O(t^{t+\delta})$ where $r_t^0 = 1 + \sum_{l=0}^{t-1} \|\Phi_l^0\|^2$ and it remains to show that for all sufficiently large t $(\geq t_0)$ we have

$$\lambda_{\min}^{0t} \geq c_0 t^a, \quad c_0 > 0.$$

In turn, this is equivalent to $\underline{\lim}_{T\to\infty} t^{-a}\lambda_{\min}^{0t} > 0$ a.s. To prove this inequality, we put $f_t = \det A(\lambda)\Phi_t^0$ where

$$\det A(\lambda) = a_0 + a_1\lambda + \cdots + a_{np}\lambda^{np}, \quad a_{np} \neq 0.$$

The last relationship holds because the matrix A_p has full rank and the degree of the polynomial $A(\lambda)$ is equal to np. If $\lambda_{\min}[\sum_{l=1}^{t} f_l f_l^T]$ denotes the minimal eigenvalue of the matrix $\sum_{l=1}^{t} f_l f_l^T$ then

$$\lambda_{\min}\left[\sum_{l=1}^{t} f_l f_l^T\right] = \inf_{\|x\|=1} \sum_{l=1}^{t}\left(\sum_{i=1}^{np} a_i(x^T)^i \Phi_{l-i}^0\right)^2 \leq (np+1)\sum_{j=0}^{np} a_j^2 \lambda_{\min}^{0t}.$$

It is clear from the above inequality that $\underline{\lim}_{t\to\infty} t^{-a}\lambda_{\min}^{0t} > 0$ a.s. We put

$$E = \left\{\omega : \lim_{t\to\infty} t^{-\alpha}\lambda_{\min}\left[\sum_{l=1}^{t} f_l f_l^T\right] = 0\right\}.$$

To prove that $\mathbf{P}\{E\} = 0$ we suppose that the contrary holds.

We write the ARMAX equation in a new form

$$x_{t-i} = A^{-1}(\nabla)B(\nabla)\nabla^i x_i + A^{-1}(\nabla)C(\nabla)\nabla^i \xi_i$$

where $B(\lambda) = B_0 + B_1\lambda^1 + \cdots + B_q\lambda^q$ (within the current proof). From this it follows that the vector φ_t^0 can be represented as follows

$$\varphi_t^0 = \begin{bmatrix} F_1(\nabla) \\ F_2(\nabla) \\ F_3(\nabla) \end{bmatrix} \begin{bmatrix} u_t \\ \xi_t \end{bmatrix} \tag{6}$$

where

$$F_1(\lambda) = \begin{bmatrix} A^{-1}(\lambda) & [B(\lambda) & C(\lambda)] \\ \lambda A^{-1}(\lambda) & [B(\lambda) & C(\lambda)] \\ \lambda^2 A^{-1}(\lambda) & [B(\lambda) & C(\lambda)] \\ \cdots & \cdots & \cdots \\ \lambda^{p-1} A^{-1}(\lambda) & [B(\lambda) & C(\lambda)] \end{bmatrix}, \quad F_2(\lambda) = \begin{bmatrix} [I_l & O] \\ \lambda[I_l & O] \\ \lambda^2[I_l & O] \\ \cdots & \cdots \\ \lambda^{q-1}[I_l & O] \end{bmatrix},$$

$$F_3(\lambda) = \begin{bmatrix} [O & I_m] \\ \lambda[O & I_m] \\ \lambda^2[O & I_m] \\ \cdots & \cdots \\ \lambda^{r-1}[O & I_m] \end{bmatrix}.$$

Here, I_k denotes the identity matrix of order k.

By our assumptions, for any $\omega \in E$ there exists a sequence of unit vectors from $\mathbf{R}^d$

$$\eta_{t_k} = \left[\alpha_{t_k}^{0^T} \cdots \alpha_{t_k}^{p-1^T} \beta_{t_k}^{0^T} \cdots \beta_{t_k}^{q-1^T} \gamma_{t_k}^{0^T} \cdots \gamma_{t_k}^{r-1^T}\right]$$

such that

$$\lim_{k\to\infty} t_k^{-\alpha} \sum_{j=1}^{t_k} (\eta_{t_k} f_j)^2 = 0. \tag{7}$$

This leads to a contradiction. To show this, let us define the following sequence of matrix polynomials (they are generated by (6))

$$
\begin{aligned}
G_{t_k}(\lambda) &= \sum_{i=0}^{p-1} \alpha_{t_k}^{i^T} \lambda^i \mathbf{Adj}\, A(\lambda)[B(\lambda)\ \ C(\lambda)] + \sum_{i=0}^{q-1} \beta_{t_k}^{i^T} \lambda^i [\mathbf{det}\, A(\lambda) I_l\ \ O] \\
&\quad + \sum_{i=0}^{r-1} \gamma_{t_k}^{i^T} \lambda^i [O\ \ \mathbf{det}\, A(\lambda) I_m] \\
&\equiv \sum_{i=0}^{s} [h_{t_k}^{i^T}\ \ g_{t_k}^{i^T}] \lambda^i, \quad s = \max(p, q, r) + mp - 1
\end{aligned}
$$

where $h_{t_k}^i$, $g_{t_k}^i$ are some sequences of bounded (in both k and ω) vectors of dimensions l and m, respectively. Obviously, in terms of this notation, equality (7) can be represented in the following form

$$\lim_{k\to\infty} t_k^{-\alpha} \sum_{i=1}^{t_k} \left(h_{t_k}^{0^T} u_i + \cdots + h_{t_k}^{s^T} u_{i-s} + g_{t_k}^{0^T} W_i + \cdots + g_{t_k}^{s^T} W_{i-s} \right)^2 = 0 \tag{8}$$

with $u_t = 0$ at $t < 0$. We consider the expression under the symbol $\lim$ on the left-hand side of (8). We have

$$
\begin{aligned}
t_k^{-\alpha} \Bigg\{ &\sum_{i=1}^{t_k} \Bigg[(h_{t_k}^{0^T} v_t)^2 + \Big(h_{t_k}^{0^T} u_i^0 + h_{t_k}^{1^T} u_{i-1} + \cdots + h_{t_k}^{s^T} u_i^{i-s} + g_{t_k}^{0^T} \xi_i \\
&+ \cdots + g_{t_k}^{s^T} t_{i-s} \Big)^2 \Bigg] + 2 h_{t_k}^{0^T} \left(\sum_{i=1}^{t_k} u_i^0 v_i^T \right) h_{t_k}^0 + 2 \sum_{j=1}^{t_k} h_{t_k}^{j^T} \left(\sum_{i=1}^{t_k} u_{i-j} v_i^T \right) h_{t_k}^0 \\
&+ 2 \sum_{j=0}^{s} g_{t_k}^{j^T} \left(\sum_{i=1}^{t_k} \xi_{i-j} v_i^T \right) h_{t_k}^0 \Bigg\} \to 0
\end{aligned}
$$

as $k \to \infty$. From Lemma 2 applied to $\sum_{i=1}^{t_k} \xi_{i-j} v_i^T$ when $\alpha = 2$ it follows that

$$\varlimsup_{k\to\infty} t_k^{-\alpha} \left[g_{t_k}^{j^T} \left(\sum_{i=1}^{s} \xi_{i-j} v_i^T \right) h_{t_k}^0 \right] \le (1 + s^2) c^2 \lim_{k\to\infty} t_k^{-\alpha} O\left(t_k^{1/2} \ln^{1/2+\eta}(t_k + e) \right) = 0$$

where we have used the uniform boundedness of the vectors $h_{t_k}^i$ and $g_{t_k}^i$. Similarly, we can verify that

$$\varlimsup_{k\to\infty} t_k^{-\alpha} \left[h_{t_k}^{0^T} \left(\sum_{i=1}^{t_k} u_i^0 v_i^T \right) h_{t_k}^0 + \sum_{j=1}^{s} h_{t_k}^{j^T} \left(\sum_{i=1}^{t_k} u_{i-j} v_i^T \right) h_{t_k}^0 \right] = 0.$$

Both these limits are equal to zero on E a.s. Now, we can rewrite (8) in the form

$$\lim_{k\to\infty} t_k^{-\alpha} \sum_{j=1}^{t_k} \left[\left(h_{t_k}^{0^T} v_t \right)^2 + \left(h_{t_k}^{0^T} u_i^0 + h_{t_k}^{1^T} u_{i-1} + \cdots + h_{t_k}^{s^T} u_{i-s} \right. \right.$$
$$\left. \left. + g_{t_k}^{0^T} \xi_i + \cdots + g_{t_k}^{s^T} \xi_{t-s} \right)^2 \right] = 0.$$

This equality is equivalent to

$$\lim_{k\to\infty} t_k^{-\alpha} \sum_{j=1}^{t_k} (h_{t_k}^{0^T} v_t)^2 = 0,$$

$$\lim_{k\to\infty} t_k^{-\alpha} \sum_{j=1}^{t_k} \left(h_{t_k}^{0^T} u_i^0 + h_{t_k}^{1^T} u_{i-1} + \cdots + h_{t_k}^{s^T} u_i^{i-s} + g_{t_k}^{0^T} \xi_i + \cdots + g_{t_k}^{s^T} \xi_{t-s} \right)^2 = 0.$$

From (5) and the assumptions it follows ($\omega \in E$) that

$$\| h_{t_k}^0 \|^2 = o\left(t_k^{-(1-\varepsilon-a)} \right), \qquad \lim_{k\to\infty} t_k^{-(1+\delta)+1-h-a} \sum_{i=1}^{t_k} \left(h_{t_k}^{0^T} u_i^0 \right) = 0,$$

$$\lim_{k\to\infty} t_k^{-(a+h+\delta)} \sum_{i=1}^{t_k} \left(h_{t_k}^{1^T} u_{i-1} + \cdots + h_{t_k}^{s^T} u_{i-s} + g_{t_k}^{0^T} \xi_i + \cdots + g_{t_k}^{s^T} \xi_{i-s} \right)^2 = 0.$$

Since $a \le 1 - (s+1)(h+\delta)$ the first of these relations means

$$\lim_{k\to\infty} \| h_{t_k}^i \|^2 = 0, \quad 0 \le i \le s. \tag{9}$$

From the second relation at $i = s + 1$ it follows that

$$\lim_{k\to\infty} t_k^{-(a+(s+1)(h+\delta))} \sum_{i=1}^{t_k} \left(g_{t_k}^{0^T} \xi_i + \cdots + g_{t_k}^{s^T} \xi_{i-s} \right)^2 = 0.$$

Acting in the same manner, we find the following pairs of equalities

$$\| h_{t_k}^s \|^2 = o\left(t_k^{-(1-h-a-j(h+\delta))} \right), \quad 0 \le j \le s,$$

$$\lim_{k\to\infty} t_k^{-a-j(h+\delta)} \sum_{i=1}^{t_k} \left(h_{t_k}^{j^T} u_{i-j} + \cdots + h_{t_k}^{s^T} u_{i-s} + g_{t_k}^{0^T} \xi_i + \cdots + g_{t_k}^{s^T} \xi_{i-s} \right)^2 = 0,$$

$$1 \le j \le s + 1.$$

By condition **C** of the theorem, we have

$$\lim_{k\to\infty} \| g_{t_k}^j \| = 0. \tag{10}$$

From (9) and (10) it follows that (the function G was defined after formula (7))

$$\lim_{k\to\infty} G_{t_k}(\lambda) = G_\infty \equiv 0.$$

Let $\{\eta_m\}$ be a converging subsequence of $\{\eta_k\}$, i.e.

$$\lim_{m\to\infty} \eta_{t_m} = \eta = (\alpha^{0^T}, \ldots, \alpha^{p-1^T}, \beta^{0^T}, \ldots, \beta^{q-1^T}, \gamma^{0^T}, \ldots, \gamma^{r-1^T}).$$

The identity $G_\infty \equiv 0$ implies

$$\sum_{i=0}^{p-1} \alpha^{i^T} \lambda^i (\mathbf{Adj}\, A(\lambda))[B(\lambda)\ \ C(\lambda)] = -\sum_{i=0}^{q-1} \beta^{i^T} \lambda^i [\mathbf{det}\, A(\lambda) I_l\ \ O]$$

$$-\sum_{i=0}^{r-1} \gamma^{i^T} \lambda^i [O\ \ \mathbf{det}\, A(\lambda) I_m]. \qquad (11)$$

According to condition $\mathbf{D}$, there are three matrix polynomials $M(\lambda)$, $N(\lambda)$, $L(\lambda)$ such that

$$A(\lambda)M(\lambda) + B(\lambda)N(\lambda) \equiv I.$$

Making use of this identity, on the left-hand side of (11) we obtain

$$\sum_{i=0}^{p-1} \alpha^{i^T} \lambda^i (\mathbf{Adj}\, A(\lambda))$$

$$= \sum_{i=0}^{p-1} \alpha^{i^T} \lambda^i (\mathbf{Adj}\, A(\lambda)) \{A(\lambda)M(\lambda) + B(\lambda)N(\lambda) + C(\lambda)L(\lambda)\}$$

$$= \mathbf{det}\, A(\lambda) \left(\sum_{i=0}^{p-1} \alpha^{i^T} \lambda^i M(\lambda) - \sum_{i=0}^{q-1} \beta^{i^T} \lambda^i N(\lambda) - \sum_{i=0}^{r-1} \gamma^{i^T} \lambda^i L(\lambda) \right).$$

We estimate the degree of the polynomial on the left-hand side of this equality

$$\mathbf{deg} \left[\sum_{i=0}^{p-1} \alpha^{i^T} \lambda^i (\mathbf{Adj}\, A(\lambda)) \right] \leq p - 1 + \mathbf{deg}\, (\mathrm{Adj}\, A(\lambda))$$

$$= p - 1 + mp - p = \mathbf{deg}\, (\mathrm{Adj}\, A(\lambda)).$$

This means that the left-hand side of the preceding equality is equal to zero and, hence, $\alpha^i = 0$ for all $i \in [0, p-1]$. From (11) it follows that $\beta^i = 0$, $i = 0, \ldots, q-1$, and $\gamma^i = 0$, $i = 0, \ldots, r-1$. Hence $\eta = 0$ which contradicts the equality $\|\eta\| = 1$. So, $\mathbf{P}\{E\} = 0$. $\qquad\qquad\qquad\qquad\qquad\qquad\qquad\qquad\qquad\qquad\qquad\qquad\quad \square$

We now turn our attention to the proof of Theorem 2 about the strong consistency of the SGM. Since this theorem does not give an estimate of the $\theta_t \to \theta$ convergence rate we only give the main steps.

We start by writing the SGM in the form

$$\theta_{t+1} = \theta_t + r_T^{-1} \Phi_t [x_{t+1}^T - \Phi_t^T \theta_t], \quad r_t = 1 + \sum_{i=1}^{t} \|\Phi_i\|^2, \quad r_0 = 1,\ t \geq 0,$$

where θ_0 is an arbitrary initial value. The history of the process up to time t (the stochastic regressor) is the vector

$$\Phi_t$$

$$= \begin{cases} [x_t^T \cdots x_{t-p+1}^T u_t^T \cdots u_{t-q+1}^T\ x_t - \Phi_{t-1}^T \theta_{t-1}\ \cdots\ x_{t-r+1}^T - \Phi_{t-r}^T \theta_{t-r}], & \text{if } t > 0, \\ 0, & \text{if } t \leq 0. \end{cases}$$

If the random actions ξ_t are observable then the real history of the process up to time t is the vector

$$\Phi_t^0 = \begin{cases} [x_t^T \cdots x_{t-p+1}^T u_t^T \cdots u_{t-q+1}^T \xi_t^T, \cdots \xi_{t-r+1}^T], & \text{if } t > 0, \\ 0, & \text{if } t \leq 0. \end{cases}$$

We put

$$\tilde{\Phi}_t = \Phi_t - \Phi_t^0, \quad \tilde{\theta}_t = \theta - \theta_t.$$

Then the initial recurrent equation for the SGM can be written as follows

$$\theta_{t+1} = \theta_t + r_T^{-1}\Phi_t(\Phi_t^T\theta_t - \tilde{\Phi}_t\theta + \xi_t - \Phi_t^T\theta_t)$$

or

$$\tilde{\theta}_{t+1} = (I - r_T^{-1}\Phi_t\Phi_t^T)\tilde{\theta}_t + r_T^{-1}\Phi_t\tilde{\Phi}_t^T - r_T^{-1}\Phi_t\theta_t^T.$$

The last equation shows the important role of the matrix $I - r_T^{-1}\Phi_t\Phi_t^T$ for studying the consistency of the SGE. We define successively the following sequence of matrices $F(t, i)$:

$$F(t+1, i) = (I - r_T^{-1}\Phi_t\Phi_t^T)F(t, i), \quad F(i, i) = I.$$

This and the previous relations lead to the following form of the SGM

$$\tilde{\theta}_{t+1} = F(t+1, i)\tilde{\theta}_i + \sum_{j=1}^{t} F(t+1, j+1)\frac{\Phi_j\tilde{\Phi}_j^T}{r_j}\theta - \sum_{j=1}^{t} F(t+1, j+1)\frac{\Phi_j\xi_{j+1}^T}{r_j} \quad (12)$$

which is convenient to analyse. Here θ denotes the true value of the parameter.

Let us find a relation between the asymptotic behavior (as $t \to \infty$) of the matrix $F(t, i)$ and the maximal and minimal eigenvalues $\mu_{\max}^t$ and $\mu_{\min}^t$ of the matrix $d^{-1}I + \sum_{i=1}^{t} \Phi_i\Phi_i^t$.

Lemma 5. *Assume that* $\lim_{t\to\infty} r_t = \infty$ *and there exist numbers* $a \in [0, 1/4]$, $b > 0, t_0 \geq 1$ *such that for all* $t \geq t_0$

$$\frac{r_{t+1}}{r_t} \leq b(\ln r^t)^a, \quad \frac{\mu_{\max}^t}{\mu_{\min}^t} < b(\ln r_t)^{1/4-a}.$$

Then $\lim_{t\to\infty} F(t, 0) = 0.$

Without going into details of the proof we only mention that considerable growth of the ratio $\mu_{\max}^t/\mu_{\min}^t$ has to be excluded. The conclusion of Theorem 2 may fail if we replace $(\ln r_t)^{1/4}$ by $(\ln r_t)^{1+\alpha}$, $\alpha > 0$.

We introduce the sequence of matrices

$$F^0(t+1, 0) = \left(I - \frac{\Phi_t^0\Phi_t^{0T}}{r_t^0}\right) F^0(t, 0), \quad F^0(0, 0) = I$$

where $r_t^0 = 1 + \sum_{i=1}^{t} \|\Phi_i^0\|^2$, i.e. instead of regressors Φ_t, we have used Φ_t^0.

Lemma 6. *Let the following conditions be satisfied for a linear ARMAX equation:*

(a) $\mathbf{E}(\|\xi_t\|^2|\mathcal{F}_{t-1}) < \infty;$
(b) *the function* $C(\lambda) - 2^{-1}I$ *is strictly positive real;*

Then the following assertions hold for SGM:

(α) $\lim_{t\to\infty} \theta_t = \theta$ *a.s. if* $\lim_{t\to\infty} F(t,0) = 0;$
(β) *at* $r = 0$ *we have* $\lim_{t\to\infty} \theta_t = \theta$ *a.s. for any initial* θ *if and only if* $\lim_{t\to\infty} F(t,0) = 0;$
(γ) $\lim_{t\to\infty} F^0(t,0) = 0$ *a.s. if and only if* $\lim_{t\to\infty} F(t,0) = 0.$

The proof of this lemma is based on relation (12), where we should show that the terms on its right-hand side tend to zero as $t \to \infty$. The detailed arguments (based on martingale considerations) are omitted but we would like to note one important fact.

The final steps of the proof of Theorem 4 are similar to the proof of Lemma 4. They differ in the two relations

$$\lim_{t\to\infty} T^{-1}\ln^h t \sum_{i=1}^{t} v_i v_i^t = \mu I, \quad r_t^0 = O(t\ln^\delta t) \quad \text{a.s.}$$

which imply (by the same arguments as in Lemma 4)

$$\frac{r_{t+1}^0}{r_t^0} = O((\ln r_t^0)^{\delta+h}) \quad \text{a.s.}$$

$$\varliminf_{T\to\infty} T^{-1}(\ln t)^{1/4-2\delta-h}\lambda_{\min}\left(\sum_{i=1}^{t} \Phi_i^0\Phi_i^{0^T} + d^{-1}I\right) \neq 0 \quad \text{a.s.}$$

It is easy to show that $C(\nabla)y_y = \tilde\theta_{t-1}\varphi_{t-1}$, where

$$y_t = x_t - \xi_t - \theta_{t-1}^T\varphi_{t-1}.$$

By condition (b) there are constants k_1, $k_2 > 0$ such that

$$S_t = 2\sum_{i=1}^{t} \xi_i^T\left(\tilde\theta_{i-1}^T\varphi_{i-1} - 2^{-1}(1+k_1)\xi_i\right) + k_2 \geq 0.$$

We put

$$V_t = \mathbf{tr}\,\tilde\theta_t^T\tilde\theta_t + \frac{S_t}{r_{t-1}} + \sigma^2\mathbf{E}\left(\sum_{i=1}^{\infty} r_i^{-2}\|\Phi_i\|^2/\mathcal{F}_t\right)$$

$$- \sigma^2\sum_{i=1}^{t-1} r_i^{-2}\|\Phi_i\|^2 + k_1\sum_{i=1}^{t-1} r_i^{-1}\|y_{i+1}\|^2.$$

This gives $\mathbf{E}V_t < \infty$. The fact that $(V_t, \mathcal{F}_t)$ is semi-martingale can be verified by standard arguments. Hence $\lim_{t\to\infty} V_t = V < \infty$ a.s. This implies the convergence of the series $\sum_{i=1}^{\infty} r_i^{-1}\|y_{i+1}\|^2$ and, according to the Kronecker lemma,

$$\lim_{t\to\infty} r_T^{-1} \sum_{i=1}^{t} \|y_i\|^2 = 0. \tag{13}$$

Similarly, one can verify that $\sum_{i=1}^{\infty} r_i^{-1}\|\tilde{\theta}_i\varphi_{i-1}\|^2 < \infty$. Therefore

$$\lim_{t\to\infty} r_t^{-1} \sum_{i=1}^{t} \|\tilde{\theta}_i\varphi_{i-1}\|^2 = 0.$$

10.7. Linear-Quadratic Problem (LQP)

We start by listing the main results on the LQP in the framework of classical theory. They will be used for the aims of adaptive control later on.

We consider a discrete time dynamical system. Let its evolution $x_t \in \mathbf{R}^n$ be modeled by the linear equation

$$x_{t+1} = Ax_t + Bu_t + C\zeta_{t+1}, \quad y_t = Dx_t \tag{1}$$

where $u_t \in \mathbf{R}^l$ is a control, $\zeta_t \in \mathbf{R}^n$ is a random disturbance (a noise). We assume that the disturbances ζ_t are unobserved and only the variables y_t are observed. The random process ζ_t is defined on a probability space $(\Omega, \mathcal{F}, \mathbf{P})$. In the simplest case ζ_t is a sequence of independent, identically distributed random variables with zero mean and finite variance. Let some non-decreasing flow of σ-algebras $\mathcal{F}_t$ be fixed. We assume that $(\zeta_t, \mathcal{F}_t)$ forms a martingale-difference and

$$\mathbf{E}(\|\zeta_t\|^b|\mathcal{F}_{t-1}) = \sigma^2 < \infty, \ \ b > 0, \quad \lim_{T\to\infty} \sum_{t=1}^{T} \zeta_t\zeta_t^T = R > 0. \tag{2}$$

The control u_t is assumed to be "admissible", i.e. to belong to the following set

$$U = \left\{ u_t, \ t \geq 0 : \overline{\lim_{T\to\infty}} \ T^{-1} \sum_{t=1}^{T}(\|x_t\|^2 + \|u_t\|^2) < \infty, \ \lim_{t\to\infty} T^{-1}\|x_t\|^2 = 0, \right.$$

$$\left. u_t \text{ is } \mathcal{F}_t\text{-measurable}, \forall t \vphantom{\sum_{t=1}^{T}} \right\} \tag{3}$$

The σ-algebra $\mathcal{F}_t$ is usually generated by ζ_j, $j \leq t$, i.e., $\mathcal{F}_t = \sigma(\zeta_j, \ j \leq t)$.

We define a bounded deterministic sequence $\{x_t^*\}$ that represents a model reference and two matrices: the n-matrix $Q_1 \geq 0$ and the l-matrix $Q_2 \geq 0$. The aim of control is:

to find a strategy u^{opt} *under which the functional*

$$W(u) = \overline{\lim_{T\to\infty}} \ T^{-1} \sum_{t=1}^{T}[(x_t - x_t^*)^T Q_1(x_t - x_t^*) + u_t^T Q_2 u_t]$$

takes the minimal value, i.e.

$$W(u^{\mathrm{opt}}) = \inf_{u \in U} W(u).$$

This aim is strong (see Chap. 1) and, in fact, it is a combination of optimal tracking after a model reference and minimization of the control.

The solution of this problem involves solving the matrix algebraic Riccati Equation:

$$S = A^T S A - A^T S B (Q_2 + B^T S B)^{-1} B^T S A + H^T Q_1 H. \tag{4}$$

If the triplet (A, B, D) is controllable and observable and the matrix D satisfies $D^T D = H^T Q_1 H$, then there is a unique non-negative definite matrix S satisfying (4). It has the property that the matrix $A + BL$ is stable, where

$$L \stackrel{\mathrm{def}}{=} -(Q_2 + B^T S B)^{-1} B^T S A.$$

The optimal control has the form

$$u_t^{\mathrm{opt}} = L x_t + d_t. \tag{5}$$

The first term on the right-hand side of (5) corresponds to the optimization problem in the case $x_t^* \equiv 0$, i.e. in absence of the model reference. The second term is a "correction" to account for the path x_t^*. The quantities d_t are defined by

$$\begin{cases} d_t = -(Q_2 + B^T S B)^{-1} B^T b_{t-1}, \\ \displaystyle b_t = -\sum_{i=0}^{\infty} (A + BL)^{i^T} H^T Q_1 x_{t+i}^* = (A + BL)^T b_{t+1} - H^T Q_1 x_t^*. \end{cases} \tag{6}$$

Due to stability of the matrix $A + BL$, the set $\{b_t,\ t \geq 1\}$ is bounded.

Under these assumptions the minimal value of the functional $W(u^{\mathrm{opt}})$ is equal to

$$W(u^{\mathrm{opt}})$$

$$= \mathbf{tr}\, SCRC^T + \varlimsup_{T \to \infty} T^{-1} \sum_{t=1}^{t} \left[x_t^{*T} Q_1 x_t^* + b_{t+1}^T B (Q_2 + B^T S B)^{-1} B b_{t+1} \right]. \tag{7}$$

This completes the solution of the stated problem. The optimal strategy proves to be *simple*, i.e. it is Markov, stationary, and deterministic. For such a strategy the calculation consists of solving the nonlinear Eq. (3). It is convenient to use the following recurrent procedure

$$S_{k+1} = A^T S_k A - A^T S_k B (Q_2 + B^T S_k B)^{-1} B^T S_k A + H^T Q_1 H, \quad k \geq 0 \tag{8}$$

with an initial $S_0 \geq 0$ being arbitrary. Under the assumption stated above this procedure converges and its iteration k_0 corresponds to the strategy $u_t^{(k_0)}$ which is suboptimal, i.e.

$$W(u^{(k_0)}) \leq W(u^{\mathrm{opt}}) + \varepsilon(k_0)$$

where $\varepsilon(k_0) \geq 0$ and $\varepsilon(k_0) \to 0$ as $k_0 \to \infty$.

The above sketch of the classical theory of LQP is based on the assumption that all parameters of Eq. (1) are known. The remainder of this section will be devoted to the case when this assumption fails.

There are two approaches to solving the LQP in the framework of the adaptive concept. The first of them is identification. It consists of solving two subproblems simultaneously. The first is the evaluation of the unknown parameters of Eq. (1). The second subproblem is to determine the control at every time t in the form $\tilde{u}_t = L_t x_t + \tilde{d}_t$ when the unknown parameters A, B, C are considered equal to their estimates A_t, B_t, C_t but the matrix S_t serves as the solution of the Riccati equation (4) calculated by using the estimates obtained. Here L_t and $\tilde{d}_t$ are defined through A_t, B_t, C_t and S_t in the same way as L and d through A, B, C and S. This results in a non-Markov, non-stationary strategy. We still have to prove that it in fact leads to the given control aim.

Another approach called "direct" uses the direct estimation of the parameters L_t, d_t of the control defined by (5). The estimation algorithm is usually based on a possibility to measure the gradient of the functional $W(u)$ by observing it after the solution x_t. In the LQP we can measure this gradient due to the linearity of the optimal control u_t^{opt} in x_t which implies the finiteness of the set of arguments (the elements of the matrix L) of the functional $W(u)$. We now consider the Riccati Equation in the adaptive version of the LQP. We assume that the matrices A, B, D are controllable and observable correspondingly. In addition, let the matrix A be stable.

Let A_t and B_t be some consistent estimates of the matrices A and B. Then, for all sufficiently large t the pair (A_t, B_t) is controllable and the pair (A_t, D) is observable. We write down the Riccati Equation

$$S_t = A_t^T S_t A_t - A_t^T S_t B_t (Q_2 + B_t^T S_t B_t)^{-1} B_t^T S_t A_t + H^T Q_1 H. \tag{9}$$

According to our assumption this equation has a unique positive definite solution S_t. The following question arises: what is the relation between the solution S_t of (9) and the solution S of (4)?

Lemma 1. *Under the stated assumptions* $\lim_{t \to \infty} S_t = S$.

Proof. First, we assume that $\sup_t \|S_t\| < \infty$. By the assumptions, there is a $c > 0$, $\rho \in (0,1)$ and t_0 such that

$$\|A_t^s\| \le c\rho^s \quad \forall s, \ t > t_0.$$

According to (9) we have

$$S_t \le A_t^T S_t A_t + H^T Q_1 H \le \cdots \le \sum_{j=0}^{\infty} A_t^{j\,T} H^T Q_1 H A_t^j.$$

Hence

$$\|S_t\| \le c^2 \|H^T Q_1 U\| \sum_{j=0}^{\infty} \rho^{2j} = c^2 \frac{\|H^T Q_1 H\|}{1 - \rho^2}$$

for any $t \ge 0$.

Next, let $R(A, B, S)$ and $R(A_t, B_t, S_t)$ denote the right-hand sides of (4) and (9) respectively. We put $E_t = R(A, B, S) - R(A_t, B_t, S_t)$. Making use of the boundedness of S_t and convergence of $A_t \to A$ and $B_t \nrightarrow B$ one can easily prove that

$$A^T S_t A - A_t^t S_t A_t = (A^T - A_t^T) S_t A + A_t^T S_t (A - A_t) \to 0,$$

$$(Q_2 + B^T S_t B)^{-1} - (Q_2 + B_t^T S_t B_t)^{-1} = (Q_2 + B^T S_t B)^{-1} (B_t^T S_t B_t - B^T S_t B)$$
$$\times (Q_2 + B_t^T S_t B_t)^{-1} \to 0$$

as $t \to \infty$. Hence

$$\lim_{t \to \infty} [A^t S_t B (Q_2 + B^T S_t B)^{-1} B^T S_t A - A_t^T S_t B_t (Q_2 + B_t^T S_t B_t)^{-1} B_t^T S_t A_t] = 0$$

and $\lim_{t \to \infty} E_t = O$. It follows that

$$\|S_t - R(A, B, S_t)\| \le \|E_t\| \xrightarrow[t \to \infty]{} 0.$$

In view of boundedness of $\|S_t\|$, there is a converging subsequence S_{t_m} such that $\lim_{m \to \infty} S_{t_m} = S'$. Then $S' = R(A, B, S')$ and the uniqueness of the solution of (4) implies $S' = S$. Hence $S_t \xrightarrow[t \to \infty]{} S$. $\qquad \square$

Another condition for the $S_t \to S$ convergence does not involve the restrictive stability condition for A.

Lemma 2. *Let (A, B) be controllable, A be non-degenerate and $(A_t, B_t) \xrightarrow[t \to \infty]{} (A, B)$. Then $\sup_t \|S_t\| < \infty$ (provided $Q_2 > 0$ and $P_0 \ge 0$). If (A, D) is observable then S_t converges to S.*

We introduce the statistical analogue of the vector b_t defined through the estimates A_t, B_t, S_t as follows

$$\hat{b}_t = -H^T Q_1 H x_t^* - \sum_{i=0}^{\infty} [A_t - B_t (Q_2 + B_t^T S_t B_t)^{-1} B_t^T S_t A_t]^{i^T} H^T Q_1 x_{t+1}^*. \tag{10}$$

Arguments similar to those in Lemma 1 lead to the following result.

Lemma 3. *If (A, B) is controllable, A is stable, $Q_2 > 0$ and $A_t \to A$, $B_t \to B$ as $t \to \infty$, then*

$$\lim_{t \to \infty} \|b_t - \hat{b}_t\| = 0.$$

We now study the adaptive version of the LQP for a concrete classes of equations. We start with the simplest case of an equation given in the state space in the form

$$x_{t+1} = Ax_t + Bu_t + \xi_t \tag{11}$$

and

$$W(u) = \varlimsup_{T \to \infty} T^{-1} \sum_{t=1}^{t} [x_t^T Q_1 x_t + u_t^T Q_2 u_t], \quad Q_1 \geq 0, \; Q_2 > 0. \tag{12}$$

As concerns the random disturbance we make two additional assumptions:

$$\sup_t \mathbf{E}(\|\xi_{t+1}\|^b | \mathcal{F}_t) < \infty, \quad \left\| T^{-1} \sum_{t=1}^{T} \xi_t \xi_t^T - R \right\| = o(T^{-\rho}) \tag{13}$$

where $b > 2$ and $\rho > 0$. We consider the class $\mathcal{K}$ of equations of the form (11) satisfying the following conditions:

(a) the triplet (A, B, D) is controllable and observable with matrix D satisfying $D^T D = Q_1$;

(b) the matrix A is stable.

The strategy described below is based on the estimates of the parameter $\theta = [A, B]^T$ by means of the LSM. At time t the estimate $\theta_t = [A_t, B_t]^T$ is obtained by the already known recurrent procedure

$$\theta_{t+1} = \theta_t + a_t P_t \left(x_{t+1}^T - \varphi_t^T \theta_t \right),$$
$$P_{t+1} = P_t - a_t P_t \varphi_t \varphi_t P_t$$

with $P_0 = dI$ and an arbitrary θ_0. Here we have used the notation

$$\varphi_t = [x_t^T u_t^T], \quad P_t = \left(\sum_{i=0}^{t-1} \varphi_i \varphi_i^T + d^{-1} I \right)^{-1}, \quad d = n + l,$$
$$a_t = (1 + \varphi_t P_t \varphi_t^T).$$

If it turns out that estimates A_t and B_t are such that the triplet (A_t, B_t, D) is controllable and observable then we construct S_t as the solution of the Riccati equation (9). Further we put

$$L_t = -(Q_2 + B_t^T S_t B_t)^{-1} B_t^T S_t A_t.$$

Let us define the following sequence of $l \times n$-matrices K_t.

$$K_t = \begin{cases} L_t, & \text{if } \|L_t\| \leq \ln^{h/2} t \text{ and } (A_t, B_t, D) \quad \text{is observable,} \\ & \hspace{8.5em} \text{and controllable} \\ K_{t-1}, & \text{otherwise.} \end{cases}$$

Here $h \in (0, 1/5)$ and K_0 is any initial matrix. We define the trial signal V_t. First, consider a sequence of independent, identically distributed random l-vectors W_t which do not depend on $\{\xi_t\}$ and have continuous distributions such that

$$\mathbf{E}W_t = 0, \quad \mathbf{E}W_t W_t^T = I, \quad \mathbf{E}\|W_t\|^4 < \infty.$$

We define the trial signal as follows

$$V_t = t^{-\delta/2} W_t, \quad V_0 = 0, \quad \delta \in \left(0, \frac{1-5\varepsilon}{2}\right).$$

We define the stopping times $3 = \tau_1 < t_1 < \tau_2 < \cdots$ by the relationships

$$t_k = \sup\left\{ t > \tau_k : \sum_{i=\tau_k}^{j-1} \|x_i\|^2 \le (j-1)^{1+h/2} + \|x_{\tau_k}\|^2 \ln^h \tau_k, \quad \forall\, j \in (\tau_k, t] \right\}$$

$$\tau_{k+1} = \inf\left\{ t > t_k : \sum_{i=t_k+1}^{t} \|x_i\|^2 \le 2^{-k} t^{1+h}, \ \|x_t\|^2 \ln^h t \le t^{1+h/2} \right\}.$$

We put

$$K_t^0 = \begin{cases} K_t, & \text{if } t \in [\tau_k, t_k), \\ O, & \text{if } t \in [t_k, \tau_{k+1}). \end{cases}$$

We can now explicitly give the form of the required adaptive control

$$u_t = K_t^0 x_t + v_t. \tag{14}$$

This strategy is denoted by $\sigma(C)$. We show that it solves the problem above, but first, we give the minimal value of the functional $W(u)$ in this situation. Instead of (7) we have (if $x_t^* \equiv 0$) $\min_u W(u) = \mathbf{tr}\, SR$.

Theorem 1. *The control strategy $\sigma(C)$ applied to the equations of the form* (11) *from the class $\mathcal{K}$ secures the minimizing of the functional* (12) *and the strong consistency of the estimates θ_t of the parameters of the equation. More precisely, we have*

$$\left| T^{-1} \sum_{t=1}^{T} [x_t^T Q_1 x_t^T + u_t^T Q_2 u_t] - \mathbf{tr}\, SR \right| = O(T^{-\min(1/2,\rho,\delta)}), \tag{15}$$

$$\|\theta_t - \theta\| = O(t^{-\beta}), \quad \beta \in [0, 1/2 - \alpha], \quad \text{a.s.} \tag{16}$$

where $\alpha \in ((\delta + 3h)(1 + h)^{-1}, 1/2)$.

Proof. We first prove (16). Using it, we prove (15) and (16). The first will follow immediately from Theorem 3, Sec. 1 if we show that the use of the strategy $\sigma(C)$ with respect to the equations from $\mathcal{K}$ leads to the relations:

(a) $\lim_{t\to\infty} r_t = \infty$,
(b) $\lambda_{\max}^t \le C r_t^{a-1}, C > 0, (\delta + 3h)(1 + h)^{-1} < a < 1.$

It is easy to verify that $r_t = 1 + \sum_{i=1}^{t-1} \|\varphi_i\|^2 \to \infty$ as $t \to \infty$. Indeed, by Lemma 1 from Sec. 6 we have[e]

$$r_{t+1} \geq \sum_{i=1}^{t} \|x_i\|^2$$

$$= \sum_{i=1}^{t} \|Ax_{i+1} + Bu_{i-1}\|^2 \left[1 + O\left(\left[\frac{\ln \sum_{i=1}^{t} \|Ax_{i+1} + Bu_{i-1}\|}{\sum_{i=1}^{t} \|Ax_{i+1} + Bu_{i-1}\|} \right]^{1/2} \right) \right]$$

$$+ \sum_{i=1}^{t} \|\xi_i\|^2 \geq \frac{1}{2} \operatorname{tr} Bt \xrightarrow[t \to \infty]{} \infty.$$

This proves (a). We obtain the estimate $r_t = O(t^{1+h} \ln^h t)$ from the above later on, but now turn to the inequality (b), where $\lambda_{\max}^t$ means the maximal eigenvalue of the matrix P_t. We prove that

$$\lim_{t \to \infty} \frac{t^{-(1+h)(1-a)}}{\ln^{h(1-a)} t} \inf_{\|x\|=1} x^T \sum_{i=0}^{t} \varphi_i \varphi_i^T x = \lambda > 0. \tag{17}$$

Suppose that (17) fails. Then we can find two sequences α_{t_m} and β_{t_m} of n-vectors and l-vectors, respectively, such that

$$\|\alpha_{t_m}\|^2 + \|\beta_{t_m}\|^2 = 1 \quad \forall m \tag{18}$$

and

$$\lim_{t \to \infty} \lim_{m \to \infty} \frac{t_m^{-(1+h)(1-a)}}{\ln^{h(1-a)} t_m} \sum_{i=0}^{t_m} \left(\alpha_{t_m}^T x_i + \beta_{t_m}^T u_i \right)^2 = 0$$

or, which is the same,

$$\lim_{m \to \infty} \frac{t_m^{-(1+h)(1-a)}}{\ln^{h(1-a)} t_m} \sum_{i=0}^{t_m} \left[\left(\alpha_{t_m}^T + \beta_{t_m}^T K_i^0 \right) x_i + \beta_{t_m}^T v_i \right]^2 = 0. \tag{19}$$

We now show that equalities (18) and (19) contradict each other.

It is clear that $\sum_{i=0}^{t} x_i v_i^T$ forms a martingale. According to Lemma 1 from Sec. 6 we have

$$\alpha_{t_m}^T \sum_{i=0}^{t_m} x_i v_i^T \beta_{t_m} = O\left\{ \left[\sum_{i=0}^{t_m} \|x_i\|^2 \cdot \ln \left(\sum_{i=0}^{t_m} \|x_i\|^2 \right) \right]^{1/2} \right\}$$

$$= O\left(t_m^{\frac{1+h}{2}} \ln^{1/2} t_m \right)$$

---

[e] In the present section we use this lemma in the following version:

$$\sum_{i=1}^{t} g_i \xi_{i+1} = O\left(\left[\sum_{i=1}^{t} \|g_i\|^2 \ln \left(\sum_{i=1}^{t} \|g_i\| \right)^2 \right]^{1/2} \right), \quad \text{a.s.}$$

as $t \to \infty$ where g_t is an $\mathcal{F}_t$-measurable random vector and ξ_t is a martingale-difference with bounded α-moments ($\alpha > 2$).

where we have used the relation

$$T^{-1}\sum_{j=0}^{t}\|x_j\|^2 = O(t^h) \tag{20}$$

which will be proved later on. Similarly

$$\beta_{t_m}^T\sum_{i=0}^{t_m}K_i^0 x_i v_i^T \beta_{t_m} = O\Big(t_m^{\frac{1+h}{2}}\ln^{\frac{1+h}{2}}t_m\Big).$$

Since the number a has been chosen such that $(1+h)(1-a) > (1+h)/2$, we can rewrite (19) as follows

$$\lim_{t\to\infty}\frac{t_m^{-(1+h)(1-a)}}{\ln^{h(1-a)}t_m}\sum_{i=0}^{t_m}\big\{\big[(\alpha_{t_m}^T + \beta_{t_m}^T K_i^0)x_i\big]^2 + \big[\beta_{t_m}^T v_i\big]^2\big\} = 0. \tag{21}$$

It follows that

$$\lim_{m\to\infty}\frac{t_m^{-(1+h)(1-a)+\delta-1}}{\ln^{h(1-a)}t_m}\frac{\beta_{t_m}}{t_m^{1-\delta}}\sum_{i=0}^{t_m}v_i v_i^T \beta_{t_m} = 0.$$

From the properties of v_i it follows that

$$\lim_{m\to\infty}t^{\delta-1}\sum_{i=0}^{t}v_i v_i^T = \frac{1}{1-\delta}I > 0$$

and, taking into account the preceding equality, we obtain

$$\|\beta_{t_m}\|^2 = O\left(\frac{t_m^{(1+h)(1-a)-1+\delta}}{\ln^{h(1-a)}t_m}\right).$$

It remains to note that $(1+h)(1-a) - 1 + \delta + 2h < 0$ due to the choice of a. Hence $\lim_{m\to\infty}\beta_{t_m} = 0$. This and (21) imply

$$t_m^{-1}\sum_{i=1}^{t_m}(\alpha_{t_m}x_i)^2 = t_m^{-1}\sum_{i=1}^{t_m}\big[\alpha_{t_m}(Ax_{i-1} + Bu_{i-1} + \xi_i)\big]^2 \xrightarrow[m\to\infty]{} 0. \tag{22}$$

Making use of Lemma 1 from Sec. 6 again we obtain

$$t_m^{-1}\sum_{i=1}^{t_m}(Ax_{i-1} + Bu_{i-1} + \xi_i) = O\Big(t_m^{-\frac{1-h}{2}}\ln^{\frac{1+h}{2}}t_m\Big) \xrightarrow[m\to\infty]{} 0.$$

This and (22) imply

$$\lim_{m\to\infty}t_m^{-1}\sum_{i=1}^{t_m}(\alpha_{t_m}x_i)^2 = 0.$$

By the conditions on the martingale-difference ξ_i, we have $\lim_{m\to\infty}\|\alpha_{t_m}\| = 0$. But from (18) it follows that $\lim_{m\to\infty}\|\alpha_{t_m}\| = 1$, since $\lim_{m\to\infty}\|\beta_{t_m}\| = 0$. This contradiction proves (17). Now, from (17), for all sufficiently large t we have

$$\lambda_{\max}^{-(t+1)} \geq \frac{1}{2}\lambda t^{(1+h)(1-a)}\ln^{h(1-a)t} \geq c^{-1}r_{t+1}^{1-a}$$

where $c > 0$ may be some r.v. This completes the verification of the relations (a) and (b) which, in turn, lead to (16). To complete the proof of the last relation we have to prove (20). To achieve this, we use the random variables τ_k and t_k from the definition of the strategy $\sigma(C)$. If $t_{k_0} = \infty$ for some $k_0 < \infty$ then (20) is obvious (according to the definition of t_k). So, we can assume that $t_k < \infty$ for all k.

In view of stability of the matrices A and $A + BL$, there is a c and a $\rho \in (0, 1)$ such that for all l

$$\|A\|^l \le c\rho^l, \quad \|(A + BL)^l\| \le c\rho^l.$$

From the definition of the matrix K^0 it follows that

$$x_{t_k+i} = A^i x_{t_k} + \sum_{j=t_k+1}^{t_k+i} A^{t_k+i-j}(Bv_{j-1} + \xi_j)$$

for $i \in [1, \ \tau_{k+1} - t_k]$. So,

$$\|x_{t_k+i}\|^2 \le 2c^2\rho^{2i}\|x_{t_k}\|^2 + \frac{2c^2}{1-\rho} \sum_{j=t_k+1}^{t_k+i} \rho^{t_k+i-j}\|Bv_{j-1} + \xi_j\|^2.$$

Hence, for any $t \in (t_k, \tau_{k+1})$ we have

$$\sum_{i=t_k+1}^{t} \|x_i\|^2 = \sum_{i=1}^{t-t_k} \|x_{t_k+i}\|^2$$

$$\le \frac{2c^2}{1-\rho}\|x_{t_k}\|^2 + \frac{2c^2}{1-\rho} \sum_{j=t_k+1}^{t} \sum_{i=j-t_k}^{t-t_k} \rho^{t_k+i-j}\|Bv_{j-1} + \xi_j\|^2 \quad (23)$$

$$\le \frac{2c^2}{1-\rho} \sum_{j=t_k+1}^{t} \|Bv_{j-1} + \xi_j\|^2. \quad (24)$$

This and the definition of τ_k imply the finiteness of τ_k for all k. So, for any n, there is a τ_k such that $\tau_k < n \le \tau_{k+1}$. From the definition of τ_j it follows that

$$\tau_{j+1}^{-(1+h)} \sum_{i=\tau_j+1}^{\tau_{j+1}} \|x_i\|^2 \le 2^{-j}$$

and this inequality is used in the following reasoning

$$t^{-(1+h)} \sum_{i=1}^{t} \|x_i\|^2 = O(1) + t^{-(1+h)} \sum_{j=1}^{k-1} \sum_{i=\tau_j+1}^{\tau_{j+1}} \|x_i\|^2 + t^{-(1+h)} \sum_{i=\tau_k+1}^{t} \|x_i\|^2$$

$$\le O(1) + \sum_{i=1}^{k-1} 2^{-j} + t^{-(1+h)} \sum_{i=\tau_k+1}^{t} \|x_i\|^2$$

$$= O(1) + \sup_{\tau_k < t \le \tau_{k+1}} t^{-(1+h)} \sum_{i=\tau_k+1}^{t} \|x_i\|^2$$

$$+ \sup_{t_k \le t \le \tau_{k+1}} t^{-(1+h)} \left[\sum_{i=\tau_k+1}^{t_k-1} \|x_i\|^2 + \sum_{i=t_k}^{t} \|x_i\|^2 \right].$$

By the definition of τ_k, t_k for $t \in (\tau_k, t_k)$ we have

$$\sum_{i=t_k+1}^{t} \|x_i\|^2 \leq \|x_{t_k}\|^2 \ln^h t_k + t^{1+h/2}, \quad \|x_{\tau_k}\|^2 \ln^h \tau_k \leq \tau_k^{1+h/2}.$$

From the listed inequalities it follows that

$$t^{-(1+h)} \sum_{i=1}^{t} \|x_i\|^2 = O(1) + \sup_{t_k \leq t \leq \tau_{k+1}} t^{-(1+h)} \sum_{i=t_k}^{t} \|x_i\|^2.$$

Now, by inequality (25) and the easily verified relation

$$\lim_{T \to \infty} T^{-1} \sum_{j=1}^{t} \|Bv_{j-1} + \xi_j\|^2 = 0, \tag{25}$$

we find

$$t^{-(1+h)} \sum_{i=1}^{t} \|x_i\|^2 = O(1) + \left(1 + \frac{2c^2}{1-\rho}\right) t_k^{-(1+h)} \|x_{t_k}\|^2$$

$$+ \frac{2c^2}{1-\rho} \sup_{t_k t \leq \tau_{k+1}} t^{-(1+h)} \sum_{i=t_k}^{t} \|Bv_{j-1} + \xi_j\|^2$$

$$= O(1) + c_1 \|x_{t_k}\|^2 t_k^{-(1+h)}.$$

If we prove that the last term here is finite we will complete the proof of (20). Indeed we have

$$\|x_{t_k}\|^2 \leq 2\|A + BK_{t_k-1}\|^2 \|x_{t_k-1}\|^2 + 2\|Bv_{t_k-1} + \xi_{t_k}\|^2$$
$$\leq O(\ln^h(t_k - 1))\left[\ln^h \tau_k \|x_{\tau_k}\|^2 + (t_k - 1)^{1+h/2}\right] + O(t_k)$$
$$\leq O(\ln^h t_k)\left[\tau_k^{1+h/2} + t_k^{1+h/2}\right] + O(t_k).$$

Thus, the consistency of the parameter estimates is proved. Moreover, the convergence rate has been obtained as well. We now want to give the upper estimate of r_t.

$$r_{t+1} = 1 + \sum_{i=0}^{t} (\|x_i\|^2 + \|u_i\|^2) = O(t^{1+h}) + O\left(\sum_{i=0}^{t} [\|K_i^0 x_i\|^2 + \|v_i\|^2]\right)$$
$$= O(t^{1+h}) + O(t^{1+h} \ln^h t) = O(t^{1+h} \ln^h t).$$

It remains to prove that the strategy $\sigma(\mathcal{C})$ is optimal with respect to $\mathcal{K}$. Consistency of the estimates gives a basis for this. Since $(A_t, B_t) \xrightarrow[t \to \infty]{} (A, B)$ we can assert that the triplet of matrices (A_t, B_t, D) is observable and controllable for all sufficiently large t. So, for the same t there exists a solution S_t of the appropriate Riccati equations but the matrices L_t (the "gain factors") are stable. According to Lemma 1, $L_t \xrightarrow[t \to \infty]{} L$. Thus $K_t = L_t$ for all sufficiently large t. We now prove a fact which will be of great significance later on. Namely, there is an integer T_0 such that

$$u_t = L_t x_t + v_t, \quad \text{a.s.} \tag{26}$$

for all $t \geq T_0$. Obviously, this is true if there is an integer k_0 such that $t_{k_0} = \infty$. Then $T_0 = t_{k_0}$. Suppose that $t_k < \infty$ for all k. According to (20) this implies $\tau < \infty$ for all k. We show that this leads to a contradiction. From the definition of t_k we have

$$\sum_{i=\tau_k}^{t_k} \|x_i\| > \|x_{\tau_k}\|^2 \ln^h \tau_k + t_k^{1+h/2}. \tag{27}$$

On the other hand,

$$\sum_{i=\tau_k}^{t_k} \|x_i\| = \|x_{\tau_k}\|^2 + \sum_{i=1}^{t_k-\tau_k} \|x_{\tau_k+i}\|$$

$$= \|x_{\tau_k}\|^2 + \sum_{i=1}^{t_k-\tau_k} \left\| (A+BL)^i x_{\tau_k} + \sum_{j=1}^{i} (A+BL)^{i-j} B(L_{\tau_k+j-1} - L)x_{\tau_k+j-1} \right.$$

$$\left. + \sum_{j=1}^{i} (A+BL)^{i-j}(Bv_{\tau_k+j-1} + \xi_{\tau_k+j}) \right\|^2$$

$$\leq c_1 \|x_{\tau_k}\|^2 + c_2 \max_{t \geq \tau_k} \|L_t - L\| \sum_{i=\tau_k}^{t_k} \|x_i\|^2 + c_3 \sum_{j=\tau_k+1}^{t_k} \|Bv_{j-1} + \xi_i\|$$

where the definition of the "gain factor" K_t^o and stability of the matrix $A + BL$ have been used. Since $\tau_k, \ t_k \xrightarrow[k\to\infty]{} \infty$ and $L_t \xrightarrow[t\to\infty]{} L$ we can choose k such that $c_2 \|L_t - L\| < 1$ for all $t > \tau_k$. Then

$$\sum_{i=\tau_k}^{t_k} \|x_i\| \leq \frac{c_1 \|x_{\tau_k}\|^2 + c_3 \sum_{j=\tau_k+1}^{t_k} \|Bv_{j-1} + \xi_i\|}{1 - c_2 \max_{t>\tau_k} \|L_t - L\|}. \tag{28}$$

Hence, for all sufficiently large k, we have the inequality

$$\sum_{i=\tau_k}^{t_k} \|x_i\| \leq \|x_{\tau_k}\|^2 \ln^h \tau_k + t_k^{1h/2},$$

which contradicts (27). So, (26) is proved. To complete the proof, it is necessary to use two auxiliary facts:

$$\textbf{(A)} \quad \overline{\lim_{t\to\infty}} \ t^{-1} \sum_{i=1}^{t} \|x_i\|^2 < \infty, \ \text{a.s.,} \quad \textbf{(B)} \quad \lim_{t\to\infty} t^{-1}\|x_t\|^2 = 0, \ \text{a.s.} \tag{29}$$

For **(A)** we can use arguments similar to those which led to (28). Then

$$\sum_{i=t_0}^{t} \|x_i\| = O(1) + O\left(\sum_{i=t_0}^{t} (Bv_{i-1} + \xi_i) \right) = O(t).$$

To prove **(B)** we note that stability of $A + BL$ implies

$$\|x_t\| \leq c\rho^t \|x_0\| + c \sum_{i=1}^{t} \rho^{t-i} \|B(K_{i-1}^0 - L)\| \|x_{i-1}\|$$

$$+ c \sum_{i=1}^{t} \rho^{t-i} \|Bv_{i-1} + \xi_i\|$$

and, using the Schwartz inequality twice, we obtain

$$\|x_t\| \leq O(1) + c \sup_{j \geq 1} \|B(K_j^0 - L)\| \left(\sum_{i=1}^{t} (\rho^2)^{t-i} \right)^{1/2} \left(\sum_{i=0}^{t-1} \|x_i\|^2 \right)^{1/2}$$

$$+ c \left(\sum_{i=1}^{t} (\rho^2)^{t-i} \right)^{1/2} \left(\sum_{i=1}^{t} \|Bv_{i-1} + \xi_i\|^2 \right)^{1/2}$$

$$= O(1) + O(t^{1/2}).$$

Thus

$$\|x_t\|^2 = O(t^{1/2}) \tag{30}$$

and **(B)** is proved.

Finally, we can prove optimality of the strategy $\sigma(C)$ with respect to $\mathcal{K}$. We put

$$\eta_{t+1} = x_t^T Q_1 x_t + u_t^T Q_2 u_t - \mathbf{tr}\, SR + x_{t+1}^T S x_{t+1} - x_t^T S x_t$$

and show that $t^{-1} \sum_{n=0}^{t} \eta_n \to 0$. If this is true then

$$t^{-1} \sum_{n=0}^{t-1} [x_n^T Q_1 x_n + u_n^T Q_2 u_n] = \mathbf{tr}\, SR + \frac{1}{2} \sum_{n=0}^{t-1} \eta_{n+1} + t^{-1} \sum_{n=0}^{t-1} [x_n^T S x_n - x_{n+1}^T S x_{n+1}]$$

$$= \mathbf{tr}\, SR + o(1) + t^{-1}(x_0^T S x_0 - x_t^T S x_t) \to \mathbf{tr}\, SR,$$

which is equivalent to optimality of the strategy $\sigma(C)$. Moreover, we want to find the convergence rate of this functional to its limit. As usual, we have to find the convergence rate of the terms entering it.

We transform η_t into a more convenient form. Note that $-(Q_2 + B^T SB)L = B^T SA$ and the Riccati equation looks as follows

$$S = (A + BL)^T S(A + BL) + L^T Q_2 L + Q_1.$$

Since $u_t = L_t x_t + v_t$ for $t \geq T_0$ we obtain

$$\eta_{t+1} = x_t^T [Q_1 + L_t Q_2 L_t + (A + BL_t)^T S(A + BL_t - S] x_t$$

$$+ (Bv_t + \xi_{t+1})^T S(Bv_t + \xi_{t+1}) - \mathbf{tr}\, SR + v_t^T Q_2 v_t + 2x_t^T L_t Q_2 v_t$$

$$+ 2x_t^T (A + BL_t)^T S(Bv_t + \xi_{t+1})$$

$$= x_t^T [\tilde{L}_t^T Q_2 \tilde{L}_t + 2\tilde{L}_t Q_2 L_t + \tilde{L}_t B^T SB \tilde{L}_t + 2(A + BL)^T SB \tilde{L}_t] x_t$$

$$+ (Bv_t + \xi_{t+1})^T S(Bv_t + \xi_{t+1}) - \mathbf{tr}\, SR + v_t^T Q_2 v_t$$

$$+ 2x_t^T L_t^T Q_2 v_t + 2x_t^T (A + BL)^T S(Bv_t + \xi_{t+1}),$$

where $\tilde{L}_t = L_t - L$. We put $\tilde{S}_t = S_t - S$. Then

$$\tilde{S}_{t+1} = (A + BK_t)^T \tilde{S}_t (A + BL) + E_t$$

where $K_t = -(Q_2 + B^T S_t B)^{-1} B^T S_t B$. Knowing the rate of $\theta_t \to \theta$ convergence we can conclude that $\|E_t\| = o(\|\tilde{\theta}_t\|) = o(t^{-1/4})$ and $K_t \to L$ as $t \to \infty$. Since the matrix $A + BL$ is stable, we have

$$\|S_t - S\| = o(t^{-1/4}), \quad \|K_t - L\| = o(t^{-1/4}).$$

Now,

$$T^{-1} \sum_{t=0}^{T-1} x_t^T [\tilde{L}_t^T Q_2 \tilde{L}_t + 2\tilde{L}_t Q_2 L_t + \tilde{L}_t B^T S B \tilde{L}_t + 2(A + BL)^T S B \tilde{L}_t] x_t = o(T^{-1/2}).$$

Next,

$$T^{-1} \sum_{t=0}^{T-1} (Bv_{t-1} + \xi_t)^T S (Bv_{t-1} + \xi_t) = \mathbf{tr}\, SR + o(T^{-\rho}),$$

$$T^{-1} \sum_{t=0}^{t-1} v_t^T Q_2 v_t = o(T^{-\delta}),$$

$$T^{-1} \sum_{t=0}^{t-1} [x_t^T L_t^T Q_2 v_t + x_t^T (A + BL)^T S B v_t] = o(T^{-\delta/2}).$$

Finally, using Lemma 1 from Sec. 6 and formula (29), we obtain

$$T^{-1} \sum_{t=0}^{T-1} x_t^T (A + BL_t)^T S \xi_t = O \left(T^{-1} \sqrt{\sum_{t=0}^{T-1} \|x_t\|^2 \ln \sum_{t=0}^{T-1} \|x_t\|^2} \right)$$

$$= O(T^{-\delta/2}).$$

From these asymptotic estimates it follows that

$$T^{-1} \sum_{t=0}^{T-1} \eta_t = T^{-1} \sum_{t=0}^{T-1} [x_t^T Q_1 x_t + u_t^T Q_2 u_t] - \mathbf{tr}\, SR$$

$$= O(T^{-1/2} + T^{-\rho} + T^{-\delta})$$

or

$$\left| T^{-1} \sum_{t=0}^{T-1} [x_t^T Q_1 x_t + u_t^T Q_2 u_t] - \mathbf{tr}\, SR \right| = O(T^{-\min(1/2, \rho, \delta)}).$$

This is the desired estimate of the convergence rate of the functional to its minimum. $\square$

The general form of the LQP for (1) can be studied by the same techniques. The difference is the form of control. If the parameters are known exactly, it is defined by (5) again, but in the adaptive version we have, instead of (14),

$$u_t = K_t^0 x_t + \tilde{d}_t + v_t$$

where

$$\tilde{d}_t = -(Q_2 + B_t^T S_t B_t)^{-1} B_t^T \tilde{b}_{t+1}$$

and $\tilde{b}_t$ is defined by (10). This leads to the strategy $\tilde{\sigma}(C)$ which is a modification of $\sigma(C)$.

Theorem 2. *The strategy $\tilde{\sigma}(C)$ applied to the equations from $\mathcal{K}$ secures minimization of the functional $W(u)$ and strong consistency of the estimates of parameters. The convergence rate for both functional and the estimates remain the same as in Theorem 1.*

The proof of this theorem is the same as that of Theorem 1. Therefore, it is left to the patient reader.

10.8. LQP for ARMAX-type Equations

In this section we deal with the ARMAX model of the form

$$A(\nabla)y_t = \triangle B(\nabla)u_t + C(\nabla)\zeta_t, \tag{1}$$

where $A(\lambda)$, $B(\lambda)$ and $C(\lambda)$ are polynomials having the same form as in Sec. 3 and the orders p, $q-1$ and r, respectively. The noise ζ_t is assumed to be a martingale-difference (under the given flow of σ-algebras $\mathcal{F}_t$) and

$$\sup_t \mathbf{E}(\|\zeta_t\|^\alpha | \mathcal{F}_{t-1}) = \gamma < \infty, \quad \alpha \geq 2,$$

$$\left\| T^{-1} \sum_{t=1}^{T} \zeta_t \zeta_t^T - R \right\| = O(T^{-\rho}), \quad \rho > 0, \ R \geq 0.$$

For an appropriate class of such equations the following problem is considered:

find a strategy minimizing the following functional

$$W(u) = \varlimsup_{T \to \infty} T^{-1} \sum_{t=1}^{T} [(y_t - y_t^*)^T Q_1 (y_t - y_t^*) + u_t^T Q_2 u_t], \quad Q_1 \geq 0, \ Q_2 > 0$$

for any equation from this class and a bounded deterministic reference path $\{y_t^*\}$.

The required strategy is based on one of the identification algorithms. The parameters of the equation form the vector $\theta = [-A_1 \cdots - A_p \ B_1 \cdots B_q \ C_1 \cdots C_q]^T$ and the vector $\theta_t = [-A_{1t} \cdots - A_{pt} \ B_{1t} \cdots B_{qt} \ C_{1t} \cdots C_{qt}]^T$ is its estimate at time t. By means of θ_t the desired adaptive control will be constructed. The first step consists of passing from Eq. (1) to some equation in the state space. It has the form

$$x_{t+1} = Ax_t + Bu_t + C\zeta_y, \quad z_t = Hx_t, \tag{2}$$

where $\dim x_t = nm$ and

$$A = \begin{pmatrix} -A_1 & I & \cdot & \cdots & O \\ -A_2 & O & I & \cdots & O \\ \dots\dots\dots\dots\dots\dots\dots \\ -A_{s-1} & O & O & \cdots & I \\ -A_s & O & O & \cdots & O \end{pmatrix}, \quad B = \begin{pmatrix} B_1 \\ \cdot \\ \cdot \\ \cdot \\ B_m \end{pmatrix}$$

$$C = [IC_1^T \cdots C_{s-1}^T]^T, \quad H = [\underbrace{I\ O\ \cdots\ O}_{nm}]\}n \tag{3}$$

$$s = \max(p, q, r-1),$$
$$A_i = O, \ B_j = O, \ C_k = O \quad \text{for } i > p, \ j > q, \ k > r.$$

From the preceding section we know that the optimal control has the linear form $u_t = Lx_t + d$, where L and d are expressed in terms of parameters of Eq. (2) and the solution of the Riccati equation. The minimal value of the functional $W(u)$ is know as well. But the case under consideration is complicated by the fact that x_t is unobservable. Hence it cannot take part in forming u_t directly. For this reason we introduce an estimate $\hat{y}_t$ of the state y_t as a solution of the following equation

$$\hat{y}_{t+1} = A\hat{y}_t + Bu_t + C(z_{t+1} - HA\hat{y}_t - HBu_t)$$

which is, in fact, an analogue of the Kalman filter. We also take into account that the matrices A, B and C in (3) are unknown.

Let A_{it}, B_{jt} and C_{kt} be the estimates of parameters of Eq. (1). Replacing A_i, B_i and C_i by these estimates in (3) we obtain the estimates $A(t)$, $B(t)$ and $C(t)$ of A, B and C, respectively. Now, we can write down the recurrent procedure of constructing the estimates of the solution of the Riccati equation

$$S_t = A^T(t)S_{t-1}A(t) - A^T(t)S_{t-1}A(t)(Q_2 + B^T(t)S_{t-1}B(t))^{-1}B^T(t)S_{t-1}B(t)$$
$$+ H^T Q_1 H$$

with an arbitrary initial $S_0 \geq 0$. The state y_t is estimated by means of the filter

$$\hat{y}_{t+1} = A(t)\hat{y}_t + B(t)u_t + C(t)(z_{t+1} - HA(t)\hat{y}_t - HB(t)u_t)$$
$$\hat{y}_0 = [x_0^T\ 0\ \cdots\ 0]^T. \tag{4}$$

At time t the control is expressed in terms of the measurable and observable quantities by the following equality

$$u_t = L_t\hat{y}_t + \tilde{d}_t,$$

where $L_t = -(Q_2 + B_t^T S_t B_t)^{-1}B_t^T S_t A_t$ and the matrix $\tilde{A}_t = A_t + B_t L_t$ is stable. The vector d_t has been defined above (see formula (6), Sec. 7). Its estimate is obtained by replacing A, B and C by their estimates in the mentioned formula. For $\hat{b}_t$ this has already been done in Sec. 7. To complete the description of the adaptive strategy it is necessary

(a) to add some trial signal v_t to the control

$$u_t = L_t \hat{y}_t + \tilde{d}_t + v_t; \tag{5}$$

(b) to define the random times t_k, τ_k, $k \geq 1$.

We do this separately for each algorithm.

The following principal condition is assumed throughout this section:

$\mathbf{C_1}$. *the triplet of matrices* (A, B, H), *where H is such that $H^T H = D^T Q D$, is controllable and observable.*

Due to this condition there exists a unique solution of the Riccati Equation in all adaptive optimization problems considered below. Hence the proposed procedure of control is correct. This condition implies that

$$m = \max(p, q) \geq r + 1. \tag{6}$$

Indeed, the requirement of controllability of the pair (A, B) means that $\mathbf{rang}\,[B\ AB \ldots A^{n-1}B] = n$ and, hence, the matrices A_m and B_m cannot be zero simultaneously. This is equivalent to (6). We note that

$$\mathbf{det}\,A = (-1)^p \mathbf{det}\,A_p,$$

for $p \geq q$ and the matrices A, A_p are invertible or degenerate simultaneously. This suggests the following condition.

$\mathbf{C_2}$. *The matrix A_p is non-degenerate.*

This condition implies the following important equalities which have been used in the proof of Theorem 3, Sec. 4.

$$\mathbf{deg}\,A(\lambda) = p, \ \ \mathbf{deg}\,[\mathbf{det}\,A(\lambda)] = np, \ \ \mathbf{deg}[\mathbf{Adj}\,A(\lambda)] = n - 1.$$

Both conditions above define a wide class of ARMAX-type equations. However, this class may be narrower in the problems of adaptive optimal control. Keeping in mind the least square method to be used later on, we introduce one more condition.

$\mathbf{C_3}$. *The matrix $C^{-1} - 2^{-1}I$ is strictly positive real and $A(\lambda)$ is stable.*

The condition on $C^{-1} - 2^{-1}I$ implies stability of $C(\lambda)$. This fact is proved in the theory of linear systems and we shall not dwell on it.

Let $\mathcal{K}'(n, l; p, q, r)$ be the class of ARMAX-type equations satisfying conditions $\mathbf{C_1}$–$\mathbf{C_3}$. The problem consists of finding an adaptive optimal strategy to minimize the functional $W(u)$. We first consider the case $y_t^* \equiv 0$.

We emphasize that the LSM is used to estimate the parameters of the equation in question. Now, let

$$\chi = \max(p, q, r) + np - 1, \quad h \in \left[0, \frac{1}{2(1 + \chi)}\right], \quad \delta \in \left[0, \frac{1 - 2h(1 + \chi)}{2\chi + 3}\right]. \tag{7}$$

We define the random times

$$t_1 = 1$$

$$t_k = \sup\left\{ t > \tau_k : \sum_{i=\tau_k}^{j-1} \|L_i \hat{y}_t\| \le (j-1)^{1+\delta} + \|L_{\tau_k} \hat{y}_{\tau_k}\|, \quad \forall j \in (\tau_k, t] \right\}$$

$$\tau_{k+1} = \inf\left\{ t > t_k : \sum_{i=t_k}^{t} \|\hat{y}_i\|^2 \le t^{1+\delta/2}, \quad \|L_t \hat{y}_t\|^2 \ge 1 + \delta, \right.$$
$$\left. \sum_{i=t_k}^{t_k-1} \|L_i \hat{y}_t\|^2 \le 2^{-k} t^{1+\delta/2} \right\}.$$

At time t the control is defined as follows

$$u_t = L_t^0 \hat{y}_t + v_t, \tag{8}$$

where

$$L_t^0 = \begin{cases} L_t, & \text{if } t \in [\tau_k, t_k) \quad \text{for some } k, \\ 0, & \text{if } t \in [t_k, \tau_{k+1}) \quad \text{for some } k. \end{cases}$$

As concerns the excitation source we assume that it is formed by a sequence of independent (of each other and of ξ_t) random vectors such that

$$\mathbf{E} v_t = 0, \quad \mathbf{E} v_t v_t^T = t^{-h} I, \quad \|v_t\| \le t^{-h} \mu.$$

This completes the description of the desired strategy which will be denoted by $\sigma(A_{CG})$.

Theorem 1. *The strategy $\sigma(A_{CG})$ applied to the equations from the class $\mathcal{K}'(n,l,p,q,r)$ ensures minimization of the functional $W(u)$ (with $y_t^* \equiv 0$) and strong consistency of the estimates of parameters. Moreover,*

$$|W_t(u^{(a)}) - \mathbf{tr}\, SCRC^T| = O(t^{-\min(\rho,h)}),$$

$$\|\hat{\theta} - \theta\| = \begin{cases} O(\sqrt{t^{-\alpha} \ln t}, & \text{if } b > 2, \\ O(\sqrt{n^{-\alpha} \ln t (\ln \ln t)^g}, & \forall g > 1 \quad \text{if } b = 2 \end{cases}$$

where

$$W_t(u^a) = T^{-1} \sum_{t=1}^{t} [y_t^T Q_1 y_t + u_t^T Q_2 u_t], \quad a \in (1/2, 1 - h(1+\chi))$$

Proof. First, we show that the estimates of parameters are consistent. From condition $\mathbf{C_1}$ it follows that the matrix polynomials $A(\lambda)$ and $B(\lambda)$ have no common left factors. To apply Theorem 1, Sec. 4 we should verify that

$$T^{-1} \sum_{t=1}^{T} \|u_t\|^2 = O(T^\delta) \quad \text{a.s.} \tag{9}$$

For this purpose, we have to prove that $\tau_{k_0} < \infty$, $t_{k_0} = \infty$ for some $k_0 < \infty$. This can be done by the same arguments as in the previous section. Then we can conclude that $L_t^0 \equiv L$ for $t \geq \tau_{k_0}$. The estimate (9) and the admissibility of the strategy $\sigma(A_{CG})$, i.e.

$$t^{-1} \sum_{n=1}^{t} (\|x_n\|^2 + \|u_n\|^2) = O(1)$$

can be obtained without difficulty.

The estimates

$$\|S_t - S\| = o(t^{-1/4}), \quad \|L_t^0 - L\| = o(t^{-1/4})$$

can be obtained by arguments similar to those in Sec. 7. To find the convergence rate for the functional $W_t(u)$, we write it in the following form

$$tW_t(u) = \sum_{n=0}^{t-1} (y_n^T Q_1 y_n + u_n^T Q_2 u_n)$$

$$= x_0^T S x_0 - x_n^T S x_n + \sum_{n=0}^{t-1} \xi_{n+1}^T C^T S C \xi_{n+1} + 2 \sum_{n=0}^{t-1} (A x_n + B u_n) S C \xi_{n+1}$$

$$+ \sum_{n=0}^{t-1} (u_n - L x_n)^T (Q_2 + B^T S B)(u_n - L x_n).$$

The properties of the noise ξ_t and the trial signal v_t imply the following estimates:

$$t^{-1}\|x_t\|^2 = O(t^{-\rho}) \quad t^{-1} \sum_{n=0}^{t} (A x_n + b u_n) S C \xi_{n+1} = O(t^{-1/2}),$$

$$t^{-1} \sum_{n=0}^{t} \xi_{n+1}^T C^T S C \xi_{n+1} - \mathbf{tr}\, S C R C^T = \mathbf{tr}\left[t^{-1} \sum_{n=0}^{t} \xi_{n+1} \xi_{n+1}^T - R \right] = O(t^{-\rho})$$

$$\sum_{n=0}^{t} (u_n - L x_n)^T (Q_2 + B^T S B)(u_n - L x_n) = O\left(t^{-1} \sum_{n=0}^{t} \|u_n - L x_n\|^2 \right) = o(t^{1-h} + t^{1/2})$$

$$\sum_{n=0}^{t} \|L_n^0(\hat{y}_n - x_n)\|^2 = o(t^{1/2}), \quad \sum_{n=0}^{t} \|(L_n^0 - L) x_n\|^2 = o(t^{1/2}).$$

These estimates lead to the final result

$$W_t(u^a) - \mathbf{tr}\, S C R C^T = O(t^{-\rho} + t^{-h} + t^{-1/2})$$

$$= O(t^{-\rho} + t^{-h}) = O(t^{-\min(\rho, h)}).$$

This completes the proof of the theorem. $\square$

In the case of the general LQP the reference path y_t^* is not zero. This case can be studied without difficulty and so we leave it to the reader. In the framework of the LSM we can require stability of the matrix polynomial $B(\lambda)$ instead of that of $A(\lambda)$. We shall not consider this problem here.

Let us discuss the problem under consideration from the standpoint of the stochastic gradient approach.

We denote by $\mathcal{K}''(n, l; p, q, r)$ the class of ARMAX-type equations satisfying the conditions $\mathbf{C_1}, \mathbf{C_2}, \mathbf{C_4}$ where the last condition means

$\mathbf{C_4}$. *The matrix $C(\lambda) - 1/2I$ is strictly positive real and $B(\lambda)$ is stable.*

We define a strategy for the class $\mathcal{K}''(n, l; p, q, r)$ which is optimal in the sense of the functional $W(u)$. Here we keep the notation introduced above. Unlike the previous case we now have two types of controls:

$$u_t^0 = L_t \hat{y}_t + \tilde{d}_t,$$

expressed explicitly in terms of the quantities to be estimated and another control u_t^1 which represents the solution of the following linear equation already familiar to us

$$B_{1t} u_t^1 + (\theta_t^T \varphi_t - B_{1t} u_t) = 0$$

where θ_t is an estimate of the parameters at time t but φ_t is the observable history

$$\varphi_t = [z_t^T \cdots z_{t-p+1}^T \ u_t^T \cdots u_{t-q+1}^T \ z_t^T - \varphi_{t-1}^T \theta_{t-1} \cdots z_{t-r+1}^T - \varphi_{t-r}^t \theta_{t-r}].$$

In Sec. 3 it has been proved that for $n = l$ this equation has a unique solution. In the case when $n < l$, the solution of this equation is defined in a non-unique way.

Before writing down the final expression for the strategy desired, we define the following random times

$$t_k = \sup\left\{ t > \tau_k : \sum_{i=\tau_k}^{j-1} \|z_i\|^2 \le (j-1)\ln^\delta(j-1) + \|z_{\tau_k}\|^2, \quad \forall j \in (\tau_k, t] \right\};$$

$$\tau_{k+1} = \inf\left\{ t > t_k : \sum_{i=t_k}^{t} \|z_i\|^2 \le 2^{-k} t \ln^\delta t, \quad \sum_{i=0}^{t} \|\hat{y}_t\|^2 \ge t \ln^\delta t, \right.$$

$$\left. \sum_{i=t_k}^{t_k-1} \|z_i\|^2 \le 2^{-k} t \ln^\delta t \right\}$$

where

$$\delta \in \left(0, \frac{1/4 - 2(m+2)mh}{2 + (n+1)m}\right), \quad m = \max(p, q), \quad h = \left(0, \frac{1}{4(n+2)m}\right).$$

The trial signals are independent of each other and of ξ_t and have continuous distributions. Moreover, $v_0 = v_1 = 0$ and

$$\mathbf{E}v_t = 0, \quad \mathbf{E}v_t v_t^T = \ln^{-h} tI, \quad \|v_t\| \le \varkappa \ln^{-h} t.$$

The required strategy has the form

$$U_t^a = u_t + v_t$$

where the control u_t has a different form on different time intervals

$$u_t = \begin{cases} 0, & \text{if } t \in [\tau_k, t_k] \cap M^c \text{ for some } k, \\ u_t^0, & \text{if } t \in [\tau_k, t_k] \cap N \text{ for some } k, \\ u_t^1, & \text{if } t \in [t_k, \tau_{k+1}) \text{ for some } k. \end{cases}$$

Here $M = \{t : \|u_y^0\| \leq t \ln^\delta t\}$. So, the desired strategy, which is denoted by $\sigma(ACG)$, is specified. The next theorem describes its main property.

Theorem 2. *The strategy $\sigma(ACG)$ applied to the equations from the class $\mathcal{K}''(n, l; p, q, r)$ implies strong consistency of the estimates of parameters (obtained by SGM) and fulfilment of the following relations:*

$$\varlimsup_{T \to \infty} T^{-1} \sum_{t=1}^{T} (\|y_t\|^2 + \|u_t\|^2) < \infty, \quad (\text{with probability one}),$$

$$W_t(u^a) = \mathbf{tr}\, SCRC^T + \varlimsup_{T \to \infty} T^{-1} \sum_{t=1}^{T} \left[y_t^{*T} Q_1 y_t^* + b_{t+1}^T B (Q_2 + B^T S B)^{-1} B b_{t+1} \right].$$

The proof of this theorem does not contain new ideas in comparison with Theorem 1. In fact this proof is even simpler because it does not obtain the estimations of the convergence rate.

We note that adaptive optimal strategies for linear stochastic difference equations are complex. They are non-deterministic, non-stationary, non-linear and non-Markov, i.e. they depend on the whole past history. Only in absence of noise, i.e. when the equations are homogeneous, these strategies have simple form. Namely, in the initial step when we have to estimate the unknown parameters, this strategy is a programmed one. Thereafter it takes the classical form defined by the algorithm studied. Apparently the fundamental difference between deterministic and stochastic problems of adaptive control is connected to this fact.

CHAPTER 11

CONTROL OF ORDINARY DIFFERENTIAL EQUATIONS

We study the basic problems of ordinary differential equations such as tracking, stabilization and optimization from the point of view of adaptive theory. We consider nonlinear equations, equations with delay and equations in Hilbert space. The Direct Lyapunov Method is the principal tool throughout this chapter. Note that the control problems mentioned above belong to the so-called "deterministic adaptive theory". This topic is undoubtedly worth a special chapter.

11.1. Preliminary Results

First of all we are interested in some stability properties of solutions of ordinary differential equations. Further, we use the terminology of the Direct Lyapunov Method (DLM) to describe them.

We consider a vector autonomous equation

$$\dot{x} = f(x(t)), \quad x(t_0) = x_0, \qquad t \geq t_0, \tag{1}$$

with $x \in \mathrm{R}^n$. In more detailed form,

$$\dot{x}_i = f_i(x_1, \ldots, x_n), \quad x_i(t_0) = x_0^i, \qquad i = 1, \ldots, n.$$

We assume that the functions $f_i(\cdot)$ satisfy the well-known conditions for the existence and uniqueness of solutions of the given equation. Let $g(x)$ be a continuously differentiable function.

Definition 1. The expression

$$\dot{g}(x) = \sum_{i=1}^{n} \frac{\partial g}{\partial x_i} \dot{x}_i = \sum_{i=1}^{n} \frac{\partial g}{\partial x_i} f_i = ((\operatorname{grad} g)^T, f)$$

is called the *derivative of the function $g(x)$ with respect to Eq. (1)*.

Definition 2. A function $V(x)$ is called a *Lyapunov function* for Eq. (1) if:

(1) it is defined for all $x \in \mathrm{R}^n$;
(2) $V(x) > 0$, $x \neq 0$, $V(0) = 0$;
(3) there exists the derivative $\dot{V}$ with respect to Eq. (1) and $\dot{V}(x) \leq 0$.

First, we focus attention on stability in the Lyapunov sense with respect to disturbances of the initial conditions. Assuming that $f(0) = 0$, we consider the stability of the trivial solution $x(t) \equiv 0$, i.e. the equipoise point.

359

Theorem 1. *If there exists a Lyapunov function $V(x)$ for Eq. (1) such that $\dot{V}(x) \leq 0$, then its trivial solution is stable.*

Theorem 2. *If there exists a Lyapunov function $V(x)$ for Eq. (1) such that $\dot{V}(x) < 0$, $x \neq 0$, then its trivial solution is asymptotically stable.*

The theorems cited above show that for either stability or asymptotic stability the initial values of the disturbed trajectories must belong to a small neighborhood of the initial value of the undisturbed trajectory. The case when initial values may be arbitrary, however large, is very important. We then say that asymptotic stability on the whole space takes place.

Theorem 3. *If there exists a Lyapunov function $V(x)$ for Eq. (1) such that*

$$\dot{V}(x) < 0, \quad x \neq 0, \qquad \lim_{\|x\| \to \infty} V(x) = \infty$$

then the trivial solution is asymptotically stable on the whole space.

Definition 3. Equation (1) is called *dissipative* for $t \geq t_0$ if there exists a number ρ such that

$$\overline{\lim_{t \to \infty}} \, \|x(t)\| \leq \rho$$

for every initial value $x(t_0)$.

Theorem 4. *If for Eq. (1) there exists a Lyapunov function $V(x)$ such that*

$$\dot{V}(x) \leq -aV + b, \quad a, b > 0, \qquad \lim_{\|x\| \to \infty} V(x) = \infty$$

then this equation is dissipative and $\lim_{t \to \infty} V(x(t)) \leq b/a$.

The previous results were concerned with stability of solutions when the initial values can deviate from the nominal values. Apart from these noises, let us consider the disturbances which influence the solution at each moment of time. The detailed description of this problem is based on the following.

Consider the equation

$$\dot{x} = f(x) + g(x). \tag{2}$$

Suppose that this equation has a unique solution just like Eq. (1). More exactly, the functions $g_1(x_1, \ldots, x_n), \ldots, g_n(x_1, \ldots, x_n)$ forming the vector $g(x)$ and being continuous guarantee the existence of the unique solution of (2) (at a given x_0). These functions need to take rather small values and not to vanish as x_i tends to zero for all i.

Definition 4. The trivial solution of Eq. (1) is called *stable under constantly acting disturbances* if for every $\varepsilon > 0$ there exist $\delta_1(\varepsilon)$, $\delta_2(\varepsilon) > 0$ such that every solution $x(t)$ of Eq. (2) with the initial value x_0 and an arbitrary disturbance $g(x)$ satisfying the conditions $\|x(t_0)\| \leq \delta_1(\varepsilon)$ and $\|g(x)\| \leq \delta_2(\varepsilon)$ on the set $\{t \geq t_0, \|x\| \leq \varepsilon\}$, respectively, satisfies the inequality $\|x(t)\| < \varepsilon$ for $t \geq t_0$.

Theorem 5. *For stability under constantly acting disturbances to take place, the following conditions should be fulfilled:*

(1) *the trivial solution of Eq. (1) is asymptotically stable;*

(2) *the equality $\lim_{t\to\infty} x(t) = 0$ holds uniformly with respect to the set $\{\|x_0\| \le c\}$.*

The above results will be used mostly for linear equations with constant coefficients. For the standard linear equation in the state space R^n

$$\dot{x} = Ax, \quad x(t_0) = x_0$$

we have the explicit form of its solution

$$x(t) = e^{A(t-t_0)}x_0.$$

In coordinate form this equation and its solution have the form

$$\dot{x}_i = a_{i1}x_1 + \cdots + a_{in}, \quad i = 1, \ldots, n,$$
$$x_0 = (x_0^1, \ldots, x_0^n),$$

and

$$x_i(t) = \sum_{j=1}^{g}(p_{ij}(t)\cos\omega_j t + q_{ij}(t)\sin\omega_j t)e^{\alpha_j t}.$$

Here, the complex numbers $\alpha_j + i\omega_j = \lambda_j$ are the eigenvalues of the matrix A, g is the number of different eigenvalues, $p_{ij}(t)$ and $q_{ij}(t)$ are polynomials in t whose degrees are one less than the multiplicity of the corresponding eigenvalues λ_j. From the last representation we have the following results.

Theorem 6. (a) *The trivial solution of a linear equation is stable if all eigenvalues of the matrix A have non-positive real parts, i.e. $\alpha_j \le 0$ and the eigenvalues having zero real parts have multiplicity one. Otherwise this solution is not stable;*

(b) *the trivial solution of a linear equation is asymptotically stable if and only if all eigenvalues of the matrix A have negative real parts, i.e. the matrix A is stable (A is a Hurwitz matrix).*

DLM allows to judge the stability of the solution of the simplest ($\dot{x} = Ax$) and more complicated linear equations without using their explicit form. As proposed first by Lyapunov, it is necessary to consider the quadratic form $V(x) = x^T P x$ with $P = P^T > 0$ some positive definite matrix. Its derivative, with respect to the equation $\dot{x} = Ax$, is equal to $\dot{V}(x) = x^T(A^T P + PA)x$ and if the matrix P satisfies the matrix equation

$$A^T P + PA = -Q < 0, \tag{3}$$

then $\dot{V}(x) = -x^T Q x < 0$, $x \ne 0$, i.e. the matrix of the quadratic form $\dot{V}(x)$ is negative definite. The following theorem gives the conditions for Eq. (3) to be solvable.

Theorem 7. *Equation* (3) *has a unique solution in the class of positive definite matrices for any Q from this class if and only if the matrix A is stable. If this is true, then $P = \int_0^\infty e^{sA^T} Q e^{sA} ds$.*

For the matrix A to have stability level $\alpha > 0$, i.e. $\operatorname{Re} \lambda_i(A) \le -\alpha$, it is necessary and sufficient that there exists a positive definite matrix P satisfying the matrix equation

$$A^T P + PA + 2\alpha P = -Q$$

for every positive definite matrix Q.

We now consider the equation containing controls $u \in \mathrm{R}^l$, $1 \le l \le n$. The simplest equations of this type have the form

$$\dot{x} = Ax + Bu \tag{4}$$

where A and B are some constant matrices of the corresponding dimensions.

Definition 5. Equation (4) is called *controllable* if for all points x_1, $x_2 \in \mathrm{R}^n$ and all t_0, $\tau > 0$, there is a piecewise continuous function $u = u(t)$ such that the corresponding solution of this equation satisfies the conditions $x(t_0) = x_1$ and $x(t_0 + \tau) = x_2$.

Equation (4) is controllable if and only if the "controllability" matrix has full rank, i.e.

$$\mathbf{rank}\,[B\ AB\ A^2 B\ \cdots\ A^{n-1}\ B] = n. \tag{5}$$

It follows that the set of all controllable pairs of matrices (A, B) is open and everywhere dense in the space $\mathrm{R}^{n^2 + nm}$, the points of which are identified with such pairs of matrices. This means that however close to an uncontrollable pair there are controllable ones (but not vice versa!).

There are other equivalent definitions of controllability. The next definition is also useful.

Definition 6. Equation (4) is called *controllable* if there does not exist a non-degenerate transformation of coordinates in R^n which leads Eq. (4) to the form

$$\begin{aligned}
\dot{y}' &= A_{11} y' + A_{12} y'' + B_1 u, \\
\dot{y}'' &= A_{22} y'',
\end{aligned} \tag{6}$$

where $y' = (x_1, \ldots, x_k)$, $y'' = (x_{k+1}, \ldots, x_n)$, $k \le n$, and the pair (A_{11}, B_1) is controllable.

In other words, a controllable equation cannot be reduced by a linear transformation to the form in which several variables are out of control.

An important fact is related to this. Suppose we have probability measures with continuous densities in the space $\mathrm{R}^{n^2 + nm}$ of all pairs (A, B). Then for any of these measures the probability of choosing an uncontrollable pair is equal to zero. This follows evidently from representation (6) because all such pairs form a set having smaller dimension than the whole space.

The most widespread (and convenient) tool of control is the linear feedback having the form

$$u(t) = Kx(t)$$

where K (the "gain factor") is a constant $l \times n$-matrix.

Definition 7. Equation (4) is called *stabilizable* if there exists a linear feedback such that the matrix $A + BK$ is stable.

As to Eq. (6) with $k < n$, the notion given above means that the uncontrollable $(n-k)$-dimensional component y'' is defined by the stable $(n-k) \times (n-k)$-matrix A_{22}. A controllable equation is stabilizable. Moreover, the following result holds.

Theorem 8. *For every polynomial* $\chi(\lambda) = \lambda^n + a_1\lambda^{n-1} + \cdots + a_n$ *with real coefficients and any controllable equation* (4) *there exists a feedback* $u = Kx$ *such that* $\chi(\lambda)$ *is the characteristic polynomial of the matrix* $A + BK$.

There are several algorithms for calculating the matrix K mentioned in the theorem. The corresponding problems belong to the so called modal control.

The following description of controlled objects

$$\dot{x} = Ax + Bu, \quad z = L^T x \tag{7}$$

is more realistic than the description in (4). The m-vector $z = z(t)$ (a vector of the "observations") entering into the considered description is interpreted as a result of some linear transformation given by the $n \times m$-matrix L. The "observability" notion is introduced because of impossibility to measure directly some components (maybe all) of the state vector.

Definition 8. System (7) is called *observable* (or, equivalently, the pair of matrices (A, L) is called *observable*) if for some $\tau > 0$ the identities $u(t) \equiv 0$, $z(t) \equiv 0$, $t \in [0, \tau)$ imply $x(t) \equiv 0$.

Observability is dual to controllability, i.e. a pair (A, L) is observable if and only if the pair (A^T, L^T) is controllable. This means that there exists a matrix K such that the matrix $A + KL$ (or $A^T + L^T K^T$) is stable. Equality (5) can easily be transformed in the present case.

The next notion is related to observability as stabilizability controllability.

Definition 9. The pair (A, L) is called *detectable* if the pair (A^T, L^T) is stabilizable.

This means that there exists a matrix K such that the matrix $A + KL$ is stable. It is said that a triplet (A, B, L) is stabilizable and detectable if the pair (A, B) is stabilizable and the pair (A, L) is detectable simultaneously.

Definition 10. A triplet (A, B, L) is called *non-degenerate* (or minimal) if the pair (A, B) is controllable and the pair (A, L) is observable simultaneously.

It is convenient to study the dependence of the solution of (7) on the control with the help of the *transfer matrix* (or function)

$$W(\lambda) = L^T(\lambda I - A)^{-1}B.$$

This matrix describes the relations between the Laplace transform of the "output" z and that of the "input" u under zero initial values. Its elements are some rational functions. Consider the scalar equation of order n

$$y^{(n)} + a_1 y^{(n-1)} + \cdots + a_n y = b_0 u^{(l)} + b_1 u^{(l-1)} + \cdots + b_l u, \quad l \le n$$

and define the following operators

$$A(p) = p^n + a_1 p^{n-1} + \cdots + a_n,$$
$$B(p) = b_0 p^l + \cdots + b_l$$

where $p = d/dt$. Then the above equation takes the form

$$A(p)y = B(p)u \tag{8}$$

and its transfer function is $W(\lambda) = B(\lambda)/A(\lambda)$. The case when this ratio presents a regular rational function is very important.

The controllability and observability criteria for the system of linear equations can be described in terms of their transfer functions.

Definition 11. A transfer function $W(\lambda)$ is called *non-degenerate* if it is represented by a ratio of two irreducible polynomials.

System (7) is observable and controllable if and only if its transfer matrix is non-degenerate.

Definition 12. Equation (8) is called *minimum phase* if the polynomial $B(\lambda)$ is stable (Hurwitzian), i.e. its roots have negative real parts.

The last definition can be extended to vector equations of type (8) whose coefficients a_i, b_i are $n \times n$-matrices. In this case we consider the roots of the polynomial $\det B(\lambda)$.

If the dimensions of the vectors of controls and states do not coincide, then the minimum phase notion can be defined by various means. For Eq. (7) we consider the transfer $m \times l$-matrix $W(\lambda)$ and form a numerical $l \times m$-matrix G. With the help of these matrices we form the new $l \times l$-matrix $Q(\lambda) = G^T W(\lambda)$. Let:

(1) the matrix $\Gamma = \lim_{\lambda \to \infty} \lambda Q(\lambda)$ be symmetric and positive definite;
(2) the polynomial $\varphi(\lambda) = \det Q(\lambda)\det(\lambda I - A)$ be stable.

In the case when the conditions stated above are satisfied the matrix $Q(\lambda)$ is called *strictly minimum phase*. For such a matrix the polynomial $\varphi(\lambda)$ has degree $n - l$ and its leading coefficient is equal to $\det \Gamma$.

The minimum phase property means neither controllability nor observability of the corresponding equation, since the numerators and denominators of the elements of the matrix can possess Hurwitzian common factors.

Definition 13. Equation (7) is called *strictly minimum phase with respect to a constant matrix G* (or just *strictly minimum phase*) if the corresponding matrix $Q(\lambda)$ is minimum phase.

We consider the stability problem for solutions of equations of type (7). On the base of DLM, the proper Lyapunov function will be sought in the class of all quadratic forms $V(x) = x^T P x$. The positive definite matrix P must be chosen so that the derivative $\dot{V}(x) = x^T P(Ax + Bu)$ should be non-negative (in the case of stability) and negative at $x \neq 0$ (in the case of asymptotic stability). We cannot do this without additional constraints on the equation in question. The desired conditions are given by the "frequency theorem" the general formulation of which exceeds the demands of further applications. For this reason we restrict attention to two corollaries from it. The second of them is given at the end of Sec. 3.

Here we are interested in solving matrix equations connected with the stability problems we shall deal with in Sec. 5. Let five matrices be given, namely, A, R be $n \times n$-matrices, B, L, G be matrices of dimensions $n \times l$, $n \times m$ and $m \times l$, respectively ($n \geq l$, $n \geq m$). The matrix R is symmetric and positive definite. It is required to determine the conditions under which there exists an $n \times n$-matrix $H = H^T > 0$ and an $m \times l$-matrix C such that

$$\left.\begin{array}{r} A^T(C)H + HA(C) + R < 0, \\ HB = LC \end{array}\right\} \tag{9}$$

with $A(C) = A + BC^T L^T$.

Theorem 9. *Assume* $\mathbf{rank}\, B = l$. *There exist matrices* H *and* C *satisfying the condition* (9) *if and only if the matrix* $GW(\lambda)$ *is strictly minimum phase.*

As for the relation between the sets of controllable and minimum phase systems we remark once more that these sets do not coincide but intersect.

In conclusion we mention one fact concerned with the differential equation of general form (1). Let $x(t, x(s))$ denote the solution of (1) for $t \geq s$ and $x(s)$ the value at time s. By the definition of a solution the following equality

$$x(t, x(s, x(\tau))) = x(t + s, x(\tau)), \qquad t \geq s \geq \tau,$$

follows. This demonstrates the so-called *semi-group property* of the solution of (1). This fact will be used in Sec. 7.

11.2. Control of Homogeneous Equations

We consider adaptive control problems for homogeneous linear autonomous equations

$$\dot{x} = Ax + Bu, \quad x(t_0) = x_0, \quad t_0 = 0 \tag{1}$$

with an initial condition $x(0) = x_0$ and matrices A, B having the dimensions $n \times n$, $n \times l$, respectively. A control $u = u(t)$ is assumed to be a piecewise continuous function with a finite number of skips on any finite time interval. We assume that these functions are continuous on the right at the points of discontinuity (i.e. $u(t) = u(t+0)$). The matrices A and B are unknown. To estimate them we use information accessible to direct measurement. Namely, it is assumed that the trajectory (the solution) x of (1) is observable, i.e. at each moment t the vector $x(t)$ is known. According to (1) there are no external disturbances (or noise) acting on the control and observations. First, we estimate the matrix A.

For this purpose, we set a number t' and put $u(t) \equiv 0$ for $t \in [0, t']$. Then Eq. (1) takes the form $\dot{x} = Ax$ and, hence,

$$x(t) = e^{At}x_0. \tag{2}$$

If $x_0 \neq 0$ then $x(t)$ is associated with some curve in the space R^n. The following two cases are possible:

(1) the vector x_0 is in **general position** (with respect to the matrix A), i.e. the vectors x_0, Ax_0, ..., $A^{n-1}x_0$ generate the whole space;
(2) there is an n_1 such that $n_1 < n$ and the vectors x_0, $Ax_0, \ldots, A^{k-1}x_0$ are linearly independent only for $k \leq n_1$.

To understand these cases we have to choose some points $0 < t_1 < \cdots < t_n$ in the interval $(0, t')$ and to consider the vectors $x(t_1), \ldots, x(t_n)$. The number of linearly independent vectors among them is equal to the dimension of the space generated by the vector x_0 and the matrix A. This follows immediately from the definition of the matrix exponent

$$e^{At} = I + tA + \cdots + \frac{t^{n-1}}{(n-1)!}A^{n-1} + \frac{t^n}{n!}A^n + \cdots.$$

This, in turn, means that the expression $e^{At}x_0$ includes the vectors x_0, Ax_0, $\ldots, A^{n-1}x_0$ for any $t > 0$.

Let case (1) take place. Then for the vector $x(t)$ we have, in the component-wise form (see Sec. 1),

$$x^{(i)}(t) = \sum_{j=1}^{g(A)} (p_{ij}(t)\cos\omega_j t + q_{ij}(t)\sin\omega_j t)e^{\alpha_j t}, \quad i = 1, \ldots, n. \tag{3}$$

Putting here $t = t_1, \ldots, t_n$ we obtain a consistent system for finding the values of the unknown quantities, namely, α_j, ω_j, $g(A)$ and the coefficients of the polynomials $p_{ij}(t)$, $q_{ij}(t)$. By using them the matrix A is determined uniquely.

To find the matrix B we have to activate the control, i.e. to set the control to nonzero values in an interval. This can be done easily when the control is one-dimensional, i.e. $l = 1$. Indeed, $B = b = (b_1, \ldots, b_n)^T$. Putting $u(t) = 1$ for $t \in [t', t'']$, $t'' > t'$, we obtain

$$x(t) = e^{At}x_0 + e^{At}\int_{t'}^{t} e^{-sA}bu(s)ds, \quad t \in [t', t'']. \tag{4}$$

Fixing t and knowing the matrix A, we obtain n linear equations defining the vector b. If $l \geq 2$ then the control should be activated l times component-wise.

Let now case (2) take place. Then the vectors x_0, $Ax_0, \ldots, A^{n_1-1}x_0$ generate a proper invariant subspace of the operator A on R^n, i.e. the vector x_0 is not that of the general position. By putting successively in the equality (3) $t = t_1, \ldots, t_n$ and using n_1 groups of the equations obtained, the matrix A is defined on an invariant subspace R^{n_1}. To find this matrix on the complementary space R^{n-n_1}, we have only l columns of the matrix $B = [b^{(1)}, \ldots, b^{(l)}]$ and their iterations. It is only the question of making the rank of the composed matrix $[B\ AB\ \ldots\ A^{n-n_1-1}\ B]$ equal to $n - n_1$ on the space R^{n-n_1}. If this is not possible then it will be impossible to obtain the estimate of all elements of the matrix A. Otherwise, we act as previously, i.e. we choose in the interval $[t'', t''')$, $t''' > t''$, some points and compose the appropriate equations by using equality (4). It remains to solve the obtained linear system to find the elements of the $n \times l$-matrix B. This completes the description of the identification procedure.

The linear operator defined by the matrix A is considered in the space R^n. Let R^{n_1} be the linearly invariant subspace generated by the vectors x_0, $Ax_0, \ldots, A^{n_1-1}x_0$ and R^{n-n_1} be its complement to R^n.

Theorem 1. *Equation (1) is identifiable at any fixed initial value x_0 if and only if there exist $n - n_1$ columns of the matrix $[B\ AB\ \cdots\ A^{n-n_1-1}\ B]$ whose projections on the subspace R^{n-n_1} form a basis.*

Now let the vector x_0 be arbitrary. Then this vector is either that of the general position (with respect to the matrix A) or it generates a non-trivial invariant n_1-dimensional subspace ($n_1 < n$) or else it belongs to the kernel of the operator (i.e. $Ax_0 = 0$). Therefore, among the projections of the columns of the composed matrix $[B\ AB\ \ldots\ A^{n-1}\ B]$ on the complementary subspace R^{n-n_1}, there must be $n - n_1$ linearly independent vectors. If we make the identification[a] at an initial value x_0 such that either $x_0 = 0$ or $x_0 \in \mathbf{Ker}\,A$ then we obtain the following

Corollary 1. *Equation (1) is identifiable at an arbitrary initial value x_0 (including $x_0 = 0$) if and only if $\mathbf{rank}\,[B\ AB\ \cdots\ A^{n-n_1-1}\ B] = n$, i.e. when this equation is controllable.*

The use of the identification procedure described requires spending time to compute the matrices A and B. That time can be made arbitrarily small because for $t > 0$ the series for e^{At} contains the matrices $A, A^2, \ldots, A^{n-1}$ with positive coefficients. Hence the vector $x(t) = e^{At}x_0$, in "general position" with respect to the invariant subspace R^{n_1} ($n_1 \leq n$) for all $t > 0$, generates the vectors $x_0, x(t), \ldots, x(t_n)$ which are linearly independent for arbitrarily small $t_1, t_2, \ldots, t_n$. For these reasons the following assertion is evident.

---

[a]Apparently, there may not be such a necessity in the stabilization and optimal control problems (with the quadratic objective function). Indeed, putting $u(t) \equiv 0$, we obtain $x(t) \equiv 0$ and, hence, the control is attained without any expenditures.

Theorem 2. *The identification procedure for Eq. (1) can be realized on an arbitrarily short time interval.*

The practical use of this result is limited by the observation accuracy of the vector $x(t)$ at close moments of time and by the methods of calculation.

Obviously, this procedure is simplified when the derivative $\dot{x}(t)$ is observed. We now consider the main problems of control connected with Eq. (1).

a. *Problem about transfering* the solution of Eq. (1) from an arbitrary initial state $x(t_1) = x_1$ into an arbitrary final state $x(t_2) = x_2$ where $t_2 = t_1 + N$, $N > 0$. We can assume that $t_1 = 0$ and $x_2 = 0$ due to linearity and stationarity of the equation considered and because it is controllable. Thus the required control (within the framework of classical theory) is represented by the following formula

$$u(t) = -B^T e^{A^T(t_1 - t)} W^{-1}(t_1, t_2)[x_1 - e^{-A(t_0 - t_1)}x_1]$$

where

$$W(t_1, t) = \int_{t_1}^{t} e^{-sA} BB^T e^{-sA^T}\, ds.$$

There are various forms of the desired controls. In the adaptive approach we use the identification procedure to start with. Hence, we take some finite time interval that has, for example, length $N/2$ and compute the matrices A and B. Then we fix the state x at time $t_1 + N/2$ and find the desired control by using the above-mentioned formula.

At present the considered problem is the only adaptive problem known of this type which can be correctly formulated and solved on a finite time interval fixed beforehand.

b. *Modal control* consists of constructing a linear feedback $u = Kx$ so that the closed loop system $\dot{x} = (A + BK)x$ has the given collection of eigenvalues $\lambda_1, \ldots, \lambda_n$. For controllable equations this can always be done and the calculation algorithm of the matrix K is known. Sometimes we are interested in changing a part of the eigenvalues, for example, one or two of them. The solution of this problem enables us to obtain both the desired dynamics of the transition processes and the required asymptotic behavior of the solution (as $t \to \infty$).

In the adaptive version the construction of modal control is obvious and we do not go into details.

c. *Control of Eq. (1) with a model reference* means that there is a desired "model" of the equation $\dot{z} = A_M z + B_M u$, where $z(t)$ is the trajectory reference, A_M and B_M are some known constant matrices, A_M is usually stable, and u is a piecewise continuous function interpreted as an external influence on the model. The controlled object is described by the equation

$$\dot{x} = (A + U(t))x + (B + W(t))\tilde{u}$$

where A and B are some unknown matrices, $U(t)$ and $W(t)$ are the parametric controlling actions on the object, $\tilde{u}$ coincides with u outside an initial finite interval which can be taken to be arbitrary. It is required to find the controls $U(t)$ and $W(t)$ so that the error $\varepsilon(t) = x(t) - z(t)$ tends to zero as $t \to \infty$.

A common approach to solving this problem (see Sec. 3) is based on the DLM and on the assumption that the matrices $U(t)$ and $W(t)$ are differentiable. If we assume that the pair (A, B) is controllable then, in fact, the solution of this problem is trivial, namely, we put $U(t) = W(t) \equiv 0$ and find the matrices A, B by the method described above. In what follows we put

$$U(t) = A_M - A, \quad W(t) = B_M - B.$$

Thus, we obtain the equation $\dot{\varepsilon} = A_M \varepsilon$ with respect to $\varepsilon(t)$ with the stable matrix A_M. Hence $\varepsilon(t) \to 0$ as $t \to \infty$.

d. *Optimal control* consists of minimizing the functional

$$\Phi(u) = \int_0^T \Psi(x(s), u(s), s)ds.$$

The case $T = \infty$ is not considered. Within the framework of classical theory the case $\Psi = x^T Q_1 x + u^T Q_2 u$ has been well studied. In the adaptive version this case creates no problems provided the pair (A, B) is controllable. If this is true, we estimate the parameters of the equation on the interval $[0, t']$ and, later on, use the methods of optimal control theory. These methods give the minimal value of the functional $\Phi'(u) = \int_{t'}^N \Psi ds$. In solving the initial problem the possible error is equal to $\int_0^{t'} \Psi ds$. For a stable matrix A this error is arbitrarily small provided identification is done in a short time. Thus the described approach is ε-optimal with respect to the functional $\Phi(u)$. Although the controllability property is, to some extent, artificial for finite values of T, it is almost necessary in the case $T = \infty$. Indeed, among conditions providing existence of optimal control in the linear-quadratic problem the stabilizability condition of the pair (A, B) near to controllability is required.

The above is also related to minimization of the "average loss per unit of time" functional

$$\Phi(u) = \lim_{t \to \infty} t^{-1} \int_0^t \Psi ds$$

and, moreover, the described control guarantees the achievement of its global minimum which is equal to zero.

The approach stated above can be extended to equations described by the "input-output" relationship

$$y^{(n)} + a_1 y^{(n-1)} + \cdots + a_n y = b_0 u^{(l)} + b_1 u^{(l-1)} + \cdots + b_l u, \quad l < n.$$

For finite time T the coefficients (a_i, b_i) are computed and, later on, stabilization or optimization control problems are solved for these equations. We shall not discuss these problems here.

11.3. Control with A Model Reference

The problem of tracking a discrete sequence of points has been thoroughly studied in the previous chapter. It has been formulated in the adaptive version for linear difference stochastic equations without any assumptions about the structure of the sequence $\{x_t^*\}$ except boundedness. Now, we consider this problem in continuous time without requiring the fulfilment of the optimality conditions.

The trajectory $x^*(t)$ in R^n is continuously differentiable and is generated by the model reference

$$\dot{x}^* = A_M x^* + B_M u. \tag{1}$$

Here A_M and B_M are matrices of dimensions $n \times n$ and $n \times l$, respectively, $u = u(t)$ is a bounded continuous vector-function of order l which signifies the control (it gives the solution $x^*(t)$ the desired properties). These are supposed to be known. The matrix A_M is also assumed to be stable. The model reference is pursued by the controlled object of the form

$$\dot{x} = A(t)x + B(t)u, \tag{2}$$

where $x(t) \in \mathrm{R}^n$ and u is an external influence which is the same as in (1), but $A(t)$ and $B(t)$ are matrices of the corresponding dimensions having the form

$$A(t) = A + \bar{A}(t) + U(t), \quad B(t) = B + \bar{B}(t) + W(t).$$

Here A and B are unknown constant matrices, $\bar{A}(t)$ and $\bar{B}(t)$ are unknown unobserved parametric disturbances and, finally, $U(t)$ and $W(t)$ are the parametric controls. The matrices $\bar{A}(t)$, $\bar{B}(t)$, $\bar{U}(t)$, $\bar{W}(t)$ are assumed to be continuously differentiable.

We denote by $\mathcal{K}(A_M, l)$ the class of equations of the form (2) with parameters satisfying the conditions stated above.

We define the control aim for the class $\mathcal{K}(A_M, l)$ as follows:

find matrices $U(t)$ and $W(t)$ so that

$$\lim_{t \to \infty} [x(t) - x^*(t)] = 0.$$

This aim does not require that $\lim_{t \to \infty} A(t) = A_M$ and $\lim_{t \to \infty} B(t) = B_M$.

We introduce the following notation:

$$e = x - x^*, \quad H_1 = A - A_M + \bar{A}(t) + U(t),$$
$$H_2 = B - B_M + \bar{B}(t) + W(t)$$

and assume that the quantities $e(t)$, $x(t)$, $u(t)$ are observed. Using this notation, from Eqs. (1), (2) we obtain the following system of differential equations

$$\begin{cases} \dot{e} = A_M e + H_1 x + H_2 u, \\ \dot{H}_1 = \dot{U} + \dot{\bar{A}}, \\ \dot{H}_2 = \dot{W} + \dot{\bar{B}} \end{cases} \qquad (3)$$

with the initial values e_0, $H_1(0)$ and $H_2(0)$. First, we assume that the parametric disturbances $\bar{A}(t)$ and $\bar{B}(t)$ are constant. We would like to find conditions ensuring asymptotic stability of equations from the class $\mathcal{K}(A_M, l)$ with respect to the variable e. Let P denote the positive definite matrix which is the solution of the Lyapunov equation $A_M^T P + P A_M = -Q$ with Q being a positive definite matrix. According to the assumptions on A_M there exists a unique matrix P satisfying the matrix equation considered.

Theorem 1. *The control of the form*

$$U(t) = -\varkappa \int_0^t Pe(s)x^T(s)ds + U_0,$$

$$(4)$$

$$W(t) = -\varkappa \int_0^t Pe(s)u^T(s)ds + W_0$$

provides stability of the trivial solution ($e = 0$, $H_1 = 0$ and $H_2 = 0$) of system (3) as well as asymptotic stability of this solution in the variable $e(t)$ with respect to the class $\mathcal{K}(A_M, l)$ under constant parametric disturbances.

Proof. We choose the following quadratic function

$$V(e, H_1, H_2) = \varkappa e^T P e + \mathbf{tr}(H_1 H_1^T + H_2 H_2^T), \quad \varkappa > 0$$

as a Lyapunov function for system (3). Its derivative, with respect to Eq. (3), is equal to

$$\dot{V} = \varkappa e^T(A_M^T P + P A_M)e + 2\mathrm{tr}[(\varkappa Pex^T + \dot{U})H_1^T + (\varkappa Peu^T + \dot{W})H_2^T].$$

If we put

$$\dot{U} = -\varkappa Pex^T, \quad \dot{W} = -\varkappa Peu^T \qquad (5)$$

and take into account the definition of the matrix P, then we obtain

$$\dot{V} = -\varkappa e^T Q e.$$

This function, treated as a function of e, H_1, H_2, satisfies the inequality $\dot{V} \leq 0$. This implies the stability declared in the theorem. Treating this derivative as a function only of e, we remark that $\dot{V} < 0$ for $e \neq 0$. Hence $e(t) \to 0$ as $t \to \infty$. $\square$

Strategy (5) is of the direct type. It has all the "good" properties of an adaptive strategy, namely, it is nonlinear, non-stationary and non-Markov (depends on the whole past history). It is interesting to take a look at the original equation by substituting the obtained parametric control (4) on the right-hand side of (2). Indeed,

$$\dot{x}(t) = \left[A - \varkappa \int_0^t Pe(s)x^T(s)ds\right]x(t) + \left[B - \varkappa \int_0^t Pe(s)u^T(s)ds\right]u(t),$$

$$e = x - x^*.$$

Although the original equation was simple, the obtained solution is rather complex.

The properties of the solutions of (3) under the control (5) can be improved if we restrict the class of the equations considered. If the parametric disturbances are constant and the control has form (5), then system (3) takes the following form

$$\begin{cases} \dot{e} = A_M e + H_1 x + H_2 u, \\ \dot{H}_1 = -\varkappa P e x^T, \\ \dot{H}_2 = -\varkappa P e u^T. \end{cases} \tag{6}$$

From the ith rows of the matrices H_1 and H_2 we form the row-vector $h_i = (h_i^{(1)}, h_i^{(2)})$ of order $n + l$ and introduce the following column vectors $\varphi = (x^*, u)^T$ and $\psi = (x, u)^T$. From Theorem 1 it follows that $\|\varphi(t)\| \le C$ and $\|\psi(t)\| \le C$ for some $C > 0$ and all t.

Theorem 2. *Assume that the matrix A_M is stable and the derivative $\dot{u}(t)$ exists. Moreover, assume that there exist numbers α, T such that*

$$\|\dot{u}(t)\| \le C, \quad \max_{t \le s \le t+T} |h\varphi(s)| \ge \alpha\|h\|$$

for all t and any vector h. Then the trivial solution of system (6) is asymptotically stable on the whole space and uniformly with respect to t_0, e_0, $H_1(0)$, $H_2(0)$ on any bounded domain.

Proof. Integrating the second and third equations in (6) we find that

$$\|h_i(t) - h_i(kT)\| \le \max_{kT \le s \le (k+1)T} \varkappa\|P\|\|e(s)\|cN = \rho_k$$

for all $t \in [kT, (k+1)T]$, $k = 1, 2, \ldots$. According to Theorem 1, $\rho_k \to 0$, and to complete the proof it is sufficient to show that $\|h_i(kT)\| \to 0$ as $k \to \infty$ and $i = 1, 2, \ldots, l$. By the assumptions of the theorem for any k we obtain

$$\alpha\|h_i(kT)\| \le \max_{kT \le s \le (k+1)T} |h_i(kT)\varphi^T(s)| \le |h_i(kT)\psi^T(t_k)| + \|h_i(kT)\|\|e(t_k)\|$$

and t_k means the moment equality is reached at, i.e.

$$\max_{kT \le t \le (k+1)T} |h_i(kT)\varphi^T(t)| = |h_i(kN)\psi^T(t_k)|.$$

The second term on the right-hand side of the inequality tends to zero but for the first one we have the following estimate

$$|h_i(kT)\psi^T(t_k)| \le |(h_i(t_k) - h_i(kT)| + |h_i(t_k)\psi^T(t_k)|$$
$$\le \rho_k C + \|\dot{e}(t_k)\| + \|A_M\|\|e(t_k)\|.$$

The first and third terms on the right-hand side of the last inequality tend to zero for known reasons. The second one vanishes according to the Landau–Hadamard theorem.[b] Indeed, $\|e(t)\| \to 0$ and from Eq. (6) it follows that $\|\ddot{e}(t)\| \le q < \infty$. □

By this result it is possible to answer not only the question of approximating the coefficients of Eqs. (1) and (2) but also of sensibility of algorithm (4) to varying parametric disturbances when the matrices $\bar{A}(t)$ and $\bar{B}(t)$ are not constant. Of course, these disturbances must satisfy some restrictions. This is true in classical control theory and, even to a greater extent, in adaptive control theory where it is either rather difficult or impossible to propose algorithms for arbitrary homogeneities. For these reasons, we demand that the functions $\bar{A}(t)$ and $\bar{B}(t)$ vary rather slowly. Now we are going to expound the above formally.

The controlled system with disturbances has the form

$$\begin{cases} \dot{e} = A_M e + H_1 x + H_2 u, \\ \dot{H}_1 = -\varkappa P e x^T + \dot{\bar{A}}(t), \\ \dot{H}_2 = -\varkappa P e u^T + \dot{\bar{B}}(t), \end{cases} \tag{7}$$

i.e. we add the terms $\dot{\bar{A}}(t)$ and $\dot{\bar{B}}(t)$ on the right-hand side of the second and third equations and also additional terms $\bar{A}$ and $\bar{B}$ have appeared in the representations of H_1 and H_2. The notion of "constantly acting disturbances" signifies that the parametric disturbances $\bar{A}(t)$ and $\bar{B}(t)$ are varied "slowly" (i.e. their derivatives are rather small). From Theorem 3, Sec. 1 and Theorem 2, the next result follows immediately.

Theorem 3. *By the assumptions of Theorem 2, control (5) provides stability of the trivial solution of Eq. (6) under constantly acting disturbances.*

Adaptive control (5), which is the same for all equations from $\mathcal{K}(A_M, l)$, allows not only to track the reference trajectory with an increasing accuracy, but to do it in the presence of "constantly acting disturbances". This means that the proposed control has the robustness property.

If the object is not described by Eq. (2) but by a scalar differential equation of order n, then the algorithm of control can be defined in a similar manner. However, when reducing an equation of higher order to a system of equations we have to observe not only the phase coordinate but all its derivatives up to $n-1$-st one. In many practical cases this demand is unrealistic but we will repeatedly meet with it within the framework of the applied concept.

Assume that the structure of the equation is known both for the model reference and for the controlled object and that the parameters entering into the description of the object are the only unknowns. It is required that the trajectories of the object

---

[b]**Theorem** (Landau–Hadamard) *Let $g(t)$ be twice differentiable and $M_0 = \sup_{t\ge 0}|g(t)|$, $M_1 = \sup_{t\ge 0}|\dot{g}(t)|$, $M_2 = \sup_{t\ge 0}|\ddot{g}(t)|$. Then we have the following inequality*

$$M_1 \le 2\sqrt{M_0 M_2}.$$

and the model reference should approach each other in time. This can be realized for scalar equations of higher orders ($n \geq 2$).

Let the object be given by the "input-output" relationship

$$\alpha(p)x(t) = \beta(p)u(t), \quad t \geq 0 \tag{8}$$

with

$$\alpha(p) = p^n + \alpha_1 p^{n-1} + \cdots + \alpha_n,$$
$$\beta(p) = \beta_1 p^{n-1} + \cdots + \beta_{n-1}p + \beta_n,$$

$p = d/dt$. Here the function $u(t) = f(t) + y(t)$, represented as the sum of some external disturbance $f(t)$ and the control $y(t)$, is assumed to be smooth so that its image under the operator $\beta(p)$ is a continuous function of t. Let the polynomial $\beta(\cdot)$ be given, i.e. the integer n and the coefficients $\beta_1, \ldots, \beta_n$ are known.

The evolution of the model reference is described by the equation

$$\lambda(p)x_M(t) = \mu(p)u(t), \quad t \geq 0, \tag{9}$$

in which the coefficients of the polynomials

$$\lambda(p) = p^n + l_1 p^{n-1} + \cdots + l_n,$$
$$\mu(p) = m_1 p^{n-1} + \cdots + m_{n-1}p + m_n$$

are unknown. No additional restrictions on Eqs. (8) and (9) are imposed.

It is required to find a control $y(t)$ so that

$$e(t) = x(t) - x_M(t) \xrightarrow[t \to \infty]{} 0$$

for every smooth function $f(t)$ and arbitrary initial values $x^{(h)}(0)$, $x_M^{(h)}(0)$ ($h = 0, 1, \ldots, n-1$).

Let the following vectors of length n

$$\alpha = (\alpha_1, \ldots, \alpha_n), \quad \beta = (\beta_1, \ldots, \beta_n),$$
$$\lambda = (l_1, \ldots, l_n), \quad \mu = (m_1, \ldots, m_n)$$

correspond to the polynomials $\alpha(p)$, $\beta(p)$, $\lambda(p)$ and $\mu(p)$ respectively. In the case when some coordinates of these vectors are equal to zero the orders of the polynomials $\beta(p)$ and $\mu(p)$ may turn out less than $n - 1$.

Let us rewrite Eq. (8) in terms of the state space. We introduce the column vectors $\nu = (\nu_1, \ldots, \nu_n)^T$ and $\delta_j = (\delta_{j1}, \ldots, \delta_{jn})^T$, $j = 1, \ldots, n-1$, (δ_{ji} is the Kroneker symbol) and define the following matrix

$$A(\nu) = [-\nu \; \delta_1 \; \delta_2 \; \cdots \; \delta_{n-1}].$$

We write down the rational function $\varkappa(p)/\nu(p)$ with $\nu(p) = p^n + \nu_1 p^{n-1} + \cdots + \nu_n$, $\varkappa(p) = k_1 p^{n-1} + \cdots + k_n$ ($\nu, \varkappa$ are the corresponding column vectors of length n) in the form

$$\frac{\varkappa(p)}{\nu(p)} = \delta^T (pI - A(\nu))^{-1} \varkappa, \quad \delta = \delta_1.$$

For the transfer function of (8) this means that

$$\frac{\beta(p)}{\alpha(p)} = \delta^T (pI - A(\alpha))^{-1}\beta.$$

It follows that $x(t)$ satisfies the following system of equations

$$\dot{z} = A(\alpha)z + \beta u, \quad x(t) = \delta^T z(t)$$

with respect to $z = (z_1, \ldots, z_n)^T$. We choose the vector $\varkappa$ so that the matrix $A(\varkappa)$ is stable (i.e. has all eigenvalues belonging to the left half-plain). Then, by using the matrix identity $A(\alpha) = A(\varkappa) + (\varkappa - \alpha)\delta^T$ we arrive at the equation

$$\dot{z} = A(\varkappa)z + \beta u + (\varkappa - \alpha)x.$$

Let us choose n linearly independent vectors $\{e_1, e_2, \ldots, e_n\}$, where $e_i = (e_{i1}, \ldots, e_{in})$, and define polynomials $e_i(p) = e_{i1}p^{n-1} + e_{i2}^p + \cdots + e_{in}$ and $\nu(p) = p^n + \nu_1 p^{n-1} + \cdots + \nu_n$ which are stable (i.e. Hurwitz) with the help of the relationships

$$\frac{e_i(p)}{\nu(p)} = \delta^T (pI - A(\varkappa))^{-1} e_i, \quad i = 1, \ldots, n.$$

From the last two equalities and Eq. (8) we have

$$x(t) = \sum_{j=1}^{n} c_j \left[\frac{e_i(p)}{\nu(p)} x(t) \right] + \frac{\beta(p)}{\nu(p)} u(t). \tag{10}$$

Further, we employ the next result which follows from the "frequency theorem" mentioned in Sec. 1.

Lemma 1. *For every Hurwitz polynomial $\nu(p)$ of degree n, there exists a Hurwitz polynomial $\rho(p)$ of degree $n - 1$ and a number $h > 0$ such that for all $\lambda \geq 0$ the following inequality (the "frequency condition") holds*

$$\mathrm{Re}\left(\frac{\rho(i\lambda)}{\nu(i\lambda)} \right) \geq h \left| \frac{\rho(i\lambda)}{\nu(i\lambda)} \right|^2.$$

Let $\rho(\lambda)$ be the polynomial given in accordance with Lemma 1. We put

$$x_i(t) = \frac{e_i(p)}{\rho(p)} x(t), \quad x_{Mj} = \frac{e_i(p)}{\rho(p)} x_M(t), \quad f_j(p) = \frac{e_i(p)}{\rho(p)} f(t),$$

$$f^*(t) = \frac{\beta(p)}{\rho(p)} f(t), \quad y^*(t) = \frac{\beta(p)}{\rho(p)} y(t)$$

and rewrite the equations of the object and model reference in the canonical form

$$x(t) = \frac{\rho(p)}{\nu(p)} \left(\sum_{j=1}^{n} c_j x_j(t) + f^*(t) + y^*(t) \right), \tag{11}$$

$$x_M(t) = \frac{\rho(p)}{\nu(p)} \sum_{j=1}^{n} (v_j x_{Mj}(t) + w_j f_j(t)). \tag{12}$$

The coefficients c_j in (11) and the coefficients v_i, w_j in (12) are the unknown components of the vectors $-\alpha + \nu$ and $-\lambda + \mu$ in the basis $(e_1, \ldots, e_n)$, respectively.

According to (11) and (12), we can find the value of the error

$$e(t) = (t) - x_M$$

$$= \frac{\rho(p)}{\nu(p)} \left\{ \sum_{j=1}^{n} [c_j x_j(t) - v_j x_{Mj}(t) - w_j f_j(t)] + f^*(t) + y^*(t) \right\}. \tag{13}$$

We now define the $3n$ functions $\xi_i(t)$, $\eta_i(t)$, $\zeta_i(t)$, $i = 1, \ldots, n$, as follows

$$\begin{aligned} \dot{\xi}_i(t) &= -\alpha_i x_i e, \quad \dot{\eta}_i(t) = -\beta_i x_{Mi} e, \\ \dot{\zeta}_i(t) &= -\gamma_i f_i e, \alpha_i, \beta_i, \gamma_i > 0. \end{aligned} \tag{14}$$

They can be rewritten in explicit form

$$\xi_i(t) = -\alpha_i \int_0^t x_i(s)e(s)ds, \quad \eta_i(t) = -\beta_i \int_0^t x_{Mi}(s)e(s)ds,$$

$$\zeta_i(t) = -\gamma_i \int_0^t f_i(s)e(s)ds. \tag{15}$$

We choose

$$y^*(t) = \sum_{i=1}^{n} [\xi_i(t)x_i(t) + \eta_i(t)x_{Mi}(t) + \zeta_i(t)f_i(t)] - f^*(t)$$

as the control law for the object or, which is the same,

$$y(t) = \frac{\rho(p)}{\beta(p)} \sum_{i=1}^{n} [\xi_i(t)x_i(t) + \eta_i(t)x_{Mi}(t) + \zeta_i(t)f_i(t)] - f(t). \tag{16}$$

We can write the equation for the error $e(t)$ in the form

$$\nu(p)e = \rho(p) \sum_{i=1}^{n} \left[(\xi_i(t) + c_1)x_i(t) + (\eta_i(t) - v_i)x_{Mi}(t) + (\zeta_i(t) - w_i)f_i(t) \right].$$

In a standard way we can verify that this equation of order n is equivalent to a system of n differential equations of the first order with respect to the collection of functions $\omega(t) = (\omega_1(t), \ldots, \omega_n(t))$, $\gamma(t) = (\gamma_1(t), \ldots, \gamma_{3n}(t))$, namely,

$$\begin{cases} \dot{\omega} = P\omega + q \sum_{i=1}^{3n} \gamma_i(t)\chi_i(t), & e = r^T \omega, \\ \dot{\gamma}_i = -\rho_i \chi_i e, & i = 1, \ldots, 3n, \quad \rho_i > 0, \end{cases} \tag{17}$$

where P is a Hurwitz matrix, q and r are some column vectors. The second subsystem in (17) corresponds to the collection of $3n$ equations of the form (14).

We can now formulate the main result.

Let $\mathcal{K}(n, \beta; F)$ denote the class of linear differential equations of order n having the form (8) with a fixed polynomial $\beta(p)$ and a set F of admissible external influence. Relation (16) together with (14), (15) defines the strategy $\sigma(B)$.

Theorem 4. *The strategy $\sigma(B)$ realizes the control aim $\lim_{t \to \infty} e(t) = 0$ for arbitrary initial values and external influence $f \in F$ with respect to the class $\mathcal{K}(n, \beta; F)$.*

Proof. The transfer function of system (17) is given by the formula $W(\lambda) = r^T(\lambda I - P)^{-1}q$. By Lemma 1 it satisfies the frequency condition

$$\operatorname{Re} W(i\lambda) \geq h|W(i\lambda)|^2, \quad h > 0, \quad \forall \lambda \geq 0.$$

The matrix $H = (H^T > 0)$, being the solution of the appropriate linear equation, exists. By using it a Lyapunov function of the system (18) can be given as follows

$$V(\omega, \gamma) = \omega^T H\omega + \sum_{j=1}^{3n} \rho_j^{-1}\gamma_j^2 \geq 0.$$

Its derivative, with respect to the above equations, is negative, i.e. $\dot{V} \leq -he^2$. Therefore for the output $e(t)$ of the system (17) the assertion of the theorem holds. $\square$

An estimate of the rate of vanishing of the function $e(t)$ must be uniform on the class $\mathcal{K}(n, \beta; F)$, i.e. it must hold for every equation of form (8) or, at least, on a large subclass of such equations. Up to now such a uniform estimate is unknown.

The adaptive control procedure described will be simplified if we know exactly, at least part of the coefficients of the equation of the object.

We now consider the problem of control with a model reference for the differential equations with constant concentrated delays.

The following equation

$$\dot{x}_M(t) = A_0 x_M(t) + \sum_{i=1}^{N} A_i x_M(t - \tau_i) + \sum_{i=1}^{L} C_i x_M(t - h_i) + Bu(t), \quad \tau_i, h_i > 0 \qquad (18)$$

with the initial condition

$$x_M(\theta) = \varphi(\theta) \in C_{[-\eta,0]}, \quad \theta \in [-\eta, 0], \quad \eta = \max_i\{\tau_i, h_i\}$$

as the model reference. As before we assume that $\dim x = n$, $\dim u = l$. The external action $u(t)$ is bounded and continuous. The control satisfies the following condition.

Condition AS: *For arbitrary fixed h_j the trivial solution is asymptotically uniformly stable on the whole space.*

Let $\lambda_{\min}(M)$, $\lambda_{\max}(M)$ denote the minimal and maximal eigenvalues of the matrix M and $\sigma_{\max}(M) = [\lambda_{\max}(M^T M)]^{1/2}$ denote the maximal singular number of the matrix M. The following theorem gives sufficient conditions for the property **AS** to take place.

Theorem 5. *Assume that $u \equiv 0$. Condition **AS** holds if:*

(a) *the matrix $\sum_{i=0}^{N} A_i$ is stable;*
(b) *the inequality*

$$\frac{\lambda_{\min}(Q)}{2\lambda_{\max}(P)} > \sigma_{\max}\left(\sum_{i=0}^{N} A_i\right)\sum_{i=1}^{N}\tau_i\sigma_{\max}(A_i) + \left[1 + \sum_{i=1}^{N}\tau_i\sigma_{\max}(A_i)\right]\sum_{i=1}^{L}\sigma_{\max}(C_i)$$

with symmetric positive definite matrices P and Q of dimensions $n \times n$ related by the Lyapunov equation

$$\left[\sum_{i=0}^{N} A_i\right]^T P + P \left[\sum_{i=0}^{N} A_i\right] = -Q$$

is satisfied.

The proof of this assertion is based on standard arguments of DLM and omitted.

Notice two special cases of condition (b). If $C_i = 0$, $i = 1, \ldots, M$, then this condition takes the form

$$\frac{\lambda_{\min}(Q)}{2\lambda_{\max}(P)} > \sigma_{\max}\left(\sum_{i=0}^{N} A_i\right) \sum_{i=1}^{N} \tau_i \sigma_{\max}(A_i).$$

If $A_i = 0$, $i = 1, \ldots, n$, and A_0 is stable then (taking into account that $A_0^T P + P A_0 = -Q$)

$$\frac{\lambda_{\min}(Q)}{2\lambda_{\max}(P)} > \sum_{i=1}^{L} \sigma_{\max}(C_i).$$

Concerning the delay, we note that $\{\tau_i\}$ and $\{A_i\}$ are included in condition (b) of Theorem 5 but the constraints on the collection $\{h_j\}$ are not imposed.

Let us consider the tracking problem of the trajectory x_M by the solution $x(t)$ of the following equation

$$\dot{x}(t) = A(t)^{(0)}x(t) + \sum_{i=1}^{N} A^{(i)}(t)x(t - \tau_i) + \sum_{i=1}^{L} C^{(i)}(t)x(t - h_i) + B(t)u(t), \tag{19}$$

$$x(\theta) = \psi(\theta), \quad \theta \in [-\eta, 0], \quad \eta = \max_{i}\{\tau_i, h_i\}$$

where the initial values are assumed to be continuous on the interval $[-\eta, \, 0]$. The matrices in the above equation have the following form

$$A(t)^{(i)} = A_i + \bar{A}_i(t) + U_i(t), \quad i = 0, 1, \ldots, N \qquad B(t) = B + \bar{B}(t) + W(t);$$

$$C^{(j)}(t) = C_j + \bar{C}_j(t) + Z_j(t), \quad j = 1, \ldots, L,$$

where the matrices A_i, B, C_j coincide with the corresponding matrices in (18) but the terms $\bar{A}_i(t)$, $\bar{B}(t)$, $\bar{C}_j(t)$ are unknown and unobserved parametric disturbances. The matrices $U_i(t)$, $W(t)$, $Z_j(t)$ are continuously differentiable parametric controls. The delays τ_i and h_j are identical in Eqs. (18) and (19) but the initial conditions may be different. To state the control aim the following notation is introduced:

$$\check{A}_i(t) = \bar{A}^{(i)}(t) + U^{(i)}(t), \quad i = 0, 1, \ldots, N_j, \qquad \check{B}(t) = \bar{B}(t) + W(t),$$

$$\check{C}_j(t) = \bar{C}^{(j)}(t) + Z_j(t), \quad j = 1, \ldots, L, \qquad e(t) = x(t) - x_m(t).$$

These quantities satisfy the equation

$$\dot{e}(t) = A_0 e(t) + \sum_{i=1}^{N} A_i e(t - \tau_i) + \sum_{j=1}^{L} C_j e(t - h_j) + \check{B}(t)u(t)$$

$$+ \check{A}_0(t)x(t) + \sum_{i=1}^{N} \check{A}_i(t)x(t - \tau_i) + \sum_{j=1}^{L} \check{C}_j(t)x(t - h_j) \qquad (20)$$

with the obvious initial condition. The solution of this equation is also obvious, namely,

$$e(t) = 0, \quad \check{A}_i(t) = 0 \ (i = 0, \ldots, N), \quad \check{B}(t) = 0, \quad \check{C}_j(t) = 0 \ (j = 1, \ldots, L).$$

Let $\mathcal{K}(A_0, \ldots, A_n; l; (\tau_i, h_j))$ be the class of equations of the form (19) with $x \in \mathrm{R}^n$, $u \in \mathrm{R}^l$ and the same values of the delays $\{\tau_i, h_j\}$. The control aim is

to design a strategy which guarantees both the stability of the solutions and the fulfilment of the condition $\lim_{t \to \infty} e(t) = 0$ *for arbitrary initial values and all equations from this class.*

Now, we study the possibilities of the following algorithm which, in principle, coincides with the one described at the beginning of the section. It is defined by

$$\begin{aligned}
\dot{U}_0 &= -Pe(t)x^T(t), & \dot{U}_i &= -Pe(t)x^T(t - \tau_i), & i &= 1, \ldots, N \\
\dot{Z} &= -Pe(t)u^T(t), & \dot{W}_j &= -Pe(t)x^T(t - h_j), & j &= 1, \ldots, L.
\end{aligned} \qquad (21)$$

Here we have used the notation

$$\varepsilon(t) = e(t) + \sum_{i=1}^{N} A_i \int_{t-\tau_i}^{t} e(s)ds.$$

We recall that the matrices A_i (the parameters of the model) are known and P is the solution of the Lyapunov equation.

If the structure of the equation of the model is simplified then some obvious simplifications can be done in the algorithm (21) as well.

Theorem 6. *The control* (21) *provides stability of the trivial solution of equation* (20) *(satisfying conditions* (a) *and* (b) *from Theorem 5) and asymptotical stability in the variable* $e(t)$ *with respect to the class of equations* $\mathcal{K}(A_0, \ldots, A_n; l; (\tau_i, h_j))$ *under constantly acting parametric disturbances and bounded continuous action* $u(t)$.

The proof of this theorem is similar to that of Theorem 1. The distinction consists of selecting a Lyapunov function and estimating its derivative. We explain these details in short.

We choose a Lyapunov function as follows

$$V(t) = \varepsilon(t)^T P \varepsilon(t) + \sum_{i=1}^{N} \alpha_i \int_{t-\tau_i}^{t} \int_{z}^{t} \|e(s)\|^2 ds\, dz + \sum_{i=1}^{L} \beta_i \int_{t-h_i}^{t} \|e(s)\|^2 ds$$

$$+ \mathbf{tr}\left[\sum_{i=1}^{N} \check{A}_i(t)\check{A}_i^T(t) + \check{B}(t)\check{B}^T(t) + \sum_{i=1}^{L} \check{C}_i(t)\check{C}_i^T(t) \right].$$

Here the values of the constants $\alpha_i > 0$, $\beta_i > 0$ should be chosen properly. The derivative $\dot{V}$, with respect to Eq. (20), is equal to

$$\dot{V} = 2e^T(t)P \sum_{i=1}^{L} C_i e(t - h_i) + 2e^T(t) \sum_{i=1}^{N} A_i^T P \sum_{i=1}^{N} A_i \int_{t-\tau_j}^{t} e(s)ds$$

$$- e^T(t)Qe(t) + 2\sum_{i=1}^{L} e^T(t - \tau_j)C_i^T P \sum_{i=1}^{N} A_i \int_{t-h_i}^{t} e(s)ds + \sum_{i=1}^{N} \tau_i \alpha_i \|e(t)\|^2$$

$$- \sum_{i=1}^{N} \alpha_i \int_{t-\tau_j}^{t} \|e(s)\|^2 ds + \sum_{i=1}^{L} \beta_i \left[\|e(t)\|^2 - \|e(t - h_i)\|^2 \right]$$

$$+ 2\mathbf{tr}\left\{ \left[P\varepsilon(t)x^T(t) + \dot{\check{A}}_0(t) \right] \check{A}_0^T(t) + \sum_{i=1}^{N} \left[P\varepsilon(t)x^T(t - \tau_i) + \dot{\check{A}}_i(t) \right] \check{A}_i^T(t) \right.$$

$$\left. + \sum_{i=1}^{L} \left[P\varepsilon(t)x^T(t - h_i) + \dot{\check{C}}_i(t) \right] \check{C}_i^T(t) + \left[P\varepsilon(t)u^T(t) + \dot{\check{B}}(t) \right] \check{B}^T(t) \right\}.$$

The term $2\mathbf{tr}\{\ldots\}$ is equal to zero due to (21). The other terms can be estimated as follows.

$$x^T Q x \geq \lambda_{\min}(Q)\|x\|^2,$$

$$2x^T(t)\left(\sum_{k=0}^{N} A_k^T \right) P A_i x(s) \leq \lambda_{\max}(P)\sigma_{\max}(A_i)\sigma_{\max}\left(\sum_{k=0}^{N} A_k \right) \left[\|x(t)\|^2 + \|x(s)\|^2 \right],$$

$$2x^T(t - h_j)C_j^T P A_i x(s) \leq \lambda_{\max}(P)\sigma_{\max}(C_j)\sigma_{\max}(A_i) \left[\|x(t - h_j)\|^2 + \|x(s)\|^2 \right],$$

$$2x^T(t)P C_j^T x(t - h_j) \leq \lambda_{\max}(P)\sigma_{\max}(C_j) \left[\|x(t)\|^2 + \|x(t - h_j)\|^2 \right].$$

The numbers α_i, β_i are chosen as follows

$$\alpha_i = \lambda_{\max}(P)\sigma_{\max}(A_i)\left[\sigma_{\max}\left(\sum_{k=0}^{N} A_k \right) + \sigma_{\max}\left(\sum_{k=1}^{L} C_k \right) \right],$$

$$\beta_i = \lambda_{\max}(P)\sigma_{\max}(C_i)\left[1 + \sum_{k=0}^{N} \tau_k \sigma_{\max}(A_k) \right].$$

Elementary (but rather combersome) transformations lead to the desired inequality

$$\dot{V} \leq -\delta\|e(t)\|^2, \quad \delta > 0.$$

Hence, using the same arguments as at the end of proof of Theorem 1, we complete the proof. $\qquad\square$

The construction of strategies for solving the problem with a model reference may be based on other principles differing from those stated in Theorem 5 but which guarantee the fulfilment of the condition **AS**. It is important to note that the property **AS** implies stability under the constantly acting disturbances both for the equations with a delay and for ordinary ones.[c] Thus control (21) will guarantee successful tracking after the "reference" if the parametric disturbances vary slowly and are not too large.

The distinctive feature of the problem considered of adaptive control is the coincidence of the delays τ_i, h_i for the "reference" and the object. This restriction is unreasonable in a number of cases. Some elements of the control should enter into the arguments of the functions $x(t - \tau + u_\tau)$ on purpose to approximate the delay of the "reference" and the object with each other in time. This problem is not likely to be solved soon.

11.4. Steepest Descent Method

We consider a general method of constructing control algorithms for many adaptive control and identification problems concerned with ordinary differential equations. The results obtained give a uniform solution of the corresponding problems provided the appropriate assumptions are satisfied.

As before, $x \in \mathrm{R}^n$ and $u \in R^l$ denote, respectively, a state of the object and a control. Let the evolution of the state of the object be described by the differential equation

$$\dot{x} = f(x, u, t), \quad x(t_0) = x_0, \tag{1}$$

where f is a continuous function with respect to all arguments and has continuous derivatives with respect to the components of the vector u. The control aim is expressed in terms of a functional Φ_t given on the trajectories of (1). Here we consider two types of such functionals:

(α) local functional

$$\Phi_t = L(x(t), t);$$

(β) integral functional

$$\Phi_t = \int_0^t \mathcal{J}(x(s), u(s), s)ds,$$

where $L(x, t)$ and $\mathcal{J}(x, u, t)$ are non-negative continuous functions and the latter has continuous derivatives with respect to the components of the vector u. Fixing

---

[c]In definitions of different types of stability for equations with an afteraction there is *an initial function* φ given on the interval $[-\eta, 0]$ in contrast with the case of ordinary differential equations when there is a fixed initial *point* x_0.

the control u, we find the derivatives of these functionals with respect to Eq. (1). In the case (α) it is equal to

$$\dot{\Phi}_t = f^T(x, u, t)\nabla_x L(x, t) + \frac{\partial}{\partial t}L(x, t)$$

and in the case (β) it is equal to

$$\dot{\Phi}_t = \mathcal{J}(x, u, t).$$

For short, in both cases we use the notation $\dot{\Phi}_t = \varphi(x(t), u(t), t)$ not to distinguish the types of functionals considered.

We define the control in the form of a differential equation

$$\dot{u} = -\Gamma\nabla_u\varphi(x, u, t) \tag{2}$$

with $\Gamma(= \Gamma^T > 0)$ a positive definite $n \times n$-matrix. Sometimes it is convenient to take $\Gamma = \gamma I$, $\gamma > 0$. The procedure of control (2) is called the *steepest descent method* (SDM for short) because in the case $\Gamma = \gamma I$ the vector u is moved in the direction of the gradient of the varying rate of the functional $\dot{\Phi}$.

Let us consider the properties of the functional Φ_t given on the paths $(x(t), u(t))$ in the space R^{n+l}, provided the SDM is used. Initially, this is not connected with adaptive problems, but later on the obtained result will be treated as a control aim secured by the SDM. The above breaks the habitual order of discussion when the control aim had been specified before the appropriate strategy was sought. Here, we study the possibility of using a strategy after it has been constructed.

Theorem 1. *Let the following conditions hold:*

(1) *there exists u_* such that $\varphi(x, u_*, t) \leq 0, \forall(x, t)$;*
(2) *$\varphi(x, u, t)$ is a convex function in u, i.e. for any u', u'', x, t the inequality*

$$\varphi(x, u', t) - \varphi(x, u'', t) \geq (u' - u'')\nabla_u\varphi(x, u'', t)$$

is satisfied. Then along any trajectory of (1) and (2) the following inequality

$$\Phi_t \leq \Phi_0 + 2^{-1}[u(t) - u_*]^T\Gamma^{-1}[u(t) - u_*] \tag{3}$$

holds.

Proof. We use the following Lyapunov function

$$V_t = \Phi_t + 2^{-1}[u(t) - u_*]^T\Gamma^{-1}[u(t) - u_*].$$

Its derivative, with respect to (1), is equal to

$$\dot{V} = \varphi(x, u, t) + [u(t) - u_*]^T\Gamma^{-1}\dot{u}(t).$$

Taking into account the assumptions we obtain the following inequality

$$\dot{V} \leq [u(t) - u_*]^T\nabla_u\varphi(x, u, t) + \varphi(x, u_*, t) + [u(t) - u_*]^T\Gamma^{-1}\dot{u}$$
$$= [u(t) - u_*]^T(\nabla_u\varphi + \Gamma^{-1}\dot{u}) + \varphi(x, u_*, t) \leq 0$$

which leads to the following

$$\Phi_t \le V_t \le V_0 = \Phi_0 + 2^{-1}[u(0) - u_*]^T \Gamma^{-1}[u(0) - u_*]. \qquad \square$$

Corollary 1. *If the assumptions of Theorem 1 hold and the local functional $\Phi_t = L(x(t), t)$ satisfies the condition*

$$\lim_{\|x\| \to \infty} \inf_{t \ge 0} L(x(t), t) = \infty, \tag{4}$$

then the paths of (1) and (2) are bounded.

Indeed, the point $(x(t), u(t))$ belongs to the domain defined by the inequality $V(x, u, t) \le V(x(0), u(0), 0)$. This domain is bounded uniformly in t.

Corollary 2. *Let the assumptions of Theorem 1 be satisfied. For the integral functional, the equality*

$$\lim_{t \to \infty} t^{-1} \int_0^t J(x(s), u(s), s)ds = \min_{u,t} t^{-1} \int_0^t J(x(s), u(s), s)ds = 0$$

holds.

This corollary follows from (3).

The next lemma will be used later on.

Lemma 1. *Let $y(t)$ be a bounded solution of the equation $\dot{y} = F(y, t)$, where $\|F(y, t)\| \le \alpha(r) < \infty$ for $y \in S_r = \{(y, t) : \|y\| \le r, \ t \ge 0\}$, and let the function $L(y, t)$ be uniformly continuous in any domain S_r so that $\int_0^\infty L(y(s), s)ds < \infty$. Then, $\lim_{t \to \infty} L(y(t), t = 0$.*

This fact follows from the Barbalat Lemma.[d]

Theorem 2. *Let the following conditions hold:*

(1) *for every $r > 0$*

$$\sup_{\|z\| \le r, t \ge 0} (\|f(z, t)\| + \|\nabla_u \varphi(z, t)\|) < \infty, \qquad z = (x, u);$$

(2) *the local functional $\Phi_t = L(x(t), t)$ is convex and uniformly continuous in (x, t) in every domain S_r, the equality (4) and inequality*

$$\varphi(x, u_*, t) \le -\delta L(x, t), \qquad \delta > 0 \tag{5}$$

holds. Then $\lim_{t \to \infty} L(x(t), t) = 0$.

Proof. Let the Lyapunov function be the same as in Theorem 1. Its derivative, with respect to Eq. (1), satisfies the inequality

$$\dot{V} \le -\delta L(x(t), t) \le 0.$$

---

[d]If $\int_0^\infty g(t)dt < \infty$ and $g(t)$ is uniformly continuous for $t \ge 0$, then $\lim_{t \to \infty} g(t) = 0$.

Hence $\delta \int_o^t L(x(s), s)ds \leq V_0 - V_t$ and $\int_0^\infty L(x(s), s)ds \leq \delta^{-1}V_0$. The assertion follows from the boundedness of the solution of the system (1) and (2) which has been proved before. $\qquad\square$

Corollary 3. *If condition (2) holds for the integral functional then* $\lim_{t\to\infty} \mathcal{J}(x(t), u(t), t) = 0.$

Let us examine the sensitivity of the SDM to the disturbances of the equation, i.e. we consider the following equation

$$\dot{x} = f(x, u, t) + h(x) \tag{6}$$

with $h(x)$ treated as an unobserved noise and satisfying the condition $\|h(x)\| \leq \chi < \infty$. The SDM is unstable and, moreover, the control $u(t)$ can increase unlimitedly in time. Therefore, it is desirable to modify this method so that it would be insensitive (robust) to, at least, small disturbances. This can be done by regularizing the functional in question, namely,

$$\hat{\Phi}_t = \Phi_t + \frac{\lambda}{2} \int_0^t \|u(s)\|^2 ds, \quad \lambda > 0.$$

Acting as before we obtain the regularized procedure

$$\dot{u} = -\Gamma(\nabla_u \varphi(x, u, t) + \lambda u) \tag{7}$$

containing negative feedback.

We consider the disturbed system (6) and (7) and study its stabilizing properties.

Theorem 3. *Let the functional* Φ_t *be locally convex and satisfy the conditions:*

$$\lim_{\|x\|\to\infty} \inf_{t>0} L(x, t) = \infty,$$

$$\|\nabla_x L(x, t)\|^2 \leq \alpha[1 + L(x, t)], \quad \alpha > 0,$$

$$\varphi(x, u_*, t) \leq -\beta L(x, t) + \gamma, \quad \beta, \gamma > 0$$

with some u_*. *Then the system* (6) *and* (7) *is dissipative.*

Proof. We choose the Lyapunov function shown above. Its derivative, with respect to (6) and (7), satisfies the inequality

$$\dot{V}_t \leq \varphi(x, u_*, t) + \nabla_x L(x, t)^T h(x) - \lambda(u(t) - u_*)u(t)$$

with $\varphi(x, u, t) = f^T \nabla_x L + \partial L/\partial t$. Making use the following simple inequalities

$$\nabla_x L^T h \leq 2a\|\nabla_x L\|^2 + \frac{1}{2a}\|h\|^2,$$

$$-(u - u_*)^T u \leq \frac{b}{2}(u - u_*)^T \Gamma^{-1}(u - u_*) + \frac{1}{2}\|u_*\|^2$$

where $a = \beta/\alpha > 0$ and b is the minimal eigenvalue of the matrix Γ, and the second and third conditions of the theorem, we arrive at the inequality

$$\dot{V}_t \leq -\frac{\beta}{2}L(x,t) - \frac{b\lambda}{2}(u - u_*)^T\Gamma^{-1}(u - u_*) + d \leq -cV_t + d$$

where

$$c = \min\left(\frac{\beta}{2}, \frac{\lambda b}{2}\right), \quad d = \gamma + \frac{\beta}{2} + \frac{\alpha\chi}{2\beta} + \frac{\lambda}{2}\|u_*\|^2.$$

From this the inequality $\lim_{t\to\infty} V_t \leq d/c$ follows. By the first condition of the theorem we obtain the required assertion. $\qquad\square$

We would like to notice that the constraints on the level of the disturbance $h(t)$ are absent. In consequence of this procedure (7) provides dissipativity for any bounded disturbance. The diameter of the limiting "sphere" increases unboundedly together with the level χ but it does not depend on the initial values of the system (6), (7).

The above can be extended to the case of random noises. Two variants of such a noise can be considered, namely,

(1) its mean is bounded, i.e. $\mathbf{E}\|h(t)\|^2 \leq \chi < \infty$;
(2) it is the white noise with bounded intensity.

For both cases the analogue of Theorem 3 holds, i.e. for any initial condition

$$\varlimsup_{t\to\infty} \mathbf{E}(\|x(t)^2 + \|u(t)\|^2) \leq d < \infty,$$

i.e. dissipativity in the mean square sense takes place.

The general results stated above allow solving many concrete problems of adaptive stabilization. We only mention two problems.

First, we turn once more to the control problem with a model reference. We examine its solution by the SDM in comparison with the solution obtained by the Lyapunov method in Sec. 3. As before (see Sec. 3) the model reference is defined by the equation

$$\dot{x}_M = Ax_M + Bu$$

with a stable matrix A and a bounded and continuous control u. The controlled object is represented by the equation

$$\dot{x} = (A_0 + U(t))x + (B_0 + W(t))u. \tag{8}$$

There are no parametric disturbances but the matrix $Z = (U(t), W(t))$ serves for the control. The quantities x, x_M, u are observed. The control aim is to secure the fulfilment of the condition $\lim_{t\to\infty} e(t) = 0$, $e = x - x_M$ or, in the form of a local functional, to minimize (in the limit as $t \to \infty$) the functional $\Phi_t = e^T(t)He(t)$, $H = H^T > 0$. According to the SDM we should first find the varying rate in time of the functional

$$\dot{\Phi}_t = \varphi(x, Z, t) = 2e^T H[(A_0 + U)x + (B_0 + W)u - Ax_M - Bu].$$

Further, we compute the derivatives φ with respect to the matrices U and W. These derivatives are some matrices composed of the derivatives $\partial/\partial \varkappa_{ij}$, where $\varkappa_{ij}$ are the elements of the matrices U and W. We obtain

$$\nabla_U \varphi = Hex_T, \quad \nabla_W \varphi = Heu_T.$$

For our problem the SDM has the following form (with $\Gamma = \gamma I$)

$$\dot{U} = -\gamma Hex_T, \quad \dot{W} = -\gamma Heu_T, \tag{9}$$

or, after integrating,

$$U(t) = U_0 - \gamma H \int_0^t e(s)x_T(s)ds, \quad W(t) = W_0 - \gamma H \int_0^t e(s)u_T(s)ds. \tag{10}$$

It remains to verify that the assumptions of Theorems 1 and 2 hold. Being linear in Z, the function φ is convex. Having the stable matrix A we can find a matrix H from the condition $A_T H + HA < 0$ and put $Z_* = (A - A_0, B - B_0)$. Then

$$\varphi(x, Z_*, t) = e^T HAe \le -\lambda e^T He, \quad \lambda > 0.$$

If the external action u is bounded then $x_M(t)$ will be bounded as well. Hence $\inf_{t>0} \Phi_t \to 0$ as $t \to \infty$.

Finally from Theorems 1, 2 and Corollary 1, the boundedness of the trajectories of the system (8), (10) follow. In other words, we have the achievement of the desired aim at arbitrary unknown parameters of the system and arbitrary initial conditions.

Another problem is concerned with nonlinear controlled objects.

Let the evolution of the object be described by the equation

$$\dot{X} = Ax + k(x)bu$$

with u a scalar control, b a constant vector of length n, and a continuous function $k(x)$. It is assumed that the function $k(x)$ is measured at each moment of time and $k(x) \ge k > 0$. The quantities $z = L^T x \in R^m$, $m < n$ are observed. The matrices A, L and vector b are unknown. Let the control u be linear in z, i.e.

$$U = C^T L^T z$$

with $C = C(t) = (C_1(t), \ldots, C_m(t))^T$. It is required to define the control law C so that the given aim should be reached. The control aim is to implement the equality $\lim_{t\to\infty} x(t) = 0$. In terms of local functionals it can be stated as minimization (in the limit) of the functional $\Phi_t = x(t)^T Hx(t)$ with $H = H^T > 0$. According to the SDM we have to find the derivative of the functional, i.e.

$$\dot{\Phi}_t = \varphi(x, c) = x^T H(Ax + k(x)bC^T z).$$

Further we obtain

$$\nabla_C \varphi(x, C) = k(x)x^T Hbz.$$

The control must depend only on the observed variables. Therefore, we have to eliminate x from the previous equality. We assume that the equality $Hb = Lg$, where g is some vector, is fulfilled. Then the desired procedure of control based on the SDM can be written in the form

$$\dot{C}(t) = -k(x)g^T z \Gamma z$$

or, after integrating,

$$C(t) = C_0 - \int_0^t k(x(s))g^T z(s)\Gamma z(s)ds.$$

The obtained algorithm of control is again nonlinear, non-stationary and non-Markov.

To realize the control aim we have to verify the assumptions of Theorems 1 and 2. This is true for Theorem 1 and for the first condition of Theorem 2. It remains to provide the implementation of condition (5). For this purpose, we choose a matrix H from the condition $A_*^T H + H A_* < 0$, where $A_* = A + bC_*^T L^T$ for some vector C_*. Adding the equality $Hb = Lg$, we obtain the modified problem of constructing a Lyapunov function (see Sec. 1). According to Theorem 8 stated there, the transfer function $W(\lambda) = L^T(\lambda I - A)^{-1}b$ must be such that the numerator of the rational function $g^T W(\lambda)$ is a stable (Hurwitz) polynomial of degree $n-1$ with positive coefficients. We can give this polynomial in explicit form, i.e. it is equal to $\chi(\lambda)g^T W(\lambda)$, where $\chi(\lambda) = \det(\lambda I - A)$.

The stated condition describes the class of equations for which the given aim of control can be attained by using the SDM. This class is specified through the parameters of the control algorithm.

11.5. Stabilization of Solutions of Minimum Phase Equations

In this section we consider two adaptive stabilization problems of solutions of linear equations

$$\dot{x} = Ax + Bu + f, \quad z = L^T x, \quad x(0) = x_0 \tag{1}$$

where $x \in R^n$, $u \in R^l$, $z \in R^m$ are the solution (the state), the control and the observation, respectively, A, B and L are some constant matrices having the dimensions $n \times n$, $n \times l$ and $n \times m$, respectively ($n \geq \max(l, m)$), and $f = f(t)$ is an external disturbance. The following $m \times l$-matrix

$$W(\lambda) = L^T(\lambda I - A)^{-1}B$$

is the transfer function of (1). Now we define the class of equations considered.

The class $\mathcal{L}(l, m, G)$ is composed of all linear equations of the form (1), so that for each of them the matrices $\delta G^T W(\lambda)$ are strictly minimum phase with respect to the matrix G given beforehand by some positive diagonal $l \times m$-matrix $\delta = \text{diag}\{\delta_1, \ldots, \delta_l\} > 0$, (i.e. $\delta_j > 0$ for all j). The dimension n of the vector x may be unknown but the condition $n \geq \max(l, m)$ is assumed. The external action $f = f(t)$ is assumed only to be continuous in $t \geq 0$.

The desired control will be sought in the form of a linear feedback

$$U = C^T(t)z = (LC(t))^T x, \quad \dot{C} = F(z, C) \tag{2}$$

where $C(t)$ and $F(\cdot)$ are an unknown $m \times l$-matrix (the gain factor) and some continuous functional, respectively. Thus the original equation takes the form

$$\dot{x} = (A + BC^T(t)L^T)x + f, \quad z = L^T x, \quad x(0) = x_0. \tag{3}$$

Further, we consider only stabilizational aims of control. Our investigation is based on using the Direct Lyapunov Method. Hence, we have to construct the appropriate Lyapunov function $V(x, C)$. For linear equations it is represented by the following quadratic form

$$V(x, C) = x^T H x + \sum_{i=1}^{l} (C_i - C_i^0)^T H_i (C_i - C_i^0), \quad H = H^T > 0, \tag{4}$$
$$H_i = H_i^T > 0,$$

satisfying the conditions

$$V(x, C) = \begin{cases} > 0 & \text{for } x \neq 0, \quad C \neq C^0, \\ = 0 & \text{for } x = 0, \quad C = C^0 \end{cases} \tag{5}$$
$$\dot{V}(x, C) \leq 0,$$

where the derivative $\dot{V}$ should be computed with respect to Eqs. (1) and (2). Obviously, the function V from (4) obeys the first of these conditions. Before verifying the second condition from (4), we calculate the derivative $\dot{V}$, where the columns of the matrices B, C, C^0, F are denoted by b_i, C_i, C_i^0, F_i, respectively. We obtain

$$\dot{V} = 2x^T H \left(Ax + \sum_{i=1}^{l} b_i C_i^T z + f \right) + 2 \sum_{i=1}^{l} (C_i - C_i^0)^T H_i F_i.$$

Making use of the equality

$$x^T H B (C - C^0)^T z = \sum_{j=1}^{l} (x^T H b_j)(C_j - C_j^0)^T z$$

and the notation $A(C^0) = A + B(C^0)^T L^T$, we transform the right-hand side of the above equality. Hence

$$\dot{V} = x^T [H A(C^0) + A^T(C^0) - H]x + 2x^T H f + 2 \sum_{i=1}^{l} (C_i - C_i^0)^T [H_i F_i + (x^T H b_i)z]. \tag{6}$$

The fulfilment of the third condition in (5) is related to the linearity in C_i of the second term in the expression for $\dot{V}$. It is possible to overcome these difficulties for the concrete forms of the function F for every problem considered. The formula (6) will be used later on (in the Sec. 2 of the next chapter).

Passing to the problems of adaptive control, we assume that matrices A, B, L are unknown and the function $f(t)$ is unobserved.

The first aim of control usually called *dissipativenes* is the existence of the solution of (3) for all $t \geq 0$ which satisfies

$$\varlimsup_{t \to \infty} \left(\|x(t)\| + \|C(t)\| \right) \leq d < \infty$$

for arbitrary initial conditions $x(0)$, $C(0)$.

In connection with this aim we consider the class of equations $\mathcal{LO}(l, m; G)$ (a subclass of $\mathcal{L}(l, m; G)$) such that the external disturbance f is bounded, i.e. $\|f(t)\| \leq \varkappa$. Here, the constant $\varkappa$ may depend on the function f.

To define the required strategy we have to give the algorithm of calculating the matrix $C = C(t)$. Let $C = (C_1, \ldots, C_l)$, $G = (g_1, \ldots, g_l)$ be the column-wise representations of the $m \times l$-matrix C and $l \times m$-matrix G, respectively. Let the columns of the matrix C be defined by the equations

$$\dot{C}_i = -\alpha_i C_i - (g_i^T z)P_i z, \quad i = 1, \ldots, l \tag{7}$$

with the matrices $P_i = P_i^T > 0$ and numbers $\alpha_i > 0$. Explicit form of the columns of the tuned matrix $C(t)$ is given by the formula

$$C_i(t) = C_i(0)e^{-\alpha_i t} - \int_0^t e^{-\alpha_i(t-s)}(g_i^T z(s))P_i z(s)ds. \tag{8}$$

The control again proves to be nonlinear, non-stationary and non-Markov. Substituting (8) in (1) we see that the final form of Eq. (1) with control (8) activated is rather complicated.

The constructed strategy is of the *direct* type. In designing it the identification is not used and, moreover, there are no estimations of parameters of the equation. The structure and the numerical coefficients of the rules of calculating the controls are known from the very beginning and not corrected during the process.

Theorem 1. *The strategy* (2), (7) *ensures dissipativity of the solutions of equations from the class* $\mathcal{LO}(l, m; G)$.

Proof. Let us construct a Lyapunov function for the system (3), (7) and show that it satisfies the inequality $\dot{V} \leq -aV + b(a, b > 0)$. Then from Theorem 4, Sec. 1 the desired assertion follows. We choose the Lyapunov function in the form (4) and, in view of (6), we obtain

$$\dot{V} = x^T \left(A_0^T H + H A_0 \right)x + 2x^T H f$$

$$+ 2\sum_{i=1}^{l} \left(C_i - C_i^0 \right)^T \left[H_i \{ -\left(g_i^T z \right)P_i z - \alpha_i C_i \} + \left(x^T H b_i \right)z \right]$$

where $A_0 = A + B(C^0)^T L^T$. We put $H_i = P_i^{-1}$, $i = 1, \ldots, l$.

For the inequality

$$\dot{V}(x, C) \leq 0 \quad \text{when } x \neq 0, \ f \equiv 0,$$

to be satisfied (we already have $V(x, C) > 0$, $x \neq 0, C \neq C^0$) it is necessary that the matrix H should be such that

$$HA_0 + A_0^T H = -Q < 0, \quad HB = LG.$$

These restrictions have been met in Theorem 8 from Sec. 1. The assumptions of that theorem coincide with those of this theorem. Hence the required matrices H and C^0 exist and for $\dot{V}$ we have

$$\dot{V} = -x^T Q x + 2x^T H f + 2 \sum_{i=1}^{l} \left(C_i - C_i^0\right)^T H_i(-\alpha_i C_i).$$

For the terms on the right-hand side of this equality the following inequalities hold

$$2x^T H f \leq \lambda x^T H x + \frac{1}{\lambda} f^T H f, \quad \forall \lambda > 0,$$

$$-2\alpha_j \left(C_j - C_j^0\right)^T H_j C_j \leq -\mu \left(C_j - C_j^0\right)^T H_j \left(C_j - C_j^0\right) + \frac{\alpha_j^2}{2\alpha_j - \mu_j} \left(C_j^0\right)^T H_j C_j^0$$

$$\forall \mu \in (0, 2\alpha_j), \quad j = 1, \ldots, l.$$

Putting here $\mu_i = \mu \in (0, 2 \min \alpha_i)$ and, moreover, choosing λ and μ so that $(\lambda + \mu)H \leq Q$, we substitute the estimates obtained in the expression for $\dot{V}$. This leads to the desired inequality $\dot{V} \leq -aV + b$, with

$$a = \mu, \quad b = \frac{\|H\|}{\lambda} \chi^2 + \frac{\alpha^2}{2\alpha - \mu} \sum_{i=1}^{l} (C_i^0)^T H_j C_j^0, \quad \alpha = \max \alpha_j$$

which completes the proof. $\hfill \square$

This result allows consideration for the adaptive stabilization problem for linear scalar equations of order n given by the "input-output" relationship ($p = d/dt$)

$$A(p)x = B(p)u + f$$

where $A(p) = p^n + a_{n-1}p^{n-1} + \cdots + a_0$, $B(p) = b_k p^k + \cdots + b_1 p + b_0$ $(k < n)$.

The control is sought in the form

$$u = c(p, t)x = \sum_{i=0}^{l-1} c_i(t)p^i x$$

which means that the vector $z = (x, \dot{x}, \ldots, x^{(l-1)})^T$ is observed. The transfer function (from u to z) is represented by the vector

$$W(\lambda) = \frac{B(\lambda)}{A(\lambda)}(1, \lambda, \ldots, \lambda^{l-1})^T.$$

If $l - 1 + k \leq n - 1$ or $l + k \leq n$ then the power of the nominator will be less than that of the denominator and we can reduce our equation to the canonical form

$$\dot{z} = Az + bu + h$$

where the $n \times n$-matrix A and the n-vector b have the following forms

$$
A = \begin{pmatrix}
0 & 1 & 0 & \ldots & 0 \\
0 & 0 & 1 & \ldots & 0 \\
\vdots & \vdots & \vdots & \ldots & \vdots \\
0 & 0 & 0 & \ldots & 1 \\
-a_0 & -a_1 & -a_2 & \ldots & -a_{n-1}
\end{pmatrix}, \quad
b = \begin{pmatrix}
0 \\ 0 \\ \vdots \\ 0 \\ 1
\end{pmatrix}.
$$

We now consider the following question. For the above equations, what does it mean to be strictly minimum phase? For an equation to have this property it is necessary that the polynomial $g^T \alpha(\lambda)$, where $\alpha(\lambda)$ is the denominator of the transfer function, should have the degree $n - 1$, positive coefficients and be Hurwitz. Since in our case

$$
\alpha(\lambda) = B(\lambda)(1, \lambda, \ldots, \lambda^{l-1})^T
$$

we obtain

$$
g^T \alpha(\lambda) = B(\lambda) \sum_{i=1}^{l-1} g_i \lambda^i = B(\lambda)G(\lambda).
$$

Hence this polynomial is stable if and only if $B(\lambda)$ and $G(\lambda)$ are stable simultaneously. The polynomial $B(\lambda)$, in turn, is stable if and only if the equation is strictly minimum phase.

In this version of the dissipativity problem the algorithm of control, i.e. the strategy has the form

$$
u = c(p, t)z, \quad \dot{C} = -G(p)zPz - \alpha C.
$$

Here $G(p)$ is an arbitrary stable polynomial, $P = P^T > 0$, $\alpha > 0$. The dissipativity of the solutions is guaranteed with respect to the class of strictly minimum phase equations which satisfy the additional conditions: $k + l = n$, the values of the polynomials $B(\lambda)$ and $G(\lambda)$ take the same sign for all λ and the external disturbance $f(t)$ is bounded. Information about the coefficients of the equation is not necessary to construct the control algorithm.

We now focus our attention on the second problem. The control aim called *stabilization* consists of the existence of the solution of (3) for all $t \geq 0$ which satisfies the conditions

$$
\lim_{t \to \infty} x(t) = 0, \quad \overline{\lim_{t \to \infty}} C(t) < \infty
$$

for arbitrary initial $(x(0), C(0))$. In connection with this aim the class of equations $\mathcal{LN}(l, m; G)$ (a subclass of $\mathcal{L}(l, m; G)$) is considered. The external disturbance f is supposed to be *damping* in the integral sense, i.e. $\int_0^\infty \|f(t)\|^2 dt < \infty$ (from this damping in the common sense, i.e. $\lim_{t \to \infty} f(t) = 0$ follows only under additional assumptions, for example, $f(t)$ is uniformly continuous for all $t \geq 0$).

To describe the strategy in accordance with definition (2) we have to specify the function F. The calculation rule for the tuned matrix $C = C(t)$ is defined as follows

$$\dot{C}_i = -\left(g_i^T z\right) P_i z, \quad P = P^T > 0 \quad \alpha_i > 0, \quad i = 1,\dots,l \tag{9}$$

where C_i and g_i are the columns of the matrices $C = (C_1,\dots,C_l)$ and $G = (g_1,\dots,g_l)$, respectively. From (9) it follows that

$$C_i(t) = C_i^0 - \int_0^t (g_i^T z(s)) P_i z(s) ds, \quad \alpha_i > 0, \quad i = 1,\dots,l. \tag{10}$$

To demonstrate the fact that the proposed algorithm guarantees adaptability with respect to the considered class (i.e. it secures stabilization with respect to this class) we have to find a Lyapunov function of the form (4) having properties (5).

Theorem 2. *We assume that* $\mathbf{rank}\, B = l$ *for all equations from the class* $\mathcal{K}$*. A Lyapunov function of the form* (4) *having properties* (5) *for the system of Eqs.* (1) *and* (2) *from the class* $\mathcal{K}$ *exists if and only if the algorithm of tuning the matrix* $C(t)$ *can be described by Eq.* (9) *with some* $m \times m$*-matrix* $P = P^T > 0$ *where the matrix* $G = (g_1,\dots,g_l)$ *is such that* $\mathcal{K} \subset \mathcal{LN}(l,m;G)$*.*

Proof. Let us assume that for any equation from $\mathcal{K}$ there exist matrices $H, H_1, \dots, H_l, C^0$ such that the Lyapunov function (4) has the properties (5). Then the matrices H, (H_i) are symmetric and positive definite. Supposing $f \equiv 0$ (see (6)) we calculate the derivative

$$\dot{V} = x^T \left(A_0^T H + H A_0\right) x + 2 \sum_{i=1}^{L} \left(C_i - C_i^0\right)^T \left[H_i F_i + \left(x^T H b_i\right) z\right].$$

For conditions (5) to be satisfied it is necessary that the following equalities

$$\left.\begin{aligned} A_0^T H + H A_0 &= -Q < 0, \\ H_i F_i(z) + (x^T H b_i) z &\equiv 0, \quad i = 1,\dots,l \end{aligned}\right\} \tag{11}$$

take place. The second of these conditions appeared due to linearity of the right-hand side of the expression for $\dot{V}$ with respect to C.

We assume that $W(\lambda) \not\equiv 0$. Then there is an x such that $z = L^T x = \mathrm{const} \neq 0$ and, in view of the second equality in (11), we obtain $x^T H b_i = \mathrm{const}$. This means, in turn, that there exist vectors g_i such that $H b_i = L g_i$, $i = 1,\dots,l$. Appealing again to the second equality in (11) and putting $P_i = H_i^{-1}$ we see that

$$F_i(z) = -\left(x^T L g_i\right) H_i^{-1} z = -(g^T z) P_i z.$$

Thus, the required derivative is defined (when $f \equiv 0$) by the formula

$$\dot{V}(x,C) = -x^T Q x.$$

Here we have applied the first equality in (11). Let us write the equality $H b_i = L g_i$ in the unified form $HB = LG$. Using Theorem 9 from Sec. 1 we conclude that the conditions of the theorem being proved are necessary.

We now focus attention on the proof of sufficiency. According to Theorem 9 from Sec. 1 there exist matrices C^0 and $H = H^T > 0$ such that the first equality in (11) and the equality $HB = LG\delta$ takes place. The algorithm of tuning the matrix C has the form $\dot{C}_i^T = -(g_i^T z)P_i z$ and a Lyapunov function can be defined by equality (4) with $H_i = \delta_i P_i^{-1}$. Since $H_i > 0$ and the second equality in (11) takes place and the mentioned function satisfies conditions (5). $\qquad\square$

Using the result above, let us consider adaptive properties of the control algorithm (2), (9) we are interested in.

Theorem 3. *The strategy* (2), (9) *ensures stabilization of solutions of equations from the class* $\mathcal{LN}(l, m; G)$.

Proof. Due to the previous theorem, for every equation from $\mathcal{LN}(l, m; G)$ there exist matrices C^0, H, (H_i) such that the Lyapunov function $V(x, C)$ having properties (5) has the form (4). Its derivative, with respect to the equation, is given by the formula

$$\dot{V}(x, C) = -x^T Q x + x^T H f.$$

Hence there are $\lambda > 0$, $\mu > 0$ such that

$$\dot{V}(x(t), C(t)) \leq -\lambda\|x(t)\|^2 + \mu\|x(t)\|\|f(t)\|.$$

Integrating this inequality from 0 to t we obtain

$$\mu I_t^2 - \lambda I_t N - V(x(0), C(0) \leq -V(x(t), C(t)) \leq 0 \qquad (12)$$

where $I_t = \int_0^t \|x(s)\|^2 ds$ and $N^2 = \int_0^\infty \|f(s)\|^2 ds$. Thus

$$I_t \leq \frac{\mu}{2\lambda} N + \sqrt{\frac{\mu^2}{4\lambda^2} N^2 + \frac{1}{\lambda} V(x(0), C(0))}.$$

It signifies the boundedness of $\int_0^t \|x(s)\|^2 ds$. It follows that $\int_0^\infty \|x(s)\|^2 ds < \infty$. Moreover, equality (12) leads to the inequality $V(x(t), C(t)) \leq V(x(0), C(0)) + \mu IN$. Therefore the solution of the system (1), (2) being bounded can be extended to the half-axis $(0, \infty)$.

On the right-hand side of (10) some quadratic forms of the components of the vector $x(t)$ appear. As seen from the above this implies the existence and finiteness of $\lim_{t\to\infty} C(t)$. According to Eq. (1) the equality

$$\|x(t)\|^2 = \|x(0)\|^2 + 2\int_0^t x^T(s)[Ax(s) + BC^T(s)L^T x(s) + f(s)]ds$$

takes place for all $t \geq 0$. From the established properties of $C(s)$ and $x(s)$ it follows that the integral in the last formula converges and, hence, $\lim_{t\to\infty} \|x(t)\|^2 = \omega \geq 0$. It remains to note that $\omega = 0$ because $\int_0^\infty \|x(s)\|^2 ds < \infty$. $\qquad\square$

The deviation of $x(t)$ and $C(t)$ from their limiting values can be estimated by using the inequality

$$V(x(t), C(t)) \le V(x(0), C(0)) + \frac{\mu}{4\lambda} N^2$$

which, in turn, follows from (12).

Let us consider once more the scalar equation of order n:

$$A(p)x = B(p)u + f. \tag{13}$$

By analogy with the previous arguments (given after Theorem 1), we conclude that in the case of a damping disturbance f, the considered equation must be strictly minimum phase, the values of the polynomials $B(\lambda)$, $G(\lambda)$ must have the same sign for all λ and the number of observed derivatives $(l - 1)$ must satisfy the inequality $l + k \le n$. Moreover, the polynomial $G(\lambda)$ enters into the description of the tuning algorithm. Indeed,

$$\dot{C}_i = -\gamma_i G(p) x p^i x - \alpha C_i, \quad i = 0, \ldots, l - 1.$$

We note the case when the class of equations can be described without difficulties. We have in mind the case of the vector equation (13) of order n for which the equalities $\dim x = \dim u = n$ hold and the matrix polynomials $A(p)$, $B(p)$ have the form

$$A(p) = p^n I + \sum_{i=0}^{n-1} A_i p^i, \quad B(p) = \sum_{i=0}^{k-1} B_i p^i.$$

The transfer matrix of the equation $A(p)x = B(p)u$ is given by

$$W(\lambda) = A^{-1}(\lambda) B(\lambda).$$

According to the condition stated above we see that

$$\det\left(\delta G^T W(\lambda)\right) = \det \delta \cdot \det G \cdot \det B(\lambda) \cdot \det A^{-1}(\lambda),$$

and the minimum phase property we are interested in signifies that $\det B(\lambda)$ is a Hurwitz polynomial. According to the equality $\lim_{\lambda \to \infty} \lambda W(\lambda) = B_{n-1}$, we conclude that the matrix $\delta G^T B_{n-1}$ must be symmetric and positive definite. If B_{n-1} is a diagonal matrix, i.e. $B_{n-1} = \mathbf{diag}\,(b_1, \ldots, b_k)$ then we can put $\delta = I$ and $G = \mathbf{diag}\,(g_1, \ldots, g_k)$. It means that we have to know only the signs of the elements of the matrix B_{n-1}.

We would like to emphasize once more that the adaptive stabilizing strategies studied above do not require to know the dimension of the state space for the controlled equations.

We consider the quadratic form $Q(x, u) = x^T P x + u^T R u$ with some non-negative definite matrices P and R and set up the linear-quadratic problem that consists of minimizing the functional

$$\Phi(x, u) = \lim_{t \to \infty} t^{-1} \int_0^t Q(x(s), u(s)) ds.$$

Theorem 4. *The control* $(2),(10)$ *gives the global minimum of the functional* $\Phi(x,u)$, *i.e.* $\min \Phi(x,u) = 0$ *for all equations from the class* $\mathcal{LN}(l,m,G)$.

Proof. By Theorem 3 the equality

$$\lim_{t \to \infty} Q(x(t), u(t)) = 0$$

holds. From this the required assertion follows immediately. $\qquad\square$

11.6. Stabilization of Minimum Phase Equations with Nonlinearities

Linear equations considered above are the simplest mathematical models of controlled objects. However, in theory and practice of automatic regulating we are forced to deal with more complicated (nonlinear) equations. In this section we are interested in controlling equations whose descriptions include two parts: a linear part with a transfer function $W(\lambda)$ and a nonlinearity with a characteristic $\Psi(\cdot)$. Within the framework of the adaptive concept, the structure of the equation is only known but not the transfer function. As concerns the nonlinearity, it is assumed that only the qualitative characteristics of the function Ψ are known. We now discuss briefly the stabilization problems for such models using the Direct Lyapunov Method introduced earlier.

We are concerned with equations which contain a nonlinearity in the "direct chain" or in the "chain of feedback". For these cases the structures of the closed loop systems are shown schematically in Figs. 4(a) and 4(b) where the notation "AR" means an "adaptive regulator" and the other notation should be clear (but will be explained below).

The above can be expressed analytically as follows

$$\dot{x} = Ax + Bv + f, \quad z = L^T x, \tag{1}$$

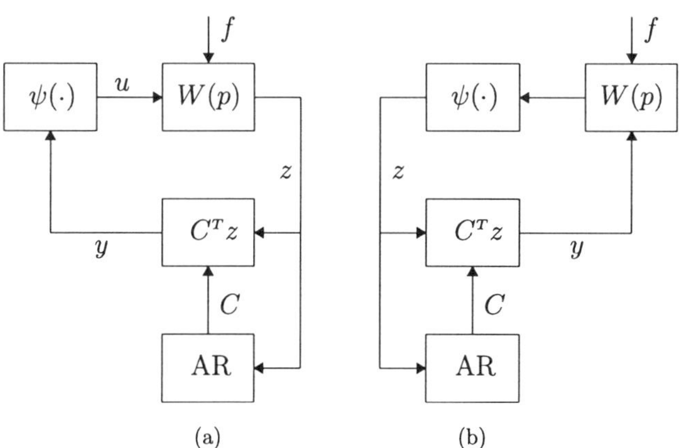

Fig. 4.

where $x \in \mathbb{R}^n$ is a state, $v = v(x, u, t) \in \mathbb{R}^m$ is the vector obtained by the nonlinear transformation of a control $u \in \mathbb{R}^l$, the variable $z \in \mathbb{R}^m$ is an observation and f is an external disturbance. The control u has the usual form

$$u = C^T(t)z = (LC(t))^T x, \quad \dot{C} = F(z, C) \tag{2}$$

where $C(t)$ is the matrix of tuned coefficients.

First, we consider equations containing a nonlinearity in the "direct chain", i.e. $v = \psi(u, t)$. The admissible types of nonlinearities are considered below. The function $\psi(y, t)$ at the moment t is assumed to depend only on the control at time t, i.e. $\psi(u(t), t)$ is an inertia-free function. This function need not be continuous in u. We assume, however, that its discontinuities (if they exist) are isolated and the function ψ is stationary (i.e. $\psi(u, t) \equiv \psi(u)$). In this case the solution might not exist in the usual sense. Here the "solution" signifies the generalized solution in the Filipov sense.

Definition 1. A scalar function $\psi(u, t)$ satisfies the *condition of linear growth* if there exist numbers $h \neq 0$ and $\lambda > 0$ such that

$$|\Psi(u, t) - hu| \leq \lambda \quad \forall u, t \geq 0.$$

A vector-function $\psi(u, t) = (\psi_1(u_1, t), \ldots, \psi_l(u_l, t))$ satisfies the linear growth condition if the above is true with respect to the functions ψ_i for some numbers h_i, λ_i.

Let $\mathcal{NLO}(l, m; G)$ denote a class of equations of the form (1) having the following properties:

(1) the disturbance f is bounded ($\|f\| \leq \chi$);
(2) the nonlinearity ψ satisfies the linear growth condition;
(3) the numerator of the rational function $\det G^T W(\lambda)$ is a stable polynomial (here $G = (g_1, \ldots, g_l)$ is some $m \times l$-matrix);
(4) $G^T L^T B = \mathbf{diag}\,(\delta_1, \ldots, \delta_l)$, $\mathbf{sign}\,\delta_j = \mathbf{sign}\,h_j$.

The dimension of the state vector $x(t)$ can be unknown.

The dissipativity is the aim of control. The algorithm of calculating the matrix $C(t) = (C_1(t), \ldots, C_l(t))$ is as follows

$$\dot{C}_j = -(g_j^T z)P_j z - \alpha_i C_j, \quad j = 1, \ldots, l \tag{3}$$

where $P_j = P_j^T > 0$, $\alpha_i > 0$ and g_i are the columns of the matrix G.

Theorem 1. *The control algorithm* (2),(3) *secures dissipativity of solutions of equations from the class* $\mathcal{NLO}(l, m; G)$.

The proof of this result is based on the Lyapunov function of the form $V(x, C) = x^T H x + \sum_{i=1}^{l} \left(C_i - C_i^0\right)^T H_i\left(C_i - C_i^0\right)$ and Theorems 4 and 9 from Sec. 1. It is analogous to Theorem 1 from Sec. 5 and hence the proof is omitted.

As concerns the scalar equations of order n

$$A(p)x = B(p)v + f, \tag{4}$$

with a nonlinearity $v = \psi(u)$, the signs of the numbers h_i in the linear growth condition are required to be known.

According to the result proved, Eq. (4) must be strictly minimum phase and the equality $k + l = n$ should take place with $k = \deg B(p)$ and l equal the number of observed derivatives of $x(t)$.

We now consider equations of the form (1) with a nonlinearity in "the chain of inner feedback" (see Fig. 4(b)). This nonlinearity is expressed analytically as follows $v = u + \Psi(z, t)$. Here the vector-function Ψ depends on $s = Q^T z = (s_1, \ldots, s_l)$ where Q is a constant $m \times l$-matrix whose columns are denoted by q_j. This means that $s_j = q_j^T z$ and we should write $\Psi(z, t) = (\psi_1(s_1, t), \ldots, \psi_l(s_l, t))$. Let us introduce the class of equations $\mathcal{NE}(l, m; Q)$ having the following properties:

(1) the disturbance f is damping, i.e. $\int_0^\infty \|f(t)\|^2 dt < \infty$;
(2) the components of the vector Ψ satisfy the inequalities

$$s_i \Psi_i(s_i, t) \geq 0, \quad i = 1, \ldots, l;$$

(3) the matrix $-\delta Q W \lambda$ is strictly minimum phase for some diagonal matrix $\delta > 0$.

The stabilization (as stated in Sec. 5) is the control aim and the algorithm of tuning the matrix $C(t)$ is chosen in the form

$$\dot{C}_i = s_i P_i z, \quad i = 1, \ldots, l \tag{5}$$

with $P_i = P_i^T > 0$ again.

Theorem 2. *The control algorithm* (2), (5) *provides stabilization of solutions of equations from the class* $\mathcal{NE}(l, m; Q)$.

The proof is similar to that of Theorem 3 from Sec. 5.

The next result is a combination of Theorems 1 and 2.

Theorem 3. *The control algorithm consisting of* (2) *and the following algorithm of tuning* $\dot{C}_j = s_j P_j z - \alpha_j C_j$ *secures dissipativity of solutions of the equations*

$$\dot{x} = Ax + B(u + \Psi(s, t)) + f; \quad z = L^T x, \quad u = C^T(t)z,$$

where the disturbance f *is bounded, i.e.* $\|f(t)\| \leq \chi$, *the nonlinearities* $\Psi_j(s_i, t)$ *satisfy the conditions* $s_i \Psi(s_i, t) \geq 0$ *and the matrix* $-\delta Q W(\lambda)$ *is strictly minimum phase for some diagonal matrix* $\delta > 0$.

11.7. Stabilization of Linear Minimum Phase Equations in Hilbert Space

First, note that at every moment t the solutions of the equations considered here are not vectors as usual but elements of some function spaces.

In the Hilbert space H we consider a one-parameter family $S(t)$ of bounded operators. It is a *semi-group* if the equality $S(t)S(s) = S(t + s)$ holds for every t,

$s > 0$. The semi-group $S(t)$ will belong to the class C_0 (or C_0-semi-group for short) if it is strongly continuous at $t = 0$, i.e. for every $x \in H$

$$\lim_{t \downarrow 0} \| S(t)x - x \| = 0.$$

Such semi-groups are strongly continuous on the right for each $t > 0$. For norms of the operators of the semi-group considered the estimate $\| S(t) \| \leq e^{\alpha t}$ for $\alpha > 0$ takes place.

Definition 1. The *generator* (or the *infinitesimal operator*) of the semi-group $S(t)$ is the linear operator A defined on the set $D(A) \subset H$ by the equality (convergence in the norm)

$$Ax = \lim_{t \downarrow 0} \frac{S(t)x - x}{t}.$$

The generator need not be bounded but for any C_0-semi-group its generator A is closed and its domain $D(A)$ is dense in H. For $t > 0$ the equality $dS(t)/dt = AS(t)$ holds. In most cases generators are differential operators. The shift-semigroup $S(\tau)x(t) = x(t + \tau)$ in the space $L_2(-\infty, \infty)$ may be regarded as the simplest example of a C_0-semi-group. Then $A = d/dt$ and its domain is given as follows

$$D(A) = \left\{ x(t) : \frac{d}{dt}x(t) \in L_2(-\infty, \infty);\ x(t) \in L_2(-\infty, \infty) \right\}.$$

Very often the generator is represented by a differential operator of the second order in the partial derivatives.

An original semi-group can be expressed explicitly by using its generator. For the finite-dimensional Hilbert space $H = \mathrm{R}^n$ the equality $S(t) = \exp\{At\}$ is valid. Without going into details we only note that this equality holds in the general case.

Now, we deal with linear equations in a Hilbert space

$$\dot{x} = Ax + f(t), \quad x(0) = x_0 \tag{1}$$

where A is the generator of a C_0-semi-group. The solution of this equation is understood in the generalized (or "weak") sense, namely,

$$\frac{d}{dt}(x(t), y) = (x(t), A^*y) + (f(t), y), \quad \text{a.s. for } t \in [0, \infty),$$

$$\lim_{t \downarrow 0} x(t) = x_0$$

for every[e] $y \in D(A^*)$. If $x_0 \in D(A)$ and $f(t)$ is a continuously differentiable function and $f(0) \in D(A)$, then $x(t)$ will be a continuously differentiable solution in the usual sense.

The solution $x(t)$ considered as a function of time can be interpreted as a curve in the infinite-dimentional space H. We will need to integrate this function (with

---

[e] Here A^* means the operator conjugate to A.

respect to t). Hence, the function $x(t)$ must satisfy the following conditions:

(1) $x(t)$ is weakly measurable with respect to t, i.e. for every $y \in H$ the scalar product $(x(t), y)$ is measurable with respect to t;

(2) $\int_0^\infty \|x(t)\|^2 dt < \infty$.

In other words, if the solution of Eq. (1) satisfies these conditions, then it will belong to the Hilbert space $\mathcal{L}_2([0, \infty); H)$ of functions taking values from H.

By the remark at the end of Sec. 1, a solution of the equation $\dot{x} = Ax$, $x(0) = x_0$, where A is a generator of a C_0-semi-group has the semi-group property and can be expressed uniquely by the formula $x(t) = S(t)x_0$. Thus the solution of Eq. (1) has the form

$$x(t) = S(t)x_0 + \int_0^t S(t - \tau)f(\tau)d\tau.$$

Here $S(t)$ denotes the strongly continuous semi-group of operators in the space H whose representation is well-known in the case of a finite-dimensional space H.

Let the Hilbert spaces H_x, H_u and H_z be the spaces of states, of controls and of observations, respectively. We consider the linear equation

$$\dot{x} = Ax + Bu + f, \quad x(0) = x_0, \quad z = L^T x \tag{2}$$

where $x \in H_x$, $u \in H_u$, $z \in H_z$ and the operators A, B, L are defined as follows.[f] Let A be the generator of the C_0-semi-group, $B \in \mathcal{L}(H_u, H_x)$, $L^* \in \mathcal{L}(H_x, H_z)$. We study the stabilization and optimization problems for the above equations. Naturally, their solution is more complicated than that in the finite-dimensional case. It requires using the functional analysis approach. The importance of these equations consist in enabling to consider problems of control for equations with a delay, for partial derivative equations and others.

Definition 2. An equation (or a pair of operators (A, B)) is called *stabilizable* if there exists an operator $K \in \mathcal{L}(H_x, H_u)$ such that any solution $x(t)$ of the equation $\dot{x} = (A + BK)x$ belongs to the space $\mathcal{L}_2([0, \infty); H_x)$.

The meaning of this notion is the existence of a linear feedback $u(t) = Kx(t)$ that generates the "damping" solution of the given equation in the sense that its norm squared is integrable on all semi-infinite intervals. It is possible to prove that in this case the estimation

$$\|x(t)\| \leq \alpha e^{-\lambda t}\|x(0)\|, \quad \alpha > 0, \ \lambda > 0$$

is valid.

Let $x_{x_0, u}(\tau)$ signify the state reached at time τ under the control $u(t) \in \mathcal{L}_2([0, \infty); H)$ provided the initial state was x_0. The set of all states reached at

---

[f]The set of all bounded linear operators mapping a Hilbert space H_1 into a Hilbert space H_2 is denoted by $\mathcal{L}(H_1, H_2)$.

time t under admissible controls u is called the *set of τ-attainable states* and denoted by $\Omega(\tau)$. The sets $\Omega = \bigcup_{\tau>0} \Omega(\tau)$ is the set of all attainable states or the attainability set.

Definition 3. Equation (2) (or the pair (A, B)) is called *controllable* if the set Ω is dense in H_x or, which is the same, the domain of values $R(B)$ of the operator B lies inside the set $\bigcap_n D(A^n)$. Moreover, the set $\bigcup_{n=1}^\infty A^n R(B)$ is dense in H_x.

Definition 4. Equation (2) (or the pair (A, L)) is called *observable* if the pair (A^*, L) is controllable.

Definition 5. The triplet (A, B, L) is called *non-degenerate* (or minimal) if the pair (A, B) is controllable and the pair (A, L) is observable, simultaneously.

We now restrict attention to a particular case of Eq. (2). Namely, let the following equation

$$\dot{x} = Ax + bu + f, \quad z = L^*x \tag{3}$$

be given. Here, the control $u = u(t)$ is a scalar function, $b \in H$ is a constant vector and the disturbance $f = f(t)$ satisfies the conditions:

$$(1)\ f(0) \in D(A); \quad (2)\ f \in \mathcal{L}([0, \infty); H_x) \cap C^1([0, \infty); H_x). \tag{4}$$

The observations z are assumed to belong to the finite-dimentional space R^l. The operator L is bounded and $L : \mathrm{R}^l \to H_x$.

The *resolvent* of a linear operator A is the operator $R_\lambda(A) = (\lambda I - A)^{-1}$ defined for any complex number λ not belonging to the *spectrum* $\sigma(A)$ — the set of all eigenvalues of the operator A. It is a quite continuous operator. We assume that this spectrum contains only a finite number of points from the half-plain $\{\lambda : \mathrm{Re}\,\lambda > -\delta\}$ for some $\delta > 0$. Under this assumption the collection of operators A described above will include strongly elliptic differential operators provided the space H is a space $\mathcal{L}_2(G)$ of functions defined on a bounded set $G \subset \mathrm{R}^n$. In this case Eq. (3) represent parabolic equations. If the spectrum $\sigma(A) = \{\lambda : \mathrm{Re}\,\lambda < -\delta\}$, for some $\delta > 0$, then, naturally, the operator A is called a Hurwitz operator.

Definition 6. The vector valued function of the complex variable λ

$$W(\lambda) = L^*(\lambda I - A)^{-1} b = L^* R_\lambda(A) b$$

is called a *transfer function* of Eq. (3).

The components of $W(\lambda)$ are meromorphic functions and $W(\lambda)$ can be written in the form

$$W(\lambda) = \chi^{-1}(\lambda) \Psi(\lambda)$$

where $\chi(\lambda)$ is an entire function and $\Psi(\lambda)$ is a vector of entire functions with $\chi(\lambda)$ and $\Psi(\lambda)$ having no common zeros.

These notions enable us to formulate the minimum phase property for Eq. (3). It has a more complicated form than in the case of a finite-dimensional space.

Definition 7. Equation (3) is called *strictly minimum phase* if one of the following two conditions is satisfied:

(1) The triplet (A, b, L) is non-degenerate and

$$W(\lambda) \neq 0 \quad \text{for Re } \lambda \geq 0;$$

(2) The triplet (A, b, L) is degenerate but there exists a vector b' and an operator L' such that the triplet (A, b', L') is non-degenerate and for $\text{Re } \lambda \geq 0$ we have

$$\chi'(\lambda)W(\lambda) \neq 0$$

where $\chi'(\lambda)$ is the denominator of the function $W'(\lambda) = L'^*(\lambda I - A)^{-1}b' = \chi'(\lambda)^{-1}\psi'(\lambda)$.

Let $\mathcal{K}_{\varphi,g}$ denote the class of equations having the form (3) where the disturbance f satisfies condition (4), the function $g^T W(\lambda)$ is minimum phase for some vector g having order l and $g^T L^* b > 0$.[g]

The control is chosen in the form of a linear feedback $u = k^T(t)z$, $k \in \mathbb{R}^l$, where the vector of the tuned parameters $k(t)$ is defined by the equation $\dot{k} = h(z)$ or

$$k(t) = k_0 + \int_0^t h(z(s))ds.$$

The choice of the strategy for any control aim consists in specifying the function $h(\cdot)$ to reach the given aim.

The control aim is

to find a strategy such that

$$(1) \int_0^\infty \|x(t)\|^2 \, dt < \infty; \quad (2) \lim_{t \to \infty} k(t) < \infty \tag{5}$$

for any equation from $K_{\varphi,g}$ and any initial values x_0, k_0.

This aim is regarded as stabilization. The tuning procedure of the parameter $k(t)$ is similar to that used in Sec. 5:

$$\dot{k} = -(g^T z)Pz \tag{6}$$

where g is the parameter of the class $K_{\varphi,g}$ and $P = P^* > 0$. The appropriate strategy with a linear feedback is denoted by $\sigma(\Phi)_{g,P}$.

Theorem 1. *The strategy $\sigma(\Phi)_{g,P}$ guarantees the attainment of the stabilizational aim (5) with respect to the class $K_{\varphi,g}$.*

---

[g]Here $L^* b = a \in \mathbb{R}^l$ and, hence, $g^T L^* b = g^T a = (g, a)$, where $(\cdot, \cdot)$ means the scalar product in $\mathbb{R}^l$.

Proof. It is based again on the Lyapunov method. The verification of the existence of a function with the desired properties is based on the frequency theorem in an infinite-dimensional space. We will not state this theorem, but we state one of its corollaries leading to the desired result.

Let $A_\alpha = A - (\alpha b)d^*$, α and d being a number and a vector, respectively.

Lemma 1. *Let the triplet (A, b', L') be non-degenerate, $d^*b > 0$ and*

$$W(\lambda) \neq 0 \quad \text{for } \mathrm{Re}\lambda \geq 0$$

where

$$W(\lambda) = \chi'(\lambda)d^T R_\lambda(A)b, \quad W'(\lambda) = L'^T R_\lambda(A)b' = (\chi'(\lambda))^{-1}\Psi'\lambda.$$

Then the operator A_α has the properties stated after the description of Eq. (3) and there is a bounded operator $Q = Q^* > 0$, mapping H onto H and a number $\varepsilon > 0$ such that

$$Qb = Lg, \quad x^* H A_\alpha x \leq -\varepsilon\|x\|^2, \quad x \in D(A).$$

The proof of this lemma is omitted.

The operators A, b', L' (which enter into the definition of the minimum phase property for the function $g^T W(\lambda)$) and vectors b, $d = Lg$ satisfy the conditions of the lemma stated above.

A Lyapunov function is chosen as follows

$$V(x, k) = x^* Q x + (k - \alpha g)^T P^{-1}(k - \alpha g).$$

Its derivative, with respect to the closed loop system, is equal to

$$\dot{V}(x(t), k(t)) = 2x^* Q A_\alpha x + x^* Q f \leq -\delta\|x(t)\|^2 + \|Q\|\|x(t)\|\|f(t)\|.$$

Integrating from 0 to t we obtain

$$\varepsilon\mu_t^2 - \|Q\|\nu\mu_t - V(x(0), k(0)) \leq V(x(t), k(t)) \leq 0$$

where $\mu_t^2 = \int_0^T \|x(s)\|^2 ds$, $\nu^2 = \int_0^\infty \|f(s)\|^2 ds$. From this it follows that

$$\lim_{t \to \infty} \mu_t^2 = \int_0^\infty \|x(s)\|^2 ds.$$

In view of (5) we have

$$k(t) = k(0) - \int_0^t (g^T z(s))Pz(s)ds,$$

and since the right-hand side of the last equality is a quadratic function in $x(t)$, we draw the conclusion that $\lim_{t \to \infty} k(t) < \infty$. $\qquad\square$

A simple example demonstrating the application of the theory above is given below.

We consider a heated rod of finite length as a concrete example of a controlled object. The heat transmission in this rod is caused by two factors. On the one hand, it is the source of heat which either adds or consumes the heat at point y proportionally to the local temperature $T(y, t)$. On the other hand, it is a source of the controlled power $u(t)$ which either adds or consumes the heat uniformly with respect to the length of the rod. The flow of heat through the ends of the rod is assumed to be equal to zero.

The mathematical model of the considered controlled object (one-dimentional with respect to the space variables) is defined as follows

$$\begin{cases} \dfrac{\partial T}{\partial t} = \theta_1 \dfrac{\partial^2 T}{\partial^2 y} + \theta_0 T + \theta_2 u, \quad z(t) = \theta_3 \displaystyle\int_0^1 T(y, t)\, dy, \\[4mm] T(y, 0) = \varphi(y), \qquad\qquad\qquad \dfrac{\partial T}{\partial y}\bigg|_{y=0, y=1} = 0 \end{cases} \tag{7}$$

where θ_0, θ_1, θ_2, θ_3 are constants satisfying the restrictions

$$\theta_0 < \pi^2 \theta_1, \quad \theta_1 > 0, \quad \theta_2 \theta_3 > 0.$$

Such a collection of parameters defines a class of equations of the form (7). These equations, in particular, have a non-stable solution (at $\theta_0 > 0$). This gives an occasion to study the stabilization problem.

For equations having the form (7) the given problem can be reduced to the control problem for equations having the form (3) in the Hilbert space $H = L_2(0, 1)$ provided that

$$Ax = \theta_1 \frac{\partial^2 x}{\partial y^2} + \theta_0 x, \quad b = \theta_2, \quad L = \theta_3$$

$$D(a) = W^{2,2}[0, 1] \cap \left\{ x : \frac{\partial x(0)}{\partial y} = \frac{\partial x(1)}{\partial y} = 0 \right\}$$

where $W^{2,2}[0, 1]$ stands for the Sobolev space.[h] Here $l = 1$ and for all collections (θ_i) the operator A is regarded as one of the operators described above.

For Eq. (7) the control is determined as follows

$$u = k(t)z, \quad \dot{k}(z) = -z^2(t). \tag{8}$$

To prove that this control leads to aim (5) we have to verify the assumptions of Theorem 1.

By the second equality in (8), $g = 1$ and, hence, the condition $g^T L b > 0$ holds. It remains to verify that the function $W(\lambda)$ is strictly minimum phase. First, let $\theta_0 = 0$, $\theta_1 = \theta_2 = \theta_3 = 1$. In this case we see that the transfer function is defined

---

[h] The Sobolev space $W^{k,p}(G)$ consists of all functions $h(x) = h(x_1, \ldots, x_n)$ defined on some set $G \subset \mathbb{R}^n$ such that both they and their generalized derivatives $D^s h$ up to the order k belong to the space $\mathcal{L}_p(G)$. The set $W^{k,p}$ is the linear normed space with norm

$$\|h\| = \left(\sum_{|s| \le k} \int_G |D^s h(x)|^p dx \right)^{1/p}.$$

The spaces $W^{k,2}(G)$ are Hilbert spaces.

as follows

$$W(\lambda) = L^T(\lambda I - A)^{-1} b = \lambda^{-1}.$$

Basing on the criteria of controllability and observability, it is possible to show that the triplet (A, b', L') is non-degenerate provided

$$b' = L' = \sum_{m=0}^{\infty} 2^{-m} \cos \pi m y.$$

The denominator of the corresponding transfer function is equal to (without going into details)

$$\chi'(\lambda) = \sqrt{\lambda} sh\sqrt{\lambda}.$$

From this it follows that

$$\varphi(\lambda) = \chi'(\lambda) W(\lambda) = \frac{sh\sqrt{\lambda}}{\sqrt{\lambda}}.$$

Hence the numbers

$$\lambda_m = -(m\pi)^2, \quad m = 1, 2, \ldots$$

are the roots of this function.

For the other values of the parameters θ_i the function $\varphi(\lambda)$ takes the form $\varphi((\lambda - \theta_0)/\theta_1)$ and its roots are $\lambda_m = \theta_0 - (m\pi)^2 \theta_1$. This means that $\varphi(\lambda) \neq 0$ for $\mathrm{Re}\,\lambda \geq 0$ under the given restriction on θ_i. Therefore for the function $g^*W(\lambda)$ the required minimum phase property, which guarantees stabilization of the solutions of (7) for any initial conditions, is proved.

The control (8) points to a solution of the linear-quadratic problem with the heating-value equation (7). Also in this case the minimum of the functional is equal to zero, as in Theorem 4, from Sec. 5.

For Eq. (7) the dissipativity problem can be considered in a similar way.

11.8. Control of Stabilizable Equations

We continue to study adaptive control algorithms for linear differential equations having the form

$$\dot{x} = Ax + Bu + f, \quad x(t_0) = x_0 \tag{1}$$

where $x \in \mathbf{R}^n$, $u \in \mathbf{R}^l$, A and B are constant matrices of the corresponding dimensions. It is assumed that the solution $x(t)$ is observed but the external disturbance $f = f(t)$ is neither known nor observed. The function $f(t)$ is assumed to be continuous and to satisfy some restrictions to be stated later on.

For Eq. (1) the stability problems considered in this section are based on a unified control strategy. The design of this strategy is based on the following simple idea. It is necessary to find a linear feedback $u = kx$ under which the solutions of the equation $\dot{x} = (A + Bk)x$ are asymptotically stable. The trajectory $x(t)$ which is observed contains necessary information for finding the matrix k ("gain matrix"). We first use this information to estimate the matrices A and B. Let $\tilde{A}(t)$, $\tilde{B}(t)$ denote the values of these estimates at time t. Then the matrix $k(t)$ is obtained as the solution of the following equation

$$\tilde{A}(t) + \tilde{B}(t)k(t) = G$$

where G is a Hurwitz matrix with a given stability level. If this equation is solvable (at least for sufficiently large t) then typical stabilization problems will be solvable as well.

We now pass to the formal consideration without going into details about the technique of calculating $\tilde{A}(t)$, $\tilde{B}(t)$. Let the eigenvalues $\lambda_i[G]$ of the matrix G be chosen so that $\max \operatorname{Re} \lambda_i[G] \leq -2v_0 < 0$. Then there exists a unique solution $H = H^T > 0$ of the matrix Lyapunov equation

$$HG + G^T H + 2v_0 H = -\Lambda < 0. \tag{2}$$

Below the matrices H and Λ will appear in the expressions of the quadratic Lyapunov function and its derivative respectively.

We define the following functional

$$T_\alpha(k, A, B) = \|A + Bk - G\|_F^2 + \alpha\|k\|_F^2, \quad \alpha > 0$$

where $\|M\|_F = (\operatorname{tr} M^T M)^{1/2}$ denotes the Frobenius' norm of a matrix M.

To define the matrix $k(t)$ we introduce some notation. Let $\mathcal{A} = \{\alpha(\delta)\}$ denote the set of positive functions defined on $[0, \delta_0]$, $0 < \delta_0 < v_0$. The functions from $\mathcal{A}$ are assumed to be continuous and increasing monotonically on $[0, \delta_0]$. Moreover, $\alpha(0) = 0$ and $\lim_{\delta \to \infty} \delta^2/\alpha(\delta) = 0$. Let $\alpha_n = \alpha(\delta_n)$, $\alpha \in \mathcal{A}$, where δ_n is defined as follows

$$\delta_n = \begin{cases} \delta_0, & \text{if } T_{\alpha(\delta_0)}^{1/2}\left[k(\alpha(\delta_0), \hat{A}(t_n), \hat{B}(t_n)), \hat{A}(t_n), \hat{B}(t_n)\right] \leq \dfrac{v_0/\mu_0 - \delta_0}{1 + \delta_0/\sqrt{\alpha(\delta_0)}}, \\ \lambda, & \text{otherwise.} \end{cases} \tag{3}$$

Here λ means the solution of the following equation[i]

$$\begin{gathered} \left(1 + \lambda\alpha(\lambda)^{-1/2}\right)T_{\alpha(\lambda)}^{1/2}\left[k(\alpha(\lambda), \hat{A}(t_n), \hat{B}(t_n)), \hat{A}(t_n), \hat{B}(t_n)\right] = \frac{v_0}{\mu_0} - \delta_0, \\ \mu_0 \in [1, v_0\delta_0^{-1}) \end{gathered} \tag{4}$$

where the moments t_n are defined by the equality

$$\left\|\tilde{A}(t_{n+1} - 0) - \hat{A}(t_n)\right\|_F^2 + \left\|\tilde{B}(t_{n+1} - 0) - \hat{B}(t_n)\right\|_F^2 = \delta_n^2. \tag{5}$$

Here the controllable pairs $(\hat{A}(t_n), \hat{B}(t_n))$ are defined as follows. If at time t the estimates $\tilde{A}(t)$, $\tilde{B}(t)$ forms a controllable pair then $(\hat{A}(t), \hat{B}(t)) = (\tilde{A}(t), \tilde{B}(t))$.

---

[i]Motivation for these cumbersome expressions will be clear from the proof of Theorem 1.

Otherwise, for some $\varepsilon > 0$ we consider the set of all controllable pairs (A, B) which are no more than ε away from the pair $(\tilde{A}(t), \tilde{B}(t))$, i.e.

$$\left\{ (A, B) : \|\tilde{A} - A\|_F^2 + \|\tilde{B} - B\|_F^2 \le \varepsilon^2, (A, B) - \text{controllable pair} \right\}. \tag{6}$$

We let $(\hat{A}(t_n), \hat{B}(t_n))$ be any element of this set.

For $t = t_n$, we define the matrix $k(t)$ as follows

$$k(t_n + 0, x(t_n)) = \begin{cases} k(\alpha_n, \hat{A}(t_n), \hat{B}(t_n)), & \text{if } \|G - \hat{A}(t_n)\| \ge \dfrac{v_0}{\mu_0} - \delta_n, \\ 0, & \text{otherwise}; \end{cases} \tag{7}$$

where we have used the notation

$$k(\alpha_n, \hat{A}(t_n), \hat{B}(t_n)) = \mathbf{argmin}\, T_{\alpha_n}(k, \hat{A}(t_n), \hat{B}(t_n)). \tag{8}$$

On the time interval $[t_n, t_{n+1}]$ this matrix is constant, i.e.

$$k(t) = k(t_n + 0, x(t_n)), \quad \forall t \in [t_n, t_{n+1}]. \tag{9}$$

The relations (2)–(9) describe the control strategy stabilizing the solutions of (1) under the appropriate conditions about the considered equation. In this description we have not touched on the estimation method for parameters A, B that will be given soon. We denote this strategy by $\sigma S(\cdot)$.

We determine the control aim. First, it is dissipativity, i.e.

$$\overline{\lim_{t \to \infty}} \left(\|x(t)\| + \|u(t)\| \right) \le d < \infty \tag{10}$$

for arbitrary initial values.

We consider the class $\mathcal{LC}(n, l)$ of controlled (in the Kalman sense) linear equations of the form (1) with a bounded disturbance ($\|f(t)\| \le \chi$), $\dim x = n$ and $\dim u = l$. We define more exactly the structure of the strategy $\sigma S(\cdot)$. Consider the following algorithm

$$\dot{\tilde{A}} = -v_0 \tilde{A} + H x x^T, \quad \dot{\tilde{B}} = -v_0 B + H x u^T \tag{11}$$

of estimating A, B with the initial values $\tilde{A}(0)$, $\tilde{B}(0)$ chosen arbitrary. The rest of the parameters of the strategy have the following sense

$$\max \operatorname{Re} \lambda_i[G] = -2v_0 < 0; \quad \frac{\lambda_{\max}[\Lambda]}{\lambda_{\min}[\Lambda]} = \mu_0 \in [1, v_0 \delta_0^{-1}]; \quad \delta_0 \in [0, v_0].$$

We denote this strategy by $\sigma_1 S(v_0, \mu_0, \delta_0)$.

Theorem 1. *The strategy $\sigma_1 S(v_0, \mu_0, \delta_0)$ secures dissipativity of solutions of equations from the class $\mathcal{LC}(n, l)$.*

Proof. For Eqs. (1) and (11) we define a Lyapunov function as follows

$$V(x, \tilde{A}, \tilde{B}) = x^T H x + \|\tilde{A} - A\|_F^2 + \|\tilde{B} - B\|_F^2.$$

Its derivative, with respect to the given equation, is equal to (after simple transformations)

$$\dot{V} = -2x^T \Lambda x - 2v_0 V + 2x^T H \tilde{A} x + 2x^T H \tilde{B} k x - x^T H G x$$
$$+ 2x^T H f - 2v_0 \mathbf{tr}\left[(\tilde{A} - A)^T A\right] - 2v_0 \mathbf{tr}\left[(\tilde{B} - B)^T B\right]$$
$$\leq -x^T \Lambda x - 2v_0 V + v_0 \|A\|_F^2 + v_0 \|B\|_F^2 + 2x^T H[\tilde{A} + \tilde{B} k - G] x.$$

We want to make sure that the sufficient condition for dissipativity of the solutions of Eqs. (1) and (11) takes place (see Theorem 4, Sec. 1). For this purpose, let us estimate the quantity $\mathcal{I}(t) = -x^T \Lambda x + 2x^T[\tilde{A} + \tilde{B} k - G] x$. We have

$$\mathcal{I}(t) \leq -\lambda_{\min}[\Lambda] \|x\|^2 + 2x^T H\left[(\tilde{B} - \hat{B}_n)k + \tilde{A} - \hat{A}_n\right] x$$
$$+ 2x^T H\left[\hat{A}_n + \hat{B}_n k - G\right] x$$
$$\leq -\lambda_{\min}[\Lambda] \|x\|^2 + 2\lambda_{\max}[H]\left\{\|(\tilde{B} - \hat{B}_n\|_F \|k\|_F + \|\tilde{A} - \hat{A}_n\|_F\right\} \|x\|^2$$
$$+ 2\lambda_{\max}[H]\|\hat{A}_n + \hat{B}_n k\|_F \|x\|^2.$$

Here the notation $\hat{A}_n = \hat{A}(t_n)$ and inequality $|x^T H y| \leq (x^T H x)^{1/2}(y^T H y)^{1/2}$ have been used. Since

$$\lambda_{\max}[H] \leq (2v_0)^{-1}\lambda_{\max}[\Lambda],$$

we obtain

$$\mathcal{I} \leq -\left\{\lambda_{\min}[\Lambda] - v_0^{-1}\lambda_{\max}[\Lambda]\left[\|\tilde{B} - \hat{B}_n\|_F \|k\|_F\right.\right.$$
$$\left. + \|\tilde{A} - \hat{A}_n\|_F\right] - v_0^{-1}\lambda_{\max}[\Lambda]\|\hat{A}_n + \hat{B}_n k\|_F\right\}\|x\|^2$$
$$= -v_0^{-1}\lambda_{\max}[\Lambda]\left\{\mu_0^{-1} v_0 - \left[\|\tilde{B} - \hat{B}_n\|_F \|k\|_F + \|\tilde{A} - \hat{A}_n\|_F\right]\right.$$
$$\left. - \|\hat{A}_n + \hat{B}_n k - G\|_F\right\}\|x\|^2.$$

Let the function $\mathcal{I}(t)$ be considered on the interval $(t_n, t_{n+1}]$ defined by the equalities (5). Then it follows that

$$\mathcal{I}(t) \leq -\frac{\lambda_{\max}[\Lambda]}{v_0}\left\{\frac{v_0}{\mu_0} - \delta_n - \left(1 + \frac{\delta_n}{\sqrt{\alpha(\delta_n)}}\right)T_{\alpha(\delta_n)}^{1/2}(k, \hat{A}_n, \hat{B}_n\right\}\|x\|^2.$$

Using the relations (3)–(9) we can verify that $\mathcal{I}(t) \leq 0$ for all t. Hence

$$\dot{V} \leq -v_0 V + v_0\left(\|A\|_F^2 + \|B\|_F^2\right) + \frac{1}{2v_0^2}\lambda_{\max}[\Lambda]\|f\|^2$$
$$\leq -v_0 V + v_0\left(\|A\|_F^2 + \|B\|_F^2\right) + \frac{\chi^2 \lambda_{\max}[\Lambda]}{2v_0^2} = -aV + b.$$

The dissipativity condition leads to the inequality

$$\varlimsup_{t \to \infty} V(t) \leq \frac{\chi^2 \lambda_{\max}[\Lambda]}{2v_0^3} + \|A\|_F^2 + \|B\|_F^2 = \rho^2 < \infty,$$

and, hence,[j]

$$\varlimsup_{t\to\infty} \|x(t)\|^2 \le \left[\frac{\lambda_{\max}[\Lambda]\chi^2}{2\lambda_{\min}[H]v_0^3} + \frac{\|A\|_F^2 + \|B\|_F^2}{\lambda_{\min}[H]} \right]^{1/2}$$

$$\le \left[(\gamma - v_0)\frac{\mu_0\chi^2}{v_0^3} + \frac{\gamma v_0}{2\lambda_{\min}[\Lambda]} \left(\|A\|_F^2 + \|B\|_F^2\right) \right]^{1/2} = \rho_0$$

as well as

$$\varlimsup_{t\to\infty} \|\tilde{A}(t) - A\| = \rho, \qquad \varlimsup_{t\to\infty} \|\tilde{B}(t) - B\| = \rho.$$

It remains to prove that $\|k(t)\| < \infty$. This follows from the boundedness of the solutions $\tilde{A}(t)$, $\tilde{B}(t)$ of the Eq. (11), from $(\hat{A}_n, \hat{B}_n)$ belonging to the ε-neighborhood of the trajectory $(\tilde{A}(t), \tilde{B}(t))$ for all t, and from continuity of the function $k(\alpha, A, B)$. Therefore it reaches a maximum value on every compact set. $\square$

It is not difficult to make sure that by the appropriate choice of parameters of the strategy $\sigma_1 S$ we can obtain arbitrarily small values of the constants ρ and ρ_0. But this results in increasing $\|k(t)\|$. In spite of the presence the estimation procedure for A and B this strategy is not identificational (the estimates *do not* convergence at all), i.e. it is of the direct type.

For the wider class of equations containing the stabilizable and not certainly controllable equations, the strategy $\sigma_1 S$ may not ensure dissipativity for any values of the parameters v_0 and $\lambda_{\min}[\Lambda]$. We define the class of equations as follows. Let the next quantities be known. Namely, χ is the upper bound of all functions $f(t)$, $e > 0$ is defined by $\mathrm{Re}\,(\lambda_i[A]) \le -e$ for $i \in I_0 = \{i : \mathrm{Re}\,\lambda_i[A] < 0\}$ and p is defined by $\|A\|_F^2 + \|B\|_F^2 \ge p^2$. Let v_0 and $\lambda_{\min}[\Lambda]$ satisfy the condition

$$4\frac{\chi^2}{e^3} \le \frac{\chi^2}{v_0^3} + \frac{p}{2\lambda_{\min}[\Lambda]}.$$

Then the strategy $\sigma_1 S$ ensures dissipativity of solutions of equations from the class of stabilizable equations under the known constants χ, e, p. It follows immediately that the norm of the vector formed by the uncontrolled components of the vector x is bounded (in the limit) by the number $4\chi^2 e^{-3}$.

In connection with Theorem 1 the following significant result used later on is valid.

Lemma 1. *Under the strategy $\sigma_1 S$ it follows that*

$$\varlimsup_{l\to\infty} \delta_l = \bar{\delta} > 0.$$

---

[j]Passing to the second inequality the following relations

$$\lambda_{\min}[H] \ge \frac{\lambda_{\min}[\Lambda]}{2(\gamma - v_0)}, \qquad \gamma = \max_i \mathrm{Re}\,(-\lambda_i[G]) \ge 2v_0.$$

were used.

Proof. Suppose this is not true, i.e. there exists a subsequence δ_{l_i} whose elements tend to 0 as $i \to \infty$. These numbers are the roots of the equation

$$\left(\frac{v_0/\mu_0 - \delta_0}{1 + \delta_{l_i}/\sqrt{\alpha_{l_i}}}\right)^2 = T_{\alpha_{l_i}}(k(\alpha_{l_i}, \hat{A}_{l_i}, \hat{B}_{l_i}), \hat{A}_{l_i}, \hat{B}_{l_i}).$$

From the inequality

$$T_{\alpha_{l_i}}(k(\alpha_{l_i}, \hat{A}_{l_i}, \hat{B}_{l_i}), \hat{A}_{l_i}, \hat{B}_{l_i}) \le \alpha_{l_i} \|\bar{k}(\hat{A}_{l_i}, \hat{B}_{l_i})\|^2$$

where $\bar{k}$ is the normal solution of the matrix equation $\hat{A}_{l_i} + \hat{B}_{l_i}\bar{k} = G$ we draw the conclusion that $\|\bar{k}_{l_i}\| \to \infty$ as $i \to \infty$. This contradicts the boundedness of the matrices k_{l_i} since $\lim_{i\to\infty} k(\alpha_{l_i}, A, B) = \bar{k}(A, B)$. $\qquad\square$

The next aim of control, called stabilization, is to provide the fulfilment of the conditions:

$$\lim_{t\to\infty} x(t) = 0, \quad \overline{\lim_{t\to\infty}} \, u(t) \le \varkappa < \infty$$

for arbitrary initial values.

We choose as the class of controlled objects the class $\mathcal{LS}(n, l)$ of stabilizable equations of the form (1) with the disturbance such that

$$\int_0^\infty \|f(t)\|^2 dt < \infty. \tag{12}$$

We denote the strategy which only differs from $\sigma_1 S$ by the estimation method of parameters of the equation by $\sigma_2 S(v_0, \mu_0, \delta_0)$. We put

$$\dot{\tilde{A}} = Hxx^T, \quad \dot{\tilde{B}} = Hxu^T \tag{13}$$

or

$$\tilde{A}(t) = \tilde{A}(0) + H\int_0^t x(s)x^T(s)ds, \quad \tilde{B}(t) = \tilde{B}(0) + H\int_0^t x(s)u^T(s)ds. \tag{14}$$

Here the initial values $\tilde{A}(0)$, $\tilde{B}(0)$ are taken arbitrarily.

Theorem 2. *The strategy $\sigma_2 S$ ensures stabilization of solutions of equations from the class $\mathcal{LS}(n, l)$.*

Proof. We consider the Lyapunov function of the form

$$V(x, \tilde{A}, \tilde{B}) = x^T Hx + \|\tilde{A} - A\|_F^2 + \|\tilde{B} - B\|_F^2$$

and, using the same arguments as in Theorem 1, we obtain the inequality

$$\dot{V} \le -\frac{v_0\lambda_{\min}[\Lambda]}{2(\gamma - v_0)}x^T x + \frac{\lambda_{\max}[\Lambda]}{2v_0^2}f^T f.$$

Integrating from 0 to t we get the following inequality

$$V(t) - V(0) \le -v_0\frac{\lambda_{\min}[\Lambda]}{2(\gamma - v_0)}\int_0^t x^T(s)x(s)ds + \frac{\lambda_{\max}[\Lambda]}{2v_0^2}\int_0^t f^T(s)f(s)ds.$$

Hence we have

$$\int_0^\infty \|x(s)\|^2 ds < \infty.$$

From (14) we obtain the inequality

$$\|\tilde{A}(t)\|_F \le \|\tilde{A}(0)\|_F + \|H\|_F \int_0^t \|x(s)x^T(s)\|\, ds \le \|\tilde{A}(0)\|_F + c \int_0^t \|x(s)\|^2 ds$$

which together with the preceding inequality gives

$$\varlimsup_{t\to\infty} \|\tilde{A}(t)\|_F < \infty.$$

Similarly $\varlimsup_{t\to\infty}\|\tilde{B}(t)\|_F < \infty$. Thus

$$\varlimsup_{t\to\infty} \|u(t)\| < \infty.$$

To complete the proof we shall integrate both sides of the equality $\frac{d}{dt}\|x(t)\|^2 = 2x^T(t)\dot{x}(t)$. We have

$$\|x(t)\|^2 = \|x(0)\|^2 + 2\int_0^t x^T(s)\big[(A + Bk(s) + f(s)\big]ds.$$

Because the integral on the right-hand side of this equality converges we see that $\lim_{t\to\infty}\|x(t)\| = \omega \ge 0$. But $\int_0^\infty \|x(s)\|^2 ds < \infty$ and, hence, $\omega = 0$. $\qquad\square$

Let us consider another set of admissible disturbances. Instead of (12) we consider the following condition

$$\lim_{t\to\infty} f(t) = 0. \tag{15}$$

Note that the sets of functions satisfying conditions (12) and (15) intersect but do not coincide.

Condition (15) together with the stabilizability requirement of the pair (A, B) defines a class $\mathcal{LS}_0(n, l)$ of equations having the form (1). The aim of control with respect to this class consists of so-called w-stabilization, i.e.

$$\varlimsup_{t\to\infty} \|x(t)\| \le w, \quad \varlimsup_{t\to\infty} \|u(t)\| < \infty \tag{16}$$

where w is an arbitrary fixed number. We describe a strategy $\sigma_w S$ which provides the w-stability.

We introduce the function $V_1(x) = x^T H x$, where H is a solution of (2) under the given matrices G and Λ. For an estimation of the parameters of the equation we use (13) where the sequence of times $\{t_n'\}$ is defined by (5) and (6). The initial conditions for the estimates on the interval $[t_{n-1}', t_n')$ are chosen as follows. Denoting $\Delta_{n-1} = w^2 \lambda_{\min}[\Lambda](s/v_0 + 1)^{-1}(t_n' - t_{n+1}')$ where $s = \max \operatorname{Re} \lambda[G] - \min \operatorname{Re} \lambda[G]$, we put

$$(\tilde{A}(t_n + 0), \tilde{B}(t_n' + 0)) = \begin{cases} \big(\tilde{A}(t_{n-1}' + 0), \tilde{B}(t_{n-1}' + 0)\big), & \text{if } V_1\big(x(t_n' - 0)\big) \le -\Delta_{n-1} \\ & \qquad\qquad + V_1\big(x(t_{n-1}' + 0)\big), \\[2mm] \big(\tilde{A}(t_n' - 0), \tilde{B}(t_n' - 0)\big), & \text{otherwise.} \end{cases}$$

$$\tag{17}$$

Let $\{t_n''\}$ be the sequence of moments when the solution of Eq. (1) crosses the boundary of the sphere S_w. These moments satisfy the following boundary conditions

$$\begin{cases} \left(\tilde{A}(t_n''+0),\tilde{B}(t_n''+0)\right) = \left(\tilde{A}(t_{n-1}''-0),\tilde{B}(t_{n-1}''-0)\right), \\ k\left(t_n''+0,x(t_n''+0)\right) = k\left(t_{n-1}'',x''(t_{n-1})\right). \end{cases} \tag{18}$$

We join the sequences $\{t_n'\}$ and $\{t_n''\}$ in one $\{t_n\}$ keeping their natural orders. The moments t_n will be the "interference" moments in the course of the process (14) when new initial conditions are defined. Integrating (13) on the interval $[t_n, t_{n+1})$ we obtain

$$\begin{cases} \tilde{A}(t_{n+1}-0) = \tilde{A}(t_n+0) + H \displaystyle\int_{t_n}^{t_{n+1}} x(s)x^T(s)ds, \\[4mm] \tilde{B}(t_{n+1}-0) = \tilde{B}(t_n+0) + H \displaystyle\int_{t_n}^{t_{n+1}} x(s)u^T(s)ds. \end{cases} \tag{19}$$

This completes the description of the strategy $\sigma_w S$.

Theorem 3. *For equations from the class $\mathcal{LS}_0(n,l)$, the strategy $\sigma_w S$ provides w-stability of their solutions.*

Proof. Let the Lyapunov function be as mentioned above, then

$$\dot{V}_1 \le -\frac{v_0 \lambda_{\min}[\Lambda]}{2(\gamma - v_0)} x^T x + \frac{\lambda_{\max}[\Lambda]}{2v_0^2} f^T f.$$

For sufficiently large t such that $\|f(t)\| \le v_0[2\mu_0(1+s/v_0]^2 w$ this inequality becomes somewhat simpler, namely,

$$\dot{V}_1 \le -\frac{\lambda_{\min}[\Lambda]}{2(1+s/v_0)}\left[x^T x - \frac{w^2}{2}\right].$$

We consider the time interval on which $x(t) \notin S_w$. Coinciding the exit moments of the trajectory $x(t)$ from the set S_w with the hitting moments into the same set and according to (19), we see that for such t the inequality

$$\dot{V}_1 \le -\frac{w^2}{2}\frac{\lambda_{\min}[\Lambda]}{(1+s/v_0)}$$

or, which is the same,

$$V_1(t_{n+1}'-0) - V_1(t_n'-0) < -\frac{w^2}{2}\frac{\lambda_{\min}[\Lambda]}{(1+s/v_0)}(t_{n+1}' - t_n') < 0 \tag{20}$$

holds.

Recalling that at the points $t \in \{t_n'\}$ the function $V_1(t)$ can have discontinuities due to the choice of the initial conditions (17), we conclude that outside the sphere S_w this function decreases on the intervals (t_n, t_{n+1}) as $t \to \infty$, $t \notin \{t_n'\}$. According to Lemma 1, the lengths of the intervals (t_n', t_{n+1}') are uniformly separated from zero. Thus, the number of intervals on which the non-negative function

$V_1(t)$ decreases by positive values (separated uniformly from zero) is finite. Hence beginning from some t the first condition in (17) fails and on the next intervals this function is continuous. It satisfies inequality (20). Thus, finally, the number of the intervals on which $x(t) \notin S_w$ is finite, i.e.

$$\overline{\lim_{t \to \infty}} \, \|x(t)\| \le w.$$

The above implies that the estimation procedure is finite-converging (see Chap. 9), i.e. the sequence $(\hat{A}_n, \hat{B}_n)$ is finite and, hence, $\overline{\lim}_{t \to \infty} \|u(t)\| < \infty$. $\qquad \square$

Notice that if the disturbances $f(t)$ are bounded uniformly by some known constant χ, then for $w > \mu_0(1 + s/v_0)\chi$ the strategy $\sigma_w S$ secures dissipativity with respect to the subclass $\mathcal{LC}(n, l)$.

We now consider the control problem called optimization. The linear-quadratic problem considered is

to construct a strategy minimizing the quadratic functional

$$W(u) = \lim_{t \to \infty} t^{-1} \int_0^t [x^T(s)Px(s) + u^T r(s)Qu(s)]ds, \quad P \ge 0, \ Q > 0 \qquad (21)$$

with respect to some class of linear equations.

The set of equations having the form (1), where a disturbance satisfies one of the two conditions

$$\int_0^\infty \|f(t)\|^2 dt < \infty, \quad \lim_{t \to \infty} f(t) = 0, \qquad (22)$$

is chosen as the class of controlled objects. We denote this class by $\mathcal{LS}_{00}(n, l)$.

Theorem 4. *For equations from the class $\mathcal{LS}_{00}(n, l)$ the strategies $\sigma_2 S$ and $\sigma_w S$ give the minimum to the functional (21).*

Proof. The solutions of equations from $\mathcal{LS}_{00}(n, l)$ are bounded and, hence, the non-negative functional W defined on them is bounded also. We prove that $\min_u W(u) = 0$ when the conditions of the theorem hold.

For the Lyapunov function used in the preceding theorems we have the inequality

$$\dot{V} \le -\frac{v_0 \lambda_{\min}[\Lambda]}{2(\gamma - v_0)} x^T x + \frac{\lambda_{\max}[\Lambda]}{2v_0^2} f^T f.$$

Integrating from 0 to t we obtain

$$\int_0^t \|x(s)\|^2 ds \le \alpha + \beta \int_0^t \|f(s)\|^2 ds.$$

Dividing both sides of this inequality by t and letting $t \to \infty$, we obtain the equality

$$\lim_{t \to \infty} t^{-1} \int_0^t \|x(s)\|^2 ds = 0,$$

where either one or both conditions in (22) hold. The boundedness of $k(t)$ and the preceding equality imply

$$\lim_{t\to\infty} t^{-1} \int_0^t \|u(s)\|^2 ds = 0.$$

Thus, for both strategies we obtain $\min_u W(u) = 0$. $\qquad\square$

In this section our study was based on the assumption that the solution $x(t)$ is observed. Of course, the special case when the derivative $\dot{x}(t)$ is observed is interesting as well. The algorithms of control are then simplified but we shall not go into details here.

Note one peculiarity of the linear-quadratic problem considered above: the minimum of the functional is equal to zero. Obviously, this is due to conditions (22) which mean that the disturbance $f(t)$ must be damped. But the condition $\|f(t)\| \le \chi$ is more natural for the problem considered. Thus, under the stabilizing controls the functional $W(u)$ proves to be finite and it remains to find a strategy minimizing it. Within the framework of adaptive control considered above, a complete solution of the linear-quadratic problem is unknown yet for deterministic equations.

It is rather interesting to compare the above assertion with the results of the preceding chapter. There, for the linear stochastic difference equations we have obtained a complete solution of the linear-quadratic problem under the quite unconstrained and natural assumptions. Due to this fact in the stochastic case, the sequence of Riccati Equations, producing the solution of the problem in the limit, can be written down. In the deterministic case there is no such possibility (due to the absence of consistent estimates of the matrices A and B) and this problem remains unsolvable so far.

11.9. Two Special Problems of Adaptive Control

We now turn to two adaptive problems which are similar to the previous ones but differ from them by a number of essential details. The first problem is concerned with two linear differential equations. These equations describe the *object* ("persecutor") and the *target* ("escapee") respectively. Both equations have the same dimension and for the sake of simplicity we restrict attention to the scalar case. The equation of the object has the form

$$a(p)x = b(p)u + c(p)f \tag{1}$$

where $a(p) = p^n + a_1 p^{n-1} + \cdots + a_n$, $b(p) = b_0 p^{n-1} + \cdots + b_{n-1}$, $c(p) = c_0 p^{n-1} + \cdots + c_{n-1}$ are the operator polynomials, u is a control and $f = f(t)$ is an unmeasured external influence. The target is described by the equation

$$\bar{a}(p)y = \bar{b}(p)v \tag{2}$$

where $\bar{a}(p) = p^n + \bar{a}_1 p^{n-1} + \cdots + \bar{a}_n$, $\bar{b}(p) = \bar{b}_0 p^{n-1} + \cdots + \bar{b}_{n-1}$ are the operator polynomials and the control v of the "escapee" is unknown and unobserved. The derivatives of the coordinates for both of the object $x(t)$ and the target $y(t)$ are not measured and the coefficients of the polynomials in (1) and (2) are assumed to be unknown.

For the class $\mathcal{K}$ of equations described by (1) and (2) we would like to find a strategy

$$u = u(x, y, \theta), \quad \dot{\theta} = h(\theta, x, y), \tag{3}$$

where θ is a vector of the tuned parameters, such that the error (mismatch) $e(t) = x(t) - y(t)$ tends to zero (in one sense or another).

First, let us define explicitly the class $\mathcal{K}$, and then the appropriate strategy.

The class $\mathcal{K} = \mathcal{K}(n;, b_0, \varphi_1, \varphi_2)$ is formed by all possible pairs of Eqs. (1) and (2) satisfying the following conditions:

   I. $\deg \alpha(p) = \deg \bar{a}(p) = n; \deg b(p) = n - 1$;

  II. $\deg c(p) < n, \deg \bar{b}(p) < n$;

 III. $b_0 > 0$ is known;

 IV. $|f(t)| \leq \varphi_1, |\bar{b}(p)v(t)| < \varphi_2$.

Now we define the strategy. The vector of the tuned parameters $\theta = (\varkappa_1, \ldots, \varkappa_{n-1}, \lambda_1, \ldots, \lambda_{n-1}, \mu_1, \ldots, \mu_{n-1})$ has $3(n-1)$ components defined by the equalities

$$\dot{\varkappa}_i = -\alpha_i e u_i, \quad \dot{\lambda}_i = -\beta_i e x_i, \quad \dot{\mu}_i = -\gamma_i e y_i, \quad \alpha_i, \beta_i, \gamma_i > 0, \quad i = 1, \ldots, n - 1.$$

Here the functions u_i, x_i, y_i are determined recurrently with the help of the Hurwitz operator polynomial $d(p) = p^{n-1} + d_0 p^{n-2} + \cdots + d_{n-2}$

$$u_1 = d^{-1}(p)u, \quad x_1 = d^{-1}(p)x, \quad y_1 = d^{-1}(p)y,$$
$$u_i = (p + \lambda)u_{i-1}, \quad x_i = (p + \lambda)x_{i-1}, \quad y_i = (p + \lambda)y_{i-1}, \quad \lambda > 0,$$
$$i = 2, \ldots, n - 1.$$

The set-valued function $\text{sign}\, z$ is defined in the following way

$$\mathbf{sign}\, z = \begin{cases} \{1\}, & z > 0, \\ \{-1\}, & z < 0, \\ [-1, 1], & z = 0. \end{cases}$$

As seen from the above it is an ordinary real function for $z \neq 0$. Let λ_0, μ_0, $c > 0$ be some constants. The control u is defined as follows

$$u(t) = \sum_{i=1}^{n-1} [\varkappa_i(t)u_i(t) + \lambda_i(t)x_i(t) + \mu_i(t)y_i(t)] + \lambda_0 x(t) + \mu_0 y(t) - c\,\mathbf{sign}\, e(t).$$

This completes the description of the strategy (3).

This cumbersome, at first sight, collection of equations describing the object, the target and strategy can be written down in a compact form (normal form[k]). For this purpose, we shall note that the variables x, y, u (input) are related with the variables $(x_1, \ldots, x_{n-1}, y_1, \ldots, y_{n-1}, u_1, \ldots, u_{n-1})^T$ (output) by the linear equation

$$\dot{w} = Aw + B(x, y, u)^T, \quad (x_1 \ldots x_{n-1} y_1 \ldots u_1 \ldots u_{n-1})^T = Cw$$

where $\dim w = 3(n-1)$. Next, the tuned parameters θ are calculated by using (3), where the $(6n-4)$-dimensional vector

$$\pi = (\varkappa_1, \ldots, \varkappa_{n-1}, \lambda_0, \lambda_1, \ldots, \lambda_{n-1}, \mu_0, \mu_1, \ldots, \mu_{n-1}, w)$$

is the state vector. Thus the normal form mentioned above for the equations of the object, the target and "regulator" is described completely as follows

$$\dot{\eta} = \varphi(\eta, t) + \psi\,\mathbf{sign}\,e, \quad e = \rho^T \eta. \tag{4}$$

All components in (4) can be expressed in terms of the functions and parameters of the original equations. We shall not go into details here. However, we notice that the presence of the nonlinearity *sign* in (4) makes "moving regimes" possible. Such a regime proves to be defined uniquely due to the conditions $\mathbf{deg}\,b(p) = n - 1$.

We shall now give some remarks about the choice of the constant c entering into the description of the control $u(t)$. For this purpose, we introduce the originals of the Laplace transformations $G_1(t)$ and $G_2(t)$ of the functions $c(p)/d(p)$ and $d^{-1}(p)$ respectively. Let $g_i = \int_0^\infty |G_i(t)|dt$, $i = 1, 2$. We demand that the inequality

$$c > \frac{g_1\varphi_1 + g_2\varphi_2}{b_0} \tag{5}$$

should be satisfied. In this connection we make the following remarks. The constant g_2 can be computed exactly since the polynomial $d(p)$ is given. But quite often for g_1 we can only find the upper estimate since $c(p)$ is unknown.[l] Every concrete choice of $d(p)$ implies its own values of g_1 and g_2. Their lower bounds are equal to zero. However, we do not strive to reach it since the "quality" of the control could get worse.

<hr>

[k]This means the usual form of the proposed equations in the state space

$$\dot{z} = Ax + Bu, \quad \nu = Cz$$

where the first equation and the vector ν correspond to the dynamic part and the observations respectively.

[l]Let $d(p) = \prod_{i=1}^{n-1}(p + \rho_i)$. The fraction $c(p)/d(p)$ can be represented in the form

$$\frac{c(p)}{d(p)} = \chi_0 + \sum_{i=1}^{n-1} \frac{\chi_i}{p - \rho_i}, \quad \rho_i > 0$$

where χ can be expressed in terms of the coefficients of $c(p)$. Then $|G_1(t)| \leq |\chi_0| + \sum_{i=1}^{n-1} |\chi_i| \exp(-\rho_i t)$. If for $|\chi_i|$ the upper bounds χ_i^* are known (it may be so in practice) then $g_1 \leq \chi_0^* + \sum_{i=1}^{n-1} \chi_i^* \rho_i^{-1}$.

Let $\sigma(B_c)$ denote strategy (3) (or (4)) with the constant c satisfying the constraint (5).

Theorem 1. *With respect to the class $\mathcal{K}(n; b_0, \varphi_1, \varphi_2)$, the strategy $\sigma(B_c)$ secures:*

(1) the boundedness of the functions $\varkappa_i(t), \lambda_i(t), \mu_i(t), \; i = 1, \ldots, n-1$;
(2) $\int_0^\infty |e(t)|dt < \infty, \quad \int_0^\infty e^2(t)dt < \infty$.

This result can be strengthened if we restrict the class of the considered equations. Namely, let $\hat{\mathcal{K}}(n; b_0, \varphi_1, \varphi_2)$ denote the subclass of $\mathcal{K}(n; b_0, \varphi_1, \varphi_2)$ for which any solution $y(t)$ of Eq. (2) is uniformly bounded and any polynomial $b(p)$ is Hurvitz.

Theorem 2. *With respect to the class $\hat{\mathcal{K}}(n; b_0, \varphi_1, \varphi_2)$ the strategy $\sigma(B_c)$ implies the uniform boundedness of both the control $u(t)$ and the function $\dot{e}(t)$.*

Corollary 1. *Under the assumptions of Theorem 2, $\lim_{t \to \infty} e(t) = 0$.*

Proof of Theorem 1. The proof is based on the possibility to write Eqs. (1) and (2) in the form of a system of equations of the first order.

$$
\begin{cases}
b_0^{-1}(\dot{x} + ax) = u(t) + \displaystyle\sum_{i=1}^{n-1}(\varkappa_i u_i(t) + \lambda_i x_i(t)) + \lambda_0 x(t) + h(t) + \varepsilon(t), \\[2mm]
b_0^{-1}(\dot{y} + ay) = \displaystyle\sum_{i=1}^{n-1}\mu_i y_i(t) + \mu_0 y(t) + \bar{h}(t) + \bar{\varepsilon}(t)
\end{cases}
\tag{6}
$$

where the functions $x_i(t)$, $y_i(t)$, $u_i(t)$ have been defined above,

$$
h(t) = \frac{1}{b_0}\frac{c(p)}{d(p)}f(t), \quad \bar{h}(t) = \frac{1}{b_0}\frac{b(p)}{d(p)}v(t),
\tag{7}
$$

and constants $\varkappa_i$, λ_i, μ_i (unknown beforehand) are uniquely defined by the equalities

$$
(p+a)d(p) - b_0\left[\lambda_0 d(p) + \sum_{i=1}^{n-1}\lambda_i(p+\lambda)^{i-1}\right] \equiv \alpha(p),
$$

$$
b_0 d(p) + b_0\sum_{i=1}^{n-1}\varkappa_i(p+\lambda)^{i-1} \equiv b(p),
$$

$$
(p+a)d(p) - b_0\left[\mu_0 d(p) + \sum_{i=1}^{n-1}\mu_i(p+\lambda)^{i-1}\right] \equiv \bar{a}(p).
$$

The functions $\varepsilon(t)$ and $\bar{\varepsilon}(t)$ are the partial solutions of the stable equation $d(p)\varepsilon = 0$ and, hence, they vanish exponentially as $t \to \infty$.

Subtracting the second equation from the first one in (6) we obtain

$$
b_0^{-1}(\dot{e} + ae) = u(t) + \sum_{i=1}^{n-1}[\varkappa_i u_i(t) + \lambda_i x_i(t)) - \mu_i y_i(t)]
$$

$$
+ \lambda_0 x(t) - \mu_0 y(t) + s(t)
\tag{8}
$$

where $s(t) = h(t) - \bar{h}(t) + \triangle\varepsilon$, $\quad d(p)\triangle\varepsilon = 0$. Let us make the following transformations in (8). Instead of $u(t)$ we substitute its representation. Denoting $\bar{\varkappa}_i(t) = \varkappa_i(t) + \varkappa_i$, $\bar{\lambda}_i(t) = \lambda_i(t) + \lambda_i$, $\bar{\mu}_i(t) = \mu_i(t) - \mu_i$ we multiply both parts of the equality obtained by $e(t)$ and on the right-hand side we substitute the expressions for the tuned parameters ($\dot{\varkappa}_i(t) = -\alpha_i e u_i$ and so on). Then we obtain the equality

$$\frac{1}{2}\frac{d}{dt}\left\{\frac{1}{b_0}e^2(t) + \sum_{i=1}^{n-1}\left[\frac{\bar{\varkappa}_i^2}{\alpha_i} + \frac{\bar{\lambda}_i^2}{\beta_i} + \frac{\bar{\mu}_i^2}{\gamma_i}\right] + \frac{\bar{\lambda}_0^2}{\beta_0} + \frac{\bar{\mu}_0^2}{\gamma_0}\right\} + \frac{a}{b_0}e^2(t) = -c|e| + s(t)e. \quad (9)$$

Let $W(t)$ denote the function bracketed on the left-hand side of (9).

The inequality (5), the Eq. (6) and the equalities (7) together with the exponential decreasing rate of $\triangle\varepsilon(t)$ mean that for any $\delta \in \left(0, c - b_0^{-1}(g_1\varphi_1 + g_2\varphi_2)\right)$ there exists some moment t_0 (depending on δ and the initial conditions) such that the right-hand side of (9) is less than $-\delta|e|$ for $t > t_0$. Taking this fact into account and integrating equality (9) from t_0 to $T > t_0$, we obtain

$$W(T) + \frac{2a}{b_0}\int_{t_0}^{T} e^2(s)ds + 2\delta\int_{t_0}^{T}|e(s)|ds \leq W(t_0).$$

Putting here $T \to \infty$ we obtain

$$\frac{2a}{b_0}\int_{t_0}^{\infty} e^2(s)ds + 2\delta\int_{t_0}^{\infty}|e(s)|ds \leq W(t_0).$$

Hence

$$\sup_{t \geq t_0}\left\{\frac{1}{b_0}e^2(t) + \sum_{i=1}^{n-1}\left[\frac{\bar{\varkappa}_i^2}{\alpha_i} + \frac{\bar{\lambda}_i^2}{\beta_i} + \frac{\bar{\mu}_i^2}{\gamma_i}\right] + \frac{\bar{\lambda}_0^2}{\beta_0} + \frac{\bar{\mu}_0^2}{\gamma_0}\right\} \leq W(t_0).$$

This inequality together with the continuity of the functions $x(t)$, $y(t)$, $\varkappa_i(t)$, $\lambda_i(t)$, $\mu_i(t)$ prove the theorem.

Proof of Theorem 2. We only give a sketch. Let the notation $[f, g, \ldots]$ mean the following statement: the "functions $f(t)$, $g(t), \ldots$ are uniformly bounded on the positive half-axis". In the course of the proof of Theorem 1 we have proved that $[e, \bar{\varkappa}_i, \bar{\lambda}_i, \bar{\mu}_i]$. Because $[y]$, we have $[x]$ and, hence, $[x_i, y_i, i = 1, \ldots, n-1]$. By using similar arguments we can prove that $[u_i, i = 1, \ldots, n-1]$. According to (8), we obtain $[\dot{e}]$. $\qquad\square$

The proof of Corollary 1 does not require any additional arguments.

The second problem of adaptive control considered in this section deals with the class of nonlinear integro-differential equations. It has been motivated by radiotechnique needs. The mathematical models studied have the following form

$$\begin{cases} \dot{x} = Ax + B\int_0^t k(s)\varphi(g(t-s))ds + Cu + f(t), \\ z = L^T x \end{cases} \quad (10)$$

where $x \in \mathrm{R}^n$, A is an $n \times n$-matrix, B and C are vectors having length n, $u = u(t)$ is a scalar control (input), $f(t)$ is an external influence, $z = z(t)$ is the quantity (the scalar output) observed. The function g is related with z as follows

$$g(t) = g_0(t) + g_1(t), \quad g_0(t) = dz(t), \quad g_1(t) = D^T \nu(t),$$
$$\dot{\nu} = P\nu + Qz, \quad \nu(0) = \nu_0$$

where d is a number, D, Q, V are vectors having length n and P is a Hurwitz matrix of order m. As concerns the functions $k(s)$, $\varphi(g)$ we assume that

$$k(s) = k_0(s) + \sum_{j=1}^{l} k_j \delta(s - \tau_i), \quad 0 \leq \tau_1 < \tau_2 < \cdots < \tau_l$$

where $\delta(s)$ is the Dirac function,[m] $k_0(s) \in C_{[0,\infty)}$ and $\varphi(g)$ is supposed to be continuously differentiable up to the order l. Moreover, there exist constants $r_i > 0$ such that $|d^j \varphi / dg^j| \leq r_i |g|$, $j = 1, \ldots l$, $l \geq n - 2$. The external disturbance f and its derivatives up to the order $n - 1$ are assumed to be square-integrable on $[0, \infty)$. The output $z(t)$ and its derivatives up to the order $s \leq \rho - 1$ are assumed to be measured. To define the number ρ we consider the system (10) with $B \equiv 0$. Let $\tilde{z} = z(t)$ be the output of the new system. Then ρ is the maximum integer such that the derivative $d^\rho \tilde{z} / dt^\rho$ is continuous. To complete the description of the mathematical model (10) we note that (10) is the equation with a delay and zero past history (i.e. $x(t) \equiv 0$, for $t < 0$).

The aim of control is

to provide the fulfilment of the following inequalities:

$$\begin{cases} \sup_{t \geq 0}[|x(t)| + |\dot{x}(t)|] + \int_0^\infty (x^2(s) + \dot{x}^2(s))ds \leq c' < \infty, \\[2mm] \sup_{0 \leq k \leq l} \left[\sup_{t \geq 0} |z^{(k)}(t) + \int_0^\infty |z^{(k)}(s)|^2 ds \right] \\[2mm] + \sup_{t \geq 0} \sum_{i=1}^{l_1 - \rho + 1} |u^i(t)| + \int_0^\infty \sum_{i=1}^{l_1 - \rho + 1} |u^i(s)|^2 ds \leq c'' < \infty \end{cases} \tag{11}$$

for some integer l_1.

This problem can be considered in the framework of Hilbert Space Theory (See Sec. 7).

We consider this problem in adaptive form. For this purpose, it is necessary to define the class of models. First, we point to the parameters in (10) which are *a priori* unknown. The functions $k_0(s)$, $\varphi(g)$, $f(t)$, the matrices A and P, the vectors B, C, D, Q, the numbers r_i, d, k_i, τ_i are assumed to be unknown. To point to the known parameters we shall rewrite system (10) in a different form excluding the state variables. This leads to the following equation with respect to the output $z(t)$

$$A(p)z = B(p)\omega(t) + C(p)u + D(p)f, \tag{12}$$

---

[m] This function is defined for the piecewice-continuous functions ψ by the equality

$$\int_{-\infty}^\infty \psi(s)\delta(s)ds = \frac{1}{2}[\psi(-0) + \psi(+0)].$$

where $\omega(t) = \int_0^t k(s)\varphi(g(t-s))ds$ and

$$A(p) = \det(pI - A), \quad B(p) = A(p)L^T(pI - A)^{-1}B,$$
$$C(p) = c_\rho p^{n-\rho} + c_{\rho-1}p^{n-\rho-1} + \cdots + c_n = A(p)L^T(pI - A)^{-1}C,$$
$$D(p) = A(p)L^T(pI - A)^{-1}.$$

We assume the order of the system is n, the degree of smoothness is ρ and the coefficient c_ρ of the polynomial $C(p)$ are known.

We denote by $\mathcal{B}(n, \rho, c_\rho)$ the class of equations of the form (10) satisfying the conditions that the pair (L^T, A) is observable, the pair (A, C) is stabilizable, the matrix P and polynomial $C(p)$ are Hurwitz and $l \geq n + \rho - 3$.

We define the structure of the strategies applied as follows

$$u(t) = u(Z, \theta), \quad \dot{\theta} = \Theta(\theta, Z), \quad s \leq \rho - 1, \tag{13}$$

where $Z = (z, \dot{z}, \ldots, z^{(s)})$ is the collection of quantities observed at each moment of time. The idea of constructing such a strategy consist in attaining the given aim (11) for Eq. (10) without the integrable summand ($B \equiv 0$). In this case we deal with an ordinary linear differential equation. The proposed strategy has a remarkable property consisting of attaining the given aim for the original Eq. (10) from the class $\mathcal{B}(n, \rho, c_\rho)$ when $B \neq 0$ as well. Let us describe this strategy.

We define it in the form adapted to Eq. (10). Let $H(p) = p^{n-1} + h_2 p^{n-2} + \cdots + h_n$ be a Hurwitz polynomial. We put

$$H(p)v_i = (p + \alpha)^{i-1}v, H(p)z_i = (p + \alpha)^{i-1}z_i, \quad \dot{\lambda}_i = -\beta_i z(t)v_i(t),$$
$$i = 1, \ldots, n-1,$$
$$\dot{\mu}_j = -\gamma_j z(t)z_j(t), \quad j = 1, \ldots, n-1, \quad z_0(t) = z(t), \tag{14}$$
$$v = \sum_{i=1}^{n-1} \lambda_i(t)v_i + \sum_{i=0}^{n-1} \mu_i(t)z_i(t), \quad u(t) = (p + \delta)^{\rho-1}v(t).$$

Here the numbers $h_2, \ldots, h_n, \{\beta_i\}, \{\gamma_i\}, \alpha$, and δ are the parameters of the adaptive regulator — they can be taken arbitrarily within the above-mentioned restrictions. We write (14) now in the normal form (12).

Let $g_2, \ldots, g_n$ be the coefficients of the polynomial $H(p)$ written down in order of the descending exponents of $(p + \alpha)$, i.e.

$$H(p) = (p + \alpha)^{n-1} + g_2(p + \alpha)^{n-2} + \cdots + g_n.$$

Equations (14) can be rewritten as follows

$$\begin{cases} \dot{v}_1 &= v_2 - \alpha v_1, \\ \dot{v}_2 &= v_3 - \alpha v_2, \\ \quad \vdots \quad \vdots \quad \vdots \\ \dot{v}_{n-2} &= v_{n-1} - \alpha v_{n-2}, \\ \dot{v}_{n-1} &= -g_n v_1 - g_{n-1}v_2 - \cdots - (g_2 + \alpha)v_{n-1} + v, \end{cases}$$

$$\begin{cases} \dot{z}_1 \;\;= z_2 - \alpha z_1, \\[4pt] \dot{z}_2 \;\;= z_3 - \alpha z_2, \\[4pt] \quad \vdots \quad\;\; \vdots \quad\;\; \vdots \\[4pt] \dot{z}_{n-2} = z_{n-1} - \alpha z_{n-2}, \\[4pt] \dot{z}_{n-1} = -g_n z_1 - g_{n-1} z_2 - \cdots - (g_2 + \alpha) z_{n-1} + z. \end{cases}$$

Together with the relations for $\lambda_i(t)$, $\mu_i(t)$ and for v, u these equations are represented in the form (12).

The vector of order $4n - 3$

$$\theta = (v_1, \ldots, v_{n-1}, z_1, \ldots, z_{n-1}, \lambda_1, \ldots, \lambda_{n-1}, \mu_0, \mu_1, \ldots, \mu_{n-1})^T$$

is the state vector. The described strategy is denoted by $\sigma(B)$.

This completes the description of the second problem. Its solution is given by the following theorem.

Theorem 3. *The strategy $\sigma(B)$ secures the attainment of the aim (11) with $l_1 = l$ with respect to the class $\mathcal{B}(n, \rho, c_\rho)$.*

In view of the complex form of the equations considered, the control aim and strategy, the proof of the above theorem is rather cumbersome. However it is not complicated and its idea is similar to the one used in Theorem 2. This proof consists in verifying the property "to be continuous and uniformly bounded (on the half-axis $[0, \infty)$) and to be square-integrable" for all components of the system studied. Beginning from the functions $z_i(t)$ this check is carried out successively up to the functions $x(t)$ and $\dot{x}(t)$. We shall not go into details here.

CHAPTER 12

CONTROL OF STOCHASTIC DIFFERENTIAL EQUATIONS

In this chapter we consider models whose dynamics are described by differential equations containing disturbances in the form of a white noise. The control of such models is, in essence, the control of stochastic Ito equations where the Wiener process is the disturbance. We consider adaptive control problems of stabilization and optimization. Both direct and identificational strategies are used. For the latter we consider the estimation procedures of the parameters of the linear Ito equation.

12.1. Preliminary Results

In this section the controlled objects in the form of ordinary differential equations depending on a control $u = u(t)$ and on an additive disturbance η_t

$$\dot{x} = f(x, u, t) + \eta_t, \quad x(0) = x_0$$

are considered. Throughout this section we assume that $f(x, u, t) = Ax + Bu$ and $\eta_t = \dot{W}_t$ is the "white noise" or, which is the same, the derivative of the Wiener process $W(t)$. The random function W_t has a peculiar feature, namely, it is continuous but not differentiable. For this reason the equation $\dot{x} = Ax + bu + \dot{W}_t$ has no mathematical sense from the point of view of classical analysis. Nevertheless, it is possible to define these equations mathematically correctly and, then, to use them as a basis to design the formal theory of control of such equations. The rest of this section is devoted mainly to stochastic equations theory.

Let $(\Omega, \mathcal{F}, \mathbf{P})$ be a probability space with an increasing flow of σ-algebras $\mathcal{F}_t$ ($\mathcal{F}_{t_1} \subset \mathcal{F}_{t_2} \subset \mathcal{F}_t$ for any $t_1, t_2, 0 \leq t_1 < t_2 < \infty$) which is continuous from the right.

Definition 1. A random process $w_t(\omega)$ taking values from R^n is called a *Wiener process* if it is continuous and $\mathcal{F}_t$-measurable for any t, and also its characteristic function has the form

$$\mathbf{E}\{\exp[i(x, w_t - w_s)] | \mathcal{F}_t\} = \exp\left[-\frac{t-s}{2}\|x\|^2\right] \quad \text{a.s.}$$

for all $x \in \mathrm{R}^n$ and $t \geq s \geq 0$.

If $w_t = \left(w_t^{(1)}, \ldots, w_t^{(n)}\right)$ then from this definition it follows that

$$\mathbf{E}\left(w_t^{(i)} - w_s^{(i)} | \mathcal{F}_s\right) = 0 \quad \mathbf{E}\left(w_t^{(i)} - w_s^{(i)}\right)\left(w_t^{(j)} - w_s^{(j)}\right) = (t-s)\delta_{ij}.$$

421

If δ_{ij} signifies the Kroneker symbol then w_t will be called the standard Wiener process.

To give an equivalent definition of the Wiener process, let us introduce the density of the normal n-dimensional distribution

$$p(x,t) = (2\pi t)^{-n/2} \exp\left[-\frac{\|x\|^2}{2t}\right]$$

and a probability measure μ onto R^n.

Definition 2. A continuous n-dimensional random process w_t is called a *Wiener process* if the probability $\mathbf{P}\{S_l\}$ of the event $S_l = \{w_{t_1} \in M_1, \ldots, w_{t_l} \in M_l\}$ for all l, $0 < t_1 < \cdots < t_l$ and all measurable sets $M_1, \ldots, M_l$ from R^n is defined by the following formula

$$\mathbf{P}\{S_l\} = \int_{\mathrm{R}^n} \mu(dx) \int_{M_1} p(x_1 - x, t_1)dx_1 \int_{M_2} p(x_2 - x_1, t_2 - t_1)dx_2 \times \cdots$$
$$\times \int_{M_l} p(x_l - x_{l-1}, t_l - t_{l-1})dx_l. \tag{1}$$

The sense of this expression is rather simple, namely, the process w_t has independent increments, i.e. the r.v. $w_0, w_{t_1} - w_0, \ldots, w_{t_l} - w_{t_{l-1}}$ are mutually independent, μ is considered as the initial distribution and the increments $w_t - w_s$ have the normal distribution $N(0, t - s)$. To determine a measure on the space $C([0, \infty], \mathrm{R}^n)$ of all continuous functions defined on the positive half-axis and taking the values from R^n, the relation (1) is used. Indeed, this relation defines an additive set function on the cylindrical sets by the method described in Sec. 1, Chap. 1 and then this function can be extended to some measure on the σ-algebra of Borel sets[a] in $C([0, \infty], \mathrm{R}^n)$. This measure is called the *Wiener measure* $\mathbf{P}_\mu$ *with the initial distribution* μ. If μ is the normal distribution then w_t will be a Gauss process.

The main properties of the Wiener process are given below.

With probability equal to one (with respect to the Wiener measure) the following hold:

(a) Hölder property

$$\overline{\lim_{0 < h \to 0}} \frac{|w_{t+h} - w_t|}{\sqrt{2h \ln h^{-1}}} = 1,$$

or, in other words, the samples of the Wiener process satisfy the Hölder condition $|w_{t+h} - w_t| \le h^a$ for $a = 1/2 - \varepsilon$ and $\varepsilon > 0$;

(b) The sample functions of the Wiener process are not differentiable;

---

[a] In the space $C([0, \infty], \mathrm{R}^n)$ the metric convergence of $w_n(t)$ to $w(t)$ means the uniform convergence with respect to t on every finite interval for is introduced. It defines a topology on $C([0, \infty], \mathrm{R}^n)$ and, hence, some Borel σ-algebra.

(c) Variation of the Wiener process is infinite on any finite interval but there exists the finite "quadratic" variation

$$\lim_{N \to \infty} \sum_{j=1}^{N} |w_{t_{j+1}} - w_{t_j}|^2 = t - s$$

where $s = t_1 < t_2 < \cdots < t_{N+1} = t$, $\max_j(t_{j+1} - t_j) \to 0$ as $N \to \infty$ and the symbol **lim** means the limit both in the "a.s." sense and in the "mean square" sense;

(d) Local law of the iterated logarithm

$$\overline{\lim_{t \to 0}} \frac{w_t}{\sqrt{t \ln \ln t^{-1}}} = 1;$$

(e) The zero set of the Wiener process, i.e. $O_w = \{t : w_t(\omega) = 0\}$ is unbounded and perfect (i.e. closed and without isolated points) and its Lebegues measure is equal to zero;

(f) Law of the iterated logarithm

$$\overline{\lim_{t \to \infty}} \frac{|w_t|}{\sqrt{2t \ln \ln t}} = 1.$$

It follows, in particular, that $\lim_{t \to \infty} t^{-1} w_t = 0$.

By using the Wiener process we now define the notion of the *stochastic Ito integral* denoted as $\int_0^T \varphi(t, \omega) dw_t(\omega)$ for a random processes from the function space $\mathcal{L}_2(T) = \{\varphi(t, \omega): \mathbf{E} \int_0^T \varphi^2(t, \omega) dt < \infty\}$. We begin with the step-functions which are defined by the partition $0 = t_0 < t_1 < \cdots < t_{n-1} < t_n = T$ and by the random $\mathcal{F}_{t_i}$-measurable variables $\varphi_i(\omega)$. For such functions, Ito defined the stochastic integral as follows

$$\int_0^T \varphi(t, \omega) dw_t(\omega) = \sum_{i=0}^{n-1} \varphi_i(w_{t_{i+1}} - w_{t_i}).$$

The step-functions are everywhere dense in $\mathcal{L}_2(T)$, i.e. for any $\varphi \in \mathcal{L}_2(T)$ there is a sequence of step-functions $\{\varphi_n\}$ converging in the mean square sense to the function φ. It follows that the sequence of random variables $\{\int_0^T \varphi_n(t, \omega) dw_t(\omega)\}$ is fundamental in the mean square convergence sense. Hence in the full space of the random variables $\{\xi\}$ with the norm $\|\xi\| = \mathbf{E}\xi^2$ there exists a random variable which is called the stochastic integral of φ denoted by $\int_0^T \varphi(t, \omega) dw_t(\omega)$ and such that

$$\text{l.i.m.}_{n \to \infty} \int_0^T \varphi_n(t, \omega) dw_t(\omega) = \int_0^T \varphi(t, \omega) dw_t(\omega).$$

This definition does not depend on the approximating sequence $\{\varphi_n\}$.

The main properties of the stochastic Ito integral are given below.

1. $\int_0^T (a\varphi(t, \omega) + b\psi(t, \omega)) dw_t = a \int_0^T \varphi(t, \omega) dw_t + b \int_0^T \psi(t, \omega) \, dw_t.$
2. $\mathbf{E} \int_0^T \varphi(t, \omega) dw_t = 0.$

3. $\mathbf{E}\left\{\int_0^T \varphi(t,\omega)dw_t \int_0^T \psi(t,\omega)dw_t\right\} = \mathbf{E}\int_0^T \varphi(t,\omega)\psi(t,\omega)dt.$

4. If $\mathbf{P}\left\{\int_0^T \varphi^2(t,\omega)dt = 0\right\} = 1$ then $\mathbf{P}\left\{\int_0^T \varphi(t,\omega)dw_t = 0\right\} = 1.$

The peculiarity of this definition implies some specific properties of the Ito integral one of which is the fact that its value differs from the value of the appropriate integral in the usual sense. For example, for the stochastic integral we obtain

$$\int_0^t w_s dw_s = \frac{w_t^2}{2} - \frac{t}{2}.$$

As seen from the above the obtained value differs from the expected one equal to $w_t^2/2$. The latter would have taken place if the corresponding integral had been treated as the Stieltjes' integral. The fact is that the considered integral $\int_0^t \varphi(\cdot)dw_s$ cannot be treated as the Lebesgue–Stieltjes integral since the variation of w_t is unbounded on any (however small) interval.

The definition of the Ito integral can be extended from the space $\mathcal{L}_2(T)$ to the set of the functions $\mathcal{P}_T = \{f : \int_0^T f^2(t,\omega)dt < \infty\}$. Now $\int_0^T \varphi(\cdot)dw_s$ is defined as the limit of the sequence $\left\{\int_0^T \varphi_n(\cdot)dw_s\right\}$ obtained with probability equal to one, where the sequence $\{\varphi_n\}$ is the same as above. But for such Ito integrals properties (2), (3) above may fail.

Using the stochastic integral we can construct a new classes of stochastic processes. Now we introduce one of them. Let $m(t,x) = (m^{(1)}(t,x),\ldots,m^{(n)}(t,x))$ and $D(t,x) = (D_{ij}(t,x))$, $x \in \mathbb{R}^n$ be a vector and an $n \times r$-matrix respectively. It is suggested that

$$\int_0^T [\|m(t,x)\| + \|D(t,x)\|^2]dt < \infty \quad \text{a.s.}$$

We consider the following stochastic integral equation ($w_t \in \mathbb{R}^r$)

$$\xi_t = \xi_0 + \int_0^t m(s,\xi_s)ds + \int_0^t D(s,\xi_s)dw_s \tag{2}$$

which is usually written in the form of a stochastic differential

$$d\xi_t = m(t,\xi_t)dt + D(t,\xi_t)dw_t$$

Definition 3. A continuous process $\xi_t(\omega)$, $t \in [0,T]$ satisfying Eq. (2) with probability equal to one is called a *diffusion process*.

In absence of the stochastic integral (2) ($D \equiv 0$) we obtain a system of ordinary differential equations of the type $\dot{x} = m(t,x)$ which defines a field of directions in $\mathbb{R}^n$. Under $D \neq 0$ the random disturbance having the intensity (or "power" D) and displacing the deterministic trajectories $x(t)$ disorderly is added. In this connection the vector m and the matrix D are called the *drift* and the *diffusion* respectively. The diffusion process is Markov and it is defined completely (except the initial distribution) by the transition functions $P(x,s;M,t) = \mathbf{P}\{\xi_t \in M|\xi_s = x\}$. By using them the drift m and the diffusion D can be defined as well.

If the drift m and the diffusion D in (2) do not depend on t explicitly then the process ξ_t is called *homogeneous*. Otherwise it is called *non-homogeneous*. In the homogeneous case $P(x, s; M, t) = P(x, t - s; M)$.

It is often desirable to know whether a function f given on the trajectories of the diffusion process ξ_t (which is defined by Eq. (2) or by a stochastic differential, i.e. the process $f(t, \xi_t)$), has the stochastic differential. In the one-dimensional case the following theorem holds.

Theorem 1. (Ito's substitution of variables formula) *If a function $f(t, x)$ is continuous together with its derivatives f'_t, f'_x, f''_{xx} then the process $f(t, \xi_t)$ has the following stochastic differential*

$$df(t, \xi_t) = \left[f'_t(t, \xi_t) + f'_x(t, \xi_t)m(t, \xi_t) + \frac{1}{2} f''_{xx}(t, \xi_t)D^2(t, \xi_t) \right]dt$$
$$+ f'_x(t, \xi_t)D(t, \xi_t)dw_t. \tag{3}$$

In the multi-variable case the continuity property of all second derivatives of the function f (with respect to x) is required. Then the one-dimensional process $f\big(t, \xi_t^{(1)}, \ldots, \xi_t^{(n)}\big)$ has the stochastic differential of the form

$$df(t, \xi_t)$$
$$= \left[f'_t(t, \xi_t) + \sum_{j=1}^{n} f'_{x_j}(t, \xi_t)m^{(j)}(t, \xi_t)) + \frac{1}{2} \sum_{i,j=1}^{n} f''_{x_i x_j}(t, \xi_t) \right.$$
$$\left. \times \sum_{l=1}^{n} D_{il}^2(t, \xi_t)D_{jl}(t, \xi_t) \right]dt + \sum_{i=1}^{n} \sum_{j=1}^{r} f''_{x_i}(t, \xi_t)D_{ij}(t, \xi_t)dw_t^{(j)}. \tag{4}$$

The application of Ito's formula is illustrated by two examples given below. We shall use them in what follows.

For the exponential function $f(w_t) = \exp(w_t + h(t))$ we obtain

$$df(w_t) = \left(h'(t)f(w_t) + \frac{1}{2}f(w_t) \right)dt + f(w_t)dw_t$$
$$= \left(h'(t) + \frac{1}{2} \right) f(w_t)dt + f(w_t)dw_t.$$

For the quadratic form $f(t, x) = x^T A(t)x$ and $d\xi_t = m(t)dt + D(t)dw_t$ we obtain

$$df = \left[\xi_t^T A'(t)\xi_t + \xi_t^T (A(t) + A^T(t))m + \mathbf{tr}\, A(t)D(t)D^T(t) \right]dt$$
$$+ (\xi_t^T A(t)D(t) + \xi_t^T A^T(t)D(t))dw_t.$$

Further, we shall focus our interest on both the existence and the uniqueness of solutions of the stochastic equation (2). For brevity, let the collection $\mathcal{E} = [\Omega, \mathcal{F}, \mathbf{P}, \xi_0, \xi_t]$ denote the set of elements satisfying the following relation

$$\mathbf{P}\left\{ \xi_t = \xi_0 + \int_0^t m(s, \xi_s)ds + \int_0^t D(s, \xi_s)dw_s, \ \forall t \right\} = 1.$$

Here m, $D \in \mathcal{P}_T$ and a random variable ξ_0 not depending on w_t and having the known distribution is the initial value.

If the collection $\mathcal{E}$ is given then the random process ξ_t is called the *strong solution* (or, briefly, the *solution*) of the stochastic equation (2).

The uniqueness of the solution of the stochastic equation in the *distribution* sense means the coincidence of the distributions of any two solutions ξ_t' and ξ_t'' under the same initial values. The distributions are interpreted as points of the space of continuous functions.

The uniqueness of solutions of the stochastic equation in the *trajectory* sense means $\mathbf{P}\{\sup_{0 \leq t \leq T} |\xi_t' - \xi_t''| > 0\} = 0$ for any two solutions ξ_t' and ξ_t'' given on the same probability space under the same Wiener process and the equality $\xi_0' = \xi_0''$ a.s., i.e. the trajectories of the solutions coincide. The uniqueness property in the trajectory sense implies obviously the uniqueness in the distribution sense.

Theorem 2. *The stochastic differential equation*

$$d\xi_t = m(t, \xi_t)dt + D(t, \xi_t)dw_t$$

has a unique strong solution if for all $t \in [0, T]$, x, $y \in \mathrm{R}^n$ *and some* $L > 0$ *the following conditions hold*

$$\|m(t, x) - m(t, y)\|^2 + \|D(t, x) - D(t, y)\|^2 \leq L\|x - y\|^2,$$
$$\|m(t, x)\|^2 + \|D(t, x)\|^2 \leq L(1 + \|x\|^2).$$

If these conditions fail, the uniqueness of strong solutions does not take place even in rather simple cases. To demonstrate the last assertion we consider the following scalar equation

$$d\xi_t = b(\xi_t)dw_t, \quad b(x) = \begin{cases} 1, & x \geq 0, \\ -1, & x < 0. \end{cases}$$

We can easily make sure that this equation has a unique solution in the distribution sense but not in the trajectory sense.

Definition 4. If for the given stochastic differential equation a collection $\mathcal{E}$ exists then we say that this equation has a *weak solution*.

The uniqueness of the weak solution means the uniqueness in the distribution sense. Therefore it is possible to say that such a solution defines a measure on the space C. A strong solution is also a weak solution but not vice versa. It is enough for a weak solution to exist that the functions m, d are continuous and bounded.

The solutions (both strong and weak) of the stochastic equation generate measures on the space C. As is usually done in the case of Markov processes, to construct these measures we have to find first the transition functions $P(x, s; M, t)$. For the Wiener process this function has the Gaussian density

$$p(x, s; y, t) = (2\pi(t - s))^{-1/2} \exp\left\{-\frac{(y - x)^2}{t - s}\right\}$$

which satisfies the following parabolic type equations in the partial derivatives

$$\frac{\partial p}{\partial t} = \frac{1}{2}\frac{\partial^2 p}{\partial y^2}, \quad \frac{\partial p}{\partial s} = -\frac{1}{2}\frac{\partial^2 p}{\partial x^2}.$$

Similar equations can be written down for any diffusion process defined by the Ito equation with the parameters m, D. We restrict attention to the homogeneous case when the vector stochastic equation has the form (the coordinate-wise representation)

$$d\xi_t^{(i)} = m_i(x)dt + D_i(x)dw_{it}.$$

Then the density $p(x,t,y)$ of the transition function $P(x,t,M)$ satisfies the *backward Kolmogorov equation* (with respect to x, t)

$$\frac{\partial p}{\partial t} + Lp = 0$$

and the *forward Fokker–Planck equation* (with respect to y,t)

$$\frac{\partial p}{\partial t} + L^*p = 0$$

where the *generative operator* L and its conjugated one L^* are defined by the equalities

$$Lp = \sum_{i=1}^{n} m_i(x)\frac{\partial p}{\partial x_i} + \sum_{ij=1}^{n} D_{ij}(x)\frac{\partial^2 p}{\partial x \partial x_j},$$

$$L^*p = \sum_{i,j=1}^{n} \frac{\partial^2[D_{ij}p]}{\partial x_i \partial x_j} - \sum_{i=1}^{n} \frac{\partial[m_i p]}{\partial x_i}.$$

Having the transition function and an initial distribution of the diffusion process ξ_t, we can define a measure $\mathbf{P}_\xi$ on the space C of all continuous functions containing the paths of the process ξ_t in the standard way. Such a measure enables us to compute or to estimate the appearance probabilities of the path set we are interested in. Sometimes it is useful to know the relation between this measure and the Wiener measure $\mathbf{P}_w$. The conditions to ensure either the equivalence of the measure $\mathbf{P}_\xi$ to the measure $\mathbf{P}_w$ or the absolute continuity of it with respect to $\mathbf{P}_w$ are very important. The absolute continuity signifies the possibility to represent the measure $\mathbf{P}_\xi$ through the measure $\mathbf{P}_w$ in the following way:

$$\mathbf{P}_\xi\{M\} = \int_M \eta(z)\mathbf{P}_w\{dz\}$$

where M is any measurable set from C and the density $\eta(\cdot)$ is the Radon–Nikodym derivative $\eta = d\mathbf{P}_\xi\{\cdot\}/d\mathbf{P}_w$ of the measure $\mathbf{P}_\xi$ with respect to the measure $\mathbf{P}_w$. For the scalar processes ξ_t described by equations with unit diffusion $(D = 1)$

$$d\xi_t = m(t,\xi_t)dt + dw_t$$

the following equality

$$\mathbf{P}\left\{ \int_0^t m^2(s,x)ds < \infty \right\} = 1$$

is a necessary condition for the measure $\mathbf{P}_\xi$ to be absolutely continuous with respect to the measure $\mathbf{P}_w$. If this condition holds then the Radon–Nikodym derivative is defined by the following formula

$$\eta_t(\xi) = \frac{d\mathbf{P}_\xi}{d\mathbf{P}_w}(t,\xi) = \exp\left\{ \int_0^t m(s,\xi_s)dw_s - \frac{1}{2}\int_0^t m^2(s,\xi_s)ds \right\}. \tag{5}$$

The random process $\eta_t(\xi)$ is a unique solution of the stochastic equation

$$\eta_t(\xi) = 1 + \int_0^t \eta_s(\xi)m(s,\xi)dw_s.$$

Formula (5) gives a basis for an estimation method of the unknown parameters of the equation and, by analogy with mathematical statistics, it is called the *maximum likelihood method* (MLM). In what follows we are interested in the identification methods of linear vector equations with the diffusion matrix D which is not equal to the identity matrix I. The representation of the Radon–Nikodym derivative for such an equation is given below. If the vector m and matrix D satisfy[b] the condition

$$\int_0^T m^T(t,\xi)[DD^T]^+m(t,\xi)dt < \infty, \quad 0 < T < \infty$$

then the measure $\mathbf{P}_\xi$ will be absolutely continuous with respect to the measure $\mathbf{P}_{\tilde{w}}$ generated by the Wiener process $\tilde{w}_t = Dw_t$ and, moreover, we have

$$\frac{d\mathbf{P}_\xi}{d\mathbf{P}_{\tilde{w}}}(t,\xi) = \exp\left\{ \int_0^t m^T(s,\xi)(DD^T)^+dw_s - \frac{1}{2}\int_0^t m^T(s,\xi)(DD^T)^+m(s,\xi)dw_s \right\}. \tag{6}$$

This formula simplifies in the case of linear equations. We consider now the linear stochastic equation for which the drift and the diffusion are equal to $A\,\xi + B$ and D respectively, i.e.

$$d\xi_t = (A(t)\xi_t + B(t))dt + D(t)dw_t. \tag{7}$$

The existence and uniqueness conditions for a strong solution are fulfilled and in the scalar case ($n = 1$) it turns out to have the following form

$$\xi_t = \exp\left\{ \int_0^t A(s)ds \right\} \left[\xi_0 + \int_0^t B(s)\exp\left\{ -\int_0^t A(v)dv \right\}ds \right.$$
$$\left. + \int_0^t \exp\left\{ -\int_s^t A(v)dv \right\} D(s)dw_s \right].$$

---

[b]The symbol M^+ denotes the pseudoinverse $r \times s$-matrix to the given $s \times r$ matrix M. That is
$$MM^+M = M, \quad M^+ = HM^T, \quad M^+ = M^TG$$
where H and G are some matrices of the corresponding dimensions.

The process ξ_t is Gaussian if the initial value ξ_0 is either Gaussian or non-random. This is not true if the diffusion is equal to $D = D(t)\xi_t$. Then

$$\xi_t = \xi_0 \exp\left[\int_0^t \left(A(s) - 2^{-1}D(s)\right)ds + \int_0^t D(s)dw_s\right]$$

where for simplicity we have assumed that $B \equiv 0$. In the general case $(n \geq 2)$ the solution of Eq. (7) has the form

$$\xi_t = \Phi(t)\left\{\xi_0 + \int_0^t \Phi^{-1}(s)B(s)ds + \int_0^t \Phi(s)^{-1}D(s)dw_s\right\}$$

where A and D are matrices now but B is a vector. Here the $n \times n$-matrix Φ, being the unique solution of the equation $\Phi(t) = I + \int_0^t A(s)\Phi(s)ds$, signifies the fundamental matrix of the equation $\dot{x} = A(t)x$.

Let us give some more results connected with asymptotic properties of solutions of stochastic equations as $t \to \infty$. In particular, we give conditions under which these solutions can be extended unlimitedly (i.e. the considered process is non-terminating).

Theorem 3. *Let*

$$d\xi_t = m(t, \xi_s)ds + D_1(t, \xi_t)dw_t^{(1)} + D_2(t, \xi_t)dw_t^{(2)}$$

be a stochastic equation such that the continuous functions m, D_1, D_2 satisfy the second condition from the previous theorem. Let a smooth function $V(x)$ exist such that $V(x) \geq 0$, $V(0) = 0$ and

$$\lim_{\|x\|\to\infty} V(x) = \infty, \quad LV \leq -aV + b,$$

$$|\nabla V(x)|^2 \leq k_1(1 + V(x)), \quad |LV(x)| \leq k_2(1 + V(x)).$$

(Here k_1, k_2, a, b are some positive constants.)

Then for any (may be random) initial value ξ_0 not depending on $w_t^{(1)}$ and $w_t^{(2)}$ there exists a unique continuous and non-terminating strong solution ξ_t. Moreover, if $\mathbf{E}V(\xi_0) < \infty$ then the following inequality

$$\mathbf{E}V(\xi_t) \leq b/a + [\mathbf{E}V(\xi_0) - b/a]e^{-at}.$$

holds for all $t \geq 0$.

Theorem 4. *Let the coefficients of the equation*

$$d\xi_t = m(t, \xi_t)dt + D(t, \xi_t)dw_t$$

have bounded derivatives (with respect to the first argument), and satisfy the second condition from Theorem 2. In addition, there is a smooth function $V(x)$ such that $V(x) \geq 0, (V(0) = 0)$,

$$\lim_{R\to\infty} \inf_{\|x\|\geq R, t\geq 0} V(x, t) = \infty$$

and

$$LV(\xi_t, t) \leq -h(t, \xi_t)$$

for a continuous function $h(t, x) \geq 0$.

Then for any (may be random) initial value ξ_0 not depending on w_t there exists a unique continuous and non-terminating strong solution ξ_t. Moreover, if $\mathbf{E}V(\xi_0, 0) < \infty$ then

$$\mathbf{E}V(\xi_t, t) \leq \mathbf{E}V(\xi_0, 0) \quad \text{and} \quad \int_0^\infty \mathbf{E}h(\xi_t, t)dt \leq \mathbf{E}V(\xi_0, 0).$$

Theorem 5. *If the coefficients m, D of the stochastic equation have continuous bounded derivatives up to the second order with respect to x and $\int_0^\infty \mathbf{E}\|\xi_t\|^2 dt < \infty$ then*

$$\lim_{t \to \infty} \mathbf{E}\|\xi_t\|^p = 0.$$

The theory of diffusion processes is, of course, not exhausted by these results.

12.2. Stabilization of Solutions of Minimum Phase Ito Equations

In this section we consider Ito equations with drifts including a control u:

$$dx_t = [Ax_t + Bu + f(t)]dt + D(t, x)dw_t, \quad x(0) = x_0. \tag{1}$$

Here $x_0 \in \mathbf{R}^n$, $u \in \mathbf{R}^l$, $w_t \in \mathbf{R}^r$ is a Wiener process, A and B are constant matrices of the corresponding dimensions, the disturbance $f(t)$ is unknown and unobserved, $D(t, x) = (d_{ij}(t, x))$ is an $n \times r$-matrix of the diffusion whose elements satisfy the conditions for the existence and uniqueness of a strong solution (under an appropriate choice of the control $u = u(t)$).

The process

$$z_t = L^T x_t \in \mathbf{R}^m \tag{2}$$

with the unknown $n \times m$-matrix L^T is supposed to be known. A Wiener process (a noise) is, of course, unobserved.

We choose the control law having the conventional form (see Chap. 11)

$$u_t = C^T(t)z_t, \quad \dot{C}(t) = F(z_t, C(t)), \quad C(0) = c_0 \tag{3}$$

where the tuned matrix $C(t)$ is computed by using the observations. To calculate the desired controls it is necessary to specify the matrix-valued function F.

It remains to introduce the transfer matrix of the system (1), (2). It is the $m \times l$-matrix $W(\lambda) = L^T(\lambda I - A)^{-1}B$. The minimum phase property associated with this matrix has been stated in Sec. 1, Chap. 11.

Here our interest is focused on two control aims corresponding to the different versions of the probabilistic stability notion of a random process. These control aims correspond to ones given in Sec. 5, Chap. 11.

Dissipativity in the mean square sense consists in the fulfilment of the following conditions:

(1) the process x_t is non-terminating;
(2) for any pair (x_0, C_0) the inequality

$$\varlimsup_{t \to \infty} \mathbf{E}\left[\|x_t\|^2 + \|C(t)\|^2\right] \le k < \infty$$

holds.

For this aim the form of the tuning law is made concrete in the following way

$$\dot{C}_i = -(g^T z)P_i z - \alpha_i C_i, \quad i = 1, \ldots, l \tag{4}$$

where $P_i = P_i^T > 0$ are $m \times m$-matrices, C_i are the columns of the matrix $C(t)$, g_i are the vectors of order m which form the matrix $G = (g_1, \ldots, g_l)$, $\alpha_i > 0$ are some numbers.

From (4) we obtain

$$C_i(t) = e^{-\alpha_i t}\left[C_i^0 - \int_0^t (g^T z(s))P_i z(s)e^{\alpha_i s}\, ds\right].$$

These equalities together with the relations (3) define a control strategy denoted by $\sigma\Phi_1$.

Finally, we give the class of equations we shall deal with. Let $\mathcal{SO}(n, l, m; G)$ denote the set of all stochastic equations (1), (2) such that the disturbance $f(t)$ and the matrix $D(t, x)$ are bounded, i.e. $\|f\| \le \chi < \infty$, $\|D(t, x)\| \le d < \infty$ and for each equation of this set the matrix $\delta G^T W(\lambda)$ is minimum phase with respect to the matrix G fixed above by some diagonal matrix $\delta > 0$. The equations considered may be regarded as equations under the influence of "input" disturbances divided into two groups. The first is the regular component $f(t)$. The second, the component Ddw_t, is a random disturbance regarded as white noise.

Theorem 1. *The strategy $\sigma\Phi_1$ guarantees dissipativity in the mean square sense with respect to the class $\mathcal{SO}(n, l, m; G)$.*

Proof. We take the following function

$$V(x, C) = x^T H x + \sum_{i=1}^{l}(C_i - C_i^0)^T H_i(C_i - C_i^0)$$

as the Lyapunov function. Here H and H_i are some symmetric positive definite matrices of the corresponding dimensions. It is easy to verify that this function satisfies the conditions of Theorem 3 from Sec. 1. For this purpose, we compute LV by using the representation of the generative operator for the equations (1), (2)

$$LV = 2x^T H(Ax + BC^T L^T x) + 2x^T H f$$

$$+ 2\sum_{j=1}^{l}\left(C_j - C_j^o\right)^T H_j(-\alpha_j C_j) + \rho(t, x)$$

where $\rho = \sum_{i=1}^{n} D_i^T(t,x)HD_i(t,x) \geq 0$. We estimate the second and third terms on the right-hand side of this equality by using the following inequalities ($\mu, \mu_i > 0$ are some positive constants):

$$2x^T Hf \leq \mu x^T HxP + \mu^{-1} f^T Hf,$$

$$-2\alpha_j\left(C_j - C_j^0\right)^T H_j C_j \leq -\mu_j\left(C_j - C_j^0\right)^T H_j\left(C_j - C_j^0\right) + \frac{\alpha_j^2}{2\alpha_j - \mu_j}C_j^{0^T} H_j C_j,$$

$$j = 1, \ldots, l.$$

The latter inequality holds for $\mu_j < 2\alpha_j$. Using Theorem 8 from Sec. 1, Chap. 11 we obtain ($Q = Q^T > 0$)

$$LV \leq -x^T Qx + \mu x^T Hx + \mu^{-1} f^T Hf - \sum_{j=1}^{l} \mu_j\left(C_j - C_j^0\right)^T H_j\left(C_j - C_j^0\right)$$

$$+ \sum_{j=1}^{l} \frac{\alpha_j^2}{2\alpha_j - \mu_j}C_j^{0^T} H_j C_j + \rho(t,x).$$

In view of the boundedness of $f(t)$ and D we can choose μ, $\mu' \in (0, \min_j 2\alpha_j)$ so that $(\mu + \mu')H \leq Q$ where $\mu' = \min_j \mu_j$. Then the previous inequality can be written as follows

$$LV \leq -aV + b, \quad a, b > 0.$$

It is not difficult to see that the remaining conditions of Theorem 3 from Sec. 1 will hold if we take $\zeta_t = (x_t, C(t))$. Hence, for all t we have

$$\mathbf{E}(\|x_t\|^2 + \|C(t)\|^2) \leq \frac{b}{a} + [\mathbf{E}V(x_0, C_0 - b/a]e^{-at}$$

and the given aim is reached. $\qquad\qquad\qquad\qquad\qquad\qquad\qquad\qquad\qquad\qquad\square$

Let us now consider the scalar equation in the "input-output" variables

$$A(p)x_t = bu + \dot{w}_t \qquad\qquad (5)$$

where $\dot{w}_t$ means the white noise, $b \neq 0$ and $A(p)$ is the operator polynomial,

$$A(p) = p^n + \sum_{i=0}^{n-1} a_i p_i.$$

Having in mind the former aim we are looking for the control in the form

$$u = \sum_{i=1}^{n-1} c_i(t)p^i x_t.$$

The vector $z = (x, \ldots, x^{(n-1)})$ is observed. By the same arguments as in Sec. 5, Chap. 11 we can write down the strategy with the help of a vector $C = (c_0, c_1, \ldots, c_{n-1})$, of a polynomial $G(p) = \sum_{i=0}^{n-1} g_i p^i$ and two positive numbers

α and β. It has the form

$$\dot{C} = -\alpha[G(p)x]z - \beta C. \tag{6}$$

As seen from the above (Sec. 5, Chap. 11), the minimum phase property will take place if $g(p)$ is a Hurwitz polynomial and $\mathbf{sign}\, g_{n-1} = \mathbf{sign}\, b$. If this is true then the control strategy defined by (6) guarantees dissipativity of solutions of Eq. (5).

We now consider another control aim which is called *stability in the mean square sense*. This means that

(1) the process x_t is non-terminating;
(2) for any pair $(x_0, C(0))$ the inequalities

$$\int_0^\infty \mathbf{E}\|x_t\|^2 dt < \infty, \quad \mathbf{E}\|C(t)\|^2 < \infty$$

hold.

To achieve the given aim we make concrete the general form of control (3) as follows

$$\dot{C}_i(t) = -(g_i^T z)P_i z, \quad i = 1, \ldots, l \tag{7}$$

or, in the explicit form,

$$C_i(t) = C_i^0 - \int_0^t (g^T z(s))P_i z(s)ds,$$

where the same notation is used as in algorithm (4). We denote this strategy by $\sigma\Phi_2$.

We consider the class $\mathcal{S}(n, l, m; \varepsilon, G)$ of equations having the form

$$dx_t = (Ax_t + bu)dt + D(t, x)dw_t \tag{8}$$

as the class of controlled equations we shall deal with. Here the matrix $D(t, x)$ satisfies the conditions $\|D(t, x)\| \leq e\|x\|$ and $e \leq \varepsilon$ where ε is an arbitrary positive number but the matrix $\delta G^T W(\lambda)$ is minimum phase for some diagonal matrix $\delta > 0$. The considered equation is said to take the influence of a "parametric" disturbance of the white noise type.

Theorem 2. *The strategy $\sigma\Phi_2$ guarantees stability in the mean square sense of the solutions of the equations from $\mathcal{S}(n, l, m; \varepsilon, G)$ when ε is sufficiently small.*

Proof. We define the Lyapunov function as follows

$$V(x, C) = x^T H x + (C - C^0)^T H'(C - C^0). \tag{9}$$

Hence we obtain

$$LV = 2x^T H(Ax + BC^T L^T x) + 2(C - C^0)^T H'F(z) + \rho(t, x)$$
$$= x^T(HA(C^0) + A^T(C^0)H)x + 2(C - C^0)^T[H'F(z) + (x^T HB)z] + \rho$$

with $A(C^0) = A + BC^{0^T}L^T$ and $\rho(t,x) = \sum_{i=1}^{k} d_i^T H d_i$. According to Theorem 8, Sec. 1, Chap. 11 and the minimum phase property, there exist matrices $H = H^T > 0$ and C^0 such that

$$HA(C^0) + A^T(C^0)H = -Q < 0, \quad HB = LG.$$

Let $P = H^{-1}$ and the function $F(z)$ in the expression of LV be replaced by its value from (7). Then we obtain

$$LV = -x^T Q x + \rho. \tag{10}$$

In view of the definition of the class $\mathcal{S}(n,l,m;\varepsilon,G)$ we have $|\rho| \leq e\|H\|\|x\|^2$. Let q' be the minimal eigenvalue of the matrix Q. If $e\|H\| < d$, then $LV(x,C) \leq -(1-q)q\|x\|^2 < 0$. It is not difficult to verify that the remaining conditions of Theorem 4 from Sec. 1 (current chapter) hold. Putting there $\xi_t = (x_t, C(t))$, $h(x) = q\|x\|^2$ we can draw the conclusion that the given aim is reached. $\qquad\square$

So, for the quadratic Lyapunov function (9) having the following properties

$$V(x,C) > 0 \quad \text{for } x \neq 0, \ C \neq C^0; \ \ LV \leq 0 \tag{11}$$

to exist it is sufficient that the conditions of Theorem 2 should be satisfied. It is interesting to know to what extent these conditions are necessary. The next theorem answers this question.

Theorem 3. *Let $W(\lambda) \neq 0$ and* **rank** $B = l$. *Then for the Lyapunov function* (9) *having properties* (11) *to exist it is necessary that the conditions of Theorem 2 should be satisfied.*

As seen from the proof of Theorem 2 for stability in the mean square sense to be provided the level of the disturbances regarded as an input of the controlled model must be chosen low (i.e. the diffusion should be small). But this serious restriction can be omitted in some special cases. We consider two of them here. Let $\mathcal{S}(n,l,m;G)$ denote the class of Ito equations of the form

$$dx_t = (Ax + Bu)dt + \sum_{i=1}^{k} d_i(t,x)dw_t^{(i)}.$$

Moreover, the matrix $\delta G^* W(\lambda)$ is minimum phase for some diagonal matrix $\delta > 0$ and the diffusion (the collection of vectors (d_i)) has either the form

$$(\alpha) \quad d_i(t,x) = \tilde{b}_i r_i^T(t)\chi \quad \text{where } \tilde{b}_i = \sum_{j=1}^{l} \beta_{ji} b_j, \ \|r_i(t)\| \leq \varkappa;$$

or

$$(\beta) \quad d_i(t,x) = d_i(t)\tilde{g}_i^T z \quad \text{where } \tilde{g}_i = \sum_{j=1}^{l} \gamma_{ji} g_j, \ \|d_i(t)\| \leq \varkappa;$$

where b_i and g_i mean the columns of the matrices B and G respectively and $z = L^T x$.

Theorem 4. *The strategy $\sigma\Phi_2$ guarantees stability in the mean square sense of solutions of equations from $\mathcal{S}(n, l, m; G)$.*

Proof. We have to verify that the proper Lyapunov function satisfies the inequality $LV(x, C) \leq -x^T R x, \; R > 0$ which guarantees attainment of the given aim. As usual, we have

$$V(x, C) = x^T H x + \sum_{i=1}^{l} (C_i - C_i^0)^T H_i (C_i - C_i^0), \quad H_i > 0.$$

Next, we obtain

$$LV(x, C) = 2x^T H(Ax + BC^T L^T x) + 2 \sum_{i=1}^{l} \left(C_i - C_i^0\right)^T H_i F_i(z) + \rho(t, x)$$

$$= x^T (HA(C^0) + A(C^0)H)x + 2 \sum_{i=1}^{l} \left(C_i - C_i^0\right)^T [H_i F_i(z)$$

$$+ (x^T H b_i)z] + \rho(t, x)$$

where $A(C^0), \rho(t, x)$ have the conventional form and the equality

$$x^T H B (C - C^0)^T z = \sum_{i=1}^{l} (x^T H b_i)(C_i - C_i^0)z$$

has been used. We now put $H_i = \delta P_i^{-1}$, then

$$LV(x, C) = x^T (HA(C^0) + A^T(C^0)H)x + \rho(t, x)$$

$$+ 2 \sum_{i=1}^{l} (C_i - C_i^0)^T [\delta_i(g_i^T z) + (x^T H b_i)z].$$

Let the diffusion satisfy the condition (α). If we choose the matrix H such that $HB = LG\delta$, i.e. $Hb_i = \delta_i L g_i, \; i = 1, \ldots, l$ then the last summand $(\sum_{i=1}^{l})$ on the right-hand of the equality above will vanish and

$$\rho(t, x) = \sum_{j=1}^{k} d_j(t, x) H d_j(t, x) = \sum_{j=1}^{k} (\tilde{b}_j r_j^T(t))^T x^T H(b_j r_j^T(t))x \leq k_1 \varkappa^2 \|H\| \|x\|^2.$$

Let the matrix $H \; (> 0)$ satisfy the equalities $Hb_i = \delta_i L^T g_i$ and

$$HA(C^0) + A^T(C^0)H = -R < 0$$

for some matrix C^0, where $R = 2k_1 \varkappa^2 I$. It exists due to the conditions on the class S. At last, we obtain

$$LV(x, V) = -x^T R x + \rho(t, x) \leq -x^T R x + k_1 \varkappa^2 \|H\| \|x\|^2 < 0.$$

This completes the first part of the proof.

Now, let $d_j(t, x)$ satisfy condition (β). Then

$$\rho(t, x) = \sum_{j=1}^{k} d_j^T(t, x) H d_j(t, x) = \sum_{j=1}^{k} d_j^T(t) H d_j(t)(d_j^T z)^T \leq k_2 \varkappa^2 \|H\| \sum_{j=1}^{k} (g_j^T z)^2.$$

Using Theorem 8 from Sec. 1, Chap. 11 again we find the matrices H and C^0 such that $Hb_j = \delta L^T g_i$ (for all i) and

$$H(A + B(C')^T L^T)^T + (A + B(C')^T L^T)H = -Q < 0.$$

We substitute both these matrices in the formula for $L(V)$ and put $C^0 = C' - \nu G$ where ν is some number to be defined later on. We obtain the following

$$
\begin{aligned}
LV(x, C) &= x^T \left[H(A + B(C')^T L^T + (A + B(C')^T L^T)H \right] x + \rho(t, x) \\
&\quad - \nu x^T H B G^T L x - \nu x^T L^T G B^T H x \\
&= -x^T Q x - 2\nu \sum_{i=1}^{l} \delta_i (g_i^T z)^2 + \rho(t, x) \\
&\leq -x^T Q x - 2\nu \sum_{i=1}^{l} \delta_i (g_i^T z)^2 + k_2 \nu \sum_{i=1}^{l} (g_i^T z)^2.
\end{aligned}
$$

If ν satisfies the condition $2\nu \min_i \delta_i < k_2$ then we obtain the required inequality $LV(x, C) \leq -x^T Q x \leq 0$. This finishes the proof of the theorem. $\qquad \square$

Corollary 1. *Under the conditions of Theorem 4*

$$\lim_{t \to \infty} \mathbf{E} \|x_t\|^2 = 0.$$

This assertion follows in an evident way from Theorem 5, Sec. 1 of the current chapter. Indeed, the conditions about the considered equations imply the existence of continuous derivatives of the diffusion coefficients up to the second order and, hence, the required assertion follows from $\int_0^\infty \mathbf{E} \|x_t\|^2 dt < \infty$.

Now we apply the results obtained to the scalar equation in the "input $-$ output" variables ($p = d/dt$)

$$p^n x + \sum_{i=1}^{n-1} [a_i + \eta_i(t)] p^i x = bu, \quad b \neq 0$$

where $\{\eta_t\}$ is a "parametric" disturbance of the white noise type such that $\mathbf{E}\eta_i(t) = 0$ and $\mathbf{E}\eta_i(t)\eta_j(t') = \varkappa_{ij}\delta(t - t')$, with the correlation matrix $(\varkappa_{ij})$ being unknown. The coefficients $a_1, \ldots, a_{n-1}, b$ are assumed to be unknown as well. The vector $z = (x, \dot{x}, \ldots, x^{(n-1)})$ is observed and the control is defined with the help of the operator polynomial $C(p, t) = \sum_{i=0}^{n-1} c_i(t) p^i$, i.e.

$$u(t) = \sum_{i=0}^{n-1} c_i(t) p^i z(t).$$

The tuned quantities $(c_i(t))$ have to be detected in the course of control. The tuning algorithm is chosen as follows

$$\dot{C}(t) = -\beta G(p) z \quad \text{or} \quad C(t) = \bar{C}_0 - \beta \int_0^t G(p) z(s) ds,$$

where $\beta > 0$ and $G(p) = g_0 + g_1 p + \cdots + g_{n-1} p^{n-1}$ is a Hurwitz polynomial such that $\mathbf{sign}\, g_{n-1} = \mathbf{sign}\, b$. The described strategy ensures stability in the mean square sense. For it to be implemented it is necessary only to know the sign of b. The considered strategies are regarded as direct. The unknown parameters of the equation are not estimated but the observations are used only to "tune" the parameters of the control law.

We pay attention to the possibility to treat the mean square problem as the optimization one or, more exactly, as a linear-quadratic problem consisting of minimization of the functional

$$W(u) = \varlimsup_{t \to \infty} t^{-1} \int_0^t \mathbf{E}\left[x_s^T Q x_s + u_s^T R u_s\right] ds$$

where $Q,\ R > 0$. Indeed, the first term under the symbol of integral is equal to zero, i.e.

$$\lim_{t \to \infty} t^{-1} \int_0^t \mathbf{E} x_s^T Q x_s ds = 0.$$

In view of the boundedness of $\|C(t)\|$, a similar equality holds for the second term as well. Hence, $W(u) = 0$ under the given algorithm of control, i.e. the functional takes its minimal value.

The specific problem of adaptive control of the Ito equations

$$dx = m(x)dt + D(u)dw_t, \quad t \geq 0 \ \ x, w \in \mathbf{R}^n \tag{12}$$

the diffusion of which depends on the control, has been left to the end of this section. The problem is to find a control such that the trivial solutions of equations from some class should be stable in the probability sense.

Definition 1. A trivial solution is *stable in the probability* sense if for any ε_1, $\varepsilon_2 > 0$ there is $\delta > 0$ such that the fulfilment of the condition $\|x_0\| < \delta$ implies $\mathbf{P}\{\sup_t \|x_t\| < \varepsilon_1\} > \varepsilon_2$.

For the trivial solution to be stable in the probability sense, the inequality $\mathcal{L}V(x) \leq 0$ is a sufficient condition which should hold in some neighborhood of the origin. Here $V(x)$ is a Lyapunov function of Eq. (12) and

$$\mathcal{L} = \sum_{i=1}^{n} m_i(x)\frac{\partial}{\partial x_i} + \frac{1}{2}\sum_{ij=1}^{n} a_{ij}(x)\frac{\partial^2}{\partial x_i \partial x_j}$$

where $a_{ij} = \sum_{k=1}^{n} d_{ik}(x)d_{jk}(x)$ is generative operator of this equation. Let us assume that in some neighborhood U of the point $x = 0$ the drift $m(x)$ and the diffusion $D(u(x))$ satisfy the Lipschitz conditions

$$\|m(x) - m(y)\| \leq \beta\|x - y\|, \quad \|D(u(x)) - D(u(y))\| \leq \gamma\|x - y\|. \tag{13}$$

Let the diffusion have the form

$$D(u(x)) = qu(x), \quad q = \text{const.}$$

The Lipschitz matrix $u(x)$ is stabilizing if there is a number q_0 such that for any $q > q_0$ the trivial solution of (12) is stable in the probability sense. An example of such a matrix $u(x) = \|u_{ij}(x)\|$ is

$$u_{i1}(x) = x_i, \quad i = 1, \ldots, n, \qquad u_{ij}(x) = 0, \quad i = 1, \ldots, n, \ j = 2, \ldots, n. \tag{14}$$

Obviously, this matrix is Lipschitz and does not depend on the properties of the function $m(x)$. Let us denote by $\mathbf{B}$ the class of Ito equations with the Lipschitz drift $m(x)$ and the control of the form (14).

Theorem 5. *The matrix $u(x)$ is stabilizing with respect to the class $\mathbf{B}$.*

Proof. We apply the generative operator $\mathcal{L}$ to the Lyapunov function $V(x) = \|x\|^{2\alpha}$, $\alpha \in (0, 1/2)$. Thus we obtain

$$\mathcal{L}V(x) = 2^{-1}q^2\|x\|^{2(\alpha-1)} \sum_{ij=1}^{n} x_i x_j \left[2(\alpha-1)\|x\|^{-2} x_i x_j + \delta_{ij}\right]$$

$$+ 2\alpha\|x\|^{2\alpha-1} \sum_{i=1}^{n} m_i(x) x_i \leq \alpha\|x\|^{2\alpha}[2\beta - (1-2\alpha)q^2].$$

As seen from the above the stability condition $\mathcal{L}V \leq 0$ holds for all $q > q_0 = \sqrt{2(1-2\alpha)^{-1}\beta}$. $\qquad\square$

We now consider robustness of the stabilizing control constructed above or, in other words, whether the matrices close to $u(x)$ are also stabilizing.

Theorem 6. *Let $\|W(x)\| \leq \varepsilon\|x\|$. Then for $\varepsilon < (2\gamma n^2)^{-1}$ the matrix $u(x) + W(x)$ is stabilizing with respect to the class $\mathbf{B}$.*

Proof. The method is clear, namely, we have to prove that the generative operator $\tilde{\mathcal{L}}V(x)$ of the equation $dx = m\,dx + qW\,dw$ differs from that of $\mathcal{L}V(x)$ of the original equation $dx = m\,dx + qU\,dw$ very little. Indeed,

$$\tilde{\mathcal{L}}V(x) \leq \mathcal{L}V(x) + \frac{q^2}{2}\|U(x)W^T(x) + W(x)U^T(x)\|n \max_{ij}\left|\frac{\partial^2 V(x)}{\partial x_i \partial x_j}\right|$$

$$\leq \mathcal{L}V(x) + 2q^2\alpha\varepsilon\gamma n^2\|x\|^{2\alpha} \leq \alpha\|x\|^{2\alpha}[2\beta - (1 - 2\alpha + 2\varepsilon\gamma n^2)q^2].$$

Therefore for $\varepsilon \in (2\gamma n^2)^{-1}$ the stabilization condition takes place. $\qquad\square$

Not every Lipschitz control can be regarded as the stabilizing control with respect to the class $\mathbf{B}$. Without going into details, note that the "non-stabilizing" controls are robust as well, i.e. the controls close to them have the same non-stabilizing properties.

The simple and elegant problem above arose due to prolonged attempts to stabilize the non-stable deterministic equation

$$\dot{x} = m(x), \quad x \in \mathrm{R}^m.$$

These attempts were based on the introduction of an artificial hindrance of the form (12). The natural interpretation of stochastic stabilization for this class of equations is likely to belong to the adaptive concept.

12.3. Identification Methods for Ito Equations

We consider the main characteristics of estimation of unknown parameters of Ito equations. We do this to solve the adaptive control problem with the help of the identification strategies. It is reasonable to consider the structure of an identification strategy while describing concrete strategies.

First, we estimate the unknown diffusion matrix D assuming it is constant. This matrix proves to be estimated exactly with probability one on any finite time interval $[T_0, T_1]$ by using the "quadratic" variation

$$\lim_{\max \triangle t_i \to 0} \sum_{i=1}^{n} (\xi_{t_i} - \xi_{t_{i-1}})(\xi_{t_i} - \xi_{t_{i-1}})^T = DD^T \tau$$

where $T_0 = t_0 < t_1 < t_2 < \cdots < t_n = T_1$, $\triangle t_i = t_i - t_{i-1}$, $\tau = T_1 - T_0$. We can assume that these calculations have been done at the very beginning of the identification procedure and, thereafter, we focus on the estimation of the drift parameters only.

Elementary solutions of identification problems exist only in simplest cases.

For example, the unknown parameter m in the equation $d\xi_t = mdt + Ddw_t$ can be found from the explicit form of its solution

$$\xi_t = mt + Dw_t$$

by using the strong law of large numbers, namely,

$$m = \lim_{t \to \infty} t^{-1}\xi_t \quad \text{a.s.}$$

We also use the same approach if the diffusion is a bounded function of time.

Below we consider the linear stochastic equations of the form

$$d\xi_t = A\xi_t dt + Ddw_t. \tag{1}$$

Even for the most simple model of this type we are forced to use the approach based on the deep results of stochastic processes theory. The Maximum-Likelihood Method (MLM) is the most important among them. It is based on the Radon–Nikodym derivative $\eta_t(\xi, \theta)$ which depends on the observable trajectory of the process and the unknown parameters θ. The value of θ maximizing $\eta_t(\cdot)$ is chosen as the parameter estimate, i.e. $\hat{\theta}_t = \operatorname{argmax} \eta_t(\xi, \theta)$. The derivative $\eta_t(\xi, \theta)$ as the function of the estimated parameter θ is called the *likelihood function*. For the linear equation (1) this function has the following form (compare with the formula (6) from Sec. 1):

$$\eta_t = \frac{d\mathbf{P}_\xi}{d\mathbf{P}_w} = \exp\left\{ \int_0^t (\delta\xi_s, d\xi_s) - \frac{1}{2}\int_0^t (A\xi_s, \delta\xi_s)ds \right\} \tag{2}$$

where $\delta = (DD^T)^+$ is the pseudoinverse matrix to DD^T and (a, b) is the scalar product in $\mathbb{R}^n$.

To find the MLM estimation $\hat{A}_t$ of A at time t we have to find a zero of the derivative of the function $\eta_t(\xi, \theta)$ with respect to the matrix A. If the matrix

$$\left(\int_0^t (\xi_s^{(i)}, \xi_s^{(j)}) ds \right) \quad i, \; j = 1, \dots, n$$

(here $\xi_i^{(i)}$ denotes the ith component of the vector $\xi_t = (\xi_t^{(i)}, \dots, \xi_t^{(n)})$) is non-degenerate we take

$$\hat{A}_t = \left(\int_0^t (\xi_s, d\xi_s) \right) \left(\int_0^t (\xi_s, \xi_s) ds \right)^{-1}. \tag{3}$$

Simple transformations of the first factor lead to two other forms of this estimate (for simplicity, let $D = I$), namely,

$$\hat{A}_t = A + \left(\int_0^t (\xi_s, dw_s) \right) \left(\int_0^t (\xi_s, \xi_s) ds \right)^{-1}. \tag{4}$$

$$\hat{A}_t = \frac{1}{2}(\xi_t \xi^T - \xi_0 \xi_0^T - tI) \left(\int_0^t (\xi_s, \xi_s) ds \right)^{-1}. \tag{5}$$

It is possible to prove consistence of the MLM estimates by using the first formula. For this purpose, the following result may be used.

Lemma 1. *Let a Wiener process w_t and some process $\zeta_t(\omega)$ satisfy the conditions*

$$\int_0^t \zeta_s^2(\omega) ds < \infty \quad \forall t \in [0, \infty), \qquad \int_0^\infty \zeta_s^2(\omega) ds = \infty \quad \text{a.s.}$$

Then with probability one we have the equality

$$\lim_{t \to \infty} \frac{\int_0^t \zeta_t(\omega) dw_s}{\int_0^t \zeta_s^2(\omega) ds} = 0.$$

Formula (5) points to a full collection of sufficient statistics for the estimates of the elements of the matrix A. Evidently, they are represented by the following random variables (provided that $\xi_0 = x$)

$$\int_0^t (\xi_s, \xi_s) ds, \quad \int_0^t \xi_k d\xi_l, \quad i, \; j, \; k, \; l = 1, \dots, n.$$

We state the properties of these estimates for the stable matrices A (their eigenvalues have negative real parts).

Proposition. *The MLM estimates $\hat{A}_t$ of the stable matrix are consistent and efficient.*[c]

---

[c]The unbiased estimate of the parameter is called *efficient* if its variance is minimal, i.e. the Rao–Kramer inequality turns into equality.

Apart from the above the matrix $\sqrt{t}(\hat{A}_t - A)$ is asymptotically normal (i.e. the $N(0, Q)$ type) with the parameter 0 (zero-matrix) and Q (some positive definite matrix) as $t \to \infty$. The convergence rate of the distributions has the order $t^{-1/2}$, i.e. it is conventional in statistics.

Different moments of the MLM estimates may be calculated and, therefore, it is possible to study the quality of the obtained estimates of $\hat{A}$. Here we omit these details.

12.4. LQP for Stochastic Ito Equations

Here we deal with the linear stochastic Ito equation

$$d\xi_t = (A\xi_t + Bu)dt + Ddw_t \tag{1}$$

where $\xi \in \mathrm{R}^n$, $u \in \mathrm{R}^l$ and A, B, D are constant matrices of the corresponding dimensions and w_t is the standard Wiener process. The process ξ_t is supposed to be observed in contrast with the process w_t. Under these conditions it is required to find the optimal control $u = u(t)$ which minimizes the non-negative quadratic functional

$$\Phi(u) = \varlimsup_{t \to \infty} t^{-1} \int_0^t [\xi_s^T Q \xi_s + u_s^T R u_s] ds \tag{2}$$

for some given matrices $Q = Q^T \geq 0$, $R = R^T > 0$. The solution of this problem is well-known in the so-called "classical" version when all parameters of Eq. (1) are known exactly.

The main problem consists of finding the solution of the matrix algebraic Riccati Equation

$$AS + SA^T + Q - SBR^{-1}B^T S = 0. \tag{3}$$

in the class of all symmetric non-negative definite matrices. If S is minimal of them (in the ordering of symmetric matrices) then the required control can be written in the form of a linear feedback

$$u_0 = K\xi_t, \quad K = -R^{-1}B^T S \tag{4}$$

where the $l \times n$-matrix K is the "gain matrix". Under the appropriate assumptions this control gives the minimal value of the functional $\Phi(u)$ equal to

$$\Phi(u_0) = \min_u \Phi(u) = \mathbf{tr}(D^T S D).$$

The matrix of the diffusion equation obtained (its drift is equal to $A + BK$) is stable.

We note two more expressions, namely, for Eq. (1)

$$d\xi_t = (A + BK)\xi_t \, dt + Ddw_t \tag{5}$$

and for its solution

$$\xi_t = e^{(A+BK)t}\xi_0 + \int_0^t e^{(A+BK)(t-s)}D\,dw_s. \tag{6}$$

To formulate solvability conditions of the Riccati Equation we introduce the matrix C defined as the factor of the decomposition $Q = CC^T$.

Proposition 1. *If the matrices triplet (A, B, C) is stabilizable and detectable then the Riccati equation is solvable.*

We remark that the required matrix S depends on the coefficients of both the equation and the functional in a continuous way, i.e. on the matrices A, B, Q, R. The minimal value of the functional is expressed by means of these coefficients and the diffusion matrix D. We can now focus on the adaptive optimization for LQP. Among the various approaches to the adaptive control, identification is the simplest but the Least Square Type Methods (or the Maximum Likelihood Type Methods) are the most known. To consider them it is necessary to specify the group of the unknown parameters of Eq. (1), on the assumption that the matrix A is only unknown. The other parameters are known exactly.[d]

So, let us have some algorithm of identification which gives the consistent estimates of the matrix A. We shall discuss now other important aspects of optimal strategies. We begin with the characteristic features of the solution of the Riccati equation. If the estimates of the matrix A are computed continuously in time then the matrices S_t are computed continuously in time too and it may happen that the pair of matrices (A_t, B_t) are non-stabilizable at some moments t. If it is true then the matrices S_t will not exist at these moments.

Proposition 2. *If the sequence of the stabilizable pairs of the matrices (A_t, B_t) converges to a non-stabilizable pair (A, B) as $t \to t_0$ then $\lim_{t \to t_0} \|S_t\| = \infty$.*

Thus, if in the course of the estimation of the matrix A its current estimates get to those regions of the parameter space where the existence conditions of the solutions of the Riccati equations fail, then the natural progress of the process of the control will be interrupted. This fact has a dramatic impact on the design of the optimal strategy for LQP.

We now describe a class of adaptive strategies suited to be the optimal ones for LQP with respect to the appropriate class of Ito equations. These strategies are differentiated one from another by the indentification methods used. For all strategies used the obtained information, i.e. the estimates of the parameters is treated in the same way. For every strategy its estimating "block" produces the estimates of the matrix A continuously but the controlling system uses them only at some discrete moments t (chosen in advance or occasionally) to solve the Riccati

---

[d]The assumption of knowing the matrix B does not seems very reasonable. However we do not know whether it will be possible to identify the matrices A and B simultaneously provided they are both unknown.

equation. These moments satisfy the following two conditions

$$\lim_{k \to \infty} t_k = \infty, \quad \triangle t_k = t_k - t_{k-1} \to 0.$$

At each moment t_k the gain matrix K_t is calculated through the matrix S_{t_k} and, thereafter, the computed matrix K_t is used up to the moment t_{k+1} and so on. The set of these strategies is denoted by σ_{VG}.

Let $\mathcal{L}(n, l; B, C, D)$ denote the class of equations of the form (5). We note that C is defined as $Q = CC^T$ where the matrix Q is given by (2). For all equations from the considered class the matrices triplets (A, B, C) must be stabilizable and detectable. The functional given by the formula (2) is chosen as the criterion of the control quality, i.e. the control aim is to minimize the following functional

$$\Phi(u) = \overline{\lim_{t \to \infty}} \, t^{-1} \int_0^t \left[\xi_s^T Q \xi_s + u_s^T R u_s \right] ds.$$

Theorem 1. *The strategies belonging to the set σ_{VG} are adaptive with respect to the class $\mathcal{L}(n, l; B, C, D)$, i.e. they guarantee minimization of the functional $\Phi(u)$ with probability one, namely,*

$$\min_u \Phi(u) = \mathbf{tr}(D^T S D).$$

Proof. We have to avoid the violation of the existence conditions of the solutions of the Riccati equation (3) to start with. These violations will appear at the moments when the $\hat{a}_{ij}(t)$ paths (the estimates of the elements of the matrix A) enter into some set M of points for which the stabilizability conditions and detectability fail.

If the equation is controllable then the violation of stabilizability will consist in reducing the rank of the controllability matrix. It is easy to understand what should have happened if the stabilizability and controllability conditions have failed simultaneously.

As known a manifold $M \subset \mathrm{R}^{n^2 + nm}$ of dimension strictly less than $n^2 + nm$ is closed and nowhere dense. Therefore we can avoid the unpleasant effects caused by the unlimited increasing of $\|S_t\|$ (as S_t approaches to M). It can be done by various means. The simplest of them is based on a transfer of S_t from the current position to a new one closed to the former the growth of $\|S_t\|$ has been noted in.

But these details are referred to the practical realization of the calculative procedure for the considered strategy σ_{VG}.

Let us consider the distribution $\mathbf{P}_t$ of the current estimates $\hat{a}_{ij}(t)$. For concreteness' sake the consistent estimates

$$\hat{A}_t = \frac{1}{2}(\xi_t \xi^T - \xi_0 \xi_0^T - tD) \left(\int_0^t (\xi_s, \xi_s) ds \right)^{-1}.$$

obtained by using the Least Square Method (LSM) are chosen.

This distribution is absolutely continuous and it has a positive density defined on $\mathrm{R}^{n(n+m)}$. Indeed, it is a function of the Wiener process and, hence the probability to enter into the manifold M at any moment t_k is equal to zero. As a result of the consistency of the estimates, the $\hat{a}_{ij}(t)$ paths enter some neighborhood (the "safety zone") of the true value of the matrix A and, thereafter, the norm $\|S_t\|$ remains finite.

Here we *do not* assert that this norm is uniformly bounded.

The $A_t \to A$ convergence a.s. and the continuity of S_t with respect to the coefficients of Eq. (1) signify the $S_t \to S$ convergence as $t \to \infty$. This, in turn, implies the $K_t \to K$ convergence a.s.

It remains to verify that the current values of the functional

$$\Phi_t(u) = t^{-1} \int_0^t \left[\xi_s^T Q \xi_s + u_s^T R u_s\right] ds = t^{-1} \int_0^t \xi_s^T (Q + K^T R K) \xi_s \, ds$$

tend to the minimal one. For this purpose, we write down the following functionals

$$\Phi_t^{(1)}(u) = t^{-1} \int_0^t \xi_s^T (Q + K^T R K) \xi_s ds,$$

$$\Phi_t^{(2)}(u) = t^{-1} \int_0^t \eta_s^T (Q + K_s^T R K_s) \eta_s ds$$

differentiated by the form of the matrices K. In the first functional it is the same as in the classical case (see the beginning of this section) but in the second one it depends on s due to identification. As a result of this two different diffusion processes ξ_t and η_t have appeared. We consider them as the discrete realizations under the same input having the form of the Wiener process w_t. Thus, these processes approach each other unlimitedly as $t \to \infty$ a.s. (These arguments can easily be done precisely.) According to properties of the Cesaro averages we can conclude that the processes ξ_t and η_t are identical. Then these functionals have the forms

$$\Phi_t^{(1)}(u) = t^{-1} \int_0^t \xi_s^T N_1 \xi_s \, ds \quad \Phi_t^{(2)}(u) = t^{-1} \int_0^t \xi_s^T N_2(s) \xi_s ds$$

where $N_1 = Q + SBRB^T S$ and $N_2(s) = Q + S_s BRB^T S_s$ stand for the exact and "experimental" values of the minimized functional.

It remains to compare the terms in both sums. They are almost identical, namely,

$$\xi_s^T N_1 \xi_s = \xi_s^T (Q + SBRB^T S) \xi_s, \quad \xi_s^T N_2(s) \xi_s = \xi_s^T (Q + S_s BRB^T S_s) \xi_s.$$

Indeed, the matrices mentioned above approach each other as $s \to \infty$. Using the properties of the Cesaro averages again we obtain the following equality

$$\lim_{t \to \infty} t^{-1} \int_0^T \xi_s^T [SBRB^T S - S_s BRB^T S_s] \xi_s ds = 0$$

which completes the proof. $\square$

If we find a method to estimate the matrices A and B simultaneously, then identification will play a more important role in constructing optimization strategies.

COMMENTS AND SUPPLEMENTS

Chapter 1

Section 1. We state the main material needed in what follows both on probability theory and on stochastic processes, in particular, the Jonescu Tulcea Theorem (see, for instance, G[20]).

Section 2. The notion of *controlled random process* in the form considered here comes from Sragovich's monograph (see S[133]) with an amendment concerned with observable processes. It differs from similar notions used in other text-books (for example, G[6], G[11]). The rest of this section is conventional. There are many text-books devoted to automata theory. The bases of this theory were founded by Mc Culloch and Pitts (see G[18]). However we use neither the abstract theory nor the structural one of automata. The interested readers may refer to the well-known book by Kalman, Falb and Arbib G[16].

Section 3. The question of the formal definition of "adaptive control" has never being touched on in the applied literature. Nor was it discussed in English-language mathematical works. But this circumstance did not impede strangely proving adaptability of strategies. For the first time the material of this section appeared S[133]. Some attempts of formalizing "adaptive control" which were undertaken by the other authors (from the former Soviet Union) were not, as seems to us, satisfactory. The definition of adaptability taken means that the given strategy can be refered to as an adaptive one only after it is proved that it ensures the attainability of the control aim stated in advance for each process from some specified class of controlled random processes.

Section 4. The "learning system" notion appeared in connection with mathematical modelling of physiological phenomena from the standpoint of bioherevizm (see Bush and Mosteller S[7]). The first mathematical problems arising here were considered by Oničescu and Mihoc as far back as in 1936 (see S[109]). The new direction of research generated by them, referred to as the theory of "systems with complete connections", became one of the main themes for Roumanian mathematicians (Iosifescu and Teodorescu S[64], Griogorescu, Iosifescu S[46]). By using another approach (the theory of Markov processes) the learning problem was studied by Norman [108]. The principal results on the asymptotic properties of the learning model (defined in the most general form) were obtained by Usachev S[146]–[149]. The relation of adaptive strategies both with automata and with learning systems were discovered by Sragovich S[133].

Section 5. We study the practice meaning of "adaptive control" by critical analysis of the problem of controlling a Markov process with unknown parameters on a finite interval and its Bayesian solution. The proposed analysis is due to Aoki (see S[1]). There, this problem was considered in context of the Bellman and Feldbaum approaches. Quite often related publications appeared in scientific periodicals. The survey of them up to 1982 can be found in Kumar S[82]. The analysis of the Bayesian approach allows to draw the conclusion (see Sragovich S[133]) about the non-adaptability of the strategy used. Unfortunately, in new publications authors continue to avoid clear statements of the control aim.

The identification methods of real objects (and the corresponding models) are described in great numbers of publications. There are many monographs devoted to this theme. Here we mention only Ljung and Söderström's book S[93] containing spacious bibliography. In connection with the identification methods, Huber developed the theory of "rough" estimates S[63] which are stable with respect to the considered class of distributions. It is called the *robust estimates theory*. The relations of the adaptive methods with the "empirical Bayesian approach" known in mathematical statistics and proposed by Robbins S[121] are of interest. A modern exposition of this approach can be found in Zaks G[25].

Chapter 2

The inertia-free controlled models describe many real phenomena. The stochastic models among them excite considerable interest both in psychology and in learning theory. The HPIV notion afford a formal basis for such models. The simplest of them have finite sets of stimuli and reactions (they were so called in stochastic learning systems of Bush and Mosteller). Therefore, it is quite reasonable to represent such models in the form of finite automata. By using automata theory one can construct control algorithms easily. Describing the control processes as an interaction of two automata gives control theory clarity and precision.

Sections 1 and 2. Here we follow, with some modifications, Sragovich S[132]. The first construction of ε-optimal automata, denoted afterwards by $\mathcal{D}_{k,n}$, was proposed and studied by Robbins S[119]. He applied them as a tool for solving statistical problems but did not use, properly speaking, the "automaton" notion. He introduced average fulfilment time of the control and established its usefulness. The other constructions of automata were considered by his followers: Isbell S[65], Smith and Pyke S[129], Samuels S[123]. In the text Theorem 2 is proved with the help of the limit theorem from Feller G[7], Vol. 2, Chap. 2.

In Russian-language scientific literature M. L. Tsetlin S[142] was the first who used the automaton principles independent of Robbins. He used names such as "stationary random surroundings" and "asymptotic-optimal sequence of automata" for a binary HPIV and an ε-optimal family of automata respectively. His purpose was to design very simple devices with the almost optimal behavior in any stationary random surroundings. The automata with linear tactics were the first devices of this kind.

Many constructions within the framework of this direction of research were rediscovered. For example, this was the case with the quasy-linear automata which were introduced simultaneously and independently of each other by Valachs S[150] and Kandelaki and Tsertsvadze S[67]. The surveys of publications devoted to this question can be found in Tsetlin S[142] and Varshawsky S[151]. The inequality (page 47) about the relations between arithmetic and harmonic means is in Hardy, Littlwood, Polya G[10].

Section 3. The idea of automata with increasing memory was discussed at the Tsetlin's seminars, but did not receive further development there. Such automata were supposed to be asymptotically-optimal without any additional conditions on the sequences $m(n)$. The first systematic investigation of this problem was done by Kolnogorov S[74].

Reasons were noticed to study automata which are, in fact, infinite. (Vavilov with colleagues S[152]). These automata were obtained as the *limit* of finite automata $A = \lim_{n \to \infty} A_n$. It was discovered, as one would expect, that the infinite limiting automata ($\mathcal{L}$, $\mathcal{D}$, $\mathcal{K}$ and others) have no useful properties with respect to the class of HPIV (see also Korolyuk and the others S[80]).

Section 4. It is based on the author's works started in 1963 (at first with U. Flerov S[134]). The exposition is close to Sragovich S[132, 133]. In comparison with these books the quantitative estimates of the mathematical expectation and variance of the hitting moments into the δ-regime are omitted.

The significance of δ-automata, especially, of the $\mathcal{G}$-type automata, was discovered in a number of practical problems of control.

The Goeffding and Petrov theorems were proved in Petrov G[121].

The asymptotic-optimal automata $\mathcal{SA}$ and $\mathcal{FP}$ were proposed by Flerov S[39]. His arguments are improved as compared with Sragovich S[133].

Another approach to designing the asymptotic-optimal automata was proposed by Varshavsky (and Vorontsowa) S[151] as far back as 1963. They were inspired by the idea of stochastic learning system (SLS) to re-distribute the probabilities in accordance with stimuli coming from the environment. This approach was realized by several recurrent procedures. Computer modelling has convinced the authors that their construction has the desired property. In United States this work caused considerable enthusiasm of many researches, see Lakshmivarahan S[87], Shapiro and Narendra S[125], Narendra and Thathachar S[103]. They proposed their own original procedures as well. But in 1974–1975 Baba and Savaragi S[13] proved that none of these procedures is, as one would expect, asymptotically-optimal with respect to the class of all binary HPIV.

Section 5. Procedure (1) was proposed by Shapiro and Narendra S[125] in the case of the constant step $a(t) = a$ but Kushner [86] and others have shown that the assertion on the asymptotic optimality of this procedure was mistaken (see Nazin and Poznyak S[100]). The modification of the initial form of the procedure that consists of introducing the changeable step was done by Nazin and Poznyak S[100]. We follow their exposition in Theorems 2 and 3. Inequalities (4) and Martingale lemma proved there were used as well.

The existence of finite automata with changeable structure which are asymptotically-optimal with respect to the class of *all* binary HPIV was questioned for a long time though public statements on that theme. Apparently, this did not appear in scientific periodicals.

Section 6. The results given here became known due to A. Kolnogorow S[72] and are published for the first time.

Chapter 3

The stochastic approximation method originated from two works: Robbins and Monro in 1951 and then Kiefier and Wolfowitz in 1952. Therafter there was a literary boom which resulted in the solution of some important problems by different authors practically simultaneously. The survey of these results (together with bibliography) can be found in Wasan S[153]. The successes which followed latter on were connected with the monographs of Nevelson and Hasminski S[104], Korostylev S[81], H.-F. Chen S[23], Kushner and Clark S[85] and with a series of L. Liung's articles S[89]–[92].

Sections 1–3. The main results on stochastic approximation are considered here as a tool of controlling the HPIV and other types of random processes, not as a technique of solving statistical problems. We have used the result obtained by D. S. Clark S[29] on necessary and sufficient conditions of convergence of the Robbins–Monro procedure as a basis of our consideration. His necessary condition (in a stronger form) was stated in Theorem 1, the proof of which is due to Chen with colleagues (see S[27]). The Clark's sufficient condition was stated in point (α). Condition (β) was formulated and proved by Shilman S[126]. This article, similar to that of Poznyak and Chikin S[114], was devoted to the behavior of the stochastic approximation procedure under dependent noises. More precisely, the mentioned articles are concerned with random sequences for which the strong law of large numbers holds. Information on processes with mixing can be found in Ibragimov and Linnik G[12]. The lemma from Sec. 2 was proved by Poznyak and Chikin S[114].

Section 4. The methodology of searching a conditional extremum and all proofs from Sec. 4 are due to Nazin and Poznyak S[100].

There are other approaches to designing recurrent procedures. For example we have the approach developed by Lakshmivarahan S[87]. Under appropriate assumptions these procedures have "absolute expediency" — a specific property leading to ε-optimality in different senses. These procedures are used for optimization of binary HPIV, for two-participant games and control of Markov processes.

Chapter 4

The minimax problems are likely to be known to specialists on adaptability. In this chapter three problems of "stochastic games" given will be discussed in a later chapter. Information on game theory is in many books.

Section 1. A deterministic two-participant game with non-opposite interests called Γ_1 is considered. It appeared within the framework of operations research developed by U. B. Germejer G[9]. His main purpose was to investigate the influence of the orders of player's moves and of their knowledge about the interests and actions of each other on the result of the game. The adaptive version of the game Γ_1 described here is due to Molodtsov S[97].

Section 3. Designing recurrent procedures to realize balanced behavior of the players in the Nash sense in multi-participant games is described in the monograph of Nazin and Poznyak S[100]. We follow mainly their exposition but treat the aim of control differently in a way which is, as it seems, closer to the true intention of the authors than it was presented in their book.

The concept of automata games as a method of describing and explaining the complex physiological phenomena was proposed by M. L. Tsetlin. Together with a group of colleagues (with I. M. Gelfand among them) he developed the theory according to which the activity of the neuron nets in the nerve-centres of the brain and in spinal cord can be formalized in the form of a game. By these ideas many phenomena (for example, changing some moves for the others) are represented as the result of changing the interaction technique of neurons from the lower regions of brain under an influence of either the stimulating flow or the inhibiting one of impulses from the upper regions. The upper regions rearrange the lower controlling links of the brain, change their functioning. According to tradition coming from McCulloch and Pitts G[18] the neuron nets are treated as automata but only with a simple structure. Their interaction entails the expediency of behavior of the whole organism. These physiological Tsetlin's views can be found in his book S[142].

The series of Tsetlin's publications (with co-authors) contains examples of games investigated by using computer simulation. These experiments led to the conclusion that automata play the Nash game. For the sake of clearness and liveliness the anthropomorphic terminology was widely used in those publications. The non-numerous analytical results were based on the Volkonski slowness hypothesis. The Varshavski's book S[151] was devoted to the concept described here.

An elegant game, which has the natural physiological interpretation, was considered by Bryzgalov, Shapiro-Pyatetsky and Shik S[8]. Different principles give a basis for another physiological model with nets of automata described by Petrov and Sragovich S[21]. The decentralized control realized by a collection of automata can be used in telemetric systems, in commutation networks, in controlling radio complexes, in controlling order of priorities in using computers and elsewhere. For applying it in communication systems one can consult Pulatov S[117].

Section 4. The results here are due to Gurvich S[58]. The slowness hypothesis was rejected in his works by constructing a contradicting example. Some results used in this section reletad to random walks and to the ruin problem can be found in Feller G[7] (Vol. 1, Chap. XIV).

There are other approaches to automata games (see Whittle S[154]).

Studying the nerve-centres depicted in the form of the neuron nets can be done by analytical methods which consist of setting up and solving the appropriate system of differential equations. Sometimes this leads to the appearence and solution of new problems in stability theory for such equations (see Sushkov S[140], Gelig S[47]). In the 1970's there were attempts to use the theory of local interaction (see Dobrushyn S[33]) for studying the behavior of automata in environments and in games. Several interesting works devoted to this theme were published. There were intentions to use this theory to investigate automata games (before the Gurvitch's works) but they ended in the stage of projects.

Chapter 5

The classical Markov chains control theory originated from Hovard's book and from a number of articles devoted to applying linear programming. Its exposition can be found in Derman G[4], Dynkin and Ushkevitch G[6]. The adaptive version of this theory was first published in Riordon S[118] and then there appeared many other algorithms (see Sragovich S[132]). The structure of the homogeneous controlled Markov chains was investigated by Zasuchin S[144]. His results were published in Sragovich S[133].

Sections 4 and 5. In describing both the identification algorithms and the automaton ones we follow the earlier Sragovich's publication S[133] with some innovations. About the method of calculating the optimal strategy one can consult Sragovich S[136].

Section 6. The recurrent procedure is stated in accordance with Nazin and Poznyak S[100] except for some simplifications. This has allowed to save space but there are some losses as well: some difficulties appear in obtaining the upper estimates of the convergence rate of current losses to optimal ones. For this reason the results were stated without proofs. Notice the references to the works of the other authors in Kumar S[85]. The researchers whose works were referred there never cited the works mentioned above though their own results were more modest.

Section 7. The conditional extremum problem is likely to be considered (following Sragovich S[134]) both in the classical version and in the adaptive one for the first time. Here the adaptive interpretation is based on identification but both automaton procedures and recurrent ones could be used as well.

The game (minimax) problems on Markov chains are older than the optimization problems. The latter were proposed by Shaplay S[125] in the form of "stochastic games" as far back as 1953. At that time the transition matrix of the chain was not supposed to be stochastic, i.e. the sums of elements of the rows could be less than one. Hence, with some positive probability the evolution of such a chain could be broken off together with the game itself. The infinite interval stochastic games were studied later on and, of course, with stochastic transition matrices. Theorems 1, 2 and 3 were proved by Gillette S[49], and Hoffman and Karp S[61] respectively. For some games the non-stationarity of optimal strategies was proved by Blackwell while investigating "Big Match".

Section 8. The results on adaptive minimax strategies on Markov chains are published for the first time.

Section 9. The optimization problems on graphs were investigated by Gössel and Sragovich S[48].

Chapter 6

Section 2. This is due to Zasuchin S[144] and was first published in Sragovich S[133]. Here this problem is presented in simpler form. For partially-observable Markov processes the general control problem is rather difficult even in the classical (non-adaptive) version. For finite Markov chains it was studied by Yushkievich (see Dynkin and Yushkievich G[6]). His solution reduces this problem to observed chains having a continuous state space. The adaptive version was first studied by Gössel and Sragovich [48]. They have obtained the optimal strategy in the class of program strategies. Chapter 8 of Sragovich's book S[133] was devoted to this theme.

Sections 3–6. For finite Markov chains with unobservable states the general statement of the adaptive control problem was considered by Konovalov S[76]–[77]. His results are presented in these sections. He extended them also to the regenerative processes.

Another approach to this problem was proposed by Hernnáder–Lerma S[60]. It was based on reducing an unobserved finite chain to an infinite one with observable states (Yushkievich's approach) and on identifying unknown parameters of the "observed" chain. Maximizing the discounted reward (as $\tau \to \infty$ where τ is an initial moment) is the aim of control.

Another form of the discussed problem was investigated by Di Masi and Stettner S[32] who supposed that the observations z_t at the moment t could be written in terms of the current states s_t by the formula $z_t = h(s_t) + \xi$ where ξ are independent identically distributed Gauss random variables. The controlled chain was supposed to be regular and the ε-optimality was considered as the aim of control.

Chapter 7

Section 1. The content is traditional. It is based on the Doob's unfading treatise G[5] and on Dynkin and Yushkievitch's monograph G[6]. One can find Theorem 1 in Taylor S[141] and Theorem 2 in Schal G[23]. The last result given is due to Gordienko S[55].

Section 2. This is a fundamental remake of Sec. 2 of Chap. 8 from Sragovich S[132].

Section 3. For classes of ergodic Markov processes, the searching control optimality in the strong sense was proposed and investigated by Gordienko S[52]. We follow his exposition.

Section 4. It contains strengthened results from Sec. 4 of Chap. 11 in Sragovich S[133] on finite semi-Markov processes.

Section 5. It is devoted to control theory of separable semi-Markov processes. Its results are due to M. Kolonko S[78] and require knowing the rather subtle topological facts on semi-continuous set-valued functions which can be found in Kuratowski G[14]. Some additional facts about constructing strongly consistent estimates are in Kolonko S[79]. A somewhat different approach to controlled jump processes was developed by Mandl with co-authors (see S[94]).

Section 6. All results from this section are due to Gordienko [50, 53, 54]. He also considered a number of applications of his theory to concrete problems.

Chapter 8

Adaptive control of stationary processes is a very popular model for practical applications. The fast development of this area left behind the clear mathematical investigation of this problem. This chapter is likely the first attempt of such a work. Its results are due to Agasandjan S[2, 3]. Before, he considered the case when the phase space was a finite set. However this simplification neither made the arguments easer nor made the result stronger.

Chapter 9

The subject-matter of this chapter is a result of the scientific activity of the Department of Theoretical Cybernetics of the Leningrad University. The most complete information can be found in the book: V. N. Fomin, A. L. Frazdkov, V. A. Yacubovich "Adaptive Control of Dynamic Object" (in Russian) (Nauka, Moscow, 1980).

Section 3. The adaptive stabilization was discussed by Lyubachevsky. The strongest results were obtained by V. Bondarko S[11]. He has considered the optimization problem of stabilizing.

Up to now his results are the most advanced investigation of finite-converging procedures. The scalar equation

$$x_t + a_1 x_{t-1} + \cdots + a_n x_{t-n} = b_1 u_{t-g} + b_2 u_{t-g-1} + \cdots + b_m u_{t-g-m+1} + c\zeta_t$$

with a delay of control is considered, where ζ_t is a bounded non-stochastic disturbance such that $|\zeta_t| \leq 1$. Let V be the collection of all sequences of this type. Define the classes of equations Θ_r such that $\Theta_r = \{(a_i),\ (b_i): |a_i| \leq r,\ |b_j| \leq r,\ b_1 \neq 0, |\lambda_i| < 1\}$, where λ_i are the roots of the characteristic equation, i.e. Θ_r consist of the minimum phase equations with bounded coefficients. For a given $\varepsilon > 0$ and the objective function $W(u) = \limsup |x_t|$, the ε-optimality consists of fulfilling the inequality $W(u) \leq W_0 + \varepsilon$, where $W_0 = \inf_u W(u)$. It was proved that there exists an ε-suboptimal strategy based on FCP.

Chapter 10

Section 1. Auxiliary information refers to the theory of linear systems, elements of random sequence theory (mainly martingales) and to identification theory. The following textbooks and monographs are, in my opinion, the best in this field: Whittle S[154], Davis and Vinter G[3], Kailath G[15], Chen Han-Fu S[23], Kumar and Varaja S[84].

Section 2. For linear homogeneous equations adaptive problems were considered in Sragovich S[135]. The finite identification turns out to be the central problem that entails solving all known control problems in the case of linear homogeneous equations for finite time.

Till now, for some reason, the asymptotical identification methods of this simple equation prevail.

Control of linear difference equations has become the main (and sometimes single) object of adaptive control. However, for us it is only part of the book which is supplemented by Chap. 9 about finite-convergence algorithms. It cannot be otherwise, since linear difference equations with constant coefficients compose a small part of all controlled objects. But it is a fact that these simplist objects do form the content of a high proportion of literature on adaptivity.

Section 3. The optimal tracking problem has a long history. It can be solved in different ways. The elegant and witty manner due to Goodwin and Sin is worth attention (see S[56]).

Sections 4–8. Both the identification problem and the linear-quadratic control problem is considered. The content of these sections is wholly based, in my opinion, on the strongest results obtained by Han-Fu Chen and Lei Guo and published in the numerous articles in 1986–1990. The brief resumes of their results can be found in two Chen's surveys S[23, 25]. Finally, more complete information on that question can be obtained from the Chen and Guo's monograph S[27].

For the linear-quadratic problems the first results are likely to be due to Mandl. Besides his own publications these results are included in Hall and Heyde's book S[59].

We would like to add some supplementary material which for a number of reasons was not included in this chapter.

We start with the "local-optimal" control proposed in 1977 by Kelmans, Poznyak and Chernitser S[68]. We study the equation

$$x_{t+1} = Ah(x_t) + Bu_t + \xi_t, \quad \xi_0 = 0, \quad x_t \in \mathrm{R}^n, \ u_t \in \mathrm{R}^m, \ \xi_t \in R^l$$

where ξ_t is a sequence of independent variables, A and B are some constant $n \times n$- and $n \times m$-matrices and $h(\cdot)$ is a Lipshitz (on the whole) function. The aim of control is to maximize the quadratic functional

$$\Phi = \varlimsup_{T \to \infty} T^{-1} \sum_{t=0}^{T-1} \left(x_t^T Q x_t + u_t^T R u_t \right), \quad Q \geq 0, \ R > 0.$$

The local-optimal (LO for short) control consists of finding the quantity

$$u_t^* = \arg\min \, \mathbf{E}\{x_t^T Q x_t + u_t^T R u_t\},$$

i.e. of minimizing the appropriate term under the symbol $\sum$ pointed to above at every moment t. We can now give the form of the LO control. Namely

$$u_t^* = -CAh(x_n), \quad C = (B^T Q B + R)^{-1} B^T Q.$$

Let the noise ξ_t be a square integrable martingale-difference and the numbers λ_Q, λ_R be the maximum eigenvalues of the matrices Q and R respectively, $\alpha = L^2 \|(I - BC)A\|^2$, $\beta = L^2 \|CA\|^2$. Here L is the Lipshitz constant.

One can prove that under the stated conditions and $\alpha < 1$ the following estimate

$$\Phi < \sigma_\xi^2 \frac{\lambda_Q + \lambda_R \beta}{1 - \alpha}$$

holds.

In the adaptive version of our optimization problem the matrices A, B are supposed to be unknown. We use the least squares' method to estimate them. Under the appropriate conditions their estimates are strongly consistent (and converge in the mean squares' sense as well). To complete the construction of the LO strategy it remains to add some random variables ν_t to the controls $\tilde{u}_t$ in keeping with the observations, i.e. to randomize the strategy. Notice that this technique was constantly used by V. H. Fomin. The final conclusion is that the described strategy implements the inequality

$$\Phi < \left(\sigma_\xi^2 + \|B\|^2 \sigma_\nu^2\right) \frac{\lambda_Q + \lambda_R \beta}{1 - \alpha}$$

with respect to the appropriate class of equations.

We also pay attention to the investigations of the Górky school belonging to this direction of research (see Kogan and Nejmark S[69]).

We may also consider optimal strategies that besides being optimal in the sense of attaining the control aim, are also optimal in the sense of the approaching rate to this aim. This means, for instance, that the functional being minimized tends to the desired minimal limit with the highest possible speed. Of course, such a "repeated" optimization is not always possible. We shall consider some examples when this is the case. These examples are recurrent stochastic approximation procedures.

In S[106] the authors seek the optimal algorithm of solving the adaptive problem of minimizing the functional $\varphi = \overline{\lim}_{t \to \infty} \mathbf{E}\|y_t\|^2$ when controlling the equations

$$y_t + \sum_{i=1}^{k} A_i y_{t_i} = u_t + \xi_t, \quad m = \dim y_t.$$

The noise ξ_t is supposed to form a sequence of independent identically distributed random variables. It turns out that

$$\min_u \Phi(u) = \mathbf{E}\|\xi_0\|^2.$$

Therefore, the optimal strategy should estimate the coefficients A_i of the equation as fast as possible. The design of the optimal strategy includes two stages:

(α) constructing information estimates in the problem considered (we have in mind the Rao–Kramer type inequalities);

(β) constructing the required optimal adaptive strategy which gives the minimum to the identification error obtained in stage (α).

By (α) and the appropriate assumptions (which are omitted here) we obtain

$$\lim_{t\to\infty} t[\mathbf{E}(y_t, h)^2 - (Dh, h)] = km\big(I_p^{-1}h', h\big), \quad \forall h \in \mathrm{R}^m,$$

where $I_p = \int p'(x)(p'(x))^T p^{-1}(x)dx$ is the information matrix of the noise density p and $D = \mathbf{E}\xi_t\xi_t^T$. The equality written above means that the convergence rate has the order ct^{-1} and $c = km(I_p^{-1}h, h)$. Stage (β) can be realized directly, i.e. a recurrent procedure can be given by which the convergence rate obtained in stage (α) is reached. Thus, under some assumptions the natural manner of regulating based on identification of the parameters of the equation by using the optimal algorithms of stochastic approximation leads to the control algorithm with the convergence rate of $\mathbf{E}\|y_t\|^2$ to $\|\xi_0\|^2$ as $t \to \infty$, which cannot be improved.

About the other problems of adaptive control and stochastic optimization where the information inequalities are used, one can consult Nazin S[98, 99], and Nazin and Juditski S[101, 102]. The mathematical interest of the problems of this kind is obvious. In applications the main problem decreasing the value of the coefficient c characterizing the convergence rate.

Choosing the optimal tracker on the basis of the classification of reference paths was discussed by Kumar and Praly S[83]. If there are no restrictions on the paths, the appropriate problem will be called the *general tracking* problem. If the reference path is generated by the output of a linear equation then it will be called the *linear model.* Further we consider only scalar equations.

We say that a numerical sequence z_t is *sufficiently rich* of order l if l is a sufficiently large integer such that there exist n and $\varepsilon > 0$ such that

$$\sum_{j=t+1}^{t+n} \big[x_{j-1}^*, \ldots, x_{j-l}^*\big]^T \big[x_{j-1}^*, \ldots, x_{j-l}^*\big] \geq \varepsilon I_l$$

for all sufficiently large t, I_l being the identity $l \times l$-matrix. Consider now a class $\mathcal{K}$ of equations of ARMAX type

$$x_t = \sum_{i=1}^{p} a_i x_{t-i} + \sum_{i=1}^{q} b_i u_{t-i} + \sum_{i=0}^{s} c_i w_{t-i}$$

leaving aside their detailed description. We only note that the assertions similar to those from Sec. 3 must hold with respect to this class. Here we are interested in a theorem of another kind.

Theorem S1. *In the general tracking problem let x_t^* be a sufficiently rich sequence of order $s + q$. Then*

$$\lim_{t \to \infty} \theta_t = \lambda \theta^0 \quad \text{a.s.}$$

for some nonzero random number λ.

For linear models this assertion remains in force irrespective of the order for all sufficiently rich sequences x_t^* which represent the reference paths.

In this theorem θ_t means the current estimate of the parameters of the equation whose explicit form is omitted. We now state an assertion on the convergence of the estimates of the parameters $\{a_i, b_i, c_i\}$.

Theorem S2. *In the general tracking problem if the sequence x_t^* is sufficiently rich having the order not less than $q + s$, then*

$$\lim_{t \to \infty} \gamma_0^{-1}(t)(\alpha_1(t) - \gamma_1(t), \ldots, \alpha_p(t) - \gamma_p(t), \beta_1(t), \ldots, \beta_q(t), \gamma_1(t), \ldots, \gamma_s(t))$$

$$= (a_1, \ldots, a_p, b_1, \ldots, b_q, c_1, \ldots, c_s), \quad \text{a.s.},$$

i.e. *the estimates are strongly consistent. In addition,*

$$\lim_{t \to \infty} \beta_1^{-1}(t)(\alpha_1(t), \ldots, \alpha_{\max(p,s)}(t), \beta_2(t), \ldots, \beta_q(t), \gamma_0(t), \ldots, \gamma_{l-1}(t))$$

$$= b_1^{-1}(a_1 + c_1, \ldots, a_{\max(p,s)} + c_{\max(p,s)}, b_2, \ldots, b_q, 1, c_1, \ldots, c_s), \quad \text{a.s.}$$

where $a_i = 0$ for $i > p$ and $c_i = 0$ for $i > s$.

From this optimality and adaptivity of the strategy follows.

These results remain in force for the linear model at $l > s$. But for $l \le s$ we have

$$\lim_{t \to \infty} \beta_1^{-1}(t)(\alpha_1(t), \ldots, \alpha_{\max(p,s)}(t), \beta_2(t), \ldots, \beta_q(t), \gamma_0(t), \ldots, \gamma_{l-1}(t))$$

$$= b_1^{-1}(a_1 + c_1, \ldots, a_{\max(p,s)} + c_{\max(p.s)}, b_2, \ldots, b_q, g_0, \ldots, g_{l-1}), \quad \text{a.s.}$$

where $a_i = 0$ for $i > p$ and $c_i = 0$ for $i > s$. Here $g_0, \ldots, g_{l-1}$ are some auxiliary parameters. This control is optimal again with respect to the given aim.

The above has given us one more adaptive strategy.

Some questions regarding the robustness of linear equations were discussed in Praly, Lin, Kumar S[116]. Recall that this means that if the strategy has some properties then all sufficiently "near" strategies have the same properties as well. In the above-mentioned article the properties of a strategy to ensure the stability of the model were mostly discussed. The Chen and Guo's article S[26] has a similar direction and is concerned with linear models considered in Secs. 4–6. Generally speaking, robustness is one of the most popular themes in recent years.

In Meyn and Caines S[96] one can find new ideas regarding adaptive strategies for linear models. The question is how to study adaptive control of stochastic systems by using ergodic theory systematically (for Markov processes). In this way it seems possible to obtain new results inaccessible by the usual methods.

Chapter 11

Section 1. It contains necessary material on linear systems and their stability. One can find more detailed information on linear systems in Kailath G[15], and on the stability theory in, for example, La Sall and Liefshetz G[19].

Section 2. This was written on the basis of Sragovich's article S[135].

Section 3. The control problem with model reference is one of the oldest in adaptive control theory and has a vast bibliography. The survey of early works with comments can be found in Carrol and Lindiff S[22].

The various versions of Theorem 1 are present in almost all works devoted to this theme. The stated version can be found both in Yuan and Wonham S[155] and in Zemlyakov and Rutkovsky S[145]. Theorems 2 and 3 are due to Zemlyakov and Rutkovsky but shorter proofs based on the Landau–Hadamard theorem (see Hardy, Littlwood, Polya G[10]) are given here. The most difficult case when the coefficients of the model are unknown was studied by V. A. Brusin by several techniques (see S[17] and S[17]).

For equations with a delay the same theme was studied by Nosov together with colleagues. These results were summarized in Kolmanovski and Nosov S[70]. However we follow here Danilin and Moiseev S[30].

Section 4. The steepest descent method was published (under different names) almost simultaneously and independently of each other by Nejmark S[105] and Fradkov S[43]. At a later date Fradkov S[45] described this method in detail together with its numerous applications. He has showed with great delight the importance of his method for future developments.

Sections 5–6. For both linear and nonlinear minimum phase equations, stabilization problems are presented on the basis of the Fradkov's thesis and a number of his articles. Before, these results were published in Sragovich S[133]. They are based on the frequency theorem that can be found in Gelig, Leonov and Yakubovich G[8].

In the case of an infinite-dimensional space (for example, a Hilbert space) the same theme requires knowing basic facts from functional analysis and, especially, from operator semigroup theory which can be found in many textbooks, but the Curtain and Pritchard's book G[2] is particularly useful.

Section 7. The results given here are due to Bondarko, Lichtarnikov and Fradkov articles S[12].

Section 8. For the quite controllable and stabilized equations the adaptive stabilization methods were studied by Sinitsin S[128]. His results together with the above-mentioned ones form a clear view of the modern state of adaptive stabilization for linear equations. It is useful to compare them with the results of other scientific schools. To make the evolution of the knowledge regarding adaptive stability obtained on the basis of the Direct Lyapunov Method one would compare directly the Parks and Shakloth's sudies, the results of Aksenov and Fomin S[5] with that of the Praly, Basin, Pomet and Ziang S[115]. The detailed article of the last authors was devoted, as usual, to the difficult stabilization problems for nonlinear equations.

They assumed that the external actions are absent, the equations are autonomous and linear with respect to unknown parameters.

Section 9. It includes two special problems coming from practical needs. They were studied by Brusin and his colleagues. It seems the exotic form of these problems, of the second in particular, makes them very interesting.

Note that the material of the present chapter was overwhelmingly based on the Second Lyapunov Method and the quite controlability notion.

Chapter 12

Section 1. It contains the traditional material about the stochastic integrals, the stochastic equations and about the properties of the linear Ito equations. More detailed information can be found in Doob G[5], Liptser and Shirjaev G[17], Arato G[1], Holewo [62], Brown and Hewitt S[15], Bellach S[9], Kazimierczyk S[66] and elsewhere. For the present the practical identification methods have been poorly worked out for the Ito equations containing the controls. This makes both theoretical investigations and practical applications difficult.

Section 2. The results regarding the stability (in the stochastic sense) of minimum phase linear Ito equations are stated, as before S[133], on the basis of Fradkov's thesis. It is useful to compare the logic of this section with that of Sec. 5, Chap. 11. The assertions connected with the stochastic stabilization of non-stable equations by means of Lipshitz control are due to Blank S[10].

For the stochastic Ito equations the history of development of the LQP in the classical version occupies a short while but it is abundant in results. It is rather difficult to ascertain the order of priority of its solution. The adaptive version of this problem was being developed (a short while) by Duncan and Pasik-Duncan S[34]–[38] and thereafter by the same authors with co-authors. The difficulties are connected both with solving the Riccati equation and with consistency of the estimates of the matrix B provided it was unknown in advance. Apparently this is the main reason the LQP has not received a complete solution yet.

Section 4. Specific problems of identification control when the estimates error appears to violate the controlability condition imposed on the parameters leads to a necessity to be careful. In other words, The Riccati equation may have no positive-definite solution (see Sragovich and Czornik S[138]). We do not know how to include the matrix B in the set of the unknown parameters of the equation yet.

Another, maybe effective, approach to constructing the optimization adaptive strategy for Ito equations was worked out by Borkar S[16]. Notice that in publications devoted to stochastic models with continuous time the problems with infinite-dimensional spaces compel attention.

GENERAL REFERENCES

[1] M. Arato, *Linear Stochastic System with Constant Coefficient. A Statistical Approach* (Springer-Verlag, Berlin, 1982).

[2] R. F. Curtain and A. I. Pritchard, Infinite dimensional linear systems theory, *Lect. Notes in Control and Inform. Sci.* Vol. 9 (Springer-Verlag, Berlin, 1978).

[3] M. H. A. Davis and R. B. Vinter, *Stochastic Modeling and Control* (Chapman and Hall, London, 1985).

[4] C. Derman, *Finite Markovian Decision Processes* (Academic Press, 1970).

[5] I. S. Doob, *Stochastic Processes* (Wiley, New York, 1953).

[6] E. B. Dynkin and A. A. Yushkevich, *Controlled Markov Processes* (Springer-Verlag, 1979).

[7] W. Feller, *An Introduction to Probability Theory and Its Application* (Wiley, New York, 1966).

[8] A. H. Gelig, G. A. Leonov and V. A. Yakubovich, *Stability of System with Non-single Equilibrium State* (in Russian) (Nauka, Moscow, 1978).

[9] J. B. Germejer, *Games with Non-opposite Interests* (in Russian) (Nauka, Moscow, 1976).

[10] C. H. Hardy, I. E. Littlewood and G. Polya, *Inequalities* (Cambridge University Press, 1934).

[11] I. I. Gihman and A. V. Skorohod, *Controlled Stoachastic Processes* (Springer-Verlag, 1979).

[12] I. A. Ibragimov and U. V. Linnik, *Independent and Stationary Connected Random Variables* (in Russian) (Nauka, Moscow, 1965).

[13] R. Z. Khasminski, *Stochastic Stability of Differential Equations* (Sijthoff and Noordoff, Alphen aan den Rijn, 1980).

[14] K. Kuratowski, *Topology*, Vol. 2 (Academic Press, New York, 1968).

[15] T. Kailath, *Linear Systems* (Prentice-Hall, New York, 1980).

[16] R. E. Kalman, P. L. Falb and N. A. Arbib, *Topics in Mathematical System Theory* (McGraw–Hill, New York, 1969).

[17] R. Sh. Liptser and A. N. Shiryaev, *Statistics of Random Processes* (Springer-Verlag, 1978).

[18] W. C. McCulloch and W. H. Pitts, A logical calculus of the ideas immanent in nervous activity, *Bull. Math. Biophys.* **5** (1943) 115–133.

[19] I. La Sall I. and S. Lefschetz, *Stability by Liapunov's Direct Method with Applications* (Academic Press, 1961).

[20] I. Neveu, *Bases Mathematiques de Calcul des Probabilites* (Masson, Paris, 1964).

[21] V. V. Petrov, *Sums of Independent Random Variables* (in Russian) (Nauka, Moscow, 1972).

[22] M. Schal, A selection theorem for optimization problems, *Arh. Mathemat.* **20** (1974) 219–224.

[23] M. Schal, Conditions for optimality in dynamic programming and for the limit of n–stage optimal policies to be optimal, *Z. Warsheinlichkeitstheorie Werb. Geb.* **32** (1975) 179–196.

[24] P. Whittle, *Optimization Over Time. Dynamic Programming and Stochastic Control.* Vol. 1, 2 (Wiley, New York, 1982, 1983).

[25] S. Zaks, *The Theory of Statistical Inference* (Wiley, New York, 1971).

SPECIAL REFERENCES

[1] M. Aoki, *Optimization of Stochastic Systems* (Academic Press, New York, 1969).

[2] G. A. Agasandian, Adaptive control for homogeneous random processes, in *Investigations in the Adaptive System Theory* (Moscow, 1976), pp. 193–219 (in Russian).

[3] G. A. Agasandian, Adaptive system for the class of homogeneous sequences, *Dokl. Ak. Nauk SSSR.* **228**(2) (1976) 329–331 (in Russian).

[4] G. A. Agasandyan, Adaptive systems for homogeneous processes with continuous states and controls spaces, *Probability Theory and its Application* (3) (1979) 515–528 (in Russian).

[5] G. S. Aksenov and V. N. Fomin, Lyapunov function method in the regulator synthesis problem, in *Cybernetics Problems Adaptive Control* (Nauka, Moscow, 1979) 69–93 (in Russian).

[6] G. S. Aksenov and V. N. Fomin, Synthesis of adaptive regulator by the Lyapunov functions method, *Avtomatika i Telemechanika* (6) (1982) 126–137 (in Russian).

[7] R. R. Bush, F. Mosteller, *Stochastic Models for Learning* (Wiley, New York, 1958).

[8] V. I. Bryzgalow, I. I. Pyatetsky-Shapiro and M. L. Shik, On two-level automata interaction model, *Dokl. Acad. Sci. SSSR* **160**(5) (1965) 1039–1041.

[9] B. Bellah, Parameter estimators in linear stochastic differential equations and their asymptotic properties, *Math. Operations Forsch. Stat. Ser. Statistics* **14**(1) (1983) 141–191.

[10] M. L. Blank, On the stochastic stabilization of the non-stable systems, *Usp. Mat. Nauk* **36**(5) (1981) 165–166 (in Russian).

[11] V. A. Bondarko, Adaptive suboptimal control of the solution of the linear difference equations, *Dokl. Ak. Nauk. SSSR* **270**(2) (1983) 301–303 (in Russian).

[12] V. A. Bondarko, A. L. Lihtarnikov and A. L. Fradkov, Design of an adaptive system for stabilizing a linear distributed plants, *Avtomatika i Telemechanika* (12) (1979) 95–103 (in Russian).

[13] N. Baba and Y. Savaragi, On the learning behavior of stochastic automata under a nonstationary random environment, *IEEE Trans. Systems. Man Cybernetics* **5**(3) (1975) 273–275.

[14] A. Benveniste, M. Métivier and P. Priouret, *Adaptive Algorithms and Stochastic Approximations* (Springer-Verlag, 1990).

[15] B. M. Brown and I. I. Hewitt, Asymptotic theory for diffusion processes, *J. Appl. Prob.* **12** (1975) 228–238 (in Russian).

[16] V. S. Borkar, Self-tuning control of diffusion without the identificability condition, *J. Optimiz. Appl.* **68** (1991) 117–138.

[17] V. A. Brusin, Synthesis of the nonsearching self-tuning system by the methods of absolute stability theory, *Avtomatika i Telemechanika* **5, 6**(7) (1978) (in Russian).

[18] V. A. Brusin, Some problems of the automatic tuning of the dynamic systems, in *Dynamic of System. Optimization and Adaptation* (in Russian) (Górky Stat. University, Górky, 1979).

[19] V. A. Brusin and M. B. Lapshyna, On one class of the continuous adaptive algorithm of control I, *Avtomatika i Telemechanika* **10** (1980) 81–90 (in Russian).

[20] V. A. Brusin and M. B. Lapshyna, On one class of the continuous adaptive algorithm of control II, *Avtomatika i Telemechanika* **12** (1980) 65–71 (in Russian).

[21] V. A. Brusin and E. Ja. Ugrinowskaya, Adaptive control of a class of nonlinear systems with an afteraction, *Avtomatika i Telemechanika* (8) (1988) 97–104 (in Russian).

[22] R. L. Carrol and D. P. Lindorff, Survey of adaptive control using Lyapunov design, *Int. J. Control* **18**(5) (1973) 428–434.

[23] H. F. Chen, *Recursive Estimation and Control for Stochastic Systems* (Wiley, 1985).

[24] H. F. Chen, *Parameter Identification and Adaptive Control*, Advances in Science of China. Mathematics (Science Press, Wiley, Beijing, 1986).

[25] H. F. Chen, Stochastic adaptive control, in *Stochastic Theory and Adaptive Control* KS, Lawrence Lecture Notes in Control and Inform. Sci. Vol. 184 (Springer-Verlag, 1992), pp. 93–120.

[26] H. F. Chen and L. Guo, A robust stochastic adaptive controller, *IEEE Trans. Automatic Control* **33**(11), (1991) 1035–1043.

[27] H. F. Chen and L. Guo, *Indentification and Stochastic Adaptive Control* (Birkhäuser, Boston, 1991).

[28] H. F. Chen and I. F. Zhang, Adaptive regulation for deterministic systems, *Acta Math. Appl. Sinica* **7**(4), (1991) 332–343.

[29] D. S. Clark, Necessary and sufficient conditions for the Robbins-Monro method, *Stochastic Processes and their Applications* **17** (1984) 359–367.

[30] A. V. Danilin and S. L. Moiseev, Design of self-adjusting system for the control of object with aftereffect based on the direct Lyapunov methods, *Avtomatika i Telemechanika* **2** (1991) 119–129 (in Russian).

[31] C. Derman, *Finite State Markovian Dessision Processes* (Academic Press, New York, 1970).

[32] G. B. Di Masi and L. Stettner, On adaptive control of a partially observed Markov chain, *Applicationes Mathematicae* **22**(2), (1994) 165–180.

[33] R. L. Dobrushin, Markov process with large number of the interaction components I,II, *Problemy Peredachy Inform.* **7**(2, 3) (1971) 149–161, 235–241 (in Russian).

[34] T. E. Duncan and B. Pasik–Duncan, Adaptive control of linear delay time systems, *Stochastics* **24** (1988) 45–74.

[35] T. E. Duncan and B. Pasik-Duncan, Adaptive control of continuous-time linear stochastic systems, *Math. Control Signal Systems* **3** (1990) 45–60.

[36] T. E. Duncan, B. Pasik-Duncan and L. Stettner, Almost self-optimization strategics for the adaptive control of diffusion processes, *J. Optimization Theory and Application* **81**(3), (1994) 449–507.

[37] T. E. Duncan, B. Pasik-Duncan and L. Stettner, On the ergodic and the adaptive control of stochastic differential delay systems, *J. Optimization Theory and Application* **81**(3), (1994) 509–531.

[38] T. E. Duncan, B. Pasik-Duncan and B. Goldys, Adaptive control of linear stochastic evolution systems, *Stochastics and Stochastics Reports* **36** (1991) 71–90.

[39] U. A. Flerov, On some class of the multi-input automata, in *Investigations in the Self-tuning Systems* (Comp. Center of Acad. Sci. USSR, Moscow, 1971), pp. 111–152 (in Russian).

[40] U. A. Flerov, On limit behavior and asymptotic optimality of the stochastic automatam in *Investigations in the Adaptive Systems Theory* (Comp. Center of Acad. Sci. USSR, Moscow, 1976), pp. 25–46 (in Russian).

[41] V. N. Fomin, *Discrete Linear Control Systems* (Kluwer Academic Publishers Group, Dordrect, 1991).

[42] A. L. Fradkov, Synthesis of the adaptive stabilization system for linear dynamic object, *Avtomatika i Telemechanika* **12** (1974) 96–103 (in Russian).

[43] A. L. Fradkov, Quadratic Lyapunov functions in the adaptive stabilization problem for the linear dynamic object, *Sib. Matem, Zurn* (2) (1976) 436–446 (in Russian).

[44] A. L. Fradkov, Speed gradient scheme and its application in the adaptive control problems, *Avtomatika i Telemechanika* (9) (1979) 90–100 (in Russian).

[45] A. L. Fradkov, *Adaptive Control in Large-scale Systems* (Nauka, Moscow, 1990).

[46] S. Grigorescu and M. Iosifescu, *Dependence with Complete Connections and its Applications* (Cambridge Univ. Press, Cambridge, 1990).

[47] A. H. Gelig, *Dynamics of the Impulse System and Neuron* (*Nets.*- Leningrad: Leningrad State University, 1982) (in Russian).

[48] M. Gössel and V. G. Sragovich, Adaptive control of the Markov chains with rewards *Dokl. Akad. Nauk. SSSR* **254**(3) (1980) 523–527 (in Russian).

[49] D. Gillett, Stochastic games with zero stop probabilities, in *Contributions to the Theory of Games*, Vol. III, *Ann. Math. Studies*, **39** (Princeton Press, 1957).

[50] E. I. Gordienko, Adaptive control of stores when the distribution of demand is unknown, *Izv. Akad. Nauk. SSSR, Techn. Kibernet* (1) (1982) 56–60 (in Russian).

[51] E. I. Gordienko, Random search in the adaptive control problem of Markov processes with discrete time, *Izv. Akad. Nauk. SSSR, Tekhn. Kibernet* (3) (1984) 26–33 (in Russian).

[52] E. I. Gordienko, Adaptive strategics for certain classes of controlled Markov processes, *Teor. Verojatnosti and Primen* **29**(3) (1984) 488–501 (in Russian).

[53] E. C. Gordienko, Controlled Markov sequences with slowly varying characteristics. I. Adaptive control problem, *Izv. Akad. Nauk. SSSR, Techn. Kibernet* (2) (1985) 53–61 (in Russian).

[54] E. C. Gordienko, Controlled Markov sequences with slowly varying characteristics. II. Adaptive optimal strategics, *Izv. Akad. Nauk. SSSR, Techn. Kibernet* (4) (1985) 81–90 (in Russian).

[55] E. C. Gordienko, Stability and existence of the canonical strategics of control for Markov sequences, Proc. 20th USSR Seminar *"Stability Problem of the Stochastic Models"* (Kuibyshev, Kuibyshev Stat. University, 1988), pp. 27–33.

[56] G. C. Goodvin and K. S. Sin, *Adaptive Filtering, Prediction and Control* (Prentice-Hall, INC, Englewood Cliffs, New Jersey, 1984).

[57] E. T. Gurvich, The asymptotic investigation method of the automata games, *Avtomatika i Telemechanika* (2) (1975) 80–94 (in Russian).

[58] E. T. Gurvich, Adaptive control for the vector random processes, in *Investigations in the Adaptive Systems Theory* (Moscow, 1976), pp. 67–118 (in Russian).

[59] P. Hall and C. C. Heyde, *Martingale Limit Theory and its Application* (Academic Press, New York, 1980).

[60] O. Hernnánder-Lerma, *Adaptive Markov Control Processes* (Springer-Verlag, 1990).

[61] A. I. Hoffman and R. M. Karp, On nonterminating games, *Management Sci.* **12** (1996) 359–370.

[62] A. S. Holevo, Estimation of the drift parameters of the diffusion process by the stochastic approximation method, In *Investigations in the Theory of Self-tunning Systems* (Computer Center of USSR Acad. Sci., Moscow, 1976), pp. 179–200 (in Russian).

[63] P. J. Huber, Robust statistical procedures (SIAM.–Philadelfia, 1977).

[64] M. Iosifescu and R. Theodorescu, *Quad Random Processes and Learning* (Springer-Verlag, 1969).

[65] J. R. Isbell, On a problem of robbins, *Annals Math. Stat.* **30** (1969) 606–610.

[66] P. Kazimercyk, Maximum likelihood approach in parametric identification of stochastic differential models on engineering systems, *Structural Safety* **8** (1990) 29–44.

[67] N. P. Kandelaki and G. N. Tsertsvadze, On behaviour of some class of the stochastic automata in random surrounding, *Avtomatika i Telemechanika* **27**(6) (1966) 115–119 (in Russian).

[68] G. K. Kelmans, A. S. Poznyak and A. V. Chernitser, Adaptive locally optimal control, *Int. J. Systems Sci.* **2**(2) (1981) 235–254.

[69] M. M. Kogan and J. I. Nejmark, *Adaptive Control* (Górky Stat University, Górky, 1987) (In Russian).

[70] V. B. Kolmanowski and V. R. Nosov, *Stability of Functional Differential Equations* (Academic Press, London, 1986).

[71] A. V. Kolnogorov, *Atomata with Finite and Increasing Memory in the Stationary Random Surrounding*, Dissertation (Wosc. Phys.-Techn. Inst., Moscow, 1984) (In Russian).

[72] A. V. Kolnogorov, Automata that are asymptotically optimal in a stationary environment and have growthing memory, *Avtomatika i Telemechanika* (9) (1981) 129–137 (in Russian).

[73] A. V. Kolnogorov, On minimax approach to the optimal expediency behaviour in the stationary surroundings on the finite time interval, *Izv. Akad. Nauk. SSSR, Techn. Kibernet* (5) (1988) 143–146 (in Russian).

[74] A. V. Kolnogorov, Estimates of the minimax risk in stationary media, *Izv. Akad. Nauk. SSSR, Techn. Kibernet* (2) (1990) (in Russian).

[75] M. G. Konovalov, Adaptive control for periodical processes with independent values, *Izv. Akad. Nauk. SSSR, Techn. Kibernet* (1) (1979) 138–144 (in Russian).

[76] M. G. Konovalov, Adaptive control of the finite automata with unobservable states, *Dokl. Akad. Nauk. SSSR* **291**(1) (1986) 59–62 (in Russian).

[77] M. G. Konovalov and I. A. Sinitzin, *To the Adaptive Control Theory* (Comp. Center. of USSR Sci., Acad., Moscow, 1988) (in Russian).

[78] M. Kolonko, The average–optimal adaptive control of a Markov renewal model in presence of an unknown parameter, *Math. Operationsforsch. u. Statist.* **13** (1982) 567–591.

[79] M. Kolonko, Strongly consistent estimation in a controlled Markov renewal model, *J. Appl. Probab.* **19** (1982) 532–545.

[80] V. S. Korolyk, A. I. Pletnev and S. D. Ejdelman, Automata. Walks. Games, *Uspehi. Mat. Nauk.* **43**(1) (1988) 87–122 (in Russian).

[81] A. P. Korostylev, *Stochastic Re current Procedures. Local Properties* (Nauka, Moscow, 1985) (in Russian).

[82] P. R. Kumar, A survey of some results in stochastic adaptive control, *SIAM J. Control Optimization* **23** (1985) 329–380.

[83] P. R. Kumar and L. Praly, Self-tuning trackers, *SIAM J. Control Optimization* **25**(4) (1987) 1053–1071.

[84] P. R. Kumar and P. Varaja, *Stochastic Systems: Estimation, Identification and Adaptive Control* (Prentice-Hall, New York, 1985).

[85] P. R. Kushner and D. S. Clark, *Stochastic Aproximation Methods for Constrained and Unconstrained Systems* (Springer-Verlag, 1987).

[86] P. R. Kushner, M. A. Thathachar and S. Lakshmivarahan, Two State Automaton — a Counter Example, *IEEE Trans. Systems, Man. Cybernetics* **2** (1972) 292–294.

[87] S. Lakshmivarahan, *Learning Algorithms. Theory and Applications* (Springer-Verlag, 1981).

[88] L. D. Landau, *Adaptive Control Systems: The Model Reference Approach* (Marcel Dekker, New York, 1979).

[89] L. Ljung, Consistency of the least square identification method, *IEEE Trans. AC* **21** (1976) 779–801.

[90] L. Ljung, Analysis of the recursive stochastic algorithms, *IEEE Trans. AC.* **22** (1977) 551–575

[91] L. Ljung, On positive real functions and the convergence of some recursive schemes, *IEEE Trans. AC.* **22** (1978) 539–551.

[92] L. Ljung, Convergence analysis of parametric identification methods, *IEEE Trans. AC.* **23** (1978) 770–783.

[93] L. Ljung and T. Söderström, *Theory and Practice of Recursive Identification* (The MIT Press, Mass, Cambridge, 1983).

[94] P. Mandl and R. Romera-Ayllon, On adaptive control on Markov processes, *Kybernetika* **23**(2) (1987) 89–103.

[95] D. E. Miller, Adaptive stabilization using a nonlinear time-varying controller, *IEEE Trans. Automatic Control* **39**(7) (1991) 1347–1359.

[96] S. P. Meyn and P. E. Caines, A new approach to stochastic adaptive control, *IEEE Trans. AC.* **32** (1987) 220–226.

[97] L. A. Molodtsov, Adaptive control in recurrent games, *Vychis. Matem. i Matem. Fizyka* **18**(1) (1978) 73–83 (in Russian).

[98] A. V. Nazin, Informational inequalities in the stochastic gradient optimization problem and optimal realizable algorithms, *Avtomatika i Telemechanika* (1) (1989) 127–137 (in Russian).

[99] A. V. Nazin, Informational inequalities in the adaptive control problem of multi-dimensional linear object, *Avtomatika i Telemechanika* (5) (1990) 127–137 (in Russian).

[100] A. V. Nazin and A. S. Poznyak, *Adaptive Choice of Variants* (Nauka, Moscow, 1986) (in Russian).

[101] A. V. Nazin and A. B. Juditsky, Informational inequalities in the adaptive control problem of linear object, *Dokl. Akad. Nauk. SSSR.* **317**(2) (1991) 323–325 (in Russian).

[102] A. V. Nazin and A. B. Juditsky, Low information boundary in the problem of adaptive tracking for a linear discrete stochastic object, *Problemy Peredachi Inform.* **31**(1) (1995) 56–67 (in Russian).

[103] K. S. Narendra and M. A. L. Thathachr, Learning automata — a survey, *IEEE Trans. Systems, Man Cybernetics* **4** (1974) 323–334.

[104] M. B. Nevelson and P. Z. Khasminski, *Stochastic Approximation and Recursive Estimation* (Nauka, Moscow, 1979) (In Russian).

[105] U. I. Nejmark, *Dynamic Systems and Controlled Processes* (Nauka, Moscow, 1976). (In Russian).

[106] A. S. Nemirowsky and J. Z. Tsypkin, On optimal algorithms of the adaptive control, *Avtomatika i Telemechanika* **12** (1984) 64–77 (in Russian).

[107] J.-F. Nash, Equilibrium point in n-person games, //*Proc. Natl. Acad. Sci., USA* **36**(1) (1950) 48–49.

[108] M. F. Norman, Markov Process and Learning Models (Academic Press, New York, 1975).

[109] O. Oniçescu and G. Mihoc, Sur les chaines statistique, *Compte Rendue Acad. Sci.* **200** (1935).

[110] B. Pasik-Duncan, On the consistency of a least squares identificational procedure in linear evolution systems, *Stochastics Reports* **39** (1992) 83–94.

[111] B. Pasik-Duncan, Asymptotic distribution of some quadratic functionals of linear stochastic evolution systems, *J. Optimization Theory and Applications* **75**(2) (1992) 389–400.

[112] P. C. Parks, Lyapunov redesign of model reference adaptive control systems, *IEEE Trans. AC.* **11** (1996) 362–367.

[113] A. A. Petrov and V. G. Sragovich, The control principles of the gas-engine, in *Investigations in the Self-tuning Systems* (Computer Center of USSR Sci. Acad., Moscow, 1967), pp. 46–66 (In Russian).

[114] A. S. Poznyak and D. O. Chikin, Gradient procedure for stochastic approximation with dependent noise and their asymptotic behaviour, *Int. J. Systems. Sci.* **16**(8) (1985) 917–949.

[115] L. Praly, G. Bastin, J.-B. Pomet and Z. P. Ziang, *Adaptive Stabilization of Nonlinear Systems: Foundations of Adaptive Control*, Lecture Notes in Control and Inform. Sci. Vol. 160 (Springer–Verlag, 1991), pp. 347–433.

[116] L. Praly, S.-F. Lin and P. R. Kumar, A robust adaptive minimum variance controller, *SIAM J. Control Optimization* **27**(2) (1989) 235–266.

[117] A. K. Pulatov, *Automata Models and their Application in Control Systems* (FAN, Tashkent, 1984) (in Russian).

[118] S. Riordon, An adaptive automation controller for discrete Markov process, *Automatica* **5** (1969) 721–730.

[119] H. Robbins, Some aspects of the sequential design of experiments, *Bulletin AMS* **58**(5) (1952) 527–535.

[120] H. Robbins, A sequential decision problem with finite memory, *Proc. Natl. Acad. Sci.* **42**(3) 920–923.

[121] H. Robbins, The optimal Bayes approach to statistical decision problem, *Ann. Math. Statist.* **35** (1964) 1–20.

[122] Y. Savaragi and N. Baba, Two ε-optimal nonlinear reinforcement schemes for stochastic automata, *IEEE Trans. SMC* (1974) 126–130.

[123] S. M. Samuels, Randomized rules for one two-armed bandit problem with finite memory, *Ann. Math. Statist.* **39**(6) (1968) 2103–2107.

[124] B. Shackloth, Design of model–reference control systems using a Lyapunov synthesis technique, *Proc. Inst. Eng.* **114**(2) (1967).

[125] I. J. Shapiro and K. S. Narendra, Use of stochastic automata for parameter self-optimization with multi-modal performance criteria, *IEEE Trans. SMC-5* **5**(4) (1968) 352–361.

[126] S. V. Shilman, Stochastic approximation of the independent noises, in *Statistical Methods and Models* (Institute of System Invest., Moscow, 1987) No. 1.

[127] L. S. Shaply, Stochastic games, *Proc. Natl. Acad. Sci., USA* **39** (1953) 1095–1100.

[128] I. A. Sinitzin, Adaptive control of the linear differential non-minimum phase equations, in *To the Aadaptive Control Theory* (Comp. Center of USSR Sci. Acad., Moscow, 1988), pp. 33–57 (In Russian).

[129] C. Smith and R. Pyke, The Robbins-Isbell two-armed bandit problem with finite memory, *Annals Math. Stat.* **36**(5) (1965) 1375–1386.

[130] M. Sohal, A selection theorem for optimizations problems, *Arch. Mathem.* **20** (1974) 219–224.

[131] M. Sohal, Conditions for optimality in dynamic programming and for the limit of n-stage optimal policies to be optimal, *Z. Wahr. Werw. Geg.* **32** (1974) 179–196.

[132] V. G. Sragovich, *Theory of Adaptive Systems* (Nauka, Moscow, 1976) (In Russian).

[133] V. G. Sragovich, *Adaptive Control* (Nauka, Moscow, 1981) (In Russian).

[134] V. G. Sragovich, Optimization with restrictions on the finite homogeneous Markov chains, *Izv. Acad. Nauk. SSSR, Techn. Kibern.* (2) (1985) 62–69 (in Russian).

[135] V. G. Sragovich, On new adaptive control problems for linear equations, *Dokl. Akad. Nauk. SSSR.* **297**(4) (1987) 812–815 (in Russian).

[136] V. G. Sragovich, On decomposition in problem of Markov chain optimal calculation by means of linear programming optimization, *Optimization* **21**(4) (1990) 593–600.

[137] V. G. Sragovich and J. A. Flerov, Construction of the class of the optimal automata, *Dokl. Akad. Nauk SSSR* **159**(6) (1964) 1236–1237 (in Russian).

[138] V. G. Sragovich and A. Chornik, On asymptotical behaviour of the solutions of the algebraic Riccati equation for continuous time, *Avtomatika i Telemechanika* (8) (1985) 90–92 (in Russian).

[139] L. Stettner, On nearly self–optimizing for a discrete–time uniformly ergodic adaptive model, *Appl. Math. Optim.* **27** (1993) 162–177.

[140] V. C. Sushkov, Qualitative investigations of the dynamics of the neuron nets, in *Investigations in Adaptive System Theory* (Moscow, 1971), pp. 223–266 (In Russian).

[141] H. M. Taylor, Markovian sequential replacement processes, *Annals Math. Stat.* **36**(19) (1965) 1677–1694.

[142] M. L. Tsetlin, *Automata Theory and Modeling of Biological Systems* (Academic Press, New York, 1973).

[143] A. M. Tsygunov, *Adaptive Control of the Objects with Afteraction* (Nauka, Moscow, 1984) (In Russian).

[144] V. S. Zasukhin, Adaptive Control of the Markov Processes, *Disertation* (Computer Center of USSR Sci. Acad., Moscow, 1975) (In Russian).

[145] S. D. Zemlyakov and V. J. Rutkovskiy, Acting conditions of the multi-dimensional self-tuning system of control with model reference under continuously acting parametrical disturbances, *Dokl Akad. Nauk SSSR* **241**(2) (1978) 301–304.

[146] E. S. Usachev, On limit distribution arising in the learning models, *Dokl. Akad. Nauk SSSR* **159**(6) (1964) 1238–1239 (in Russian).

[147] E. S. Usachev, The stochastic learning model and its properties, in *Investigations in the Theory of Self-tuning Systems* (Computer Center of USSR Sci. Acad., Moscow, 1967), pp. 8–26 (in Russian).

[148] E. S. Usachev, Asymptotical properties and approximation of the stochastic learning models, *Dokl. Akad. Nauk. SSSR* **182**(2) (1968) 282–284 (in Russian).

[149] E. S. Usachev, On asymptotical properties of random processes arising in the learning models, in *Investigations in the Theory of Self-tuning Systems* (Computer Center of USSR Sci. Acad., Moscow, 1971), pp. 153–206 (In Russian).

[150] V. Ja. Valah, On behaviour of automata with selective tactics in stationary surroundings, *Kibernetika* (4) (1968) 10–14 (in Russian).

[151] V. I. Varshavsky, *Collective Behaviour of the Automata* (Nauka, Moscow, 1973) (in Russian).

[152] E. I. Vavilov, S. D. Ejdelman and A. I. Ezrohi, Asymptotic behaviour of stochastic automata in stationary surroundings, *Kibernetica* (5) (1975) 3–21 (in Russian).

[153] M. T. Wasan, *Stochastic Approximation* (Cambridge University Press, 1969).

[154] P. Whittle, Co–operative effects in assemblies of stochastic automata, in *Proc. Sympo. Honor Jerzy Neyman* (PWP, Warszawa, 1977).

[155] J. S.–C. Yaan and W. M. Wonham, Probing signals for model reference identification, *IEEE Trans. AC.* **22**(4) (1977).

ADDITIONAL REFERENCES

In Chapter 11, the control models to be described by the ordinary differential equations were studied. However, only some of the problems associated with such models were considered. In fact, bibliography on such question is rather specious and, moreover, increase constantly. We would now like to help, to a certain extent, to the reader who is interested in studying these models in detail. The brief list annexed below gives an idea of the existent publications on the adaptive control problems of the mentioned class of models which were not entered into present consideration. Besides, these publications contain additional references on the questions pointed out.

[A1] A. J. Astrovsky and I. V. Gajshun, The uniform and approximate observability of linear nonstationary systems, *Avtomatika and Telemechanika* (7) (1988) 3–13.

[A2] K. Balachandran and P. Balasubramaniam, Remarks on the controlability of nonlinear pertubatuous of volterra integro-differential systems, *J. Appl. Math. Stochastic Analysis* **8**(2) (1995) 201–208.

[A3] S. V. Burnosov and R. I. Kozlov, Investigation of the dynamics of nonlinear systems with uncertainty and perturbations on the basis of the method of vector Lyapunov functions I, *J. Comp. Syst. Sci. Int.* **33**(5) (1985) 75–81.

[A4] S. V. Burnosov and R. I. Kozlov, Investigation of the dynamics of nonlinear systems with uncertainty and pertubations on the basis of the method of vector Lyapunov functions II, *J. Comp. Syst. Sci. Int.* **34**(2) (1986) 82–90.

[A5] N. V. Druzhinina, V. O. Nikiforov and A. L. Fradkov, Adaptive control method for object with nonlinearity in output, *Avtomatika Telemechanika* (2) (1996) 3–33.

[A6] G. Kreisselmeir and R. Lozano, Adaptive control of continuous-time overmodeled plants, *IEEE Trans. Automatic Control* **44**(2) (1966) 1779–1794.

[A7] C.-H. Lee, Robust stabilization of linear continuous systems subjected to time-varying state delay and pertubations, *J. Franklin Inst.* **333**(B)(5) (1996) 707–720.

[A8] R. Lozano and G. A. Suárez, Adaptive control of non-minimum phase systems subject to unknown bounded disturbances, *Int. Series Numerical Math.* (Birkhäuser Verlag, Basel, 1996), Vol. 121, pp. 125–133.

[A9] H. Logemann and A. Ilchamann, An adaptive servomechanism for a class of infinite-dimentional systems, *SIAM J. Control Optimization* **32**(4) (1994) 917–936.

[A10] W.-M. Lu, K. Zhou and J. C. Doyle, Stabilization of uncertain linear systems: an LFT approach, *IEEE Trans. Automatic Control* **41**(1) (1966) 50–65.

[A11] D. E. Miller, Adaptive stabilization using a nonlinear time-varying controller, *IEEE Trans. Automatic Control* **39**(7) (1994) 1349–1359.

[A12] G. Nahapetian and W. Ren, Uncertainty structures in adaptive and robust stabilization, *Automatica* **31**(11) (1995) 1565–1575.

[A13] K. S. Narenda and J. Balakrishnan, Adaptive control using multiple models, *IEEE Trans. Automatic Control* **42**(2) (1997) 171–187.

[A14] J.-B. Pomet and L. Praly, Adaptive nonlinear regulation: estimation from the Lyapunov equation, *IEEE Trans. Automatic Control* **47**(6) (1992) 729–740.

[A15] L. Praly, S.-F. Lin and P. R. Kumar, A robust adaptive minimum variance controller, *SIAM J. Control Optimization* **27**(2) (1989) 235–266.

[A16] S. A. Quesada and M. De la Sen, Robust adaptive tracking with pole placement of first-order potentially inversely unstable continuous-time systems, *Informatica* **9**(3) (1998) 259–278.

[A17] E. P. Ryan, An integral invariant principle for differential inclusions with applications in adaptive control, *SIAM J. Control Optimization* **36**(3) (1988) 960–980.

[A18] A. V. Savkin and I. R. Petersen, Robust stabilization of discrete-time uncertain nonlinear systems, *J. Optimization Theory Appl.* **96**(1) (1988) 87–107.

[A19] S. Triantafillidis, J. A. Leach, D. H. Owens and S. Townley, Limit systems, limit gains and non-genetic behaviours arising in adaptive stabilizing control, *Dynamics and Stability Systems* **10**(3) (1995) 233–254.